THERMOPHOTOVOLTAIC GENERATION OF ELECTRICITY

Previous Proceedings in the Series of Conferences on Thermophotovoltaic Generation of Electricity

	Year	Held in	Publisher	ISBN
4th	1998	Denver, CO	AIP Conf. Proceedings vol. 460	1-56396-828-2
3rd	1997	Colorado Springs, CO	AIP Conf. Proceedings vol. 401	1-56396-734-0
2nd	1995	Colorado Springs, CO	AIP Conf. Proceedings vol. 358	1-56396-509-7
1st	1994	Copper Mountain, CO	AIP Conf. Proceedings vol. 321	1-56396-353-1

Other Related Titles from AIP Conference Proceedings

654 Space Technology and Applications International Forum - STAIF 2003: Conf. on Thermophysics in Microgravity; Conf. on Commercial/Civil Next Generation Space Transportation; 20th Symp. on Space Nuclear Power and Propulsion; Conf. on Human Space Exploration; 1st Symp. on Space Colonization
Edited by Mohamed S. El-Genk, February 2003, 0-7354-0114-4
CD-ROM: 0-7354-0115-2

608 Space Technology and Applications International Forum - STAIF 2002: Conf. on Thermophysics in Microgravity; Conf. on Innovative Transportation Systems for Exploration of the Solar System and Beyond; 19th Symp. on Space Nuclear Power and Propulsion; Conf. on Commercial/Civil Next Generation Space Transportation
Edited by Mohamed S. El-Genk, February 2002, 0-7354-0052-0
CD-ROM: 0-7354-0053-9

552 Space Technology and Applications International Forum - 2001: Conf. on Space Exploration Technology; Conf. on Thermophysics in Microgravity; Conf. on Innovative Transportation Systems for Exploration of the Solar System and Beyond; Conf. on Commercial/Civil Next Generation Space Transportation; 18th Symp. on Space Nuclear Power and Propulsion; Space Radiation and Environment Effects Track
Edited by Mohamed S. El-Genk, February 2001, 1-56396-980-7
CD-ROM: 1-56396-981-5

504 Space Technology and Applications International Forum - 2000: Conf. on International Space Station Utilization; Conf. on Thermophysics in Microgravity; Conf. on Enabling Technology and Required Scientific Developments for Interstellar Missions; Conf. on Commercial/ Civil Next Generation Space Transportation; 17th Symp. on Space Nuclear Power and Propulsion
Edited by Mohamed S. El-Genk, January 2000
2 vol. print set, hard cover: 1-56396-919-X
CD-ROM: 1-56396-920-3

462 NCPV Photovoltaics Program Review
Edited by M. Al-Jassim, J. P. Thornton, and J. M. Gee, March 1999, 1-56396-836-3

404 Future Generation Photovoltaic Technologies: First NREL Conference
Edited by R. D. McConnell, 1997, 1-56396-704-9

394 NREL/SNL Photovoltaics Program Review: Proceedings of the 14th Conference - A Joint Meeting
Edited by C. E. Witt, M. Al-Jassim, and J. M. Gee, May 1997, 1-56396-687-5

To learn more about these titles, or the AIP Conference Proceedings Series, please visit the webpage
http://proceedings.aip.org/proceedings

THERMOPHOTOVOLTAIC GENERATION OF ELECTRICITY

Fifth Conference on Thermophotovoltaic Generation of Electricity

Rome, Italy 16–19 September 2002

EDITORS
Timothy J. Coutts
National Renewable Energy Laboratory, Golden, CO
Guido Guazzoni
U.S. Army Central Electronic Command, Fort Monmouth, NJ
Joachim Luther
Fraunhofer Institute for Solar Energy Systems, Freiburg, Germany

SPONSORING ORGANIZATIONS
U.S. Army, CECOM
Army Research Office (European Office)
National Renewable Energy Laboratory
U.S. Department of Energy
National Aeronautics and Space Administration
Glenn Research Center

AMERICAN INSTITUTE OF PHYSICS

Melville, New York, 2003
AIP CONFERENCE PROCEEDINGS ■ VOLUME 653

Editors:

Timothy J. Coutts
National Renewable Energy Laboratory
1617 Cole Boulevard
Golden, CO 80401
USA
E-mail: tim_coutts@nrel.gov

Guido Guazzoni
CECOM-ASD
Myer Center
Fort Monmouth, NJ 07712
USA
E-mail: Guido.Guazzoni@mail1.monmouth.army.mil

Joachim Luther
Fraunhofer Institute for Solar Energy Systems
Oltmannsstr. 5
D-79100 Freiburg
GERMANY
E-mail: luther@ise.fhg.de

L.C. Catalog Card No. 2002116779
ISBN 0-7354-0113-6
ISSN 0094-243X
Printed in the United States of America

CONTENTS

SESSION 3: EMITTERS/FILTERS
CHAIR: CHRISTOPHER HEBLING

SESSION 4: EMITTERS/INNOVATIVE APPROACHES
CHAIR: ROBERT NELSON

SESSION 7: CELL TECHNOLOGY
CHAIR: M. N. PALMISANO

Preface

The Fifth Conference on Thermophotovoltaic Generation of Electricity was originally planned for October 2001 but was unavoidably delayed because of the events of September of that year. The interval between the Fourth and Fifth conference was longer than has typically been the case and was almost three years, instead of the more usual two years. Not surprisingly, there was a great deal of new information to report in this meeting, given the length of time since the previous conference. The meeting was held at the Hotel Villa Pamphili in Rome, Italy, during the period September 16-19, 2002. There were 80-90 attendees (about one third fewer than the conferences in the U.S.) with 54 Invited and Contributed papers being presented. By far the majority of contributors have chosen to submit a written version of their paper for this volume and we continue to believe that the archives of these meetings is useful.

The Co-Chairs of the meeting were Timothy J. Coutts (NREL), Guido Guazzoni (U.S. Army CECOM),, and Joachim Luther (Fraunhofer Institute for Solar Energy Systems). The first four TPV conferences were chaired by the first of these individuals and one or more colleagues from NREL. The previous meetings were all held in the vicinity of Denver, primarily because NREL is located in this area and was the primary sponsor of the conferences. On this occasion, the primary sponsor was the U.S. Army Communication Electronic Command (CECOM) and one of the motivations of this organization was to learn about TPV activities in Europe. Dr. Robert Hamlen of CECOM, provided introductory remarks (not included in this volume) in which he outlined the possible role of TPV systems for the U.S. military of the future. Interests in TPV in the U.S. have mainly been concerned with military applications and there has been little interest in non-military applications. This has not been the case in Europe where TPV is seen as being a potential means of saving energy, through applications such as recovery of industrial waste heat, and as an enabling component in new applications such as stand-alone domestic gas furnaces.

It is the intention of the community, and the Co-Chairs of this meeting, to hold the next TPV conference (TPV6) in Freiburg in approximately eighteen months. At that time, it should be clearer whether or not commercial and military interest in TPV is increasing or decreasing. Professor Joachim Luther and his colleagues at the Fraunhofer Institute for Solar Energy will take the leading role in organizing the next meeting and correspondence/suggestions etc. regarding it should be sent to him.

In this meeting, there was a change in the organization of earlier conferences and papers on Systems were placed first. This was followed by papers dealing with emitter (radiator) technologies, and then with those dealing with the semiconductor converters themselves. Arguments for and against this sequence may be found but, generally, this organization was popular amongst the attendees.

Most of the papers presented on systems and utilization, dealt with stand-alone power generator units intended for the commercial market. Typically, these were either directly fueled by gas, by heat recovered from a domestic gas-fired space heating unit, or by heat from a high-temperature industrial process. These attempts to promote the use of TPV in the private sector are almost entirely being pursued in Europe and Japan. Almost all of the investigations are quite innovative and, if successful, have the potential to open a larger commercial market for the specific semiconductor cell technology utilized.

The main topics in the section of the meeting dealing with emitters were concerned with selectively radiating emitters, optimized optics of photovoltaic cells and the electromagnetic coupling between emitter and photovoltaic converter. The introduction of concepts based on radiation in the near-field limit occurred at an earlier conference but this was the first at which the concept was demonstrated. It was surprising and regrettable that there were very few contributions dealing with the design and implementation of sophisticated filters systems used to return sub-bandgap radiation to the emitter. Filters beyond the quartz technology could still, in principle, play an important role in TPV photon management. There remain many problems associated with the materials used in optical systems that have to be solved to implement systems that are stable in the long-term and are highly efficient.

Timothy J. Coutts
Guido Guazzoni
Joachim Luther

SESSION 1

Systems and Utilization

Chair: T. J. Coutts

TPV Systems and State-of-Art Development

Robert E. Nelson

Quantum Group Inc.
San Diego, CA 92121

Abstract. Even though TPV energy converters may have weight, size, and cost advantages over other direct energy conversion techniques, TPV shares a low conversion efficiency with its direct energy conversion counterparts. Consequently, the virtues of TPV may be confined to low power applications where fuel costs are not significant. A number of commercial organizations are developing portable power systems embracing the full breath of TPV options. Fuels, which govern many system characteristics, range from methane to diesel fuel. Photoconverters are narrow band GaSb or InGaAs to wide band Si. Emitters are either broadband black body types, selective, or matched. Filters, particularly resonant filters, play a role in spectrum control. Optical and thermal recuperation are important. Fuel delivery, combustion control, ignition means, and thermal management impact overall performance. The design details of a recently developed TPV system will be covered.

BACKGROUND

The origin of TPV is generally traced by most observers[1] to Dr. Pierre Aigrain who was a visiting professor at MIT in late 1960 and early 1961. During one of many lectures, Aigrain proposed a direct energy conversion process comprised of a flame-heated radiant emitter and a semiconductor photoconverter. Dr. George Turner of Lincoln Laboratory called my attention to a short reference[2] in a Quarterly Research Report of that organization where H.H. Kolm used a Coleman camping lantern with a Welsbach mantle to illuminate a silicon solar cell. This experiment was performed in 1956, but apparently, no subsequent effort at Lincoln Laboratory developed. Regardless of the location of the origin, a series of TPV development programs began in the early 1960's and continued to the middle 1970's. General Motors Corporation was an active participant with contributions from their Defense Research Laboratory, Delco Radio Division, Allison Division, and Harrison Radiator Division. MIT also participated in advancing the technology with work done by Professors David White, John Blair and Bruce Wedlock. Prof. Richard Schwartz, who started his TPV work at MIT, continued the study of photodiodes under high injection conditions at Purdue University. Fort Monmouth was the site of the U.S. Army's major TPV developmental effort by Drs Kittl and Guazzoni. Dr. Guazzoni,[3] for example, was the first to report high-temperature spectral emittance of four rare earth oxides. By the middle 1970's, however, interest in the conventional applications of TPV faded.

The energy crises of the 1970's generated some interest in TPV for solar energy

CP653, *Thermophotovoltaic Generation of Electricity: 5th Conference*
edited by T. J. Coutts, G. Guazzoni, and J. Luther

conversion. Dr. Richard Swanson at Stanford performed a number of studies for the Electric Power Research Institute and reported on silicon photoconverters.[4] Ed Horne, then at Boeing developed a solar TPV concept.[5] Researchers in Europe also participated during this interval. P. Wurfel and W. Ruppel performed analytical studies of TPV systems at the University of Karlsruhe. In Italy, F. Demichelis and. E. Minetti-Mezzetti worked on TPV concepts at the Politecnico di Torino.

After a period of dormancy, interest in TPV reignited in the late 1980's. This renewal may have been prompted by the increased capabilities of photoconverters, advancements in emitters, or special military interests. The development activity and participants in this phase are recorded in the four volumes of the Thermophotovoltaic Generation of Electricity conferences sponsored by NREL.

As is the case in the development of many technologies, early predictions of performance are optimistic and zeal for participation is evident until realism sets in and modest reports of actual performance dampens the enthusiasm. TPV has clearly entered this phase as we now observe another decline in interest. This pessimism has been caused primarily by the realization that conversion efficiencies of TPV are low and are currently in the single digit range. Substantial inventive steps are required to restore the levels of past interest.

With the current prospect of low efficiencies, TPV has no ability to compete with prime movers like gas turbines or diesel engines. Even small internal combustion engines are in many cases more attractive than TPV options. There may be applications, however, where low efficiency is acceptable. Low power systems where fuel cost is inconsequential is one possibility. Power levels from a few watts up to about 500 watts satisfy our notion of low power. Furthermore, the user may require the attributes of a direct energy conversion technique such as low noise, no significant moving parts, and reliability. The direct energy conversion candidates include

Thermophotovoltaic
Thermoelectric
AMTEC (Alkali Metal Thermal-to-Electric Converter)
Thermionic

Generally speaking, all of the aforementioned direct energy conversion methods have about the same efficiency. We look to TPV for smaller size, lighter weight, and possibly lower costs as compared with the other candidates.

Most current low-power applications are serviced by batteries, and this sets the stage for a potential competition between TPV and batteries. Table 1 catalogs the energy densities (volumetric and gravimetric) of some common batteries. Rechargeable secondary and single-use primary batteries are included. Also included in Table 1 are the energy densities of common liquid hydrocarbon fuels.

Liquid hydrocarbon fuels have a significant energy density advantage over batteries. Liquid hydrocarbon fuels are also far less costly on a per unit energy density basis than batteries. The energy density stored in a battery (electrical), however, is more valuable than the energy stored in a fuel (chemical). Batteries power any number of light-generating, heating, or electronic appliances. While liquid fuels are best used in heating applications, the thermal energy derived from combusting these fuels may be used in a thermodynamic heat engine to generate the more valuable shaft power or electrical power. Of course, the TPV energy converter must meet size, cost, efficiency, and reliability criteria. There are additional advantages of liquid hydrocarbon fuels in that most of the fuels are stable and do not degrade during storage. Thus, there is no shelf life limitation for these fuels in appropriate containers whereas all electrochemical batteries degrade immediately after manufacture or activation. Unlike batteries, TPV systems provide rated power at low temperatures providing the proper fuel is employed. The energy density of all electrochemical batteries drops with temperature, and some batteries, such as carbon zinc, lose half their available energy at 0° C.

TABLE 1

ENERGY DENSITY[6]

	Volumetric Energy Density (Watt-hour / Liter)	Gravimetric Energy Density (Watt-hour / Kilogram)
SECONDARY BATTERIES		
Lead Acid	70	35
NiCd	90	37
NiMH	175	50
Li Ion	200	90
PRIMARY BATTERIES		
Carbon Zinc	100	65
Alkaline	330	125
Silver Oxide	500	120
Lithium	550	230
HYDROCARBON FUELS		
Propane	6896	15200
Isobutane	7567	13700
N-Butane	7872	13600
Gasoline	9215	13200
Diesel Fuel	10862	12400

There is a potential TPV application that may exceed slightly our self-imposed limit of 500 watts and where a low efficiency is tolerated. This is a cogeneration application where the primary output is thermal energy. An excellent example is a self-powered residential furnace,[7] which generates its own electrical power to provide for ignition, controls, and a motor for an air mover or water circulator. The electrical power required to operate the typical furnace (about 500 watts) is far less than the thermal output (about 30,000 watts). Full use of the conversion losses in the generation of electrical power is realized because this loss is added to thermal output of the furnace. This is a rare application where full recovery of all conversion losses is possible. This type of furnace may exhibit lower operating costs while benefiting from reliable service in areas experiencing electrical utility outages.

Now that we have defined a special niche for TPV, we note that there is another technology focussing on the same application. Small fuel cells are currently under active development[8] for portable electronic devices such as cell phones and laptop computers. Unlike TPV, which is a thermodynamic heat engine, fuel cells are electrochemical devices that operate isothermally. In addition, fuel cells have been under development for many years and far longer than most direct energy conversion methods. For low power applications, fuel cells have challenges. First, all fuels cells operate best on hydrogen. Hydrogen, however, is very difficult to store with an acceptable energy density. Very high pressures or very low temperatures are required to compress hydrogen. Storage in hydrides offers a slight advantage, but this approach has other limitations. Consequently, a search for a liquid fuel to power fuel cells directly led to methanol (also ethanol). Methanol, however, has a problem called "crossover," where the fuel migrates through the fuel cell membrane, which disables the device. "Crossover" can be eliminated by diluting the methanol with water (20 volumes of water for each volume of methanol), but this dilution reduces substantially the energy density of the stored fuel and, furthermore, the water has to be eliminated by processes such as evaporation, which requires energy. Some current development is aimed at recovering and recirculating the dilution water, which is desirable. Microminiature pumps and fans, however, are a challenge and require development. Another approach to the fuel problem is to select a fuel with a high energy density and reform the fuel to hydrogen. Generally, the reforming occurs at high temperatures with catalysts. Methanol is a candidate fuel for portable applications because it reforms at a relatively low temperature (about 200° C). Fuel reforming is undergoing significant development at automobile companies looking forward to fuel-cell-powered automobiles powered by gasoline or diesel fuel. Clearly, fuel cells are a formidable challenge to TPV technology.

CURRENT STATUS OF TPV TECHNOLOGY

The recent commercial developmental work in the USA has been reported mostly at the many NREL meetings. A brief review here of the current status should suffice. Lew Fraas at JX Crystals Inc. has developed GaSb photoconverters for illumination by silicon carbide emitters. Fraas has developed also what he calls matched emitters for GaSb cells. His matched emitters have a broader emission band than selective emitters in order to generate a higher power density. He employs chromia stained alumina[9] and antireflection coated refractory metals[10] as well as other material systems.

Ed Horne at EDTEK Inc. also employs GaSb cells for TPV systems. In addition, Horne employs inductively resonant mesh metal film filters,[11] which are selective and reflect emitter radiation with wavelengths above and below the peak responsivity wavelength of the photconverter. The resonant filters exhibit a very high reflectivity (high 90's %) at long wavelengths. Since the resonant filter consists of an etched pattern of openings on a metal film, this approach offers an added benefit of current collection if the metal film is deposited directly on the photoconverter.

The Tecogen Division of Thermo Power Corporation, a Thermo Electron Company recently completed a developmental program[12] for Fort Monmouth. A combination flashlight and TPV generator (3 watts) developed by Tecogen satisfies our criteria for a suitable TPV device---a low power portable device. The Tecogen system employs a light emitting mantle for the flashlight function and a selective fibrous erbia emitter illuminating a GaSb array for the TPV section.

The Quantum Group Inc. has developed a similar concept for a flashlight/battery charger for Fort Monmouth.[13] A discussion of the features of this device will outline the technical approach employed in the development of this cost-effective, low-power, portable appliance.

Fuel Choice

A significant first step in this design is the choice of fuel to power this device. Even though military interests prefer diesel fuel or jet fuel, there are system complexities with these fuels, and we elected to use propane, which is readily available and low cost ($0.458/gallon, Wall Street Journal, 6Sept02, p. B-7). More importantly, propane volatilizes at atmospheric pressure and at temperatures over which the power device is expected to operate. Propane boils at – 42.1° C, and there are other volatile candidates such as isobutane (boiling point: – 11.7° C) and normal butane (boiling point: - 0.5° C). Consequently, propane is chosen for applications requiring cold weather operation. Since propane is so volatile, it must be stored in appropriate hermetic containers. Fortunately, there are many consumer aerosol products, which dispense shaving cream, hair spray, cologne, air freshener, paint, etc. by employing propane-butane blends as propellants. Of course, in our application we fill the canister only with the propellant--propane.
The packaging for these products include impact-extruded aluminum canisters and standard aerosol valve assemblies that cap off the canisters. Figure 1 illustrates a representative array of small impact-extruded canisters with a rolled lid on the open end. The canisters come in various sizes. We have selected one size that is very close to the size of a standard "D" cell battery. Other canisters shown are similar in size to "AA" batteries, to two "AA" batteries in tandem, and to two "C" batteries in tandem. Since canisters and aerosol valves are

manufactured at rates that are measured in the billions annually, these components are mature in design, reliable, and low in cost.

Fuel/Air Delivery

The vapor pressure of propane above the boiling point rises with temperature. We require that the fuel rate be controlled over the full operating temperature range (-40° C to 50° C). In this application the pressure of the propane vapor is reduced and controlled by a diaphragm type of pressure regulator. A very simple atmospheric burner that provides a premix of fuel and primary combustion air that is approximately stoichiometric (about 24 volumes of air for each volume of propane vapor). If the propane vapor issues from a small diameter orifice and the high velocity vapor stream is directed down the throat of a venturi, the proper amount of primary combustion air can be entrained and mixed with the fuel. A photograph of the assembly is shown in Figure 2, and an assembly sketch and piece-part drawings of this fuel/air delivery system are depicted in Figure 3. The fuel air premix is delivered to the emitter and burned. The high velocity propane vapor stream is obtained with a regulated propane vapor pressure of 10 to 30 Kpa behind a small orifice with a diameter of 40 to 80 μm. A cost effective orifice is a drilled sapphire or ruby disc that is a major component in jeweled Swiss watches.

Ignition

Ignition of the fuel/air mixture at the emitter is accomplished with a high voltage arc discharge at electrodes located near the emitter. A piezoelectric igniter, which is a low cost component found these days in a disposable butane cigarette lighter, generates the high voltage pulse. In this component a piezoelectric material is deformed momentarily by a hammer blow and a high voltage (measured in kilovolts) is generated. The piezoelectric igniter contains a moveable plunger, which, when depressed, stores energy in an internal spring. Near the end of the plunger stroke, the hammer is released and propelled to the piezoelectric crystal by the spring. Contacts at the ends of the piezoelectric material are connected to the electrodes near the emitter. When the plunger is released, it returns to its original position, and the igniter is reset and ready for the next actuation cycle. The piezoelectric igniter is a reliable component with a cycle life measured in the tens of thousands of actuations.

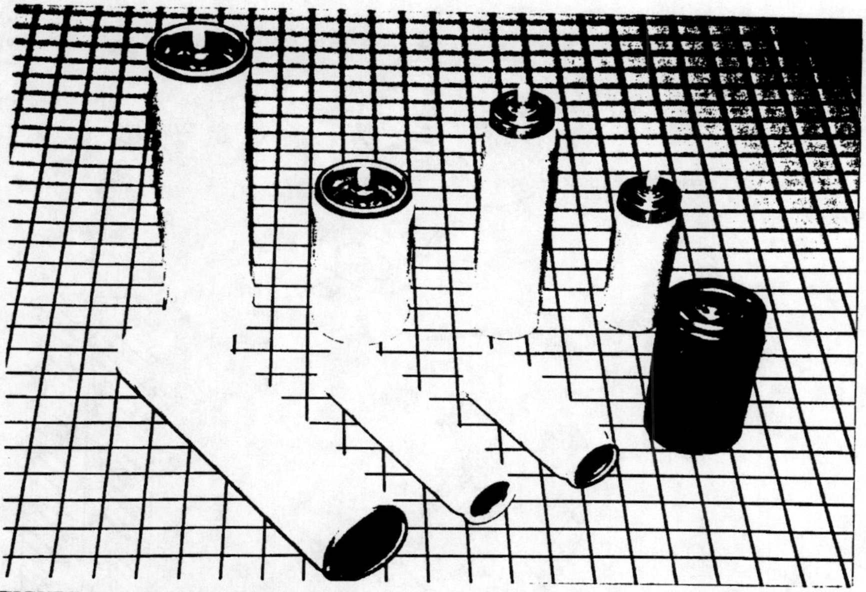

FIGURE 1. Open canisters are horizontal. Vertical canisters have aerosol valve lids. Vertical canister second from left is the same size as D cell battery shown on right.

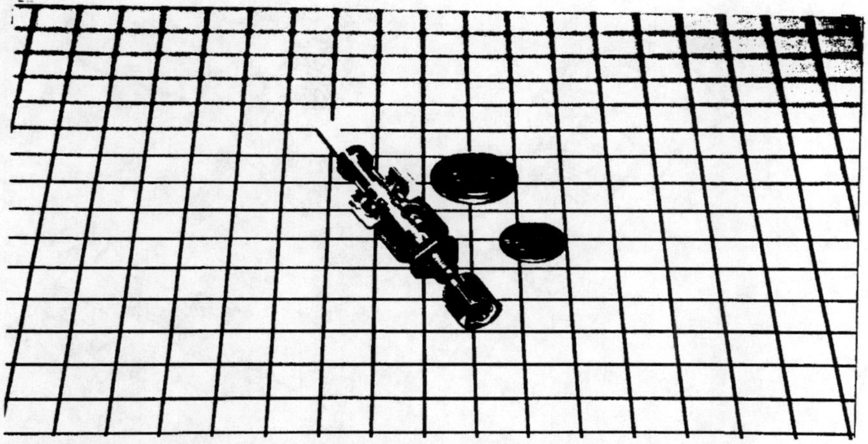

FIGURE 2. Atmospheric burner test fixture with recuperator for use in TPV battery charger.

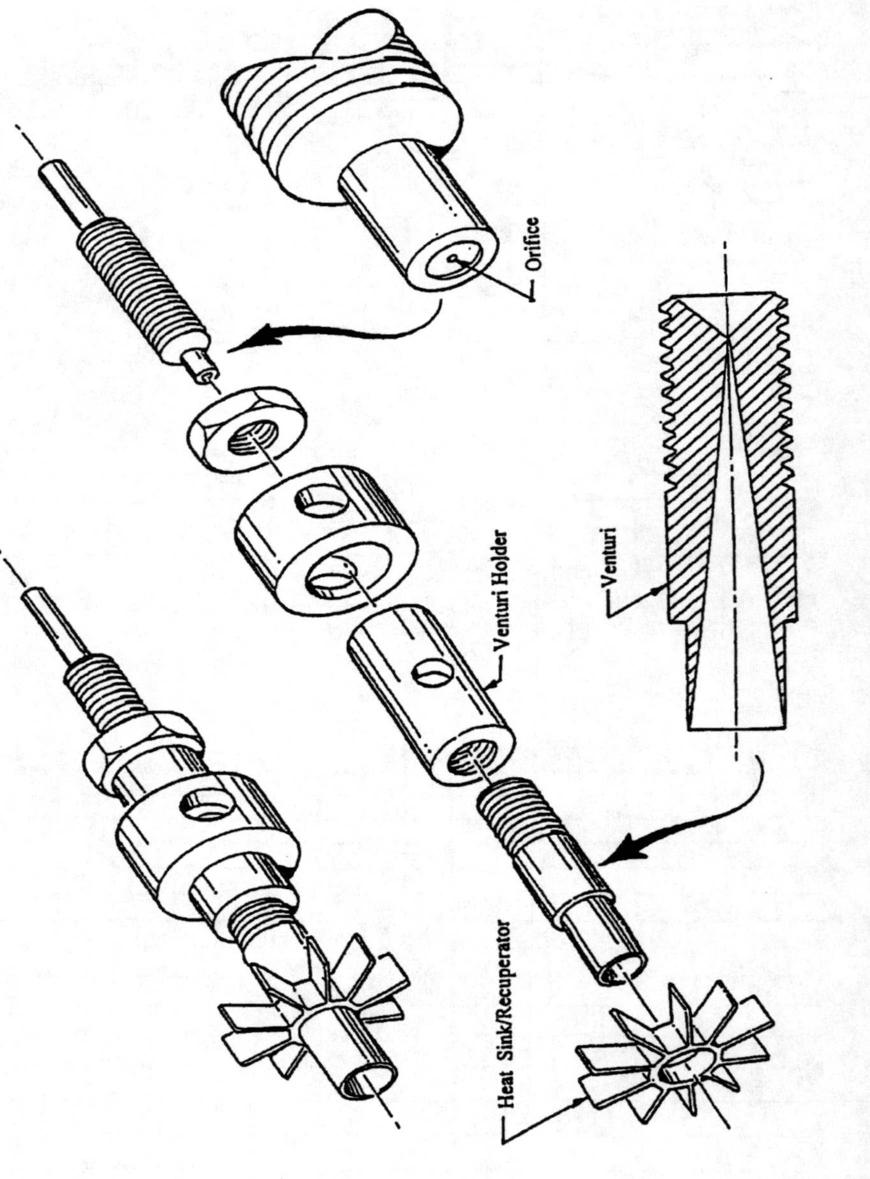

FIGURE 3. Components and assembly of atmospheric burner system.

Emitters

Mantle emitters are selected for both applications---for visible light generation and for IR radiation for TPV. The classical mantle geometry of a closed end cylinder yields an efficient emitter along with the potential for mechanical strength.

For visible light emission, we employ a network of filaments comprised of thorium oxide (ThO_2) doped with an optical modifier, which is 0.7% by weight of cerium oxide (CeO_2). This well-known composition was employed first by the Austrian chemist, Carl Auer, who invented this most efficient converter of gas combustion into visible light. This very refractory ceramic composition is a concern in some applications inasmuch as it is radioactive. The dominant component is a heavy element, which has no stable isotopes. Natural thorium Th^{232} is an alpha emitter with a half-life of 10^{10} years.

The TPV emitter is comprised of ytterbium oxide (ytterbia Yb_2O_3), which exhibits optically a single emission at 0.98 μm with a full width at half maximum of 150 nm. There is very low off-band emission in these structures since the ceramic filaments are very small in diameter (about 10 μm) and are optically transparent. The 0.98 μm emission is very dominant and arises from electronic transitions among vacancies in the inner electron shells, which is a property common in rare earth elements.

Specialized ceramic processing[14] of these emitters yield mechanically robust mantles that survive impacts of 1500 to 2000 g's.

Photovoltaic Array

The aforementioned ytterbia emitter illuminates a crystalline silicon cell array. The silicon cells have back contacts, which require no shadowing front surface grids. An etched front surface and a back reflector serve to trap 0.98 μm photons in the cell for enhanced generation of hole-electron pairs. All cell surfaces are treated and passivated to minimize surface recombination velocity.

Optical Components

In both systems the emitter is located approximately at the focus of a parabola of revolution. The parabolic reflector is formed from polished aluminum and has a silicon monoxide coating for passivation. The parabola's aperture is about 70 mm in diameter. Visible light from the flashlight passes through a lens system and emerges as a uniform beam. Infrared radiation is intercepted by the array of crystalline silicon cells in the battery charger. A single layer of transparent quartz (or borosilicate) glass resiliently mounted protects the emitter from accidental contact. The planar glass lens is vulnerable and needs protection in the flashlight application. A domed polycarbonate lens with a proper antireflection coating and spaced apart from the planar lens provides superb mechanical protection for the glass lens without compromising the optical performance. The silicon photovoltaic array mounted on its heat sink serves also to protect the glass lens in the TPV battery charger.

Recuperator

A recuperator is desirable in these applications because the exhaust gases leaving the emitter are at very high temperatures. If some of the exhaust gas enthalpy can be transferred (recuperated) to the incoming primary combustion air, system efficiency is enhanced. Unfortunately, in our simple systems, it is difficult to employ a recuperator in the flashlight because we depend on free convection to disperse the exhaust gases. Since we anticipate that the flashlight will be used in all orientations, this rules out a flashlight recuperator in our design. For the TPV battery charger, however, we recognize that the charging process takes hours, and the user will leave the equipment unattended. Consequently, it is relatively simple to design the package so that the user is encouraged to place this equipment in the desired emitter-pointing-down orientation. The small heat exchanger (with radial fins) attached to the fuel/air delivery conduit in Figure 2 is an effective and very low cost recuperator. With the TPV battery charger in the proper orientation, the efficiency is increased by 30% with the recuperator. If the user disregards the orientation requirement, the equipment will still function but more time and fuel are required to recharge.

Complete Systems

Assembly sketches of the flashlight and the TPV battery charger are shown in Figures 4 and 5. Figures 6a and 6b are photographs of the prototypes. Propane fuel contained in a single canister/aerosol valve package is inserted into the device similar to battery insertion in a flashlight. The fuel is fed to the input port of the pressure regulator, which is adjusted for the proper pressure at the outlet port. A small stem containing the orifice is connected to the outlet port to direct the fuel vapor stream down the venturi to mix the fuel and entrained primary combustion air as discussed earlier. The fibrous mantle emitters are mounted on ceramic mantle support tubes, which serve as conduits to deliver the combustible fuel/air mixture to the mantle. Ignition is accomplished by an electrical spark at the ignition electrodes connected to a piezoelectric igniter (not shown). The optical components, reflectors and lenses, have been described. Note in the photograph of

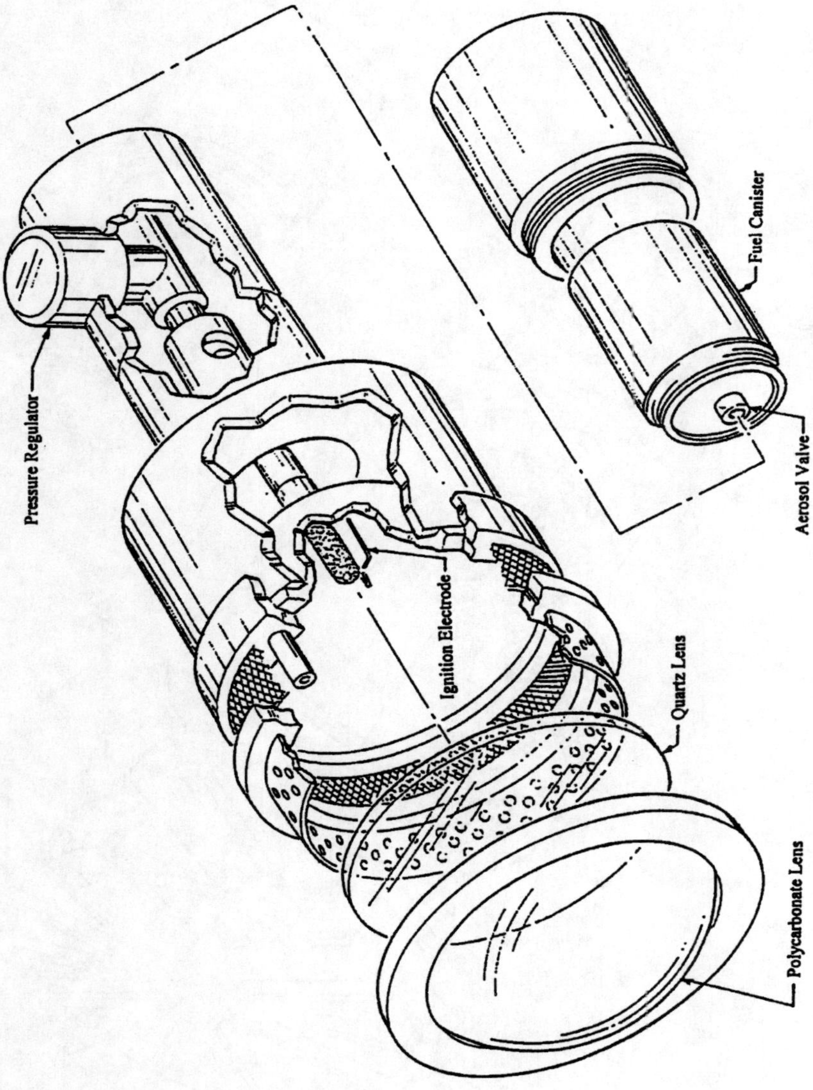

FIGURE 4. Prototype flashlight assembly

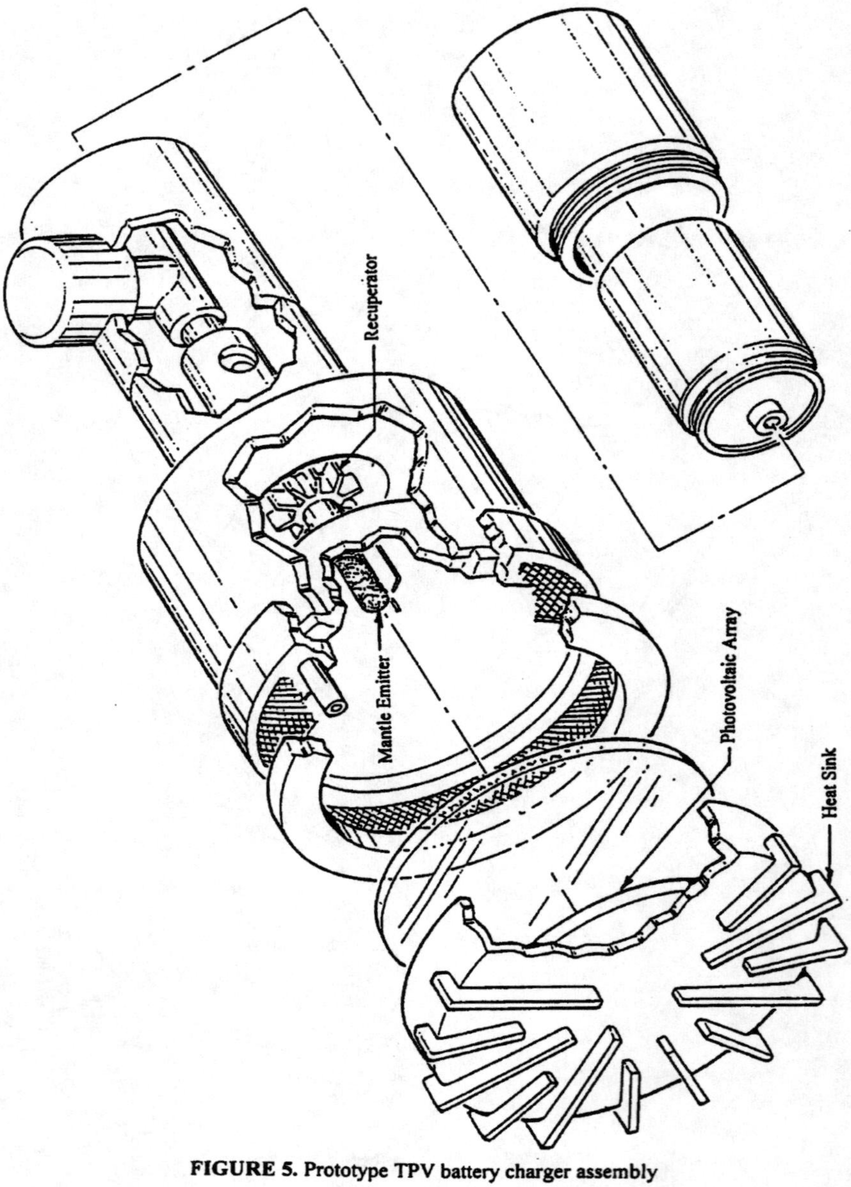

FIGURE 5. Prototype TPV battery charger assembly

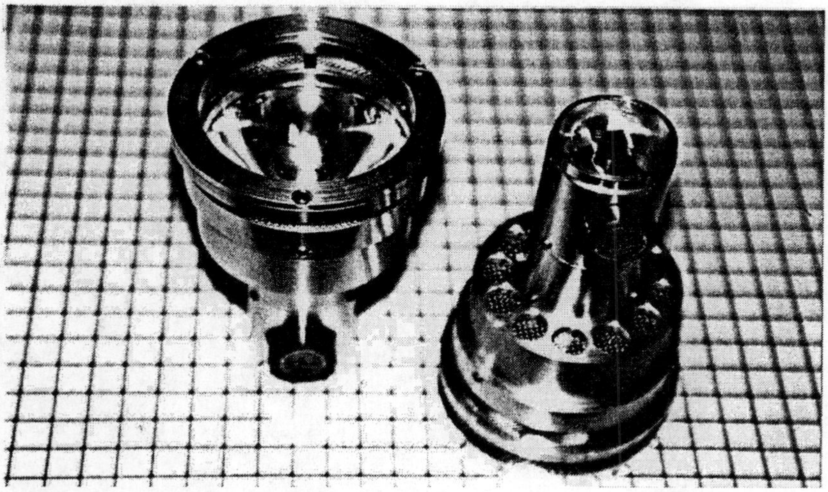

FIGURE 6a. End views of prototypes. TPV battery charger on left is shown without PV array.

FIGURE 6b. Side views of prototypes. Flashlight shown in front.

the prototypes in Figure 6a that all air inlets/exhaust ports are covered with metallic screens that serve as flame arresters. That is, this equipment is designed so as not to serve as an ignition source in the event that these devices are used in locations where there may be spilled flammable liquids or combustible gas leaks. The flame arresters serve also another function in that they tend to attenuate weather effects such as wind and rain. Since mantle emitters require a small amount of secondary combustion air, some of the screened ports are air inlets while other screened ports are exhaust outlets. Air inlets and exhaust ports will interchange as the orientation of the device changes.

Operational Results

Flashlight

Luminous Output	18 lumens
Fuel Rate (propane)	1 g/hr
Fuel Content ("D" size canister)	25 g
Color Temperature	2450 K

Color Rendition (Comparable to tungsten filament @ 2450 K)

TPV Battery Charger

Efficiency	3-5%
Power Output	1-3 watts
Fuel Rate	2-6 g/hr

Voltage/current (Depends on cell size, number and interconnections)

References

1. D.C. White and R.J. Schwartz, *Proceedings of NATO AGARD Conference*, Cannes, France, 1964, Gordon & Breach Science Publishers, London, 1967, pp.897-922.

2. H.H. Kolm, Quarterly Progress Report, Solid State Research, Group 35, MIT-Lincoln Laboratory, Lexington, MA, 1May1956, p. 13.

3. G. Guazzoni, *Applied Spectroscopy*, **26**, No. 1, 1972, pp. 60-65.
4. R.M. Swanson, *Proceedings of the International Electron Devices Meeting*, Washington, D.C., 8-10December1980, pp. 186-189.

5. W.E. Horne, U.S. Patent No. 4,313,024, Filed 5April1977, Issued 26January1982.

6. For battery data, see D. Linden, *Handbook of Batteries, 2nd ed.,* Editor-in-Chief David Linden, McGraw-Hill Inc., New York, 1995, p. 7.8 and p. 23.10. For gross heat of combustion of hydrocarbon fuels, see C.F. Taylor, *The Internal Combustion Engine in Theory and Practice, Vol. 2, Combustion, Fuels, Materials, Design*, MIT Press, Cambridge, MA, 1968, pp. 120-121.

7. R.E. Nelson, "Grid-Independent Residential power Systems," *The Second NREL Conference on Thermophotovoltaic Generation of Electricity,* AIP Conference Proceedings, **358**, Eds. J.P. Benner,T.J. Coutts, and D.S. Ginley, 1995, pp221-237.

8. D. Voss, *Technology Review*, **104,** No. 9, 2001, pp. 68-75.

9. L. Ferguson and L. Frass, "Matched Infrared Emitters for Use with GaSb TPV Cells," *The ThirdNREL Conference on Thermophotovoltaic Generation of Electricity*, AIP Conference Proceedings, **401**, Eds. T.J. Coutts, C.S. Allman, and J.P. Benner, 1997, pp. 169-179

10. L. Frass, J. Samaras, J. Avery, and L. Minkin, "Antireflection Coated Refractory Metal Matched Emitters for Use with GaSb Thermophotovoltaic Generators," *Proceedings of the 39th Power Sources Conference,* Sponsored by U.S. Army Communications Electronics Command and U.S. Army Research Laboratory, Cherry Hill, NJ, 12-15June2000, pp. 367-370.

11. W. E. Horne, M.D. Morgan, and V.S. Sunduram, *Proceedings of the 31st Intersociety Energy Conversion Engineering Conference*, 1996, pp. 995-1000.

12. U.S. Army CECOM Contract No. DAAB07-99-C-D251.

13. U.S. Army CECOM Contract No. DAAB07-99-C-D252.

14. W.J. Diederich and R.E. Nelson, U.S. Patent No. 4,883,619, Filed 24Feb1986, Issued 28Nov1989.

Record Electricity-to-Gas Power Efficiency of a Silicon Solar Cell Based TPV System

Bernd Bitnar[1], Jean-Claude Mayor[1], Wilhelm Durisch[1], Andreas Meyer[2], Günther Palfinger[1], Fritz von Roth[1], Hans Sigg[1]

[1] Paul Scherrer Institut, CH-5232 Villigen PSI, Switzerland
[2] Solaronix SA, CH-1170 Aubonne, Switzerland
Author for correspondence: Bernd Bitnar, email: bernd.bitnar@psi.ch

Abstract: In this paper we report on the development and characterisation of a small TPV prototype system, which uses silicon photocells and a rare earth selective emitter. A simulation model of this system was developed, which allows studying the system theoretically. The fabrication of the selective incandescent mantle emitter from Yb_2O_3 and detailed measurements of its radiation power and emissivity are presented. The maximum emissivity was 0.85 at 1.27 eV. An emitter temperature of 1735 K was obtained for an approximately 75 cm^2 large emitter heated by a butane burner. A SnO_2 filter tube was developed. The photocell generator is composed of monocrystalline silicon solar cells and a water-cooling circuit. The prototype system reached, without a selective filter and without preheating of the combustion air, a record electricity-to-gas power efficiency of 2.4 %. We compare the experimentally achieved system efficiency with simulations using our model. The possibilities to further increase the system efficiency are discussed.

INTRODUCTION

Up to now research in thermophotovoltaics (TPV) is mainly focused on the development of low bandgap photocells that convert infrared emission of a broadband radiator with a typical temperature of 1500 – 2000 K with a high efficiency. Possible materials are Ge [1], GaSb [2] or different GaSb compounds [3]. However, photocells from these materials are produced only in a laboratory scale up to now. Therefore, they remain expensive and most of them contain the toxic element Sb.

A promising application of TPV is the integration of a TPV generator into a residential heating system [4]. The TPV system supplies electricity that is used to run the heating system independently of the electrical grid. Typically an electrical power of about 150 W is necessary to supply sufficient electricity to run a 20 kW_{th} gas heating system [4]. If the electrical power of the TPV generator exceeds the electricity consumption of the heater, the additional generated electricity could be fed into the house-grid or into the utility mains, e.g. during high electricity-demand periods. If TPV heating systems should not be delimited to the niche-market of off-grid sites, the most important requirement of the TPV system is its cost, which has to be considerably lower than that of the heater itself. Furthermore, the use of non-toxic and ecologically unhesitating components is important for a broad acceptance by the customers.

CP653, *Thermophotovoltaic Generation of Electricity: 5th Conference*
edited by T. J. Coutts, G. Guazzoni, and J. Luther

We favour therefore silicon photocells to be used in TPV systems. Monocrystalline silicon solar cells are available in large quantities and with relatively low costs, so that even large photocell modules are not expensive and, of course, silicon is non-toxic. However, because of the high bandgap of silicon ΔE=1.12 eV, most of the radiation of a grey body emitter with a temperature below 1800 K (emission maximum at E = 0.75 eV) can not be converted by silicon photocells. To reach a sufficient high efficiency of a silicon TPV system, a selective emitter matched to the silicon bandgap is indispensable.

We designed a TPV generator, which uses an Yb_2O_3 selective mantle emitter and silicon photocells. To study the system theoretically, a simulation model was developed, which allows investigating the influence of different components on the system efficiency. We discuss the individual components of our system. At the end of this paper, experimental results of the prototype system are compared to simulations. Technical details of the prototype as well as next steps towards a TPV gas heating system are described in our forthcoming paper in this volume (in the following text referred to as Ref. A):

Bitnar B., Durisch W., Meyer A., Palfinger G., „New flexible module for thermophotovoltaic applications".

SIMULATION MODEL

We developed a simulation model, which allows a detailed study of a TPV system theoretically. Fig. 1 shows the structure of the model. The model accounts for the cylindrical geometry of our system, as shown in Fig. 1. For the calculation we used the following assumptions: All surfaces are diffuse reflectors, heat transport by convection as well as optical absorption in the gas between emitter and filter are neglected.

The system is described by the following physical parameters (Fig. 1):

$$\vec{q} = (q_1, \ldots, q_{10}) \quad \text{vector of internal radiation fluxes,}$$

$$\vec{P} = (P_{em}, P_{fi}, P_{pc}, P_{el}, P_{r1}, \ldots, P_{r4}) \quad \text{vector of external energy fluxes,} \quad (1)$$

$$\vec{E}(E,T) = (\varepsilon_1(E)L_1(E,T), \ldots, \varepsilon_{10}(E)L_{10}(E,T),) \quad \text{vector of emission characteristic.}$$

Here ε_i means the emissivity of the surface i and L_i the blackbody radiation power at the temperature of the surface i. T is the temperature and E the photon energy. The electrical power P_{el} is given by:

$$P_{el} = \int_E \eta_{el} \cdot q_6 dE \quad (2)$$

with the photocell spectral efficiency $\eta_{el}(E)$ and the irradiation power $q_6(E)$. The emitter radiation power P_{em} is calculated from a combustion model [5]:

$$P_{em} = (T_{ad} - T_{em}) \cdot (c_{p,g}\dot{m}_g + c_{p,a}\dot{m}_a) \quad (3)$$

with the adiabatic combustion temperature T_{ad}, the emitter temperature T_{em}, the specific heat capacities for gas $c_{p,g}$ and air $c_{p,a}$ and the mass flows of fuel gas $\dot{m}_g$ and air $\dot{m}_a$. With the lower heat value ΔH of the combustion gas, the thermal power P_{th} of the burner is given by:

$$P_{th} = \Delta H \cdot \dot{m}_g \quad . \tag{4}$$

The geometrical relations of the radiation exchange between the internal surfaces i and j are defined by the radiation configuration factors F_{ij}, which were determined for our specific geometry using [6, 7, 8].

Our model uses the net radiation method, which is described in [9]. The internal radiation balance is given by the equation:

$$[M(E)] \cdot \frac{\partial}{\partial E}\vec{q} = -\vec{E} \tag{5}$$

and the external energy balance by

$$\frac{\partial}{\partial E}\vec{P} = [N(E)] \cdot \frac{\partial}{\partial E}\vec{q} \quad . \tag{6}$$

The array [M(E)] contains the radiation configuration factors F_{ij} and the transmittance $T_k(E)$ and reflectance $R_k(E)$ spectra of the emitter, the filter, the reflectors and the photocells as input parameters. The elements of [N(E)] are the radiation configuration factors and $\eta_{el}(E)$.

The differential equations (5) and (6) can be numerically solved to determine $\vec{q}, \vec{P}, \vec{E}$ and therewith the temperature of all components. Internal multiple reflections are automatically taken into account.

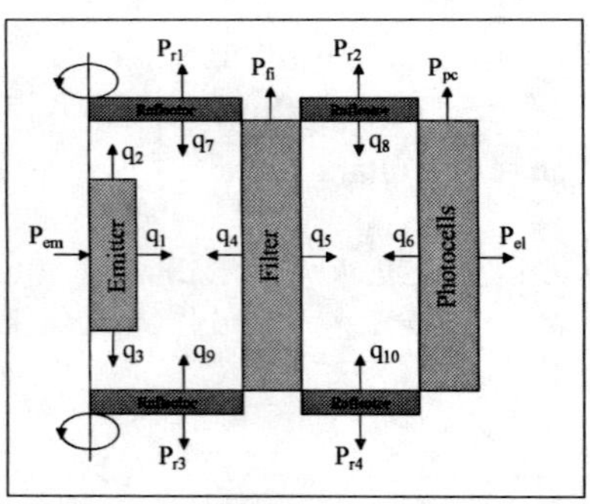

FIGURE 1: Schematically drawing of the simulation model. The cylindrical system consists of the radiation emitter, a selective filter, the photocell generator and reflectors in axial direction.

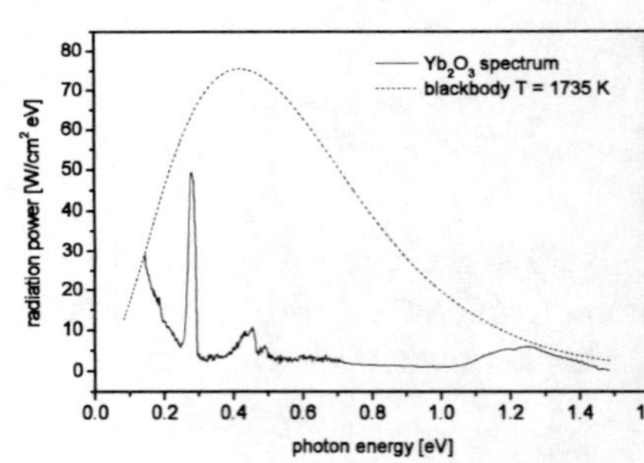

FIGURE 2: Radiation spectrum of the Yb_2O_3 mantle emitter in comparison to a blackbody spectrum with a corresponding temperature of 1735 K.

An opening in the upper reflector, which is necessary to let the exhaust gas flow out in a real system, is considered by calculating an effective T and R of the reflector. In a similar way, the view factor of the photocell module, which is defined as the relative coverage of the module area with cells, is taken into consideration by correcting P_{el} and the cell reflectance.

In practice, the spectra $T_k(E)$ and $R_k(E)$ of the system components and $\eta_{el}(E)$ of the photocells are measured experimentally and used as input parameters for the model. Together with the parameters of a specific combustion gas and P_{th}, P_{el} is determined and can be compared to experimental results with a TPV system.

YB$_2$O$_3$ MANTLE EMITTER

Various rare earth oxides show selective emission bands at different photon energies [10]. The emission band of Yb_2O_3 is matched very well to the bandgap of silicon. Yb_2O_3 is stable at high temperatures in air.

We investigated Yb_2O_3 incandescent mantle emitters. Mantles are formed by thin porous structures. The gas penetrates through the emitter during burning. Therefore the contact area between the hot flame gases and the emitter is large and the temperature difference between the flame close to the emitter and the emitter itself is supposed to be very small. The optical transmittance of the emitter mantle due to its pores was found to be typical 6 % and does not reduce the emitting area significantly.

Our emitters were heated with Butane and Methane burners with a thermal power of 1.2 – 2.0 kW.

Emitter Fabrication

In the past, the incandescent mantle emitters were fabricated from raw mantles, which were impregnated with Yb_2O_3 from Auer (Berlin, Germany). Our new emitters are impregnated by ourselves in the facility of Solaronix.

The cotton raw mantle is impregnated with Yb-nitrate, followed by a hydrolyses in Yb-hydroxide. The impregnated mantles were fixed either directly in an elliptical shape to a burner tube, or with ceramic discs at the top and the bottom into a cylindrical shape to a metal rod above a tubeless burner. In a baking process at about 500°C the Yb_2O_3 ceramic structure is formed and the cotton mantle is burnt. We used emitters with an area of about 75 cm^2 for a small prototype system and 500 cm^2 large emitters for a demonstration system.

The disadvantage of the described Yb_2O_3 mantle structure is its poor mechanical strength. In future we plan to apply our process to create an overcoat of a ceramic mesh substrate to combine the excellent optical behaviour of the mantle with the mechanical strength of the ceramic mesh.

Optical Characterisation

An Yb_2O_3 mantle emitter with an area of approximately 75 cm^2 was fabricated and its temperature and emission spectrum were measured. The emitter was mounted open, so that no radiation from the environment was back reflected, and it was heated with a 1.2 – 2.0 kW tubeless butane burner. To determine its temperature, we used a Flash assisted Multiwavelength Pyrometry (FAMP) [11], which measures the spectrum of a flashlight reflection from the emitter together with its emission spectrum. The emitter transmittance was determined at the cold mantle using a Perkin Elmer Lambda 19 spectrometer. To measure the emission spectrum in the range 0.15 –1.50 eV the emitter radiation was coupled into a Bruker 113 IV FTIR spectrometer. A detailed description of our optical characterisation method including the calibration of the FTIR spectrometer is given in [12].

We measured an emitter temperature of 1735 K. Fig. 2 shows the radiation spectrum of the emitter in comparison with a corresponding blackbody spectrum at 1735 K.

The emitter spectrum shows the selective emission peak with its maximum at $E = 1.27$ eV, two emission bands at 0.45 eV and 0.28 eV and a continuous emission at $E < 0.2$ eV, which is assumed being blackbody emission for $E < 0.15$ eV. The two bands at 0.45 eV and 0.28 eV were also visible in a spectrum of the butane flame without emitter and we identified the relatively sharp line at 0.28 eV with CO_2 emission and the 0.45 eV band occurs most probably from O-H bonds of water in the exhaust gas [11].

As the result of our measurements, Table 1 summarizes some characteristics of the emitter. The selective peak has its maximum emissivity ε_{max} at the photon energy E_{peak}. The selectivity $\Delta_{1.1}$ is defined as the radiation power integrated over photon energies above the silicon bandgap of 1.12 eV relative to the total radiation power P_{rad}. For comparison, a blackbody with $T = 1735$ K has a selectivity $\Delta_{1.1}$ of only 5 %.

TABLE 1. Results of the optical characterisation of an Yb_2O_3 mantle emitter

T [K]	ε_{max}	E_{peak} [eV]	P_{rad} [W/cm^2]	$\Delta_{1.1}$ [%]
1735 ± 10	0.85	1.27	8.4	20

Therefore, the Yb_2O_3 mantle emitter has an emission spectrum suitable to illuminate silicon photocells. The convertible radiation power at the emitter surface is 1.7 W/cm^2. Compared to sunlight, the emitter radiation is concentrated by about a factor of 20.

SELECTIVE FILTER

To further increase the selectivity of the radiation spectrum, a selective filter is used, which transmits the convertible radiation and reflects sub-bandgap radiation back to the emitter. Transparent conducting oxides (TCOs) show a high transmittance for radiation with a frequency higher than the plasma frequency of the charge carriers and a high reflectance for lower frequency radiation. The plasma frequency can be varied with the doping of the film. Therefore the optical characteristics of a TCO filter can easily be matched to a given emitter-photocell combination in a TPV system. TCOs are already commercially applied for several filter applications, i.e. lamp housings, and widely available on the market.

A disadvantage of TCO filters is free carrier absorption in the film at photon energies around the plasma frequency. This absorption decreases with increasing carrier mobility.

We developed a special CVD process to deposit F doped SnO_2 films on flat substrates or onto the inner surface of quartz or heatproof borosilicate tubes.

A filter sample was deposited onto a borosilicate-wafer and characterised in detail. In Fig. 3 the transmittance and reflectance spectra are shown, which were measured with a PE lambda 19 spectrometer and (at the low photon energies) with the Bruker 113 IV. The transmittance at the emitter radiation maximum of 1.27 eV (vertical line in Fig. 3) is 76 %. The absorption, which was calculated from T and R, appears in a spectral region, where the emitter radiation power is low (see Fig. 2).

To determine the physical film parameters, we measured the sheet resistance of the SnO_2 film with four-point-measurement, its thickness with profiling an etched step in the film and the carrier density and mobility by Hall measurements. The contacts for the hall measurement were formed with screenprinted silver paste on the film, which have been wire bonded to the sample holder. Table 2 gives a list of the resulting film characteristics. The error in sheet resistivity gives the inhomogeneity over the 4" wafer, the uncertainty in thickness is caused by the accuracy of the profiling technique. The difference in sheet resistivity from 4-point and Hall measurement results from the non-uniform film thickness.

TABLE 2. **Characteristics of the SnO_2 filter**

Sheet Resistivity [Ω/□] from 4-point-probe	**13.6 ± 2.7**
Layer thickness [nm]	**468 ± 23**
Carrier density [cm^{-3}]	**$7.1*10^{20}$**
Mobility [cm^2/Vs]	**17.8**
Sheet Resistivity [Ω/□] from Hall measurement	**10.5**
Layer thickness [nm] from simulation	**490**

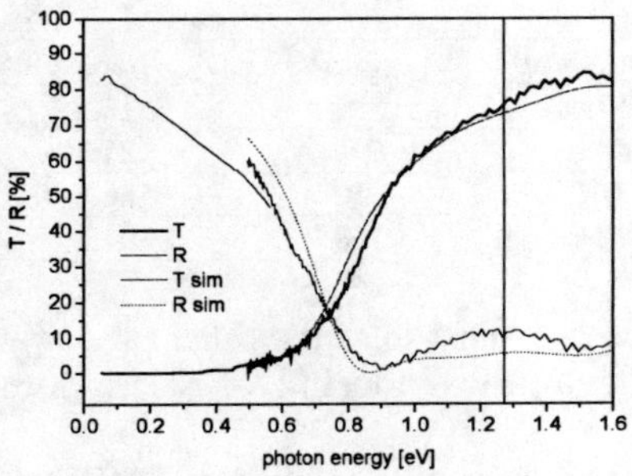

FIGURE 3: Transmittance T and reflectance R of the SnO_2 filter. The result of a simulation (sim) of the T and R spectra is shown, dashed lines.

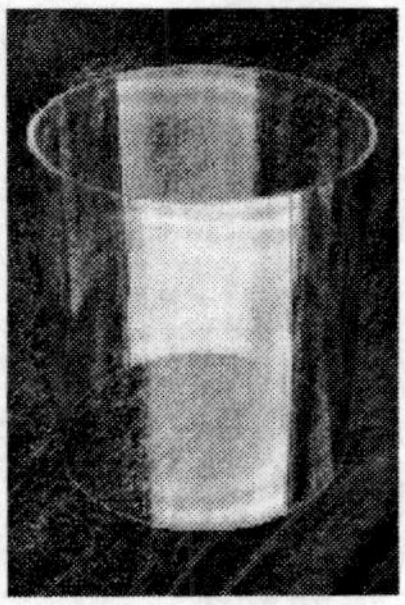

FIGURE 4: TPV filter tube. The SnO_2 layer was deposited on the inner surface of a quartz tube.

With these experimental filter parameters, T and R spectra were calculated, which are although shown in Fig. 3. A dielectric constant ε_∞ of 3.70 from [13] was used for the calculation. The effective mass of electrons in the n-doped film was fitted to the spectra to 0.48 m_0 and agrees with the values given in [13]. The thickness of the film was varied for an optimum agreement to the measurement and the resulting thickness is given in Table 2, too.

The TPV filter consisting of this SnO_2 film on the inner surface of a quartz tube is shown in Fig. 4. The tube has a radius of 6.5 cm and a length of 13 cm. The deposition is uniform over almost the whole tube length. For technical reasons, a border of about 2 cm is non-uniform, as is appearent from the interference rings shown in Fig. 4. This border-area can easily be cut off.

SILICON PHOTOCELL MODULE

To convert the Yb_2O_3 emission, silicon photocells are well suited. Important cell parameters for a sufficient high efficiency are a high external quantum efficiency (EQE) close to the bandgap and a small series resistance, because of the high radiation power of the emitter. For the construction of TPV prototype systems, we selected two monocrystalline silicon solar cells, which fulfills these requirements: the SH2 cell from ASE (Heilbronn, Germany) and the high efficiency cell type C from the University of New South Wales UNSW (Australia). Both cells show a high EQE at photon energies of the selective emission peak. The EQE of the ASE cell is compared with the Yb_2O_3 radiation spectrum in [14]. The results of IV-measurements, carried out under a sun simulator and under Yb_2O_3 irradiation, are given in Table 3. The cell temperature T, the short circuit current I_{SC}, the open circuit voltage V_{OC} and the maximum power P_{max} of the cells are shown. For the Yb_2O_3 illumination, the cells were mounted onto a metal cooling plate in a distance of about 70 mm to the Yb_2O_3 mantle emitter, heated with the butane burner. The efficiency of the UNSW cell is quite higher than that of the ASE cell. But the ASE cell is much less expensive, so that it can be used to build larger photocell modules to a reasonable price (see [4]).

TABLE 3. IV-characteristics of photocells

Cell	radiation	T [°C]	I_{SC} [mA/cm^2]	V_{OC} [mV]	P_{max} [mW/cm^2]
ASE	Sun AM1.5	25	37.4	595	16.0
UNSW	Sun AM1.5	25	40.6	694	21.1
ASE	Yb_2O_3	28	146	630	50.8
UNSW	Yb_2O_3	24	171	738	81.2

From both photocells modules were built. These consist of cells glued onto water cooled metal blocks, which were mounted into a cylindrical housing. A description of the module construction is given in [15].

Table 4 shows the result of an IV-measurement of the module with the UNSW cells under sun irradiation, mounted onto a suntracker. The module consists of 22 cells with a total area of 480 cm^2. The sun global irradiation for the measurement was 805 W/m^2. Corrected to a cell temperature of 25°C, a module efficiency of 20.7 % was achieved. The low resistive cell interconnection in the module as well as the use of almost identical cells result in a module efficiency close to that of a single cell.

The spacing between the cells was minimised, so that the view factor of the module is 0.91 for the ASE cell generator and 0.81 for the UNSW cell generator, respectively. In the ASE module all 16 cells are series connected. The UNSW module has 2 series connected strings with 11 cells each.

TABLE 4. IV – characteristic of UNSW cell module, 22 series connected cells

I_{SC} [A]	V_{OC} [V]	FF [%]	η [%]	T [°C]
0.72	13.8	73.4	18.9	47

RESULTS

First, experiments with the butane TPV prototype system (see Ref. A) were carried out. The results are given in Table 5. The prototype was tested with both photocell generators. The quartz tube from the standard configuration was changed to the SnO_2 filter tube. The combustion air had room temperature for all experiments.

The system efficiency of **2.4 %** with the quartz tube and the UNSW cell module is to our knowledge a **world record for a silicon TPV system**. The experiment with the ASE cell module achieved a lower efficiency of 1.6 %. The reduced η_{sys} of the ASE module is consistent with the lower performance of the single ASE cell, given in Table 3.

The SnO_2 filter, unfortunately, achieved a lower η_{sys} compared to the quartz filter. The transmission loss of the SnO_2 filter seems to overcompensate the effect of increased emitter temperature due to back-reflected sub-bandgap radiation.

TABLE 5. Experimental results with the butane TPV prototype system

Photocells	Filter	P_{th} [kW]	I_{SC} [A]	V_{OC} [V]	FF [%]	P_{el} [W]	η_{sys} [%]
ASE	quartz	1.85	4.0	10.4	69	29.0	1.6
UNSW	quartz	1.98	7.9	8.1	75	47.9	2.4
UNSW	SnO_2	2.12	7.4	8.0	76	44.7	2.1

For a detailed analysis, the butane TPV experiments were simulated using the above-described model. Table 6 shows the simulation results. The agreement between simulation and experiments with the quartz filter is excellent. The increased T_{em} of the UNSW system compared to the ASE system by about 20 K results from the higher reflectivity for sub-bandgap radiation of 50 % of the UNSW cell compared to only 20 % of the ASE cell.

The simulation of the SnO_2 filter, which includes the filter absorption losses, result in an increased η_{sys} by about 10 % compared to the quartz tube. We give two possible reasons for the contradiction between experiment and simulation:

- The assumption of neglecting absorption in the exhaust gas between the emitter and the filter is not allowed, cf. the two gas emission lines in the emitter spectrum (shown in Fig. 2). This may lead to re-absorption of the reflected radiation.
- The air number is not well defined in the experiments with the self-sucking butane burner. Due to the higher emitter temperature with SnO_2 filter, the amount of sucked air, maybe, was changed compared to the quartz filter experiment.
- P_{th} of the burner was determined by weighting the mass loss of butane in the cylinder during experiment. An exact control of the power is not possible.

TABLE 6. Simulations of the butane prototype system

Photocells	Filter	P_{th} [kW]	T_{em} [K]	P_{el} [W]	η_{sys} [%]
ASE	quartz	1.85	1756	28.6	1.5
UNSW	quartz	1.98	1777	46.9	2.4
UNSW	SnO_2	1.98	1803	50.6	2.6

For a more detailed study of this problem, we used the methane TPV system (see Ref. A), which allows the exact definition of the air number. Table 7 gives the results of simulations corresponding to the experimental results given in Table 8. The ASE photocell module was used. P_{el} was measured with a power tracker. λ was 1 for all investigations. The experiments achieved only lower efficiencies compared to the simulations. The top reflector increases the emitter temperature resulting in an efficiency enhancement. In the experiment, the SnO_2 filter achieved the same efficiency as the quartz tube. This disagrees with the strong increase in efficiency due to the SnO_2 filter predicted from the simulation.

TABLE 7. Simulations of the methane prototype system

P_{th} [kW]	Filter	Top reflector	T_{em} [K]	P_{el} [W]	η_{sys} [%]
1.8	quartz	-	1725	19.7	1.1
1.8	quartz	yes	1739	24.0	1.3
1.8	SnO_2	yes	1779	34.4	1.6

TABLE 8. Experimental results with the methane prototype system

P_{th} [kW]	Filter	Top reflector	T_{cell} [°C]	P_{el} [W]	η_{sys} [%]
1.8	quartz	-	22	15.4	0.9
1.8	quartz	yes	26	19.1	1.1
1.8	SnO_2	yes	25	19.1	1.1

Fig. 5 shows P_{el} as a function of P_{th} resulting from a simulation. The corresponding experimental results are shown in Fig. 6. We summarise our results from these measurements:

- From the simulations, an almost linear function of P_{el} versus P_{th} is expected. The enhancement due to the SnO_2 filter should be almost constant over P_{th}.
- The experiments with the quartz filter agree well with the simulations for P_{th} less than 1.4 kW. The increase in P_{el} with P_{th} is sub linear in the experiment. This is also visible in I_{SC} versus P_{th}, which is not influenced by errors in the power tracking (see Ref. A) and insensitive to the cell temperature.
- The experiments with the SnO_2 filter achieved a slightly higher efficiency and I_{SC} for low P_{th} (Fig. 6) and a reduction of both for high P_{th}.

From these studies we deduce, that the emitter is not sufficiently heated from sub-bandgap radiation, reflected back from the SnO_2 filter, to overcompensate the loss in transmittance through the filter. Only at low P_{th} we could demonstrate the positive effect of back-reflected radiation in principle. The simulation overestimates this heating, because absorption in the gas is not considered.

In the methane system, only for low P_{th} smaller 1.4 kW the emitter reached the theoretical temperature. For higher P_{th} the combustion probably happens in a distance to the emitter surface resulting in a temperature difference between flame and emitter. For the future, other emitter sizes has to be investigated to get the experiment and the simulation into agreement in a wider range of P_{th}.

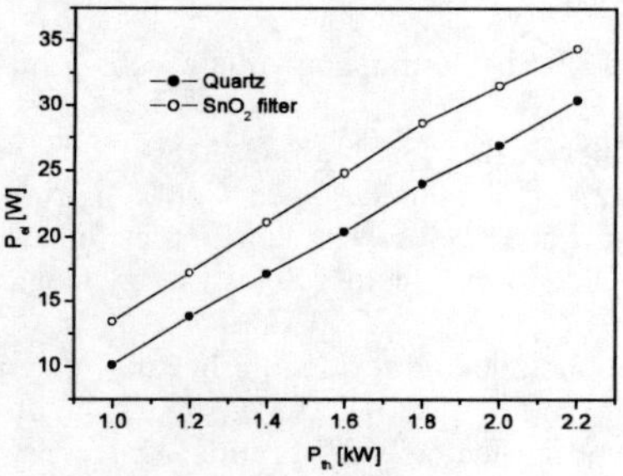

FIGURE 5: Simulation P_{el} as a function of P_{th} with quartz and SnO_2 filter tube.

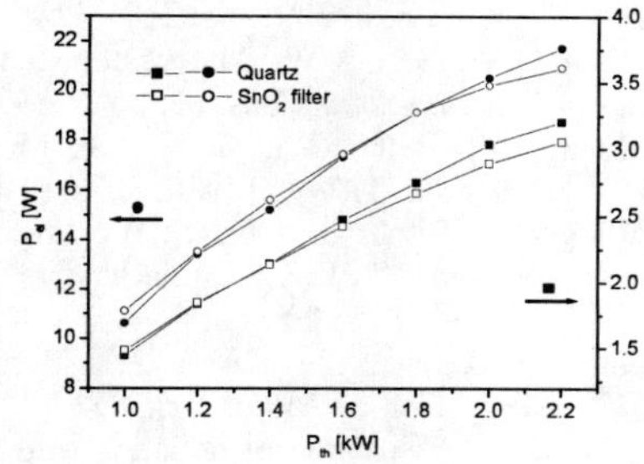

FIGURE 6: P_{el} (left axis) and I_{SC} (right axis) as a function of P_{th} for experiments with quartz and SnO_2 filter tube.

CONCLUSION

With a TPV prototype system, which achieved a record system efficiency of 2.4 %, we successfully demonstrated that silicon photocells are a high efficient and cost effective component for TPV. We developed a simulation model, which describes the prototype system in detail and can be used for its optimisation. The disadvantage of the butane self-sucking burner, which does not allow a definition of the air number, was removed in a new methane gas flow-controlled burner. In this system, too low emitter temperatures were observed at higher P_{th} than 1.4 kW. We suggest the use of larger emitters to solve this problem. The influence of heat reflective filters were studied and showed a slightly increase in efficiency, but only at low P_{th}. We plan an improvement in transmittance of these filters to achieve a sufficient high efficiency gain.

For technical applications of TPV, our system has to be up-scaled and the components have to be simplified to allow an industrial production of TPV systems in the future. Such a system is described in our forthcoming paper, Ref. A.

ACKNOWLEDGEMENT

We thank Stefan Stutz for the FTIR measurement of the filter. The project was founded by the Swiss "Kommission für Technologie und Innovation" KTI, contract No. 5692.2 EBS.

REFERENCES

1. Andreev V.M., Khvostikov V.P., Khvostikova O.A., Oliva E.V., Rumyantsev V.D., Shvarts M.Z., "Thermophotovoltaic cells with sub-bandgap photon recirculation", Proc. 17th Europ. Photov. Solar Energy Conf. & Exh. 2001, pp. 219 –222.
2. Sulima O.V., Bett A.W., "Fabrication and simulation of GaSb thermophotovoltaic cells", *Solar Energy Mat. and Solar Cells.* **66**, pp. 533 - 540 (2001).
3. Sulima O.V., Bett A.W., Dutta P.S., Mauk M.G., Mueller R.L., "GaSb-, InGaAsSb-, InGaSb-, InAsSbP- and Ge-TPV cells with diffused emitters", Proc. 29th IEEE Photov. Spec. Conf., in print.
4. Palfinger G., Bitnar B., Durisch W., Mayor J.C., Grützmacher D., Gobrecht J., "Cost estimates of electricity from a TPV residential heating system", Proc. 17th Europ. Photov. Solar Energy Conf. & Exh. 2001, pp. 117 - 120.
5. Schubnell M., Benz P., Mayor J.-C., "Design of a thermophotovoltaic residential heating system", *Solar Energy Mat. and Solar Cells.* **52**, pp. 1-9 (1998).
6. Howell J.R., "A catalog of radiation heat transfer configuration factors", 2nd Edition, University of Austin, http://mohican.me.utexas.edu/~howell/intro.html.
7. Rea S.N., "Rapid method for determining concentric cylinder radiation view factors", *AIAA Journal* **13** (8), pp. 1122 – 1123.
8. Mahbod B., Adams R.L., "Radiation view factors between axisymmetric subsurfaces within a cylinder with spherical centerbody", *J. Heat Transfer* **106** (1), pp. 244 – 246 (1984).
9. Siegel R., Howell J.R., "Thermal radiation heat transfer", Hemisphere Publ. Corp., New York, 2nd Ed. (1981).
10. Krishna M.G., Rajendran M., Pyke D.R., Bhattacharya A.K., "Spectral emissivity of ytterbium oxide-based materials for application as selective emitters in thermophotovoltaic devices", *Solar Energy Mat. and Solar Cells.* **59**, pp. 337-348 (1999).
11. Tschudi H.R., Schubnell M., "Measuring temperatures in the presence of external radiation by flash assisted multiwavelength pyrometry", *Rev. of Scientific Instr.* **70** (6), pp. 2719-2727 (1999).
12. Bitnar B., Durisch W., Mayor J.-C., Sigg H., Tschudi H.R., „Characterisation of rare earth selective emitters for thermophotovoltaic applications", *Solar Energy Mat. and Solar Cells.* **73**, pp. 221 - 234 (2002).
13. Rakhshani A.E., Makdisi Y., Ramazaniyan H.A., "Electronic and optical properties of fluorine-doped tin oxide films", *J. Appl. Phys.* **83** (2), pp. 1049 – 1057 (1998).
14. Bitnar B., Durisch W., Grützmacher D., Mayor J.-C., Müller C., von Roth F., Anna Selvan J.A., Sigg H., Tschudi H.R., Gobrecht J., „A TPV system with silicon photocells and a selective emitter", Proc. 28th IEEE Photov. Spec. Conf. 2000, pp. 1218 – 1221.
15. Kushch A.S., Skinner S.M., Brennan R., Sarmiento P.A., "Development of a cogenerating thermophotovoltaic powered combination hot water heater/hydronic boiler" in *Thermophotovoltaic Generation of Electricity: Third NREL Conference*, edited by Benner/Coutts/Allman, AIP Conference Proceedings 401, New York: American Institute of Physics, 1997, pp. 373-386.

Cost Estimates of Electricity from a TPV Residential Heating System

Günther Palfinger, Bernd Bitnar, Wilhelm Durisch, Jean-Claude Mayor, Detlev Grützmacher, Jens Gobrecht

Paul Scherrer Institut, CH-5232 Villigen PSI, Switzerland

Abstract. A thermophotovoltaic (TPV) system was built using a 12 to 20 kW_{th} methane burner which should be integrated into a conventional residential heating system. The TPV system is cylindrical in shape and consists of a selective Yb_2O_3 emitter, a quartz glass tube to prevent the exhaust gases from heating the cells and a 0.2 m^2 monocrystalline silicon solar cell module which is water cooled. The maximum system efficiency of 1.0 % was obtained at a thermal input power of 12 kW_{th}. The electrical power suffices to run a residential heating system in the full power range (12 to 20 kW_{th}) independently of the grid.
The end user costs of the TPV components - emitter, glass tube, photocells and cell cooling circuit - were estimated considering 4 different TPV scenarios. The existing technique was compared with an improved system currently under development, which consists of a flexible photocell module that can be glued into the boiler housing and with systems with improved system efficiency (1.5 to 5 %) and geometry. Prices of the electricity from 2.5 to 22 EURcents/kWh_{el} (excl. gas of about 3.5 EURcents/kWh), which corresponds to system costs of 340 to 3000 EUR/$kW_{el,peak}$, were calculated. The price of electricity by TPV was compared with that of fuel cells and gas engines. While fuel cells are still expensive, gas engines have the disadvantage of maintenance, noise and bulkiness. TPV, in contrast, is a cost efficient alternative to produce heat and electricity, particularly in small peripheral units.

INTRODUCTION

Residential heating systems are one of the prime targets for a TPV mass market as they are not necessarily very demanding in respect to efficiency, e.g. as in self powering systems. The 'losses' of the electricity production can in principle totally be used by the heating system. In most modern heating systems the water of the heating circulation has a temperature of around 30 to 40 °C so that it can be used for cell cooling. In these cases, 100 % of the additional fuel that a TPV-boiler - compared to a conventional boiler without TPV - uses is converted into electricity. Moreover, the electricity does not need to be transported over long-distance lines as it is produced right where it is needed. For comparison, conventional gas power plants bring only 40 % of the input power to the end user. Therefore, TPV can contribute to reduce the demand of primary energy. As the price of a product decreases with the increase of its production volume, a TPV mass market also lowers the price for solar cells.

Although the cost of electricity produced by TPV is one of the most important factors for marketability, very little has been published so far. In this work we take a similar approach as in [1], but now considering a new technological development, i. e. a flexible photocell module that fundamentally simplifies the cell cooling. This

CP653, *Thermophotovoltaic Generation of Electricity: 5th Conference*
edited by T. J. Coutts, G. Guazzoni, and J. Luther

FIGURE 1. TPV system mounted into residential boiler by Hoval AG, Liechtenstein

module demonstrates how the photocell generator and its cooling can be realized with less expenditure.

A practical realization of a domestic gas heating system with TPV cogeneration and further steps using the improved flexible photocell module are presented in the next two sections. The thermal power of the system of 12 to 20 kW is suited to heat a one or two family home. The electrical output power of 120 to 164 W suffices to run the system independently of the electrical grid. In the section thereafter, the cost of electricity for the presented system is estimated and compared with future TPV systems, gas engines and fuel cells.

EXISTING TECHNIQUE

This cost estimate refers to a TPV system which is described in detail in [1, 2] so that that the system is only briefly described here: The essential parts are a **heat source** (here a gas flame) heating an **emitter** which illuminates **photocells.** Figure 1 shows our TPV system built into a residential boiler (Hoval AG, FL), with condensate heat exchanger. In Figure 1, the emitter is not shown and has to be inserted from the top. The cell cooling water can be taken from the heating circulation. A **(quartz) glass tube** protects the photocells from the hot fume gases. The cylindrical geometry of the system reduces the geometrical loss of radiation since it allows for a high view factor of the cells. A reflector can optically close the system at the gas inlet. However, an opening is needed to allow the fume gases to leave the system.

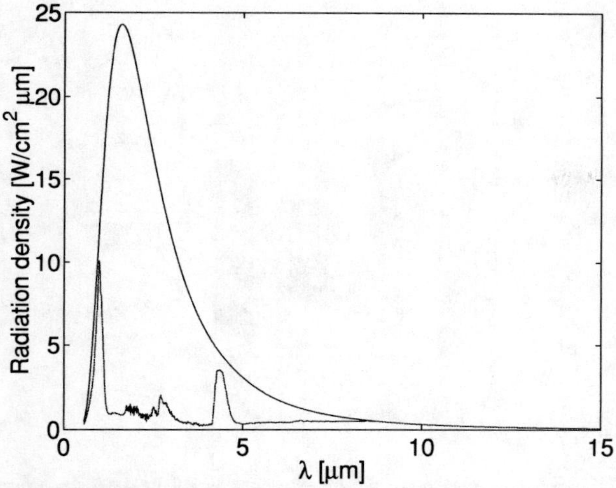

FIGURE 2. Blackbody and Yb_2O_3 spectrum at 1800 K

The Selective Emitter

The emitter has the same structure as an incandescent mantle of a gas lamp. It is fabricated by coating of textile material with a solvent containing Yb. After the textile raw mantle has been pyrolysed, a fragile Yb_2O_3 structure remains [3].

The emitter has a cylindrical surface of 360-500 cm^2 (depending on the used burner tube) delivering a radiation density of about 100 kW/m^2 (100 times solar irradiation AM1.5) at 1800 K of which 24 % has a wavelength below 1.1 µm so that it can be converted by silicon photocells. Figure 2 compares the narrow emission spectrum of the selective emitter to the broad blackbody spectrum at 1800 K.

The Quartz Tube

The quartz tube protects the photocells from the hot fume gases. Heat resistant glasses like DURAN® work as well, which was demonstrated in a smaller system under similar conditions. Its slightly higher absorption is practically negligible.

The Photocells

ASE SH2 solar cells with an efficiency of 16 % under AM 1.5 irradiation were used [3]. If the emitter is thermally heated with 12 kW, the radiation density at the surface of the cells is 9400 W/m^2 (about 10 times AM1.5) producing 530 W_{el}/m^2 which is 3.3 times as much as under solar irradiation (AM 1.5). The cells are not optimized for concentrated light, thus, the fill factor decreases from 0.72 at AM1.5 irradiation to 0.64 at TPV irradiation with 12 kW_{th} and further to 0.50 at TPV irradiation with 20 kW_{th} thermal input power. This is also the reason why the system efficiency dropped from 1.0 % at 12 kW_{th} to 0.83 % at 20 kW_{th} thermal input power, although the emitter temperature increases with the thermal input power.

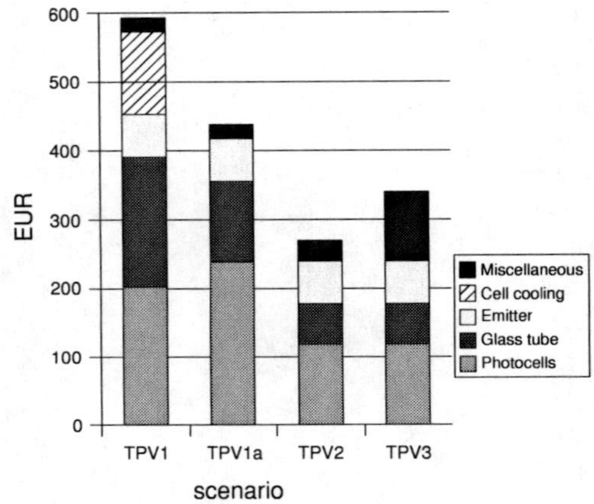

FIGURES 3. Comparison of the investment cost of the 4 TPV scenarios

The Cell Cooling Elements

As the radiation density at the cells is around 10 suns, it is necessary to cool them actively. For that, the cells are glued onto water cooled aluminum blocks. The cooling effect can be further enhanced by replacing the aluminum with copper.

IMPROVED TECHNIQUE: FLEXIBLE PHOTOCELL MODULE

To avoid the above mentioned cell cooling blocks of which the production involves a high amount of manual labor, a flexible photocell module was developed [2]. It is a stack of an aluminum foil, a thin isolation layer, the series connected ASE SH2 cells and a plastic foil. As a single cell has a width of only 12-16 mm, the module can be glued on the inner wall of the boiler, keeping the gap between the curved wall and the cell small. This allows for a good thermal contact to the water cooled boiler wall. The fabrication of the module is much easier than that of the cooling blocks and, therefore, it is expected that the TPV system becomes more cost efficient.

COST ESTIMATES

To determine the production cost of the TPV system, the costs of the individual components of the system presented in the previous section - emitter, glass tube, photocells or photocell module and cell cooling elements (if applicable) - were estimated and 100 EUR/$kW_{el,peak}$ were added to account for assembly and miscellaneous parts. The cost of electricity was calculated so that it amortizes the investment over the lifetime of the system.

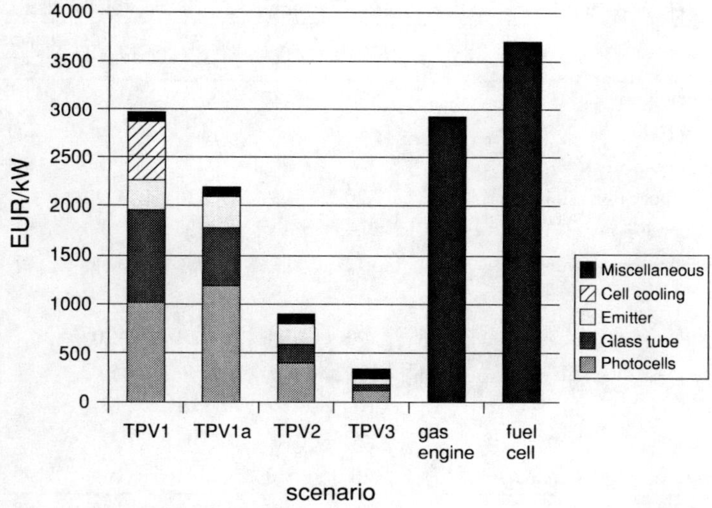

FIGURES 4. Comparison of the investment cost per kW electrical output power

The Scenarios

4 TPV-scenarios are compared to 2 heat and electricity producing systems working with either fuel cells or a gas engine.

In **scenario TPV1**, the figures are based on the performance of the currently realized system (1 % system efficiency). The cost of labor for the producing and assembling the cell cooling is not accounted for.

In **scenario TPV1a** the flexible photocell module is used to largely simplify the cell cooling. The quartz tube is replaced by a cheaper tube made of heat resistant DURAN® glass. All other components and the system performance is kept as in scenario TPV1.

Scenario TPV2 is an optimization of the system described in scenario TPV1a using short term available technology. The system is improved by reducing its diameter by 50 % and using cells optimized for the prevailing concentration (about 10 suns). This cuts the system cost by 40 % and improves the system efficiency to 1.5 %.

Scenario TPV3 increases the system efficiency to 5 %, but keeps the costs as in scenario TPV2. This increase in efficiency can be achieved by either developing an inexpensive low bandgap cell or by increasing the emitter temperature. The latter requires either an NO_x filter or the fuel has to be burnt with pure oxygen instead of air to keep the NO_x emissions within legal limits. NO_x filters are too expensive to date, but might get low cost if they are frequently used in diesel engines of cars due to the new EU and US exhaust gas regulations around 2008. Whereas oxygen might be a byproduct obtained from the electrolysis of water for hydrogen production.

TABLE 1. Cost estimate for 4 TPV scenarios and a gas engine and a fuel cell for comparison.

scenario	TPV1	TPV1a	TPV2	TPV3	gas engine [4]	fuel cell[1]
area of the photocells (m^2)	0.20	0.20	0.10	0.10		
system efficiency (%)	1	1	1.5	5	25	40
peak thermal input power (kW)	20	20	20	20	18.8	10
peak electrical output power (kW)	0.20	0.20	0.30	1.0	4.7	4.0
average thermal input power (kW)	13	13	13	13	12	6.5
average electrical output power (kW)	0.13	0.13	0.2	0.65	3.1	2.6
photo cells (EUR)	200	240	120	120		
quartz/DURAN-glass tube (EUR)	190	120	59	59		
emitter (EUR)	62	62	62	62		
cell-cooling (EUR)	120	0	0	0		
100 EUR/$kW_{peak,el}$ for miscellaneous	20	20	30	100		
total investment cost (EUR)	**590**	**440**	**270**	**340**	**14000**	**15000**
total investment per $kW_{peak,el}$ (EUR)	**3000**	**2200**	**900**	**340**	**2900**	**3700**
yearly electricity production (kWh)	230	230	350	1200	5500	470
yearly electricity production per unit cell area (kWh/m^2)	1200	1200	3500	12000		
EURcents/kWh without capital costs	13	9	3.8	1.5	12	16
EURcents/kWh including interest, maintenance, running expenses	**22**	**16**	**6.6**	**2.5**	**21**	**27**
--"-- incl. Gas[2]	**25**	**19**	**10**	**6.0**	**25**	**31**

The Investment Cost

The investment cost is the sum of the costs of main components plus the 100 EUR/$kW_{el,peak}$ for miscellaneous purposes. The cost of the components are derived from retail prices. Hence, the calculated result is the end user investment cost. Figure 3 shows a bar graph of the investment cost of the TPV systems for the 4 different TPV-scenarios. The TPV systems have a thermal input power of 20 kW. To illustrate the influence of the system efficiency, a bar graph of the investment cost per unit peak electrical output power is shown in Figure 4. See Table 1 for details of the calculations.

The **emitter cost** is deduced from the retail price per unit weight of a 1.5 g of standard Yttrium oxide incandescent mantle emitter for lighting purposes. The result is about 20 times the cost of the textile mantle and the 10 g Yb_2O_3, the final emitter consists of. Because of the similar production process, the Yb_2O_3 incandescent mantle can be used to estimate the end user price of an Yb_2O_3 mantle emitter.

For the **glass tube**, the retail price of a quartz tube in scenario TPV1 and a DURAN® glass tube in scenario TPV1a, TPV2 and TPV3 was used.

[1]Accuracy of cost estimate ± 30 % [6]
[2]Average end user price for gas in Switzerland in 1999 of 3.5 EURcents/kWh [7]

The existing system contains ASE SH2 solar **cells** bought for 1000 EUR/m^2, which is about 3 times the price of standard solar cells. This high price allows for improved cells also in scenario TPV2 and TPV3. The cost of module is the material cost of 23 EUR/m^2 plus 150 EUR/m^2 for assembly, cell cutting and -tabbing [5].

Note that if a low bandgap cell together with a suited emitter has the ability of doubling the system efficiency, the cells can be five times as expensive without increasing the cost of electricity.

The **cell cooling blocks** were handmade for the prototype. In our cost estimate we accounted for the cooling elements by taking the material price of 1 cm thick copper blocks, its production and assembly is not fully accounted for. Therefore, the price for cell cooling is scenario TPV1 is an underestimate. The cooling blocks are not needed for scenario TPV1a, TPV2 and TPV3.

The Cost of Electricity

The investment cost was written off over a presumed lifetime of the system of 20 years and with an interest rate of 4.25 % per year. Yearly maintenance cost of 1 % of the investment was taken into consideration. The amount of produced electricity was deduced from the system efficiency and an average thermal output power of 13 kW_{th} for 1800 hours per year.

To get the net price of heat driven, cogenerated electricity, the cost of a conventional boiler (scaled to the thermal power) was deducted from the price of the fuel cell and the gas engine and 700 EUR/kW_{el} for an inverter was deduced from the price of the fuel cell. Furthermore, it was assumed that the system efficiency is independent of the input power for all systems.

Some Notes on Filters and Recuperators

In the following we consider a residential heating system optimized to deliver the amount of heat demanded. This heating system is supplied alternatively with a TPV generator using a filter or not. The emitters of both optimized systems are kept at the same temperature. Since a filter reflects the low wavelength radiation back to the emitter, either a larger emitter area can be used or less gas has to be burnt to keep the emitter at the same temperature. In the first case, the emitter area can be increased by increasing the height or the radius. Assuming that in an optimized system the photocells are mounted as close as possible to the emitter, the area of the photocells increases accordingly. Thus, the accountable radiation density per unit photocell area remains constant and the investment cost per kWh is about the same. In the latter case, the investment cost increases slightly due to the addition of the filter, but the same amount of electricity is produced with and without the filter. Thus, the filter does not reduce the cost of electricity per kWh.

Clearly, it has to be pointed out that in both cases the use of a filter increases the system efficiency drastically, multiplying all other advantages mentioned in the introduction. Moreover, a higher system efficiency allows to feed electricity into the grid and opens up a host of other applications. The same arguments hold for recuperators.

AC/DC and Inverters

In Table 1, the cost of electricity natively produced by each system, i. e. AC for the gas engine and DC for the others, is compared. Although there are some applications in which DC is favored, like self powering boilers and island applications storing in batteries, the mass market is feeding into the grid. This makes inverters necessary which increases the cost of electricity by 5 EURcents/kWh. In case the PV system is combined with a TPV boiler, which is advantageous because of their complementary electricity production, the inverter can be shared between each other.

A Cost estimate based on comparison with PV

Cost estimates on solar cells give as a rule of the thumb that the cell-, encapsulation- and mounting cost contribute to the total cost about one third each. TPV has no inherent cost of mounting as the boiler has to be put up anyway. Presuming that the cost of encapsulating and mounting of PV cells is priced similarly as emitter, glass tube, cell cooling and assembly for TPV, the costs of PV and TPV per unit cell area are about the same. However, the sun yields about 1.2 MWh/m^2 per year (in Central Europe), while the TPV-emitter irradiates 180 MWh/m^2 per year (if burning 1800 h/year at 1800 K). Even in the more experimental setup of scenario TPV1 and TPV1a, the (for AM1.5 radiation optimized) photocells produce 8 times the amount of electricity they could produce with solar irradiation on a roof in Zürich, Switzerland. This factor increases to 23 for scenario TPV2 and to 80 for scenario TPV3, showing the potential for cutting cost. This can even be pushed further by bringing the photocells closer to the emitter or by concentrating the light with mirrors [8].

CONCLUSION AND OUTLOOK

Our study shows that TPV can become a cost efficient technology for the decentralized production of electricity in the very near future. Electricity from fully developed gas engines is still more expensive than electricity from TPV. The comparison with fuel cells, which have drawn a lot of attention from funding agencies recently, shows that fuel cell electricity is produced at a cost 60 % higher than that of TPV, even when compared with a non-optimized TPV system as outlined in scenario TPV1a. Moreover, TPV shows a huge potential, for both, to increase the efficiency and to cut cost. However, the mix of heat and electricity differs significantly for TPV on the one hand and gas engine and fuel cells on the other hand so that they cover two different market segments.

A first application might be a self powering boiler with high heat supply security. This extra comes for free as it pays off with the electricity savings. Being economic, it opens up a market of 6 million boilers per year in the EU.

REFERENCES

1. G. Palfinger, B. Bitnar, W. Durisch, J.-C. Mayor, D. Grützmacher, J. Gobrecht, "Cost Estimates of Electricity from a TPV Residential Heating System", *Proceedings of the 17th European Photovoltaic Solar Energy Conference and Exhibition 22-26 October 2001*, Vol. I, 2001, pp. 117-120
2. B. Bitnar, W. Durisch, A. Meyer, G. Palfinger, "New Flexible Module for Thermophotovoltaic Applications", *this volume*
3. B. Bitnar, J.-C. Mayor, W. Durisch, A. Meyer,G. Palfinger, F. von Roth, H. Sigg, "Record Electricity-to-Gas Power Efficiency of a Silicon Solar Cell Based TPV System", *this volume*
4. ecopower energy solutions ag, Biel, Switzerland.
5. Andreas Meyer, Solaronix SA, Aubonne, Switzerland, personal communication, 2002
6. Burri Bruno, Vaillant GmbH, Switzerland, personal communication, 2001.
7. Schweizerische Gesamtenergiestatistik 1999 BFE.
8. http://vri.etec.wwu.edu/tpv.htm

ACKNOWLEDGEMENT

The work was supported by the Swiss "Kommission für Technologie und Innovation" KTI, contact No. 5692.2 EBS. The photocell module was contributed by Solaronix SA, Aubonne, Switzerland and the boiler by Hoval AG, Vaduz, Liechtenstein.

TPV Tube Generators for Apartment Building and Industrial Furnace Applications

Lewis M. Fraas, James E. Avery, Wilbert E. Daniels, Huang X. Huang, Enrico Malfa*[‡], Matteo Venturino*, Giandomenico Testi*, Gianni Mascalzi[*%] Joachim G. Wuenning**

JX Crystals Inc., 1105 12th Ave. NW, Suite A2, Issaquah, WA 98027, USA
**ABB Service S.R.L., Via C. Arconati, 1- 20135 Milano, MI, ITALY*
[‡] Presently with Centro Sviluppo Materiali S.p.A, Piazza Caduti 6 luglio 1944, 1 Dalmine (BG), ITALY
[%] Presently with Environmental & Power Technologies, v.le Edison 50, Sesto S. Giovanni (MI), ITALY
***WS Wärmeprozesstechnik GmbH, Dornierstrasse 14, D-71272 Renningen, GERMANY*

Abstract. Major changes in the regulation of electric and natural gas industries during recent years have forced energy companies to explore opportunities in small-size Combined Heat and Power systems. These differ fundamentally from the traditional model of central generation and delivery since small, modular electric generators can be located very close to end-users inside a building or a single house within an industrial area, combined with the production of heat and cold. In particular, interest is growing in the new technologies for sub-100kWe units, including systems based on thermophotovoltaic (TPV) technology. TPV generator tubes can be inserted into hot furnaces to generate electricity and low-grade heat. In this generator tube, a water-cooled GaSb photovoltaic converter array inside the tube faces outward toward an infrared emitter liner mounted on the inside surface of the closed-end tube. Each tube can be sized to generate several kW and a given furnace can heat several tubes. We have conducted pilot experiments on key components in order to develop the concept just described. This includes a pilot scale array tested in an electrical furnace that heat a 3" diameter alumina tube with an infrared emitting liner. Also, a silicon carbide tube with a water-cooling system was tested in a ceramic fiber lined furnace equipped with a commercial 200 kW flameless regenerative burner, simulating a TPV generator tube in such a system.

INTRODUCTION

Large furnaces such as those used in glass or melting industrial operations or alternatively in apartment buildings already have efficient burners and large high temperature spaces. TPV generator tubes such as the one shown conceptually in figure 1 can be inserted into these hot furnaces to generate electricity and low-grade heat (as in figure 2). In this generator tube, a water-cooled photovoltaic array inside the tube faces outward toward an infrared emitter liner mounted on the inside surface of the closed-ended tube. Each tube can be sized to generate several kW and a given furnace can heat several tubes.

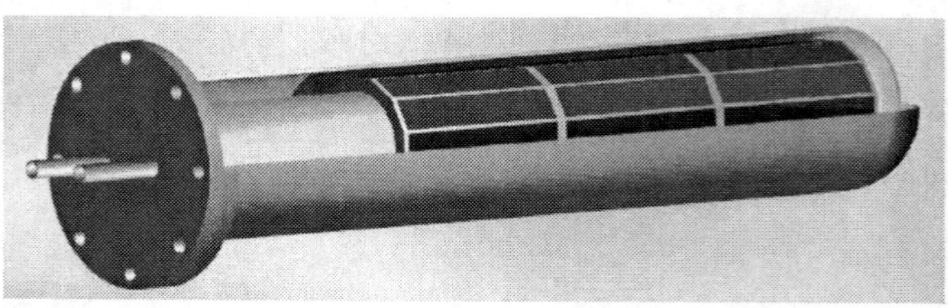

FIGURE 1. Multi-kW TPV tube concept for industry – 5 kW from a two meter tube.

CP653, *Thermophotovoltaic Generation of Electricity: 5th Conference*
edited by T. J. Coutts, G. Guazzoni, and J. Luther

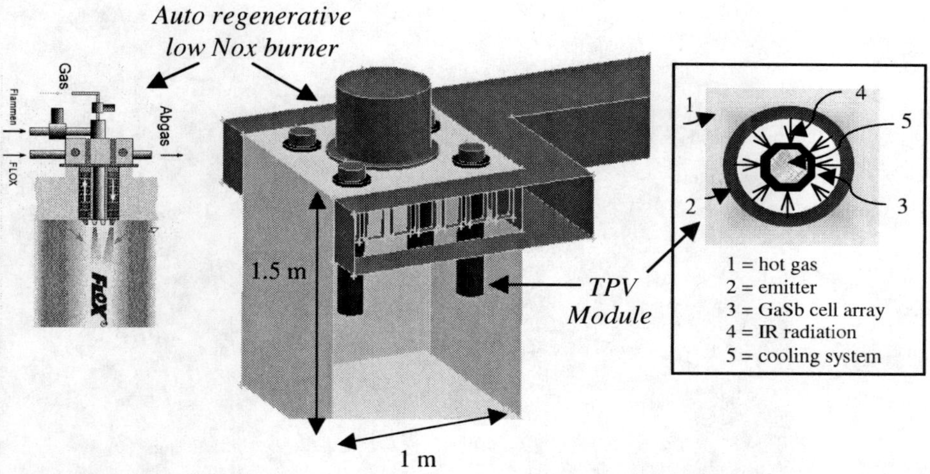

FIGURE 2. ABB's concept for insertion of four tubes into a furnace with a regenerative burner.

JX Crystals has a patent pending on the TPV tube generator design shown in fig. 1 and is developing these tubes. The three key TPV components used here are the GaSb infrared sensitive photovoltaic cell, antireflection-coated refractory metal emitters, and simple dielectric filters. These three key components and how they work together are described in more detail in a companion paper presented at this conference. These TPV generator tubes can potentially be incorporated into furnaces in apartment buildings as is shown in the figure 2 ABB furnace design.

These concepts originated approximately three years ago. However, at that time, no work had been done to demonstrate this concept. ABB then funded JX Crystals to conduct some pilot demonstration experiments on the TPV converter tubes intended to lead to the apartment building combined heat and power application. JX Crystals also obtained funding for pilot experiments from the US DOE directed toward industrial applications of these concepts. Tests have been carried out by ABB to investigate the possibility to design a mini co-generator around a flameless regenerative burner. Both sets of experimental findings are presented in the following sections.

PILOT EXPERIMENTS ON TPV CONVERTER TUBES

In our pilot experiments, an electrical furnace is used to heat a 7.7 cm diameter alumina tube with an infrared liner. PV circuits are mounted on a water-cooled paddle and the paddle is inserted into the heated tube. Each circuit contains an array of GaSb cells. Antireflection-coated tungsten foil is used as the emitter liner. Dielectric filters are located between the cells and the emitter. Figure 3 shows a 3D drawing of our pilot experimental configuration.

Over the course of a year, we fabricated five test paddles, four of which have live circuits (pictured in fig. 4). The fifth paddle is a gold plated reference paddle used for background absorption measurements. The paddles shown in figure 4 all contain circuits with 15 series-connected GaSb cells shingle mounted in rows. Each paddle in is 18 cm long, with an active circuit length of 16.5 cm, using 1.65 cm wide cells.

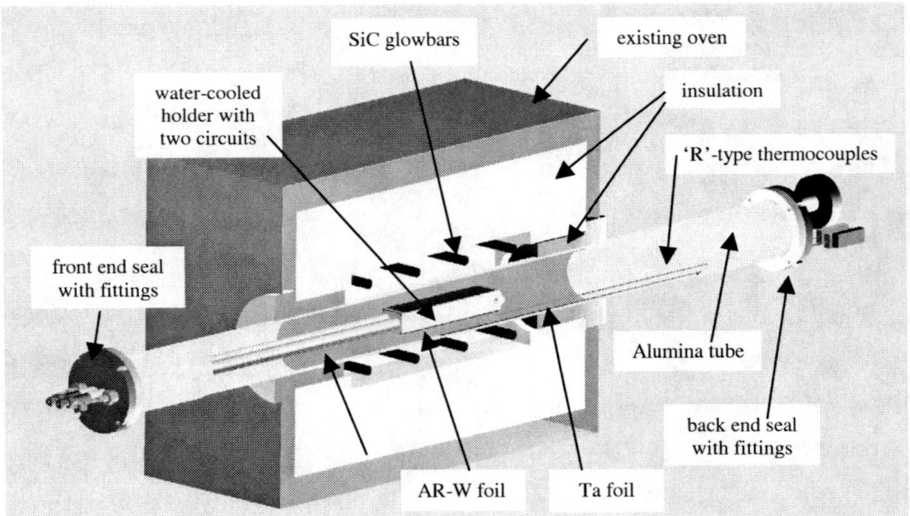

FIGURE 3. Tube furnace with photovoltaic paddle - oven and tube cut out to show interior features.

The paddles in figures 4a and 4d have 30-cell circuits mounted on their top and bottom faces, with gold plated side walls; each circuit has two rows of 15 cells mounted in parallel (note that figure 4d has filter cover slides over the circuits). The paddle in figure 4b has 30-cell circuits mounted on all four faces. Finally, the paddle shown in figure 4c has eight 15-cell single-row circuits mounted in an octagonal array.

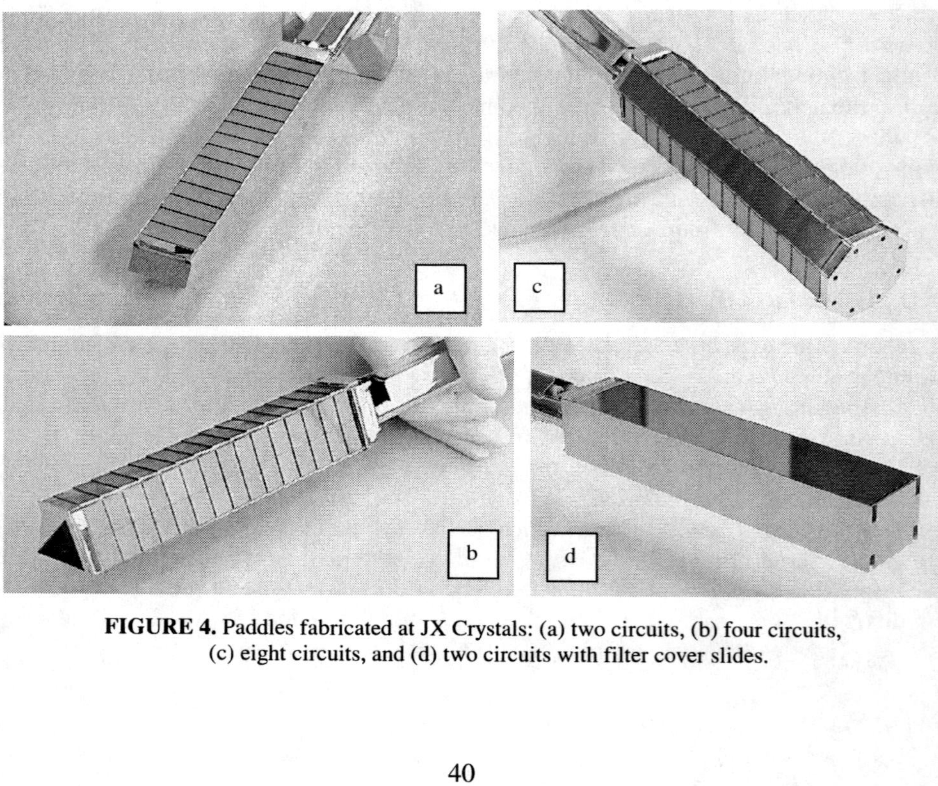

FIGURE 4. Paddles fabricated at JX Crystals: (a) two circuits, (b) four circuits, (c) eight circuits, and (d) two circuits with filter cover slides.

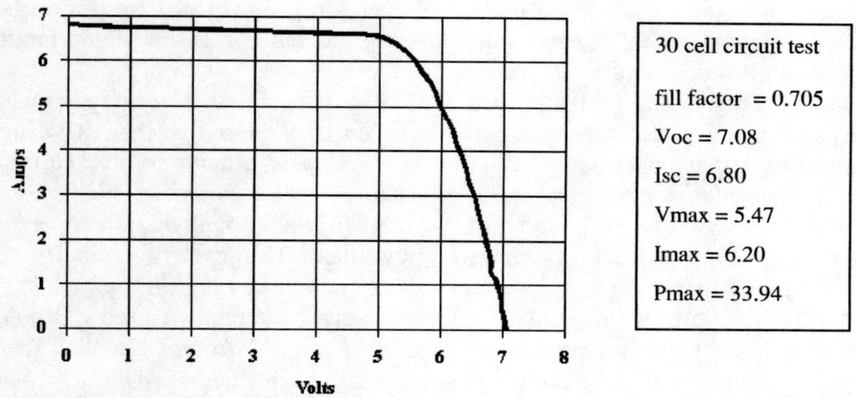

FIGURE 5. Current vs. voltage for a 30 cell circuit, such as the ones used in 4a, 4b, and 4d above.

Figure 5 shows a typical flash test result for a 30-cell circuit in the last paddle fabricated. In addition to the various paddles used in our experiments over the last year, we have also experimented with various emitter liner configurations as will be described in the next section.

TPV EXPERIMENTAL RESULTS

Over the course of a year, we conducted 61 high temperature furnace runs. Each run takes a full day with a slow furnace ramp up followed by a dwell for several hours at temperature and then a slow ramp down. In all cases, the tube was evacuated, the paddle was water-cooled with thermister probes located at the paddle inlet and outlet, the water flow was monitored including provision for back up water supply if required, and I vs. V curves were recorded rapidly and periodically at temperature. In the following, we first describe the results qualitatively. Then, toward the end of this section, we present quantitative results.

Our work over the last year can best be described in terms of the identification and resolution of three different problems that were initially intertwined:

1) Slow contaminant deposit buildup on the cold paddle emanating from the hot side elements.
2) Cracks and steps between the cells and circuits producing additional absorption losses degrading efficiency.
3) Poor heat transfer to the emitter liner resulting in poor emitter temperature uniformity and low emitter temperature. Both poor uniformity and low temperature degrade system performance.

In the following, we describe the origin of each of these problems, its solution, and the effect that it had on our experiments. As these problems are intertwined, we highlight our key experiments in historical order. In our first two-circuit paddle experiment (run 9), we measured an efficiency of 11.2%. This result repeated in runs 11 and 12 as witnessed by engineers from ABB. This result was obtained at an emitter temperature of 1150° C. A subsequent recalibration of our thermisters reduced the efficiency number to 10.9%. This emitter temperature was derived from modeling based on the measured cell current since the thermocouple reading of the emitter temperature proved unreliable.

We wanted a higher emitter temperature. Also, at this point, the four-circuit paddle had been built and was ready for test. So prior to run 16, we removed the AR/W liner. We planned to coat the backside of this foil with WC in an attempt to blacken it and improve heat transfer to it from the furnace heater elements. However, this first liner was tightly fitting in the tube and had become brittle through high temperature operation. So, it broke and we fabricated a new liner. The results for run 16 were disappointing and there was a pronounced contamination coating on the paddle. It turned out that unknown to us at the time, the WC source material we used was bound together using cobalt and this produced the contaminant. Unfortunately, both the 4-circuit paddle and the initial 2-circuit paddle were irreversibly degraded in runs 16 and 18 by cobalt film deposition.

Run 19 started us on a search for the origins of contaminants. Fortunately, we found that we could monitor contaminant buildup by using the gold reference paddle and watching the rate at which heat input into this paddle increased with time at temperature. Also, fortunately after each run, the contaminant layer could be wiped off the gold paddle to return it to its original condition. The contaminant could also be analyzed chemically. In runs 20 through 48, we found and eliminated the contaminant sources. There were two. The first and more extreme came from the Cobalt and could be eliminated by eliminating the use of Cobalt bonded WC. The second and more minor contaminant source came from a reaction between the silica binder in the alumina tube and the backside of the W foil. The silica reacted with W to produce tungsten oxide and silicon. We eliminated this problem by welding the tungsten seam with a proprietary welding method. This produced a tube where nothing from the backside could find its way to the inside paddle.

By run 48, the first problem, the contaminant problem, had been resolved. On runs 49 through 52, we collected data on the 8-circuit paddle. The measured conversion efficiency for this paddle of 5.6% was very disappointing. We now believe that this poor efficiency was a result of problems 2 and 3. We decided that we could attack problem number 2, i.e. absorption at cracks and steps, by building a 2-circuit paddle using planar filter-coated glass slides adhesive bonded to the circuits thereby covering the cracks and steps. A photograph of this last 2-circuit paddle is shown in figure 4d. A visual inspection of this photograph should convince the reader that problem 2 is now resolved. We measured the performance of our last 2-circuit paddle in run 61. Our results for runs 9 and 61 are summarized in table 1.

The measured efficiency for run 61 was actually somewhat lower at 9.5% than we had measured in run 9. This was at first surprising since we believe that the elimination of cracks and steps should have produced a higher efficiency paddle. We believe that problem 3, i.e. heat transfer to the emitter foil, is now the remaining problem. As we noted earlier, the foil in run 9 was in good mechanical contact with the alumina tube whereas in run 61, the foil was a welded tube that could be slid in and out of the alumina tube. Hence in run 61, problem 3 was more pronounced. This issue is discussed in more detail in the next section.

TABLE 1. Paddle Runs 9 and 61			
		First paddle (Run 9)	Second paddle (Run 61)
Circuit Data	Isc	7.8 A	7.5 A
	Voc	6.5 V	6.5 V
	FF	0.66	0.66
	Pmax	33.3 W	32.3 W
Paddle Data	Pelectric	66.6 W	64.6 W
	Pwater	840 W	940 W
	Pgold	230 W	260 W
Derived	Temitter	1150° C	1150° C
	Tcell	50° C	50° C
	Pwater-Pgold	610 W	680 W
Efficiency & Power Density	Pelec./(Pwater-Pgold)	10.9%	9.5%
	Power Density (elec.)	0.76 W/cm^2	0.74 W/cm^2

TPV MODELING AND INTERPRETATION

JX Crystals used TracePro software to develop a model that reproduces the measured results from run 61. TracePro is a 3D Monte Carlo Ray Trace program that can generate heat transfer predictions given geometry and surface and material properties. Figure 6 illustrates the setup procedures for TracePro model calculations.

How can we explain the differences between runs 9 and 61? Our thesis is that the difference comes from enhanced non-uniformity in the emitter temperature profile. We begin with visual observation made during run 61. There is a window at the end of the alumina tube that allows us to look at the paddle and emitter during a run. In run 61, it was observed that the emitter was hotter at its ends than in the region over the circuit. Our thesis is that the emitter temperature is drawn down in its center by heat transfer to the circuit whereas at its ends where less heat is lost, it is hotter. This thesis is strongly supported by inspection of the actual furnace geometry shown in the vertical cross section drawing depicted in figure 7. Note that the 34 cm furnace hot zone is double the length of the 17 cm paddle. Note also that only four of the furnace heating elements (shown as circles) directly heat the emitter in the paddle region, while the other four heater elements heat the end zones.

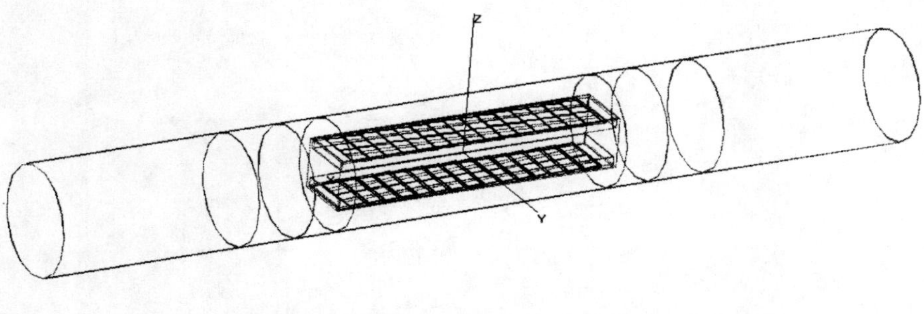

FIGURE 6. TracePro model of paddle and tube.

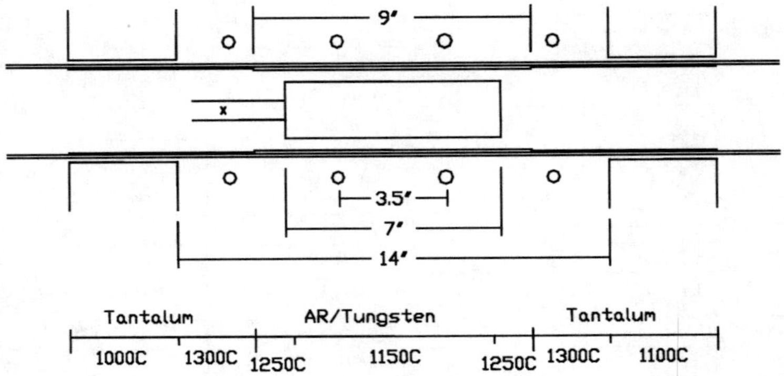

FIGURE 7. Test configuration with assumed temperature profile for modeling, accounting for both poor heat distribution and poor heat coupling to tungsten emitter foil.

The thesis that the emitter foil temperature can be drawn down by heat loss is further supported by an additional observation made during a secondary experiment done during run 61. When we back filled with Argon, we noted an interesting transient. The circuit current went up from 7.5 A to 8.6 A for a short time and then settled back down at a new equilibrium value of 7.8 A. We explain this as follows. In vacuum, the coupling between the alumina tube and the foil is poor. So the foil temperature is initially appreciably lower than the alumina tube temperature. When Argon fills the space between the two tubes, the heat transfer rate improves and the foil temperature rises toward the alumina temperature. Because of the stored heat in the alumina tube, it takes a short time for the alumina tube temperature to fall to a new equilibrium value. After establishing the new equilibrium in Argon, the foil temperature is higher and the alumina temperature is a little lower than in vacuum. Calculation suggests a foil temperature of 1150° C for 7.5 A and a foil temperature of 1180° C for 8.6 A or a change in temperature of 30° C.

All of the above serves as justification for the temperature profile shown in fig. 7, where the emitter foil ends are assumed to be at 1250° C and the center region of the foil is assumed to be at 1150° C. This profile then produces additional power losses in the end segments of the circuit as is shown in figure 8.

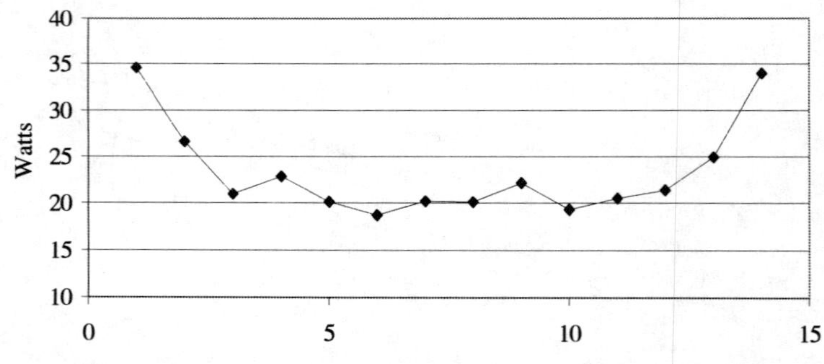

FIGURE 8. Power distribution into 14 circuit segments in Watts per segment.

Since the cells with the lowest illumination level control the current from a circuit, the excess power coupled into the circuits at each end is simply an additional loss. Our model suggests that this is an additional loss of approximately 72 Watts. Our model results reconciling runs 9 and 61 are presented in table 2. It has to be noted that our arguments here are simply plausible but certainly not rigorously proven.

TABLE 2. TracePro analysis of run 61 (see table 1)		
Given	Emitter Center Temperature	1150° C
	Emitter End Temperature	1250° C
	Cell Temperature	50° C
Prediction	Pwater	936 W
	Pgold	252 W
	Isc	7.5 A
	Pelectric	65.2 W
	P(excess at circuit ends)	72 W
	Pwater-Pgold-Pend	612 W
	Efficiency	**10.6%**

The fact remains that problem 3, i.e. poor heat coupling to the emitter foil, needs to be addressed. For our pilot experiments, the problem can be resolved in three steps. First, since it is easier to radiation-couple to a black tube than to a white tube, we propose to replace the alumina white tube with a SiC black tube. This will have additional advantages in that SiC will be the material of choice in real systems and SiC can be more rapidly cycled. Second, we need to bond the emitter foil to the SiC. This is where some research will be required. Perhaps silicon powder can be used as a bonding agent. In any case, we note that the thermal expansion coefficient for SiC is less than that of W or Mo. This means that the emitter foil will expand and lock into the SiC. This is not the case for the alumina tube where the thermal expansion coefficient for alumina is larger and the gap widens between the two tubes with an increase in temperature. Finally, third, we propose to build a longer paddle that is more nearly the same length as the furnace hot zone length. This will improve uniformity and decrease the importance of end effects.

Returning to our modeling results, the fact that we are able to fit our measurements with our model allows us to use our model to make projections regarding potential future performance improvements. Table 3 presents some model predictions. The first row of this table simply represents the fact that the model agrees with experiment. Subsequent rows report the predicted efficiencies when specific changes are made in the model. In the second row, we add perfect end mirrors in the model, eliminating the hot end effects, and the efficiency rises to 12.4%. In the third row, we hypothesize an improvement in cell quantum efficiency. The effect is minor. A more significant improvement to 14.2% results from an assumed reduction in circuit series resistance. Finally, rows 5 and 6 indicate that an efficiency greater than 15% can result from an emitter temperature exceeding 1200° C.

Our model predicts that a major gain in both power density and TPV efficiency will come with increases in emitter temperature as can be seen in figure 9.

TABLE 3. Path to Higher Efficiencies	
TracePro Model	**Efficiency**
1) Modeled with emitter at 1150° C per table	10.6%
2) Emitter at 1100° C with paddle end	12.4%
3) #2 with improved cell QE	12.8%
4) #2 with lower circuit series resistance	14.2%
5) #4 with emitter at 1200° C	16.6%
6) #4 with emitter at 1300° C	18.9%
7) Cell only with emitter at 1300° C	23.0%

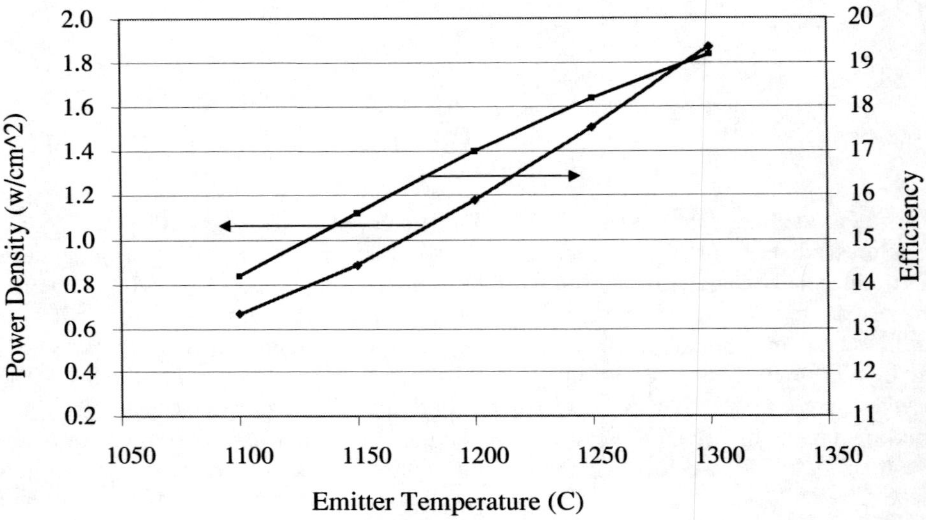

FIGURE 9. Efficiency and array power density vs. temperature.

PILOT FURNACE EXPERIMENTS

A series of tests have been carried out at WS (Wärmeprozesstechnik GmbH) ceramic fiber lined furnace (figure 10) to investigate the possibility to design a miniature co-generator around a flameless regenerative burner.

In particular, the Regemat® 200kW manufactured and commercialized by WS for direct heating in industrial application has been selected. It is composed of a series of nozzles, surrounding a central NG gun. Each ceramic nozzle is fed alternatively with hot combustion gases and air, acting as regenerative air preheater. In this way the combustion air can reach temperatures of about 800-1000°C, required for the FLOX behavior. A series of valves provides the switching between suction and injection for each nozzle. The burner should start up in "flame mode" and, when the furnace temperature reach 850°C, the burner can be switched to FLOX mode.

The furnace is 2000 mm long and has a cross section of 1200x1200 mm for the front dimensions. During FLOX mode, part of the flue gases are exhausted through the front of the furnace after their pass through the regenerative heat exchanger.

In the WS test rig during furnace operation, there are four stainless steel air-cooled pipes arranged along the furnace, as depicted in figure 10, that usually provide heat extraction. In order to reproduce the operating condition for the TPV generator tube, a SiC tube has replaced one of these cooling pipes with a water-cooling probe inside it (figure 10).

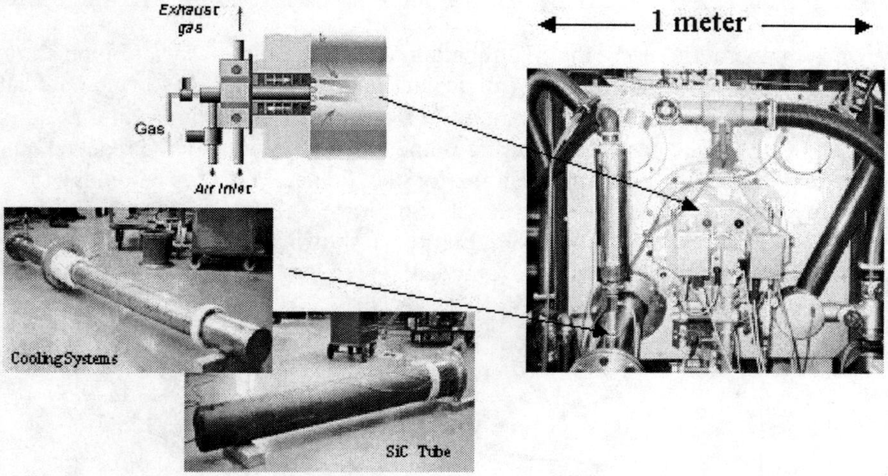

FIGURE 10. Furnace experimental set-up.

The SiC tube selected, the commercial one used in WS radiant burner (C200/1650), has an external diameter of 195 mm (internal 185 mm) and a active length inside the furnace 1600 mm. The external surface has been equipped with thermocouples to measure the temperature distribution. Surface temperature has been recorded along the tube axis following three lines on the cylindrical pipe. The SiC tube surrounds the external pipe (ϕ_o=141.3 mm, active length 1400 mm) of the water cooling system.

FURNACE TEST RESULTS

The goals of the tests were to verify the impact of FLOX operation mode on temperature level and uniformity along the SiC water cooled pipe at different furnaces operating conditions and the corresponding percentage of available heat that it is possible to transfer from the fuel to the SiC tube. During the test the burner efficiency and the pollutant emissions have also been recorded.

Several tests have been performed by varying the furnace temperature starting from 850 °C to 1250 °C at 100% of burner power. In order to increase the furnace temperature the heat extraction has been adjusted reducing the flow rate to the air-cooled pipes. The furnace operation has been tested also at partial load and measurements have been recorded for 50% burner load. Table 4 shows the furnace test results.

TABLE 4. Combustion Tests Matrix									
		Flame 900	FLOX 950	FLOX 1000	FLOX 1100	FLOX 1200	FLOX 1250	FLOX 1200	Flame 1200
Test number		2	3	4	5	6	7	8	9
Nominal load	%	100	100	100	100	100	100	100	100
NG Power input	kW	200.2	191.2	200.2	202.2	200.2	203.2	199.2	204.2
Total Power input	kW	202.2	193.2	202.2	204.3	202.1	205.2	201.3	206.4
Fuel-to-emitter η	-	52%	79%	78%	75%	72%	67%	73%	41%
Furnace Temperature	C	906	957	1009	1109	1207	1259	1207	1162
Nox @ 3% O2	ppm	53	14	25	52	55	80	55	62

Surface temperature along the SiC tube containing the internal water pipe showed an average temperature gap with the furnace in the range of 80-110°C. No appreciable ΔT difference was detected from flame to FLOX modes. The main reason for that is in the high mixing of flue gases even during flame mode promoted by the reduced length of the furnace and the high momentum for the flame. The dimensionless (T/T_{ref}) temperature profiles along the tube axial coordinate (x_{TC}/x_{ref}) for both FLOX and flame mode varied from 82% to 94%. Figure 11 shows surface temperature profiles along the silicon carbide tube in FLOX mode.

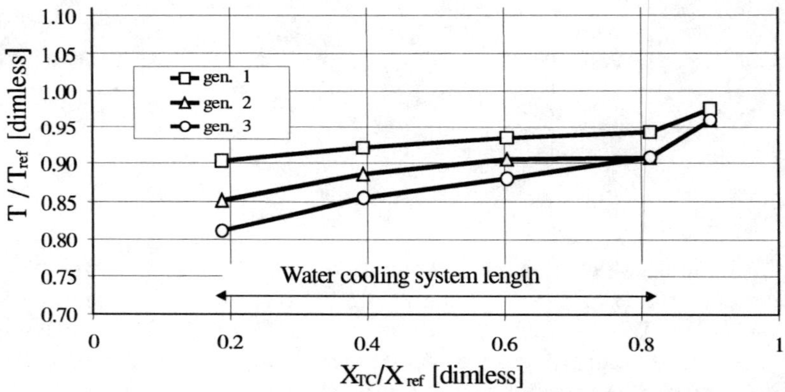

FIGURE 11. Surface temp. profiles along the SiC tube in FLOX mode (furnace temp. = 1200 C).

CONCLUSIONS

Herein, we have described a new TPV geometry suitable for use in industrial furnaces or central heating furnaces in large buildings. In this geometry, the TPV array is located inside an emitter tube and the tube is heated from outside. As it turns out, this geometry is also very suitable for pilot experiments where the interactions between full size circuits and full size emitters can be explored. In the experiments reported here, we have measured a TPV efficiency of over 10% for the first time using complete circuits and full size emitters. We have also developed a model that quantitatively fits our measurements and allows us to predict efficiencies over 15% after additional development. The test campaign conduced at WS has confirmed that the autoregenerative burner operating in FLOX mode can provide suitable conditions in the combustion furnace for the development of a mini-CHP TPV unit.

Electric Power Generation Using Low Bandgap TPV Cells in a Gas-fired Heating Furnace

K. Qiu and A.C.S. Hayden

CANMET Energy Technology Centre-Ottawa
Natural Resources Canada, 1 Haanel Drive, Ottawa, Ontario, Canada K1A 1M
Email: kqiu@nrcan.gc.ca

Abstract. Low bandgap TPV cells are preferred for electric power generation in TPV cogeneration systems. Recently, significant progress has been made in fabrication of low bandgap semiconductor TPV devices, such as InGaAsSb and InGaAs cells. However, it appears that only limited data are available in the literature with respect to the performance of these TPV cells in combustion-driven TPV systems. In the research presented in this paper, power generation using recently-developed InGaAsSb TPV cells has been investigated in a gas-fired space heating appliance. The combustion performance of the gas burner associated with a broadband radiator was evaluated experimentally. The radiant power density and radiant efficiency of the gas-heated radiator were determined at different degrees of exhaust heat recuperation. Heat recuperation is shown to have a certain effect on the combustion operation and radiant power output. The electric output characteristics of the InGaAsSb TPV devices were investigated under various combustion conditions. It was found that the cell short circuit density was greater than 1 A/cm^2 at a radiator temperature of 930°C when an optical filter was used. An electric power density of 0.54 W/cm^2 was produced at a radiator temperature of 1190°C. Furthermore, modeling calculations were carried out to reveal the influence of TPV cell bandgap and radiator temperature on power output and conversion efficiency. Finally, the design aspects of combustion-driven TPV systems were analyzed, showing that development of a special combustion device with high conversion level of fuel chemical energy to useful radiant energy is required, to improve further the system efficiency.

INTRODUCTION

Thermophotovoltaic (TPV) generation of electricity has application in electric grid independent or self-powered gas appliances, cogenerators of electricity and heat, portable generators, hybrid electric vehicles, recovery systems of high temperature waste heat from industrial processes, etc. In TPV power generation systems, combustion-driven radiant sources are utilized. The photons emitted from the sources at a temperature of practical interest are distributed at much lower energies or longer wavelengths in comparison to solar radiation. Thus, low bandgap (e.g. E_g<0.75) TPV cells should be used in the TPV systems to optimize the power density and efficiency. InGaAsSb and InGaAs cells are two types of low bandgap TPV devices that presently are of considerable interest. Significant progress has been made in fabrication of these TPV cells. The diode behavior and electric characteristics of InGaAsSb and InGaAs TPV devices have been reported and related mechanisms discussed extensively (1-3). In order to provide the basis for the design of TPV power generation systems, it is necessary to investigate the performance of these TPV cells in combustion-driven radiant sources, specifically in gas-fired furnaces. In the present work, power generation using InGaAsSb

CP653, *Thermophotovoltaic Generation of Electricity: 5th Conference*
edited by T. J. Coutts, G. Guazzoni, and J. Luther

TPV cells was studied in a gas-fired heating furnace where radiation was emitted by a broadband thermal radiator. The combustion operation of the burner/radiator was also studied. The electric output characteristics of the InGaAsSb TPV cells were examined at various combustion conditions. The study was aimed at system testing. Modeling predications were performed, revealing the effect of TPV cell bandgap and radiator temperature on the power output and cell efficiency. Furthermore, thermal energy flow over the TPV unit was analyzed, in an effort to elucidate the design aspects of TPV systems.

EXPERIMENTAL SECTION

The low bandgap TPV cells used in this study are quaternary semiconductor InGaAsSb cells (AstroPower, USA). Currently, mainly two designs of InGaAsSb cells are being developed. One involves a thick (3-5 μm) low-doped epitaxial p-emitter with an AlGaAsSb window layer. The other design involves a relatively thin (<0.7 μm) high-doped emitter without a window layer. The InGaAsSb cells used in this study were fabricated from the latter design. Figure 1 depicts the InGaAsSb cell structure. An epitaxial n-type $In_{0.15}Ga_{0.85}As_{0.17}Sb_{0.83}$ layer is grown on an n-type GaSb substrate by liquid phase epitaxy. A p-type $In_{0.15}Ga_{0.85}As_{0.17}Sb_{0.83}$ layer is then developed as emitter on the epitaxial n-doped $In_{0.15}Ga_{0.85}As_{0.17}Sb_{0.83}$ layer by a Zn-diffusion process, forming p-on-n InGaAsSb devices. The cell size used in this work is 0.4 cm in diameter. At 300K, the nominal bandgap of the cells is 0.53 eV, which corresponds to a cutoff wavelength of 2.4 μm.

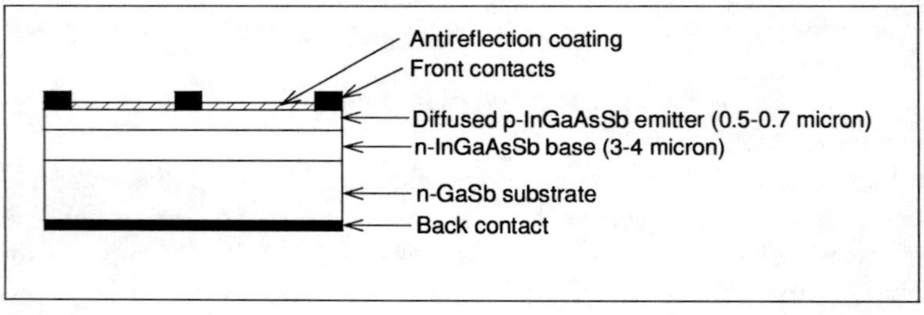

Figure 1. Schematic representation of InGaAsSb cell structure

The system testing was performed in an experimental set-up of a gas-fired heating furnace, shown in Figure 2. The experimental set-up consisted of a gas burner, a thermal radiator, an optical filter, a cooling system, an air recuperator, various measuring devices and a data acquisition system. The radiant power output from the radiator was measured by means of a radiometer (Land 2π Ellipsoidal Radiometer, Land Combustion Limited, UK). The gas-fired burner in the furnace was fully aerated, where gas and combustion air are entirely premixed before entering the burner. The broadband radiator with reticulated structure was made of SiC, having an emissivity of approximately 0.9. The optical filter in the experimental assembly has a high transmission below the cutoff wavelength of InGaAsSb cells and a relatively high reflectivity above 2.5 μm.

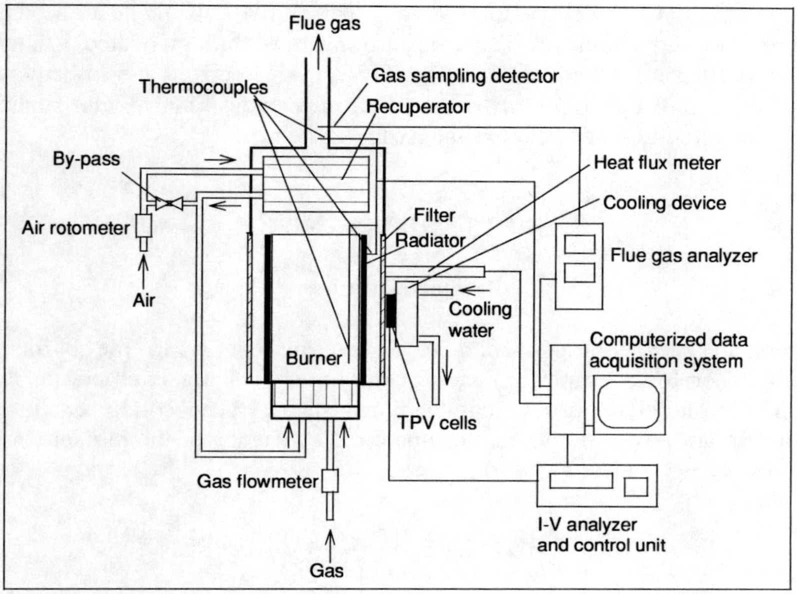

Figure 2. Schematic diagram of the TPV experimental set-up

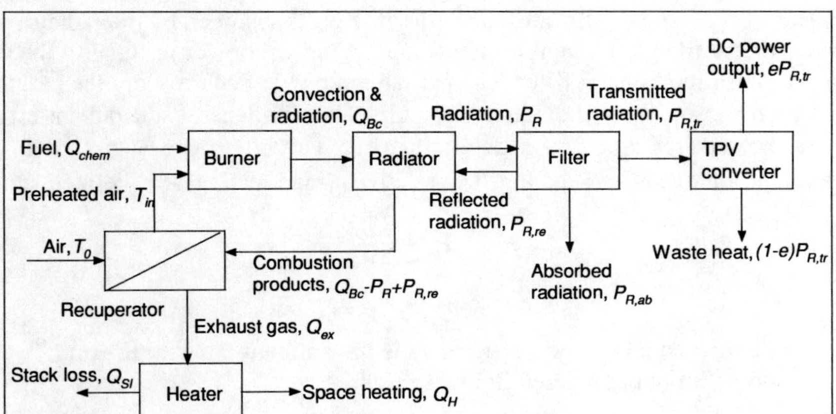

Figure 3. Energy flow diagram for combustion-driven TPV cogeneration systems

Figure 3 schematically depicts the components of the experimental TPV unit and thermal energy flow over the system. The heat flux transferred from combustion products to the radiator is emitted effectively. The radiation energy from the radiator is subsequently filtered by the optical filter. The combustion products leaving the burner/emitter enter an air recuperator, by which part of the sensible heat of the products is transferred to combustion air. The combustion air is thus preheated from ambient temperature to a desirable temperature. The TPV devices convert the in-band radiation incident on them into electricity at specific cell efficiency. Finally, the exhaust gases leaving the recuperator enter a heat exchanger for space heating.

RESULTS AND DISCUSSION

Burner/radiator

Fuel chemical energy is converted into radiation energy in the burner/radiator combination. Therefore, the performance of the burner/radiator is crucial to the entire TPV system. Initially, a series of combustion tests has been conducted, in order to measure the radiant power output and estimate the efficiency of the radiant source. The radiant efficiency is expressed as

$$\eta_{rad} = \frac{p_{rad} A_r}{m_{fuel} \Delta H_L} \tag{1}$$

where p_{rad} is the net radiant power density, A_r is the radiator area, m_{fuel} is the fuel mass flow rate and ΔH_L is the fuel lower heating value. The net radiant power refers to the total radiation minus the amount of radiation reflected by the filter and absorbed by the radiator. Here, it is assumed that all reflected radiation is absorbed by the radiator and the absorptivity of the filter is negligible. Thus, the net radiant power is equal to the amount of radiation transmitted by the filter. The net radiant power density from the radiator was measured by means of the radiometer. The radiant power density and radiant efficiency of the gas-heated radiator were determined at varying degrees of exhaust heat recuperation or relative air preheating. The relative air preheating, ε, is defined as

$$\varepsilon = \frac{T_{in} - T_0}{T_{ex,in}} \tag{2}$$

where T_{in} is the preheated air temperature, T_0 is the ambient temperature and $T_{ex,in}$ is the exhaust gas temperature before entering recuperator.

Table 1 shows the major results of combustion operation of the experimental set-up. The radiant burner performance was affected by the combustion variables. The heat recuperation was observed to have a significant influence on the combustion operation and radiant power output. In general, the radiation source efficiency achieved in the combustion unit is modest (Table 1). This is ascribed mainly to the relatively low radiator temperature and also to the fact that the burner/radiator used in this work was not designed for maximization of radiant efficiency. In a previous paper (4), it has been shown that the reticulated radiator geometry and combustion mode affect on the radiator temperature and thus the radiant efficiency.

It should be pointed out that the filter used in the test facility is not ideal, and therefore transmits a considerable amount of infrared radiation with longer wavelengths, especially beyond 3.8 μm. As a result, the filtered radiant power contains a significant portion of non-convertible radiation. This will be discussed below.

Table 1 Results of combustion performance of the experimental set-up

Heat release rate[a], kW	Relative air preheating	Air temperature, °C	Radiator temperature, °C	Net radiant power density, W/cm²	Radiant efficiency, %	Total net radiant output[b], W
4.8	0	24	930	3.5	21.7	1040
5.5	0	24	970	3.9	21.1	1153
6.3	0	24	1005	4.4	20.7	1307
6.9	0	24	1050	4.7	20.6	1396
6.9	0.18	265	1131	5.6	24.1	1663
6.9	0.31	472	1190	6.1	26.3	1812

[a] Fuel: natural gas; Excess air: 10%.
[b] Total effective area of the SiC radiator: 297 cm².

Cell Electric Output Characteristics in the combustion-driven radiant source

The electric output characteristics of the InGaAsSb TPV cells were measured, with the cells illuminated by the gas-heated radiator in the furnace under various combustion conditions. Figure 4 shows the variations of open circuit voltage (V_{oc}) with short circuit current density (J_{sc}) at different cell temperatures. At a low cell temperature (approximately 25°C), the open circuit voltage increases with short circuit current density as the radiator temperature or radiant power density is raised. Apart from the radiant power density, the short circuit current and open circuit voltage are affected by the Zn concentration profile in p-$In_{0.15}Ga_{0.85}As_{0.17}Sb_{0.83}$ emitter. Also, the surface treatment and front contact fabrication have an influence on the short circuit current. Superior light trapping and/or higher absorption coefficients enhance the short circuit current density.

At a given level of illumination, the short circuit current is governed primarily by the spectral response, or is proportional to the quantum efficiency. The results presented in Figure 4 indicate that the open circuit voltage is higher than 300 mV at a short circuit current density greater than 1 A/cm². Interestingly, the dependence of open circuit voltage on the circuit density was altered when the cell temperature was elevated (Figure 4). As mentioned earlier, the filtered radiation contained a certain amount of non-convertible radiation energy. This energy became waste heat when it was incident on the cells. Although the waste heat was dissipated with the water-cooled heat sink, it was found that a certain rise in the cell temperature appeared to be inevitable after the cells had been exposed to the radiant source in the furnace for some time. It is known that for the very low bandgap InGaAsSb cells, the cell temperature greatly influences the open circuit voltage, especially under high current density conditions. For instance, in the range 25 to 60°C, the temperature coefficient of V_{oc} was reported ranging from –1.79mV/K to –43mV/K at current densities of 0.3-6.0 A/cm^2 (1). In consequence, the open circuit voltage decreased with short circuit current density at elevated cell temperatures in the system testing (Figure 4). This is ascribed to a greater temperature coefficient of V_{oc} at a higher current density. Figure 5 presents the short circuit current density and open circuit voltage as a function of radiator temperature at different cell temperatures.

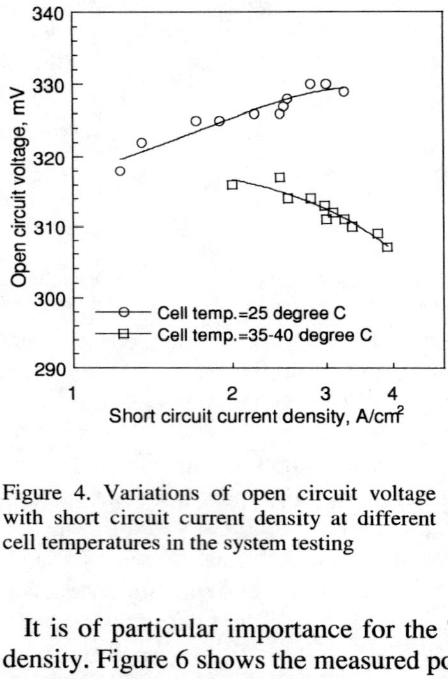

Figure 4. Variations of open circuit voltage with short circuit current density at different cell temperatures in the system testing

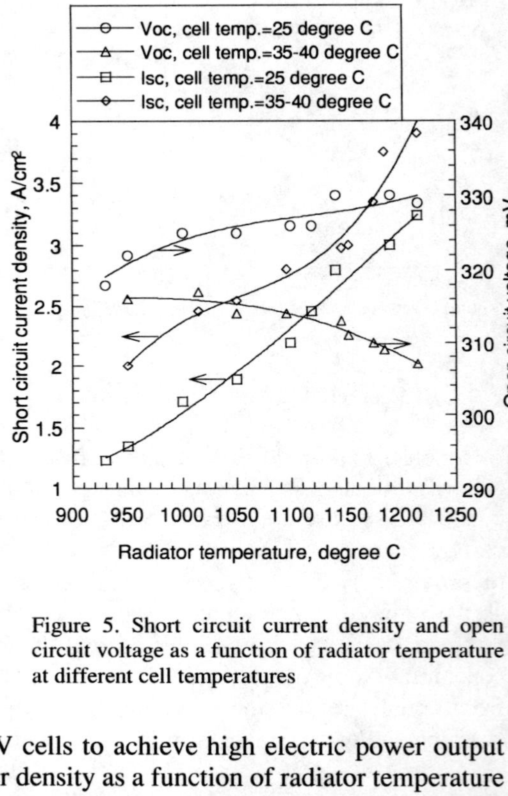

Figure 5. Short circuit current density and open circuit voltage as a function of radiator temperature at different cell temperatures

It is of particular importance for the TPV cells to achieve high electric power output density. Figure 6 shows the measured power density as a function of radiator temperature during the system testing. As expected, the power output noticeably increases with the radiator temperature. Obviously, the cell temperature has an influence on the maximum power output. On the one hand, increasing cell temperature gives rise to a reduction in open circuit voltage (Figure 5) as well as fill factor. On the other hand, the cell spectral response increases with rising cell temperature, thereby resulting in a significant rise in the current density (Figure 5). As a consequence of these combined effects, the electric power density drops off only slightly over the cell temperature range being considered. As shown in Figure 6, a reasonably high power density has been produced by the low bandgap TPV cells in the combustion-driven radiant source, despite a certain rise in the temperature of cells in the heating furnace.

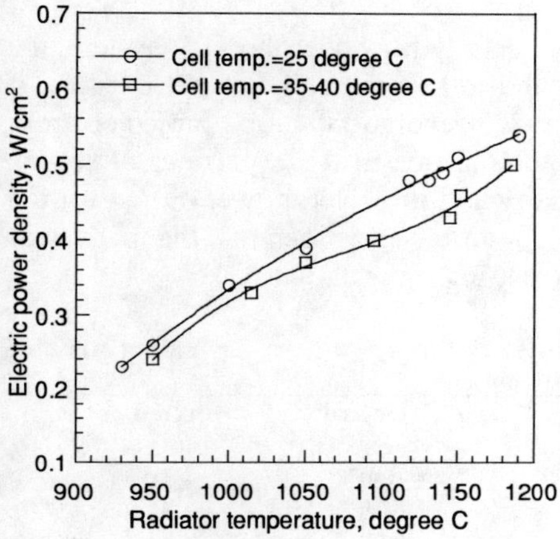

Figure 6. Electric power out density vs. radiator temperature at different cell temperatures

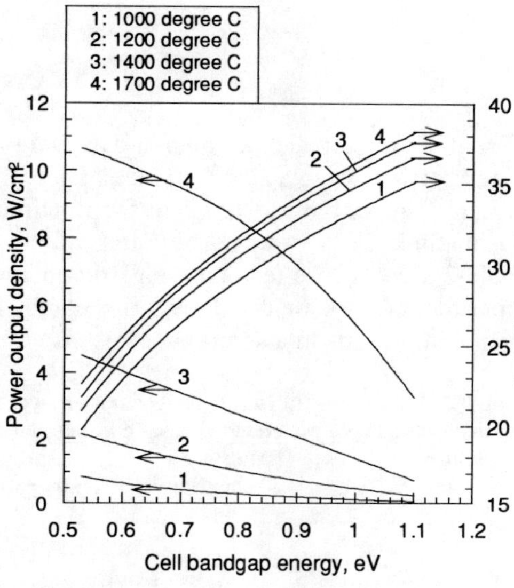

Figure 7. Variations of power output density and cell efficiency with cell bandgap energy for various radiator temperatures

Modeling

Due to rising environmental concerns, gas combustion temperature should generally not exceed 1500°C. In this case, a relatively low temperature of radiator is required. Besides, if a low temperature radiator is employed , the design of TPV systems can be greatly simplified and it will be less problematic with regard to the material lifetime and reliability of the radiator. As a result, lower bandgap TPV cells are more suitable, since there is more photon convertible radiation energy available for these cells at a lower radiator temperature and ,accordingly, a higher power output density can be achieved. However, lowering the cell bandgap decreases the open circuit voltage and fill factor, thus limiting the attainable cell efficiency. This must be taken into account in the designing of TPV systems.

In order to look into the influence of both TPV cell bandgap and radiator temperature on power output and cell efficiency, modeling calculations have been made. Figure 7 presents the variations of power output density and cell efficiency with the bandgap energy for various radiator temperatures. The predictions were made on the assumption that the radiator was a blackbody and the optical filter was an idealized edge filter. Detailed descriptions of the mathematical model will be given elsewhere (5). As shown in Figure 7, there is a trade-off for the efficiency and power density for varying cell bandgap energies. In spite of this, the lower bandgap cells are still preferred for the TPV generation systems where radiators operate at low temperatures. This is due to the fact that the power density of high bandgap cells (e.g. 1.1 eV cells) is too low at a relatively low radiator temperature.

TPV System

From the obtained experimental data and thermal energy flow over the TPV system, the system efficiency may be analyzed. The cells tested in this research are not capable of covering the whole radiator, but the maximum power generation can be projected from the obtained cell power density and the actual radiator area in the experimental set-up. Table 2 presents the results of electric power density and projected power output for the experimental TPV unit under various combustion conditions. It is seen that the modest air preheating results in a significant increase in the electric power output.

Table 2 Experimental data of electric power density and projected power output for the experimental TPV unit under various combustion conditions

Heat release rate, kW	Relative air preheating	Radiator temperature, °C	Cell electric power density[a], W/cm²	Projected power output[b], W
4.8	0	930	0.23	68.3
5.5	0	970	0.28	83.2
6.3	0	1005	0.34	101.0
6.9	0	1050	0.39	115.8
6.9	0.18	1131	0.48	142.6
6.9	0.31	1190	0.54	160.4

[a] The cell power density was obtained at a cell temperature of approximately 25°C.
[b] Total cell area is assumed to amount to the radiator area (297 cm²).

The electric efficiency of TPV system is defined as the ratio of electric output power, P_{el}, to chemical energy input rate, Q_{chem}:

$$\eta_{TPV} = \frac{P_{el}}{Q_{chem}} \tag{3}$$

or

$$\eta_{TPV} = \frac{p_{el} A_r}{Q_{chem}} \tag{4}$$

where p_{el} is the cell power density. Calculations show that the system electric efficiency is fairly low for this experimental facility. This is mainly due to the low conversion levels of fuel chemical energy to useful radiant energy. Conceptually, TPV generation systems seem quite simple. However, the development of these systems has proved to be much more challenging than anticipated (6). During the processes of TPV conversion, the three primary components, heat source/thermal radiator, optical filter and bandgap-matched TPV cells, must function as a well integrated system. The system efficiency can also be expressed as the product of efficiencies of the three components (see Figure 3), i.e.

$$\eta_{TPV} = \eta_{rad}\eta_{spe}\eta_{cell} \tag{5}$$

where η_{spe} is the spectral efficiency and η_{cell} is the cell efficiency. The spectral efficiency refers to the ratio of the radiant power output with TPV cell convertible wavelengths to the total net radiant power output. The cell efficiency is here defined as the ratio of maximum electric power to photon convertible radiant power. Note that Equation (5) implies that all net radiation output is incident on the cells. The component efficiencies for the experimental system have been obtained and are given in Table 3. Since it was impossible to measure the cell efficiency precisely for the present experimental facility, the cell efficiency was estimated from the modeling calculations (5).

Table 3 Component efficiencies for the experimental TPV system

Heat release rate, kW	Radiator temp., °C	Radiant efficiency (η_{rad}), %	Spectral efficiency (η_{spe}), %	Conversion of fuel to photon convertible radiation ($\eta_{rad}\eta_{spe}$), %	Cell efficiency (η_{cell}), %
4.8	930	21.7	34.8	7.5	19.1
5.5	970	21.1	37.0	7.8	19.4
6.3	1005	20.7	39.1	8.1	19.7
6.9	1050	20.6	40.8	8.4	20.0
6.9	1131	24.1	41.5	10.0	20.6
6.9	1190	26.3	42.2	11.1	20.9

As shown in Table 3, the major energy loss in the system results from the low conversion of fuel energy to photon convertible radiation. Therefore, in order to realize high-efficiency TPV systems and thus approach commercial readiness, further work should be directed to developing a special gas combustion device with a high conversion level of fuel chemical energy to useful radiant energy. For instance, a gas-fired self-recuperative SiC radiant-tube burner can operate at an outer tube surface temperature of 1250°C, with the temperature uniformity being within ±10°C (7). The thermal efficiency of the radiant-tube burner reaches the level of 70% when using a built-in ceramic recuperator with relative air preheating above 0.65 (e.g. air preheating temperature = 845°C). The radiant tubes for higher temperatures are under development. Such radiant-tube burners may offer the potential for high system efficiency for TPV power production and will be the subject of our future work.

Here, it is of interest to note that the TPV generation in gas-fired heating appliances is a cogeneration system where exhaust heat is in large part utilized for space and/or water heating. Hence, in spite of a low system electric efficiency, the overall fuel utilization efficiency in such a system can be sufficiently high, e.g. higher than 90%. This is superior to conventional heating equipment. Also, for the cogeneration system, the required electricity share is usually small, as compared with the thermal load. Therefore, the simultaneous production of electric power and heat in residential homes and commercial buildings using the TPV converter is still very attractive, even when the system electric efficiency is not very high at the present time.

CONCLUSIONS

Performance of InGaAsSb TPV cells was investigated in a gas-fired furnace where a thermal radiator was heated directly by a combustion flame. The radiant power and radiant efficiency of the thermal radiator were determined under various combustion conditions. It has been shown that exhaust heat recuperation has a significant effect on the burner/radiator operation. The cell short circuit density was greater than 1 A/cm^2 at a relatively low radiator temperature of 930°C. The electric output was found to reach a power density of 0.54 W/cm^2 at a radiator temperature of 1190°C. Modeling calculations have shown that there is a trade-off with regard to the cell efficiency and power density for varying cell bandgap energies. In spite of this, the lower bandgap cells are still preferred for the combustion-driven TPV systems. Finally, the analyses of energy flow over the TPV unit revealed that the major energy loss arose from the low conversion of fuel chemical energy to photon convertible radiation. Therefore, a gas burner/radiator system with a high conversion level of fuel energy to useful radiant energy is required for high-efficiency TPV units.

ACKNOWLEDGEMENTS

The authors would like to thank Dr. M. G. Mauk of AstroPower, Inc., USA, for his valuable help and discussions. Funding for this activity is primarily from the Canadian Program on Energy Research and Development (PERD).

REFERENCES

1. Sulima, O. V., Beckert, R., Bett, A. W., Cox, J. A. and Mauk, M. G., InGaAsSb photovoltaic cells with enhanced open-circuit voltage, *IEE Proc.-Optoelectron.*, 2000, Vol.147, No. 3, pp. 199-204
2. Wojtczuk, S., Colter, P., Charache, G. and DePoy, D., Performance status of 0.55eV InGaAs thermophotovoltaic cells, *4th NREL Conf. on Thermophotovoltaic Generation of Electricity*, 1998, pp. 417-426
3. Wang, C. A., Choi, H. K., Oakley, D. C. and Charache, G. W., Extending the cutoff wavelength of lattice-matched GaInAsSb/GaSb thermophotovoltaic devices, *4th NREL Conf. on Thermophotovoltaic Generation of Electricity*, 1998, pp. 256-265
4. K. Qiu and A. C. S. Hayden, Thermophotovoltaic generation of electricity in a gas-fired heater: influence of radiant burner configurations and combustion processes, Submitted to *Energy Conversion and Management*, 2002
5. K. Qiu and A. C. S. Hayden, in progress
6. Schroeder, K. L., Performance optimization of thermophotovoltaic systems. Ph.D. thesis, Auburn University, Auburn, Alabama, USA, 1998
7. Wunning, J. A. and Wunning J. G., Ceramic burner and radiant tubes for high-temperature processes, Tile & Brick International, 12(5), 1996

SESSION 2

Systems and Utilization

Chair: Joachim Luther

Thermophotovoltaics for Combined Heat and Power Using Low NOx Gas Fired Radiant Tube Burners

Lewis Fraas & James Avery, Enrico Malfa*, Joachim G. Wuenning**, Gary Kovacik†, Chris Astle†

JX Crystals Inc., 1105 12th Ave. NW, Suite A2, Issaquah, WA 98027, USA
**ABB Service S.R.L., Via C. Arconati, 1- 20135 Milano, MI, ITALY*
***WS Wärmeprozesstechnik GmbH, Dornierstrasse 14, D-71272 Renningen, GERMANY*
†Alberta Research Council, 250 Karl Clark Rd., Edmonton, Alberta, CANADA

Abstract: Three new developments have now occurred, making economical TPV systems possible. The first development is the diffused junction GaSb cell that responds out to 1.8 microns producing over 1 W/cm^2 electric, given a blackbody IR emitter temperature of 1250 C. This high power density along with a simple diffused junction cell makes an array cost of $0.50 per Watt possible. The second development is new IR emitters and filters that put 75% of the radiant energy in the cell convertible band. The third development is a set of commercially available ceramic radiant tube burners that operate at up to 1250 C. Herein, we present near term and longer term spectral control designs leading to a 1.5 kW TPV generator / furnace incorporating these new features. This TPV generator / furnace is designed to replace the residential furnace for combined heat and power for the home.

INTRODUCTION

Solar cells can produce electric power from sunlight without burning a fuel. If very low cost solar cells can be made, a homeowner can generate his own electricity at rates below the electric utility rate. However, while this can be true during summer months, there is a problem in winter months when the sun does not shine. Low bandgap photovoltaic or "solar" cells can solve this problem. A homeowner can put solar cells on his roof for electric power in the summer months and "solar" cells in his furnace for electric power during the winter months. The idea of "solar" cells in a heating furnace is called ThermoPhotoVoltaics or ThermoPV or TPV. The idea is that a ceramic element is placed in the flame in the furnace and this element then glows like the coals in a fireplace. "Solar" cells near by then convert the glow into electricity. Using TPV, the homeowner generates electricity whenever heat is needed. Therefore, it is not necessary to burn additional fuel.

While this TPV concept is simple, the problem has been that the two types of solar cells (or more accurately photovoltaic cells) are not the same. While the solar cells on the roof convert visible light into electricity, the TPV cells in the furnace need to convert infrared radiation into electricity. JX Crystals has invented and developed the required GaSb TPV cells.

OVERVIEW OF KEY COMPONENTS

After inventing the required IR sensitive cells, JX Crystals began to develop complete TPV systems. This effort then required us to invent and integrate several key components into a practical generator that can be manufactured economically.

CP653, *Thermophotovoltaic Generation of Electricity: 5th Conference*
edited by T. J. Coutts, G. Guazzoni, and J. Luther

While the TPV idea was first proposed in 1960, three new developments have taken place in the last ten years that now make economical TPV systems possible. The first new development is the diffused junction GaSb cell [1,2,3,4]. Personnel now at JX Crystals were the first to fabricate GaSb cells in 1989. In 1991, our work was presented in a paper entitled "Fundamental Characterization Studies of GaSb Solar Cells" [1]. This new GaSb cell responds out to 1.8 microns. The process used to fabricate this cell replicates the silicon solar cell fabrication process. Diffusions are known to be much less expensive than epitaxy and, unlike in complex epi cells, no toxic gases are used in our process. JX Crystals has issued patents and licensed patents on the GaSb cell and circuits [2]. Two TPV circuits incorporating these new cells are shown in figure 1a along with a power curve in figure 1b showing 2 Watts per cell. The power density is over 100 times more than the traditional planar solar cell making cost of $0.50 per Watt possible.

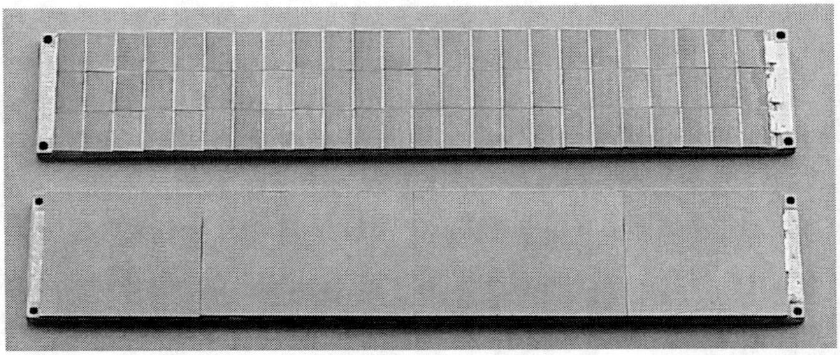

FIGURE 1a. 72 GaSb cells shingle mounted on a circuit measuring 5 cm by 26 cm. Top circuit shows shingles with steps and cracks; bottom circuit is covered with planar filter covers.

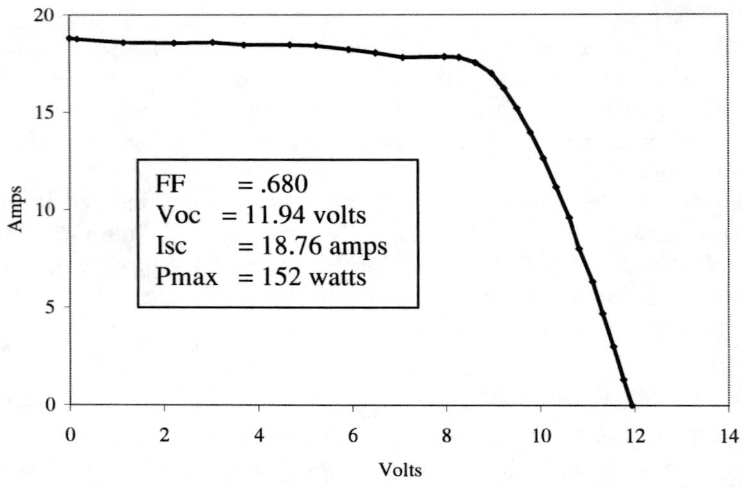

FIGURE 1b. Illuminated current vs. voltage test for 72-cell circuit showing over two Watts per cell and over 1 Watt per cm^2.

The second new development is new IR emitters and filters that put 75% of the radiant energy in the cell convertible band. These spectral control elements [5,6,7] along with the TPV cells and circuits have been described in more detail previously [1-7]. However, since these elements are key to the success of TPV systems, we will describe near term and longer term spectral control options in the next section.

The third new development is the commercial availability of low NOx SiC radiant-tube burners that operate at up to 1250 C. Figure 2a shows a schematic of a radiant-tube burner from WS Inc. and figure 2b shows a photograph of several of these burner tubes in operation in an industrial furnace. According to the WS Inc. literature, "In over a decade of practical experience with ceramic radiant tubes, material failure due to thermal wear has not yet occurred."

In a subsequent section of this paper, we will describe a TPV generator / furnace built around these three key components. This unit is designed for combined heat and power (CHP) for the home.

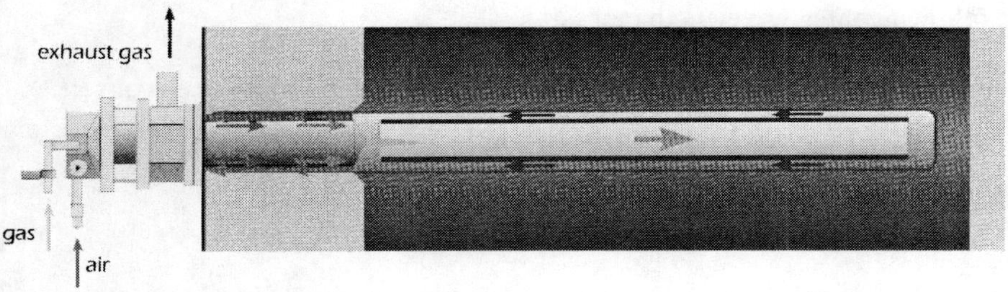

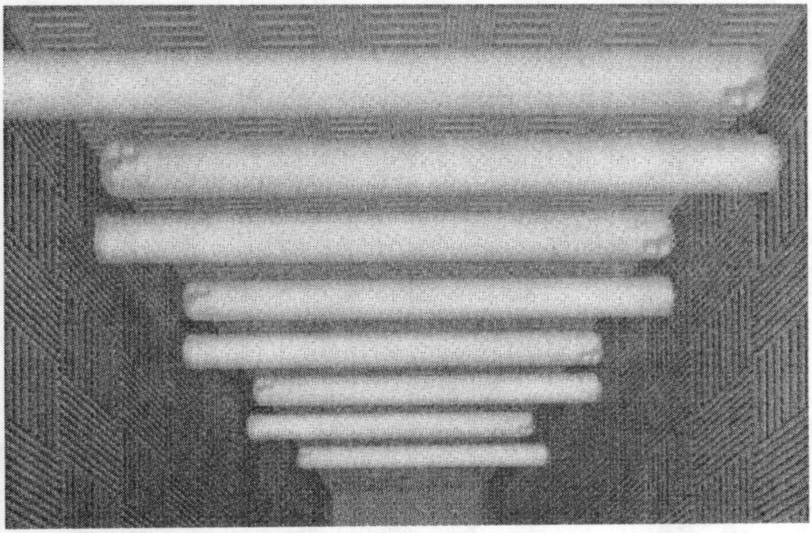

FIGURE 2. (a) Schematic of a low NOx radiant tube burner as manufactured by WS Inc. (b) Photograph of several of these burners in operation in an industrial furnace.

SPECTRAL CONTROL

Given cells and circuits that work and a developed high-temperature radiant tube burner with a recuperator, the key to the success of TPV is in the spectral control. The optical path for TPV begins with the IR emitter and ends with the PV cells. The components in between such as a spectral filter or an emitter foil are designed to control the spectrum of the radiation. Ideally, the emitted spectrum would consist only of photons with wavelengths in the PV cell convertible band. However in practice, this is not achieved. Ideally, the IR filter would reflect all non-convertible photons back to the emitter and transmit all convertible photons to the PV cell array. However in practice, this is also not achieved. Here, we describe spectral control in terms of three optical elements, the IR emitter, the IR filter, and the PV cells. Two compromise solutions for spectral control are described in this section.

The two alternate spectral control systems we describe here differ primarily in the tupe of emitter utilized. The PV cells are common and both use dielectric filters. To qualitatively understand spectral control, it is convenient to divide the emitted spectrum into the following three wavelength ranges:

1. Short wavelength (<1.8 microns) radiation that the PV cells can convert.
2. Mid-wavelength (< 3.6 microns and >1.8 microns) radiation where a simple dielectric filter easily reflects and recycles non-useful radiation.
3. Long wavelength (> 3.6 microns) radiation which must somehow be suppressed.

As the quantum efficiency shows in figure 3, a GaSb PV cell efficiently converts short wavelength photons into electricity. Similarly, as the reflection spectra in figure 4 shows, simple dielectric filters efficiently reflect mid-wavelength photons back to the emitter. The two spectral control systems we describe here represent alternate means of suppressing the long wavelength radiant energy.

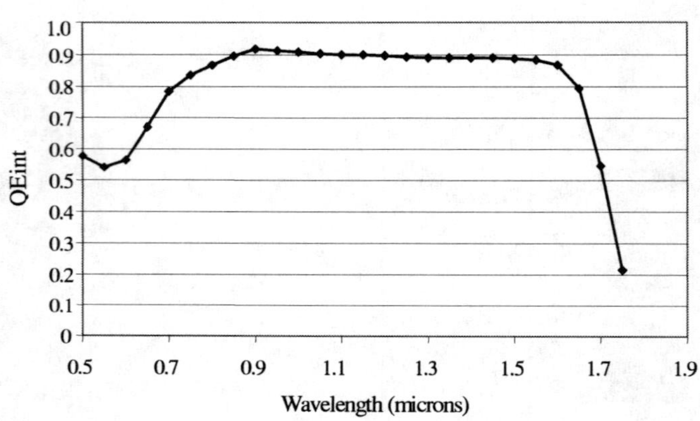

FIGURE 3. GaSb cell internal quantum efficiency.

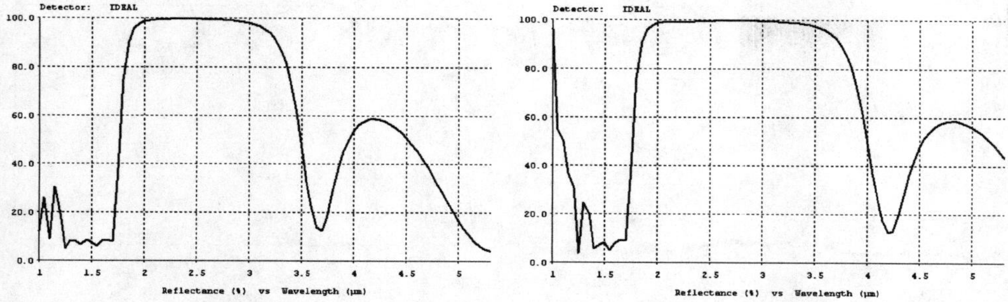

FIGURE 4. Reflectance for (a) 9 layer and (b) 18 layer dielectric filters.

Our first method of suppressing the longer wavelength radiation uses a double quartz heat shield around a SiC emitter. Since quartz absorbs longer wavelengths beyond 4 microns, the inner most quartz shield absorbs energy from the SiC emitter and re-radiates half of that energy back toward the emitter. The second heat shield does the same thing with the net result that the long wavelength emitted energy is reduced by 3 times. In effect, as is shown in figure 5(a), the emittance is reduced by 3 times at wavelengths beyond 4 microns. We first described this double quartz heat shield concept in the 2nd TPV conference in 1995 [8]. Pierce and Guazzoni presented further information on this concept in the 4th TPV conference in 1998 [9].

The higher performance longer-term option uses an antireflection-coated tungsten foil around the heated SiC tube [5,6,7]. Tungsten has a low emittance (high reflection) at long wavelengths and using the anti-reflection (AR) coating can enhance its emittance at cell-convertible shorter wavelengths. The measured emittance of AR/W is shown in figure 5(b). A patent on this emitter has been issued to JX Crystals Inc. [6]

Referring to figure 5, note that the emittance at shorter wavelength for AR/W is comparable to SiC and its emittance at longer wavelengths is lower than for the double quartz shield case yielding better spectral control. However, the negative for AR/W is that it can not operate in air. As in a light bulb, a noble gas backfill is required. Figure 6 shows the radiant power spectra arriving at the PV cells for the two spectral control cases just described.

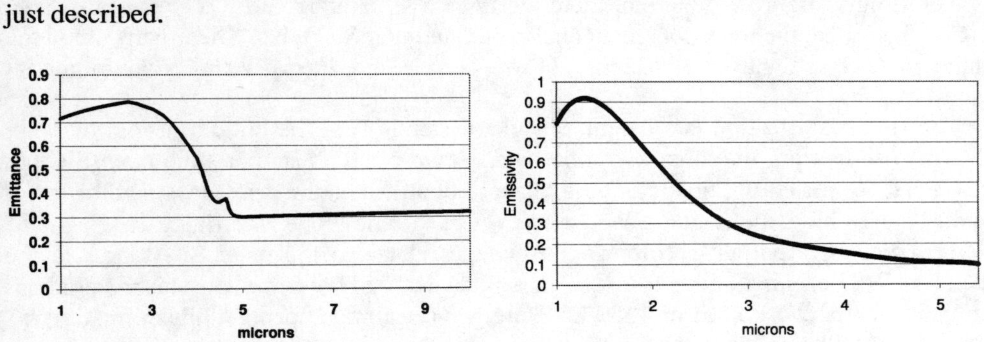

FIGURE 5. (a) Effective emittance for double quartz shield around SiC emitter. (b) Emittance of AR coated tungsten foil.

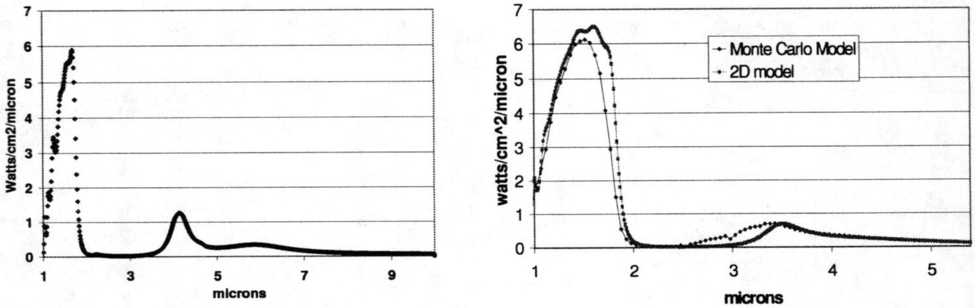

FIGURE 6. Emitted power spectra for (a) SiC / q^2 at 1200 C & (b) AR/W at 1300 C.

In figure 6(a), the emittance for SiC with two-quartz shields from figure 5(a) is convolved with a blackbody at 1200 C and the 18 layer filter spectra from figure 4(b). The 18-layer filter is used because it has reflectivity extending out to 4 microns where the quartz shields become effective at reducing the SiC emittance. In figure 6(b), the AR/W emittance is convolved with a blackbody at 1300 C and the 9-layer filter spectra.

Qualitatively, figure 6 shows that both spectral control strategies are effective. We have used TracePro to analyze the performance of each of these options and these results will be presented in the next section. First, however, we describe our proposed generator / furnace design for home combined heat and power. This design begins with a commercially available WS radiant tube burner.

1.5 KW TPV GENERATOR / FURNACE DESIGN AND PERFORMANCE

We have previously described the design, fabrication, and testing of a 500 W cylindrical TPV generator [7] in which a TPV array surrounds a custom radiant tube burner. All of the parts for this 500 W TPV generator were built at JX Crystals. We have now adapted this earlier design to use a small WS Inc radiant tube burner resulting in a detailed design of a larger 1.5 kW TPV generator / furnace. Figure 7 shows an overview of this design.

Referring to figure 7, note that there are three SiC ceramic parts. They are the large SiC radiant tube, the inner SiC recuperator, and an inner SiC tube. These parts are in the standard WS radiant tube design. However, we are incorporating five important modifications.

The first modification is a ceramic insulator stand-off at the bottom inside the outer SiC tube supporting the inner SiC tube. This stand-off both supports the inner SiC but also insulates the end of the outer tube to avoid or at least reduce heat losses at the lower end. This is unlike the standard WS design where the end glows (see figure 2b).

The second important recommended change relates to the inner SiC tube. In the standard WS design, all three SiC tubes are made using Si bonded SiC. This material has an upper temperature limit of 1350 C. Here, we recommend using a higher quality SiC for the inner tube with a higher temperature capability.

The third important change is the addition of the fold-back section in the outer SiC tube. This allows the addition of the secondary stainless steel recuperator coaxial with the first higher temperature SiC recuperator.

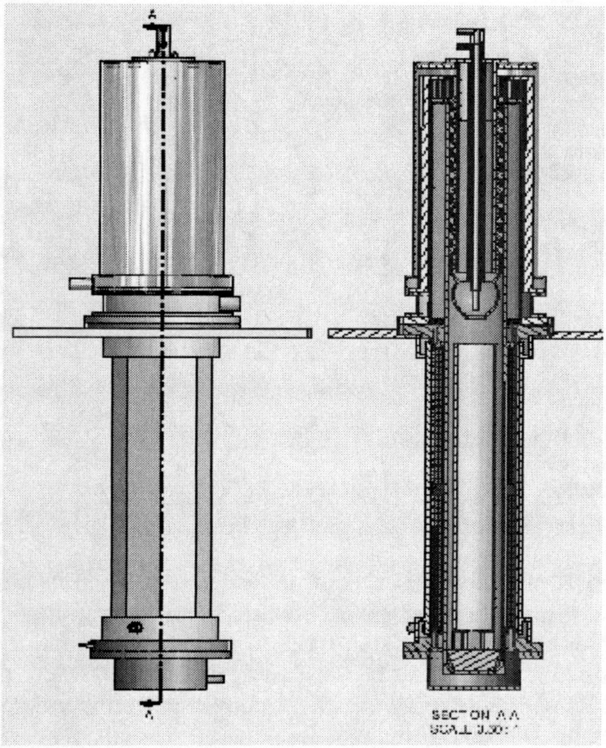

FIGURE 7. Design of TPV 1.5 kW electric generator / 12.2 kW thermal furnace for home combined heat and power.

In the longer term, the fourth important feature in our design is the use of the AR coated refractory metal foil IR emitter wrapped around the outer SiC tube as a means of controlling the IR spectra for maximum conversion efficiency. There are also refractory metal foil heat shield wrappings.

In the longer term, the fifth and final important feature in our design is the use of o-ring seals at the top and bottom PCA flanges. These seals are required so that the space between the PCA and the radiant tube can be evacuated and back filled with a noble gas such as krypton. This is done for two reasons. First, it is necessary to protect the emitter foil against oxidation and second, krypton is used to reduce heat transfer losses to the PV array via thermal conduction through the gas. This noble gas fill feature appears daunting but is really no different than the conventional incandescent tungsten filament light bulb.

We have projected the performance of this TPV generator design based on data from WS Inc. and our own experience with burners and based on modeling of the emitter and PV array performance using TracePro. TracePro is a Monte Carlo based software package by Lambda Research that has been used for TPV modeling by Richard Thomas at Bechtel Bettis [10]. Table 1 summarizes the resultant predicted system performance for a near term design using SiC with a double quartz shield and a longer term design using an AR/W foil emitter with a hermetic seal.

TABLE 1. Predicted TPV generator / furnace performance.		
Modeled with: Cell Temp = 50 C 16 circuit array with 63 cell (21x3) circuits # cells = 1008 Emitter Area = 1480 cm^2 Circuit Area = 1900 cm^2		
	Case 1 AR/W emitter	Case 2 SiC emitter with double quartz shield
Temitter	1260 C	1250 C
Pmax	1.5 kW	1.1 kW
Ptotal-Precycle	9729 W	11221 W
Pwater	8229 W	10121 W
Pmax/(Ptotal-Precycle)	15.4%	10%
Burner/Recup Effic	80%	80%
System Electric Effic	12.3%	8%
Fuel Burn Rate	12.2 kW	14 kW
Emitter heat transfer rate	6.6 W/cm^2	7.6 W/cm^2

As Table 1 shows, TracePro calculation indicates that our TPV generator / furnace should be capable of producing 1.5 kW of electricity for a burn rate of 12.2 kW with an overall electrical conversion efficiency of 12.3% for the AR/W emitter case. Similarly, our TPV generator / furnace should be capable of producing 1.1 kW of electricity for a burn rate of 14 kW with an overall electrical conversion efficiency of 8% for the SiC emitter case with a double quartz shield.

Note that the unit that we are describing is designed to produce both power and heat. Therefore, the cells are operated at 50 C in the Table 1 calculation. This higher cell temperature operation reduces the electrical efficiency relative to room temperature operation but allows for hot water and space heating for the home. In a combined heat and power (CHP) system, the waste exhaust heat can also be used. The CHP efficiency can then be over 90%.

TPV TEST STATION

A TPV test station using a WS SiC radiant tube burner is being assembled at the Alberta Research Council in Canada. Figure 8 shows photographs of this system. Experimental data from this test station will be used to verify model calculations and to evolve the TPV key components.

ECONOMICS

From the point of view of the homeowner and the country, our strategy is to conserve fuel and save money by combining an energy efficient TPV furnace (winter) with a renewable energy source, solar cells (summer). While the TPV furnace does burn fuel, it utilizes the fuel energy with over 90% efficiency (referred to the fuel lower heating value, 100% efficiency is possible). Our distributed combined heat and power residential system saves on the need for additional central power plants and additional transmission lines. While fuel energy utilization efficiency in a central power plant might be 40%, the other 60% will be thrown away as waste heat. Co-generating electricity and heat in the home with TPV avoids burning fuel in the central power plant.

A very interesting advantage for our TPV generator / furnace is that it is both a generator and a furnace. Therefore, for a new home purchase, one gets a credit for the furnace. In other words, a new homeowner can chose to buy a furnace for \$2700 for example or a TPV generator / furnace for \$2700+\$1500 = \$4200.

Figure 8: TPV experimental test station with glowing WS SiC radiant-tube burner (left) and water cooled filter array (right).

CONCLUSIONS

From the point of view of technical design and market potential for this technology, the conclusions in this paper are very positive. We present a design capable of producing 1.5 kW electric with an array efficiency of 15.4%, a burner / recuperator efficiency of 80%, an overall electrical efficiency of 12.3% and a combined heat and power efficiency of over 90%. The cost analysis of this design suggests a total cost of approximately \$3200. JX Crystals' analysis for the US market suggests an installed cost to a homeowner of \$4200. Given a furnace credit of \$2700 (for heating), this corresponds to a cost for the electric generator of \$1500/1.5kW or \$1000/kW. This corresponds to an energy payback time of approximately 4 years at 10 cents per kWh.

REFERENCES

1. L. M. Fraas, J. E. Avery, V. S. Sundaram, et. al. "Fundamental Characterization Studies of GaSb Solar Cells", 22nd IEEE PVSC (1991) p. 80.
2. US Patents #5,217,539, #5,389,158, #6,057,507, #6,232,545 B1, and PCTs assigned to JX Crystals Inc.
3. L. M. Fraas, H. X. Huang, et. al., "Low Cost High Power GaSb Photovoltaic Cells", TPV Generation of Electricity, 3rd NREL Conference, AIP Proceedings #401 (1997) p. 33.
4. J. E. Avery, L. M. Fraas, et. al., "Lightweight Concentrator Module with 30% AM0 Efficient GaAs/GaSb Tandem Cells", 21st IEEE PVSC (1990) p. 1277.
5. L. M. Fraas, J. E. Samaras, J. E. Avery, and L. Minkin, "AR Coated Refractory Metal Matched Emitter for Use with GaSb TPV Generators", 28th IEEE PVSC (2000) p. 1020.
6. US Patent # 6,177,628 and PCT assigned to JX Crystals Inc.
7. L. M. Fraas, J. E. Samaras, et. al., "TPV Generators Using the Radiant Tube Burner Configuration", Proceedings of 17th European PV Solar Energy Conference, Munich, Germany 22-26 October 2001.
8. L. M. Fraas, U. C. Pernisc, et. al., "SiC IR Emitter Design for TPV Generators", TPV Generation of Electricity, 2nd NREL Conference, AIP Proceedings #358 (1995) p. 488.
9. D. E. Pierce and G. Guazzoni, "High Temperature Optical Properties of TPV Emitter Components", TPV Generation of Electricity, 4th NREL Conference, AIP Proc. #460 (1998) p. 177.
10. Richard M. Thomas and Bernard R. Wernsman, "Thermophotovoltaic Devices and Photonics Modeling", Optics & Photonics News, (August 2001) p. 40.

Small Thermophotovoltaic Prototype Systems

Wilhelm Durisch, Bernd Bitnar, Fritz von Roth, Günther Palfinger

Paul Scherrer Institute, PSI, CH-5232 Villigen PSI, Switzerland
phone: 0041 56 310 2625, fax: 0041 56 310 2199
Author for correspondence: Wilhelm Durisch, email: wilhelm.durisch@psi.ch

Abstract. In an earlier paper [1], we reported on a small grid-connected thermophotovoltaic (TPV) system consisting of an ytterbia mantle emitter and silicon solar cells with 16 % efficiency (under solar irradiance at Standard Test Conditions, STC). The emitter was heated up using a butane burner with a rated thermal power of 1.35 kW (referring to the lower heating value). This system produced an electrical output of 15 W, which corresponds to a thermal to electric (direct current) conversion efficiency of 1.1 %. In the interim, further progress has been made, and significantly higher efficiencies have been achieved. The most important development steps are: 1) The infrared radiation-absorbing water filter between emitter and silicon cells (to protect the cells against overheating and against contact with flue gasses) has been replaced by a suitable glass tube. By doing this, it has been possible to prevent losses of convertible radiation in water. 2) Cell cooling has been significantly improved, in order to reduce cell temperature, and therefore increase conversion efficiency. 3) The shape of the emitter has been changed from spherical to a quasi-cylindrical geometry, in order to obtain a more homogeneous irradiation of the cells. 4) The metallic burner tube, on which the ytterbia emitter was fixed in the initial prototypes, has been replaced by a heat-resistant metallic rod, carrying ceramic discs as emitter holders. This has prevented the oxidation and clogging of the perforated burner tube. 5) Larger reflectors have been used to reduce losses in useful infrared radiation. 6) Smaller cells have been used, to reduce electrical series resistance losses. Applying all these improvements to the basic 1.35 kW prototype, we attained a system efficiency of 1.5 %. By using preheated air for combustion (at approximately 370 °C), 1.8 % was achieved. In a subsequent step, a photocell generator was constructed, consisting of high-efficiency silicon cells (21% STC efficiency). In this generator, the spaces between the cells were minimized, in order to achieve as high an active cell area as possible, while simultaneously reducing radiation losses. This new system has produced an electrical output of 48 W, corresponding to a system efficiency of 2.4 %. This is the highest-ever-reported value in a silicon-cell-based TPV system using ytterbia mantle emitters. An efficiency of 2.8 % was achieved by using preheated air (at approximately 500 °C).

An electronic control unit (fabricated of components with low power consumption, and including a battery store) was developed, in order to make the TPV system, described in [1] self-powered. This unit controls the magnetic gas supply valve between gas supply cylinder and burner as well as the high-voltage ignition electrodes. Both the control unit's own power consumption and the battery-charging power are supplied directly by the TPV generator. A small commercial inverter is used to transfer excess power to the 230 V grid.

INTRODUCTION

Thermophotovoltaic devices convert radiant heat directly into electrical power. The integration of this principle into residential heating systems would open up a major potential for combined heat and power generation. Part of the electrical power produced can be used to make heating systems self-powering. Such systems would continue to operate during grid failures (blackouts). Excess power can be fed into the

CP653, *Thermophotovoltaic Generation of Electricity: 5th Conference*
edited by T. J. Coutts, G. Guazzoni, and J. Luther

household's-grid to supply appliances, and/or transferred to the utility's-grid for external consumers.

In Switzerland, about 1.5 million oil-burning and gas-fired domestic central heating systems could be equipped with thermophotovoltaic generators [1]. Assuming a fuel-to-alternating-current (AC) conversion efficiency of 5 %, the potential for household TPV co-generation systems in Switzerland then amounts to 2.3 billion kWh of electricity per year. Furthermore, if the domestic space heating and hot water production that is presently based on electricity were replaced by oil-burning and gas-fired TPV co-generation systems, an additional contribution of about 0.5 billion kWh/year could be produced, resulting in an effective TPV potential of 2.8 billion kWh/year. This corresponds to 6 % of the total annual electricity consumption in Switzerland, which means that decentralized TPV co-generation systems could - in the long term - provide a valuable contribution to the Swiss electricity supply. However, further research, development and demonstration work will be required to achieve a fuel-to-AC efficiency of 5 %. Based on the latest economic considerations [2], it is concluded that competitive electricity prices should be anticipated in comparison with prices for household electricity.

15 W PROTOTYPE

The purpose of developing TPV prototype systems was: 1) To investigate the influence of different components and configurations on electrical output, 2) To improve system design and enhance device efficiency, 3) To provide experimental data for the modeling of larger systems, 4) To demonstrate the feasibility for continuous operation, 5) To demonstrate self-powered operation 6) To demonstrate self-powered operation plus the simultaneous feeding of excess power to the 230 V grid.

A first prototype system was presented in [1]. This consists of a conventional butane burner (rated thermal power 1,35 kW), which was equipped with a spherical mantle emitter made of ytterbia, Yb_2O_3. Gaseous butane is provided by evaporation of liquid butane contained in a steel cylinder. Eight commercial silicon solar cells, measuring 5cm x 10cm, made by TESSAG, Heilbronn, Germany, with STC-efficiency of 16 %, were used to fabricate the photocell generator. In order to obtain as high a voltage as possible, the cells were connected in series. An infrared radiation-absorbing water layer between the emitter and photocell-generator protects the photocells from overheating and against direct contact of the cells with flue gasses The cell temperature was limited to about 55 °C by additional free convection cooling. Upper and lower reflectors prevent axial radiation losses. Under emitter radiation, the photocell generator produced a maximum power point (mpp) current of 4.6 A at an mpp voltage of 3.3 V, corresponding to a direct current (DC) output of 15.2 W and a thermal input to DC output conversion efficiency of 1.1 %. The short circuit current was 5.0 A and the open circuit voltage 4.5 V, resulting in a fairly low fill factor of 0.67.

A high efficiency DC/DC converter (voltage multiplier), developed at the University of Lausanne, Switzerland, and an inverter specially developed by the Swiss company, LEC, Küsnacht were used to feed the generated power into the local 230 V utility grid [1]. An overall converter/inverter interface efficiency of up to 89 % was achieved. Using commercially available DC/DC converters led to unsatisfactory results (very low interface efficiency due to low efficiency of a series connection of two industrial DC/DC converters).

30 W PROTOTYPE

A second prototype, based on the same butane burner, was developed in order to increase the electrical output and system efficiency. The same ytterbia mantle structure as in the first prototype was used as emitter. A detailed simulation model [3] was used to study the effect of the geometrical form of the emitter on the illumination of the photocells. Spherical emitters in a cylindrical cell-generator lead to a non-homogeneous irradiation distribution on the photocells, and hence a reduction in their efficiency. The emitter geometry was therefore changed from spherical to ellipsoidal. Furthermore, the perforated burner tube was replaced by a metallic diffuser/rod system with ceramic mantle holders in order to eliminate oxidation and clogging of the tube. In [4], it was shown that the water filter also absorbs some of the convertible emitter radiation. In order to reduce these losses, a glass tube replaced the water filter. However, this requires much more efficient cell cooling. Therefore the cells were glued onto metallic support plates that were cooled by tap water flowing through boreholes. In spite of the higher heat load on the cells, this produced a reduction in the cell temperature, resulting in higher cell efficiency. A glass tube between emitter and photocell generator is necessary to avoid direct contact between the cells and flue gasses, and to prevent the condensation of water (contained in the flue gas) onto the cell surface. We installed larger lower and upper reflectors to reduce axial radiation losses. The result of all these measures is shown in Fig. 1.

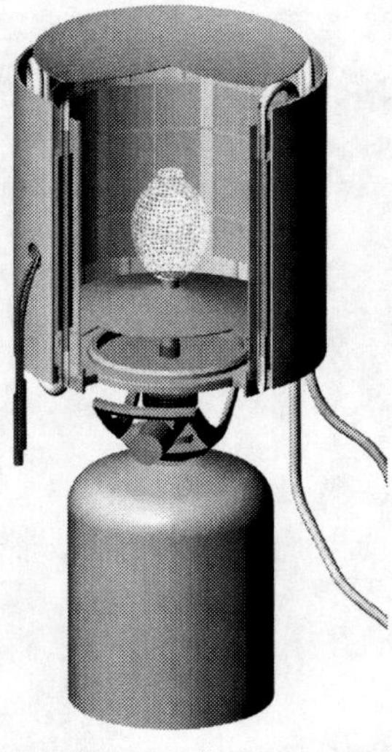

FIGURE 1. 30 W TPV prototype system. The incandescent mantle emitter made from ytterbia, Yb_2O_3, is positioned in the center. Upper and lower reflectors prevent axial radiation losses. The silicon photocells can be seen in the background. Cell temperature is kept at an acceptable level by efficient cell cooling, using tap water. A glass tube between the emitter and photocell generator prevents contact between the cells and exhaust gases.

In this second prototype, we again used silicon solar cells made by TESSAG, Heilbronn, Germany, with an STC efficiency of 16 %. We used smaller cells, producing lower currents, in order to minimize series resistance losses in the cell circuitry. A total of sixteen cells were installed, each measuring 2.4 cm by 9.8 cm. The active cell area amounts to 376 cm^2. All the cells were connected in series to achieve as high an output voltage as possible.

An electrical output of 29 W was obtained at a thermal input of 1905 W, corresponding to a system efficiency of 1.5 %. The short circuit current and open circuit voltage were measured to be 4.0 A and 10.4 V respectively, leading to a fill factor of 0.7. At the maximum power point, the current and voltage were 3.7 A and 7.8 V respectively.

The electrical data for the photocell generators was acquired using a current/voltage (I/V) test device developed by PSI [5]. This consists of a dynamic load, electronic load control, three precise digital multimeters and a notebook. Suitable precision resistors are used to measure the current. Surface temperature probes (Pt-100) are arranged on the rear side of the photocells to measure their temperatures. The signals from the multimeters are transmitted via an IEEE bus to the notebook. Specially developed software allows to evaluate the measured data on line. The results are stored, and can be printed out immediately in the form of a condensed test report, containing all the relevant electrical data and the temperature of the generator under test.

50 W Prototype

A third prototype was fabricated, Fig. 2, based on the experience gained with the previous prototypes.

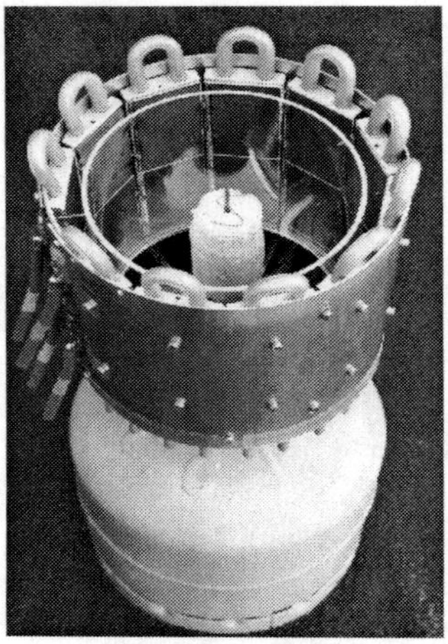

FIGURE 2. 50 W TPV prototype system. A cylindrical ceramic mantel emitter made from ytterbia, Yb_2O_3, was placed in the center of a photocell generator. Butane combustion heats up the emitter. A glass tube is positioned between the emitter and photocells to protect the photocells from the hot exhaust gases. Tap water flows through the cell support plates to provide an efficient cell cooling. Lower and upper reflectors (upper reflector not shown in this picture) reduce radiation losses in axial direction.

The main improvements are: 1) High efficiency monocrystalline silicon solar cells from the University of New South Wales, UNSW, Sydney, Australia, with an STC efficiency of 21.1 %. 2) Cylindrical Yb_2O_3 emitter for more homogeneous illumination of the photocells. 3) Active cell area increased from 376 to 477 cm^2. Two strings of eleven cells each were glued onto metallic support plates and built into a stainless steel cylinder. Each string has separate electrical connection leads, so that the strings can either be connected in parallel or in series. The axial position of the emitter was chosen in such a way that the two strings deliver almost the same short circuit current. Active cell cooling was obtained by the flow of tap water through the cell support plates, and again a glass tube protected the cells from contact with the flue gases.

Infrared mirrors (glass with 1 μm gold coating) at both ends of the glass tube reflect useful emitter radiation back to the photocells and emitter. The upper reflector (not shown in Fig. 2) has a hole in its center to allow the flue gas to escape. The flow rate of butane was determined by measuring the time and butane mass before and after each experiment.

An electrical output of 47.9 W was attained at a thermal input power of 1985 W, corresponding to a system efficiency of 2.4 %. The short circuit current and open circuit voltage were measured to be 4.0 A and 16.0 V respectively (all cells in series). From I/V tests, the mpp-data was found to be: I_{mpp} = 3.76 A and V_{mpp} = 12.73 V, leading to a fill factor of 0.75. Using tap water at 14°C, the cell temperature was kept at 20.6°C in the lower string and at 23.6°C in the upper string. By using preheated air (at approximately 500°C) for combustion, a system efficiency of 2.8 % was achieved.

A series of I/V-tests was carried out under solar irradiation before mounting the photocell generator into the steel cylinder. The result shown in Fig. 3. was obtained after evaluation of the tests.

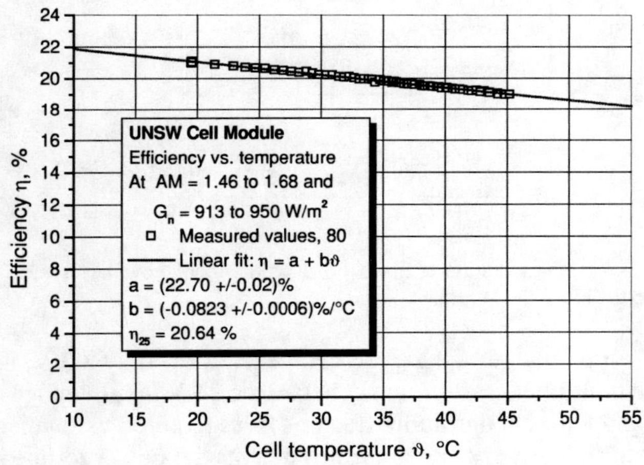

FIGURE 3. Efficiency η vs. cell temperature ϑ at an air mass AM of approximately 1.5, and at an almost constant global normal irradiance G_n of 930 W/m^2. The test module consists of 22 series-connected high efficiency monocrystalline silicon solar cells from the University of New South Wales, UNSW, Sydney, Australia.

According to Fig. 3, the STC efficiency of the UNSW cell module used as TPV-generator is found to be 20.64 %. As expected for a series-connected string, this value is somewhat lower than the supplier's specification for the single cells, which is in the range of 21.0 to 21.2 %. Using the STC efficiency, STC power is found to be 9.85 W. This means that the generator, in comparison with sun illumination, produces 5.7 times more output power under the emitter radiation in the TPV system.

SELF-POWERED GRID-CONNECTED SYSTEM

In order to demonstrate the feasibility of self-powered operation of TPV systems, the 15 W prototype presented in [1] was supplemented by an electronic control unit developed by PSI. This operates the magnetic gas supply valve as well as the high-voltage ignition electrodes. Furthermore, a commercial DC/AC grid inverter from NKF, Delft, The Netherlands, was added in order to demonstrate simultaneous back-supply of excess power into the 230 V utility grid. The complete TPV system is schematically represented in Fig.4.

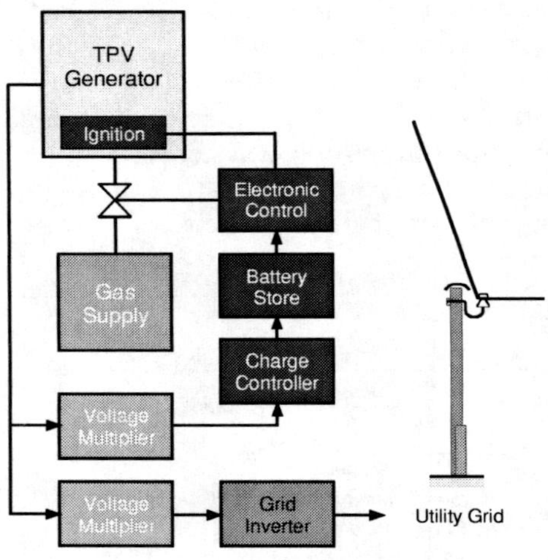

FIGURE 4. Schematic representation of a self-powered thermophotovoltaic system with back-supply of excess power to the local 230 V grid.

Because the mpp voltage of the photocell generator is only 3.3 V, it was necessary to use a voltage multiplier (factor 8) to generate the 24 V input voltage required by the grid inverter. An additional multiplier (factor 4) was required to obtain enough voltage for charging the 12 V battery store, used to deliver the power to open the gas valve, supply the ignition electrodes and power the control unit.

Measurements showed that power consumption for the gas valve during continuous operation of the burner was 0.71 W, and that for the control unit was 1.36 W. In total, 2.07 W (or 14 %) of the photocell output is required for self-powered operation. The remaining part is fed into the grid.

In order to reduce investment costs, the necessity for the multipliers mentioned above can be avoided by replacing the 8 large photocells with 60 smaller cells, all series-connected, and by changing the control unit's supply voltage from 12 to 24 V.

A series type of charge controller is required to charge the battery store, since shunt controllers short circuit the photocell-generator at fully charged store and therefore impede simultaneous back-supply to the utility grid.

SUMMARY AND OUTLOOK

Two new small thermophotovoltaic prototype systems were successfully realized and tested. System efficiencies of up to 2.8 % were achieved. Analogous system efficiencies in the relevant literature [6,7,9] do not exceed the value of 2.1 %.

Electrical characterization of the UNSW high efficiency photocell-generator under solar illumination and under selective emitter radiation [10] in a TPV-prototype revealed an almost six-fold electrical power density of the photocells installed in TPV-systems, reducing the specific generator cost considerably as compared with solar photovoltaics.

We have successfully demonstrated the self-powered operation of a small TPV-system, including simultaneous back-supply of excess power to the 230 V utility grid. For this purpose an electronic control unit was developed by PSI using components with low power consumption. The principle can easily be applied to larger TPV-systems.

In future, emphasis will have to be placed on the development, investigation and practical use of appropriate emitters, taking account of thermal emission characteristics as well as the chemical, thermal and mechanical properties. However, simplified, efficient, cost-effective and reliable photocell-generators will also be important. Finally, system's development and testing will be indispensable.

In a next important step a photocell-generator and a burner/emitter system developed by PSI will be installed in a commercial boiler produced and sold by HOVAL AG, Vaduz, Fürstentum Liechtenstein. In lab tests the generator produced an electrical power of 164 W, which is sufficient for self-powered operation of such boilers.

Our goal is to investigate in detail the thermophotovoltaic conversion of radiant heat into electricity, in order to provide the basic understanding of the physical processes involved. This is required to transfer this new technology successfully into industry, as an efficient, economical and sustainable new technique for combined heat and power generation.

ACKNOWLEDGMENTS

We thank the Swiss "Kommission für Technologie und Innovation" KTI for supporting this project, contract No. 5692.2 EBS.

REFERENCES

1. Durisch W., Grob B., Mayor J.-C., Panitz J.-C., Rosselet A. Interfacing a small thermophotovoltaic Generator to the grid. Proceedings of the 4th Conference on Thermophotovoltaic Generation of Electricity, AIP, 1999, p. 403-414.
2. Palfinger G., Bitnar B., Durisch W., Mayor J.C., Grützmacher D., Gobrecht J. Cost estimates of electricity from a TPV residential heating system. Proc. 17th Europ. Photov. Solar Energy Conf. & Exh. 2001, pp. 117 - 120.
3. Mayor J.-C. Projekt Thermophotovoltaik - Mathematische Modellierung. Internal Technical Report, PSI-TM-50-00-04, November 2000.
4. B. Bitnar, W. Durisch, D. Grützmacher, J.-C. Mayor, F. von Roth, J.A. Anna Selvan, H. Sigg, H.R. Tschudi, J. Gobrecht. A TPV system with silicon photocells and a selective emitter. Proceedings of the 28th IEEE Photovoltaic Specialists Conference, Anchorage, September 2000, p. 1218 – 1221.
5. Durisch W., Tille D., Wörz A., Plapp W. Characterisation of photovoltaic generators Applied Energy 65 (2000), 273-284.
6. Crowley C. J., Elkouh N. A., Magari P.J. Thermal spray approach for TPV emitters. Proceedings of the 4th Conference on Thermophotovoltaic Generation of Electricity, AIP, 1999, p. 197-313.
7. Ferber J., Aschaber J., Helbling C., Heinzel A., Wiehle R., Zenker M., Luther J. Microstructured tungsten surfaces as selective emitters in thermophotovoltaic systems. Proceedings of the 16th European Photovoltaic Solar Energy Conference and Exhibition, Glasgow, p. 77-81.
8. Johnstone S.J., "The rare earth industry", in: Manuals of chemical technology II, edited by G. Martin, London: Crosby Lockwood and Son, 1915.
9. Fraas L. M. Combined heat and power for home using thermophotovoltaics. Proceedings of the 8th International Energy Forum, Las Vegas,July 2000, p. 289-294.
10. Bitnar B., Durisch W., Mayor J.-C., Sigg H., Tschudi H. R., Palfinger G., Gobrecht J. Development of a small TPV prototype system with an efficiency >2 %. Proc. 17th Europ. Photov. Solar Energy Conf. & Exh. 2001, pp. 33 – 36.

Additional references related to this paper are to be found in the present proceedings under: Bitnar B., Durisch W., Meyer A., Palfinger G., "New flexible module for thermophotovoltaic applications".

The Challenge of Realistic TPV System Modeling

J. Aschaber*, C. Hebling† and J. Luther†

*Freiburger Materials Research Center, Stefan-Meier-Str. 21, D-79104 Freiburg, Germany
johannes.aschaber@fmf.uni-freiburg.de

†Fraunhofer Institute for Solar Energy Systems, Heidenhofstr. 2, D-79110 Freiburg, Germany

Abstract. Realistic modeling of a TPV system is a very demanding task. For a rough estimation of system limits many of assumptions simplify the complexity of a thermophotovoltaic converter. It's obvious that real systems can not be described by this way. An alternative approach that can deal with all these complexities like arbitrary geometries, participating media, temperature distributions etc. is the Monte Carlo method (MCM). This statistical method simulates radiative energy transfer by tracking the histories of a number of photons beginning with the emission by a radiating surface and ending with absorption on a surface or in a medium. All interactions in this way are considered. The disadvantage of large computation time compared to other methods is not longer a weakness with the speed of todays computers. This article points out different ways for realistic TPV system simulation focusing on statistical methods.

RADIATION BETWEEN NONDIFFUSE NONGRAY OPAQUE SURFACES

Net Radiation Method in an Enclosure

Consider an enclosure composed of N discrete surface areas as shown in Fig. 1. We assume, as usually in enclosure theory, that the outgoing spectral radiation intensity $i_{\lambda o}$, temperature, and surface properties are uniform over each surface. If A_k and A_j describe the k-th and j-th surfaces, then the spectral outgoing intensity in direction $\theta_{r,k}, \phi_{r,k}$ from dA_k is then given by [1]

$$i_{\lambda o,k}(\lambda,\theta_{r,k},\phi_{r,k},\vec{r}_k) = \varepsilon_{\lambda,k}(\lambda,\theta_{r,k},\phi_{r,k},\vec{r}_k)\, i_{\lambda b,k}(\lambda,\vec{r}_k)$$

$$+\frac{1}{A_k}\sum_{j=1}^{N}\int_{A_k}\int_{A_j}\rho_{\lambda,k}(\lambda,\theta_{r,k},\phi_{r,k},\theta_k,\phi_k,\vec{r}_k)\, i_{\lambda 0,j}(\lambda,\theta_j,\phi_j,\vec{r}_j)\frac{\cos\theta_k\cos\theta_j}{|\vec{r}_j-\vec{r}_k|^2}\,dA_j\,dA_k \quad (1)$$

where $\varepsilon_{\lambda,k}$ is the directional spectral emissivity, $\rho_{\lambda,k}$ the bidirectional spectral reflectivity and $i_{\lambda b}$ the spectral blackbody intensity. The dependance of these properties on temperature T_{dA} is omitted to simplify the notation. The first term in Eq.(1) is the spectral intensity emitted by dA_k in the direction $\theta_{r,k}, \phi_{r,k}$. The second term is the intensity reflected from dA_k in the same direction. The sum regards to contributions from all the dA_j surfaces.

CP653, *Thermophotovoltaic Generation of Electricity: 5th Conference*
edited by T. J. Coutts, G. Guazzoni, and J. Luther

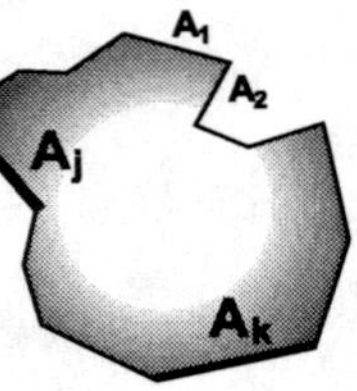

FIGURE 1. Enclosure composed of N discrete areas including surfaces A_j and A_k.

All the k-equations in Eq.(1) are in terms of outgoing intensities in each direction $\theta_{r,k}, \phi_{r,k}$. Separated for each k, one obtains a set of N simultaneous equations for $i_{\lambda o,k}(\lambda, \theta_{r,k}, \phi_{r,k}, \vec{r}_k)$ for $k = 1, \ldots, N$. An iterative numerical solution is required. After calculating all $i_{\lambda o,k}(\lambda, \theta_{r,k}, \phi_{r,k}, \vec{r}_k)$ the total energy per unit time and per unit surface can be determined. It must be supplied to each surface to keep its specific temperature. This is found from the $i_{\lambda o}$ as the difference between the total emitted and absorbed energies. For A_k this leads to

$$\frac{Q_k}{A_k} = \int_{\lambda=0}^{\infty} \int_{\theta_{r,k}=0}^{\pi/2} \int_{\phi_{r,k}=0}^{2\pi} i_{\lambda 0,k}(\lambda, \theta_{r,k}, \phi_{r,k}, \vec{r}_k) \cos\theta_{r,k} d\omega_{r,k} d\lambda$$

$$-\frac{1}{A_k} \sum_{j=1}^{N} \int_{\lambda=0}^{\infty} \int_{A_k} \int_{A_j} i_{\lambda 0,j}(\lambda, \theta_j, \phi_j, \vec{r}_j) \frac{\cos\theta_k \cos\theta_j}{|\vec{r}_j - \vec{r}_k|^2} dA_k dA_j d\lambda. \qquad (2)$$

for $k = 1, \ldots, N$ where the differential solid angle is given by $d\omega_{r,k} = \sin\theta_{r,k} d\theta_{r,k} d\phi_{r,k}$. For diffuse-gray surfaces, so that all $i_{\lambda o,k}$ are independent of λ, θ, and ϕ, this simplifies to

$$\frac{Q_k}{A_k} = \pi i_{0,k} - \pi \sum_{j=1}^{N} i_{0,j} F_{k-j}, \qquad (3)$$

where F_{k-j} is known in literature as the (geometric) configuration factor or the view factor. It is defined as that fraction of energy which leaves diffuse surface area A_k and arrives at surface A_j,

$$F_{k-j} = \frac{1}{A_k} \int_{A_k} \int_{A_j} \frac{\cos\theta_k \cos\theta_j}{\pi |\vec{r}_j - \vec{r}_k|^2} dA_k dA_j. \qquad (4)$$

If Q_k rather than T_k is specified, T_k must be determined and the solution of Eq.(2) becomes more difficult. A temperature is assumed for A_k, and the enclosure equations of the form Eq.(1) are solved to find the $i_{\lambda o}$. The outgoing intensities are substituted into Eq.(2) and the computed Q_k is compared to the given value. The T_k is then adjusted and the procedure repeated until a desired convergence.

The conclusion from the section above is that although the formulation of radiation-exchange problems involving directional and spectral property effects is not conceptually difficult, it is often tedious to obtain solutions to the resulting integral equations.

In order to simplify the equations, it is usually necessary to invoke assumptions and approximations such as non-directional and/or non-spectral surface properties. Numerical techniques such as iteration can be used for directional spectral problems, since closed-form analytical solutions can rarely be obtained. For complicated directional and spectral effects, the Monte Carlo method is often a better approach than the integral equation method.

The Monte Carlo Method for Radiative Exchange in an Enclosure

As discussed in the section above the formulation of radiation heat balances in enclosures leads to integral equations for unknown surface temperatures or heat flux distributions. These equations can be difficult to solve and are the consequence of using a deterministic viewpoint when deriving the heat flow quantities. By invoking a probalistic model of the radiative exchange process and applying the Monte Carlo technique, it is possible to avoid many of the inherent difficulties in the integral equation formulation.

Considering again an enclosure of N discrete surface areas. The idea of the Monte Carlo method is to use probability concepts to model stochastic physical events such as photon emission, reflection and absorption. The paths of individual photon "bundles" are tracked beginning with emission at a certain location and ending with absorption at a different location. To determine which a photon bundle should go and wether it should be absorbed or reflected when it hits a surface, random numbers are generated and compared with appropriate probability functions. For further details see references [1, 2, 3].

The energy per bundle ω is defined as the power emitted from a given surface ΔA divided by the number $N_{\Delta A}$ of bundles emitted per unit time giving

$$\omega_{\Delta A} = \frac{\Delta A}{N_{\Delta A}} \int_0^\infty \int_{\frown} \varepsilon_\lambda(\lambda, \theta, \phi, T_A) i_{\lambda,b}(\lambda, T_A) \cos\theta \, d\omega \, d\lambda = const. \tag{5}$$

The value $\omega_{\Delta A}$ is kept constant and therefore the number $N_{\Delta A}$ has to be calculated for each surface element ΔA. If $S_{\Delta A}$ describes the number of photon bundles absorbed by area ΔA, the net flux of by ΔA is given by

$$Q_{\Delta A}^{net} = \omega_{\Delta A}(N_{\Delta A} - S_{\Delta A}). \tag{6}$$

The flowchart in Fig. 2 describes the algorithm. The steps outlined by a dashed line involve random number selections.

The value of Monte Carlo methods is that program complexity increases in proportion to problem complexity, while the difficulty of carrying out conventional solutions increases as the square of the complexity as shown in Fig. 3. Because a Monte Carlo method is somewhat more difficult to apply to the simplest problems, it is most effective when complex geometries and variable properties must be considered. In complex geometries, a Monte Carlo method has the advantage that simple relations will specify the path of a given energy bundle, whereas most other techniques require integrations over surface areas.

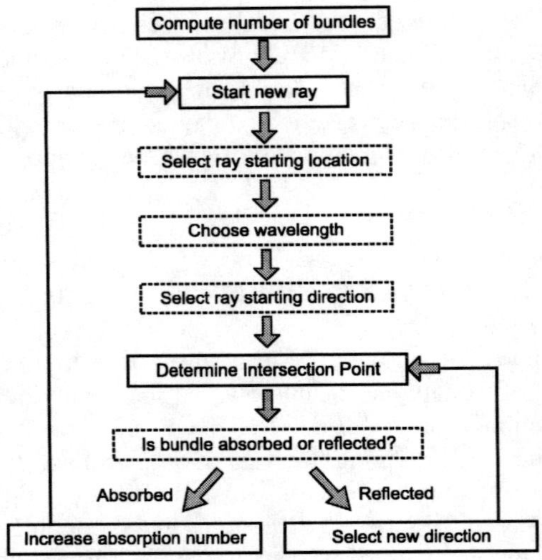

FIGURE 2. Flowchart of collision-based Monte Carlo simulation for a single emitting surface element.

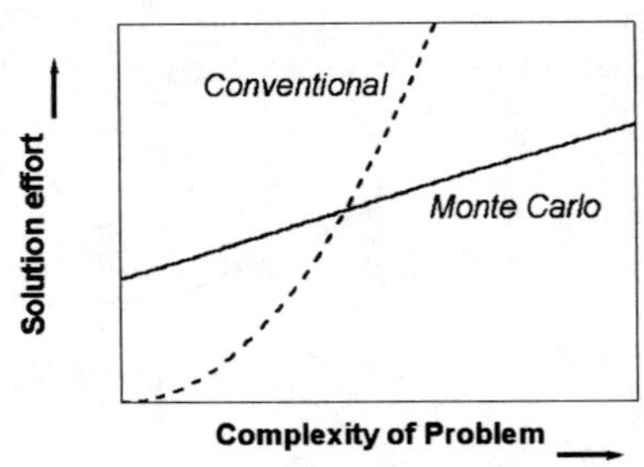

FIGURE 3. Comparison of Monte Carlo and conventional solution techniques [1, 2].

RADIATION IN PARTICIPATING MEDIA

Equation of Transfer

The radiation of intensity $i_\lambda(S)$ within an absorbing, emitting and scattering medium with spectral absorption coefficient a_λ and the spectral scattering coefficent $\sigma_{s\lambda}(S)$ is shown in Fig. 4. The change in intensity with S in the solid angle $d\omega$ about the S direction for a medium with an index of refraction $n = 1$ is given by [1]

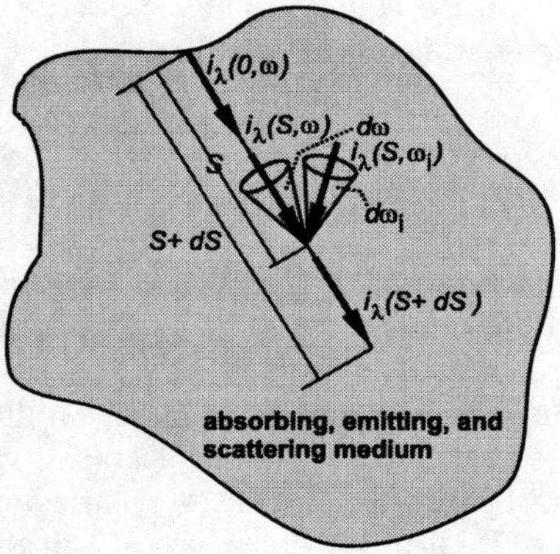

FIGURE 4. Drawing describing situation for the equation of transfer

$$\frac{di_\lambda}{dS} = \underbrace{-a_\lambda(S)i_\lambda(S)}_{\substack{\text{loss by absorption}\\ \text{(including induced}\\ \text{emission)}}} + \underbrace{a_\lambda(S)i_{\lambda b}(S)}_{\substack{\text{gain by emission}\\ \text{(not including}\\ \text{induced emission)}}} - \underbrace{\sigma_{s\lambda}(S)i_\lambda(S)}_{\text{loss by scattering}} + \underbrace{\frac{\sigma_{s\lambda}(S)}{4\pi}\int_{\omega_i=0}^{4\pi} i_\lambda(S,\omega_i)\Phi_\lambda(\omega,\omega_i)\,d\omega_i}_{\text{gain by scattering in direction } S} \tag{7}$$

where $\Phi_\lambda(\omega,\omega_i)$ is called the *phase function*:

$$\Phi_\lambda(\omega,\omega_i) = \frac{di_\lambda(S,\omega)}{(1/4\pi)\int_{\omega_i=0}^{4\pi} di_\lambda(S,\omega_i)\,d\omega_i} \tag{8}$$

The net outflow of radiant energy per unit volume is the emitted energy rate minus the absorbed energy rate at the location S. This describes the desired divergence of the radiant flux vector and can be derived from Eq.(7) with the optical depth

$$\kappa_\lambda(S) = \int_0^S K_\lambda(S^*)\,dS^* \tag{9}$$

to

$$\nabla q_r\vec{(S)} = 4\int_0^\infty a_\lambda(\lambda)[e_{\lambda b}(\lambda) - \frac{1}{4}\int_0^{4\pi} i_\lambda(\lambda,\omega,\kappa_\lambda)\,d\omega]\,d\lambda, \tag{10}$$

which is valid for both anisotropic and isotropic scattering. K_λ in the Eq.(9) is the extinction coefficient given by

$$K_\lambda = a_\lambda + \sigma_{s\lambda}. \tag{11}$$

The expression $-\nabla q_r(\vec{S})$ is the so called radiative source term in the general energy conservation equation, which will be discussed later.

Monte Carlo Method for Participating Media

The Monte Carlo method was introduced as a technique for modeling radiative transfer between surfaces in an enclosure with a nonparticipating medium inside the cavity. This technique can also be extended to include radiative transfer in a participating medium. There is no limit (except computer time) on the complexity of problems that can be modeled by this method. The effects of anisotropic scattering, nonuniform properties, nongray (spectral) properties, and even polarization can be included by this method when it is essential [3]. Another big advantage of the Monte Carlo method is that additional effects can be included easily while keeping the basic algorithm. Thus Monte Carlo technique is expandable, too.

Photon Bundles Energy Values - The basic equation for accounting radiant energy in the participating media is a balance of radiant energy for a differential volume element. This relation is given by Eq.(10). The idea of the MCM in this case is to calculate the second (absorption) term as the product of the number of absorbed bundles $S_{\Delta V}$ and the bundle volume energy $\omega_{\Delta V}$ divided by ΔV. This energy portion is kept constant for all of the bundles regardless of which subvolume they are emitted from. Thus the bundle energy emitted from subvolume ΔV with an index of refraction $n = 1$ is given by

$$\omega_{\Delta V} = \frac{\Delta V}{N_{\Delta V}} 4 \int_0^\infty a_\lambda(\lambda) e_{\lambda b}(\lambda)\, d\lambda = const. \tag{12}$$

Multiplying Eq.(10) by ΔV and using Eq.(12) results in the net transfer rate

$$\nabla \vec{q}_r = \frac{\omega_{\Delta V}}{\Delta V}(N_{\Delta V} - S_{\Delta V}) \tag{13}$$

Assuming non radiating boundary areas the absorbed power per unit of elemental area ΔA is then given by $q_{\Delta A}^{abs} = \omega_{\Delta V} / \Delta A\, S_{\Delta A}$.

For a known temperature distribution the solution scheme is quite simple. The number of bundles $N_{\Delta V}$ that should be emitted from each volume element is determined from Eq.(12). Each bundle is tracked through the medium until it is absorbed at some volume or surface element. The number of bundles absorbed at each volume element $S_{\Delta V}$ is counted and the source term (net transfer rate) is calculated from Eq.(13). If the temperature field is not known the source term is substituted into the energy equation which is solved for a new temperature field. Iteration of this procedure leads to a converged solution [3].

Speeding up [4, 3] - The algorithm described above (sometimes called "collision based" in literature) is simple and powerful. Its weakness is that each ray only contributes to the results of the medium or surface regions where the ray experiences collisions events. Since a ray is either scattered (reflected) or completely absorbed at each collision it can contribute to only one region. If there are many regions for which results must be obtained a large number of rays must be used to ensure statistically precise and accurate results. An alternative to these techniques is the pathlength method. For a nonscattering optically thick medium the bundle energy content is changed along its path. No random number is drawn to determine the distance traveled until absorption. Rather the energy content is depleted due to gradual absorption. That amount is added to the absorption rates of the subvolumes which are parsed by the bundle. The bundle is traced until it leaves the enclosure or until its energy is depleted. For highly reflective surfaces energy partitioning at each surface reflection can also increase convergence. The energy of a bundle will be partitioned at each reflection into the fraction α which is absorbed and the fraction $\rho = 1 - \alpha$ which is reflected. Therefore a random number to decide whether a bundle is absorbed or not is not necessary. The bundle is then traced until it either leaves the cavity or its energy is depleted (e.g. below a certain fraction of the original energy content).

ENERGY TRANSFER BY RADIATION COMBINED WITH CONDUCTION AND/OR CONVECTION

Radiative transport of energy in a participating media is usually accompanied by conductive energy transfer or convective transfer or both. With multiple modes of energy transport possible the question is what are the relative magnitudes of the various modes of energy transport. If radiative transfer dominates radiative equilibrium establishes the temperature distribution. If radiative and conductive and/or convective transport are of similar magnitudes a combined balance of all modes establishes the temperature distribution. The occurence of combined modes significantly increases the mathematical complexity of the essential equations. In particular the energy and radiative transfer equations are coupled and must be solved simultaneously. This coupling introduces nonlinearity into the mathematical formulation in addition to the complex radiative transfer formulation. As a result of the increased mathematical complexity it is reasonable to introduce various simplifications.

The governing relation concerning multimode heat transfer is conservation of energy. The energy equation for a single-component fluid is given by

$$\rho c_p \frac{DT}{Dt} = \beta T \frac{Dp}{Dt} + \nabla\left(k\nabla T - \vec{q}_r\right) + q_S + \Phi_d \tag{14}$$

where D/Dt is the substantial derivative[1], β is the thermal coefficient of volume expansion of the fluid, q_S is the local heat source per unit volume and time, and Φ_d is the heat production by viscous dissipation.

The equation requires the term $-\nabla\vec{q}_r$ which is the energy supplied locally by the radiative intensities. This term is obtained by using Eq.(10). A general solution requires the spectral intensities at each location and in each direction. Since the temperature is in the blackbody spectral intensity a *simultaneous solution* with the energy equation (14) is required.

---

[1] $\frac{D}{Dt} \equiv \frac{\partial}{\partial t} + \vec{u}\nabla$, where $\vec{u}$ is the fluid velocity vector

This section shows some solution techniques for combined heat transfer modes. First, coupled conduction and radiation are discussed when convection is negligible. At the end some ideas for including convection problems are shortly discussed.

Coupled Radiation and Conduction

Consider a participating medium that can be described as a continuum for both radiative and conductive energy transport. If there are no sources or sinks of energy, no convective enery transport and no unsteadiness the energy balance given by Eq.(14) on a differential volume element of the medium simplifies to

$$\nabla(\vec{q}_c + \vec{q}_r) = 0 \tag{15}$$

where the conductive heat flux vector with the thermal conductivity k is given by Fourier's law

$$\vec{q}_c = -k\nabla T. \tag{16}$$

The following simplifications can be used to obtain the radiation source term:

Optically Thick Systems - In this case the so called *diffusion approximation*[2] holds and it can be shown [1, 3] that the net radiation flux is given by

$$\vec{q}_r = -k_r(T)\nabla T \tag{17}$$

where $k_r(T)$ is the radiative conductivity defined by

$$k_r(T) \equiv \frac{16\sigma T^3}{3K_R}. \tag{18}$$

The variable σ is the Stefan-Boltzmann constant and the K_R is called the Rosseland mean extinction coefficient. It is defined by

$$\frac{1}{K_R} = \int_0^\infty \frac{1}{K_\lambda(\lambda)} \frac{\partial e_{\lambda b}}{\partial e_b} d\lambda, \tag{19}$$

where the $\frac{\partial e_{\lambda b}}{\partial e_b}$ is found by differentiating Planck's law after setting $T = (e_b/\sigma)^{1/4}$. The energy flux vector resulting from combined radiation and conduction can be expressed as

$$\vec{q} = \vec{q}_r + \vec{q}_c = -(\frac{16\sigma T^3}{3K_R} + k)\nabla T. \tag{20}$$

Integrating the simplified energy equation $\nabla\vec{q}$ once gives the net flux q and twice gives the temperature distribution. The constants of integration can be determined by applying

[2] valid for $l_m/H \ll 1$, where H is the path length, over which radiant energy density changes appreciably and l_m is the extinction mean free path given by $l_m = 1/K_\lambda$ [1]

the no-slip conditions to the conductive term and the slip conditions are applied to the radiation term[3].

Optically Thin Systems - Under optically thin conditions the coupling between radiative and conductive heat flux terms in the energy equation is weak. The two contributions can be decoupled, calculated seperately and added to obtain the total flux as given by the equations [3]

$$\begin{aligned} \nabla \vec{q}_r &= 0 \\ \nabla \vec{q}_c &= 0 \\ \vec{q} &= \vec{q}_r + \vec{q}_c. \end{aligned} \tag{21}$$

An additive solution cannot be used to predict temperature profiles except in the extreme cases where either radiative transfer or conductive transfer dominates. If conduction dominates obviously a nearly linear profile would be established. If radiation dominates a nearly isothermal profile would be established.

Coupled Radiation, Conduction and Convection

For problems including convective heat transfer the conservation equations of mass, momentum and energy have to be considered. Therefore it's again reasonable to introduce simplifications. It is prudent to characterize the radiative transfer as being either optically thin or optically thick and use either an uncoupled, isothermal formulation or the diffusion approximation. This was discussed in the previous section for coupled radiation and conduction. However, the initial designation of optical thickness should be made on the basis of the thermal boundary layer thickness which is established by the convective, conductive and radiative balance.
When free convection is significant buoyancy appears in the momentum equation while the continuity and energy equations are unchanged. As an example for fully developed tube flow the momentum equation is [1]

$$\frac{dp}{dx} + \rho(T)g = \mu \frac{1}{r}\frac{d}{dr}(r\frac{du}{dr}). \tag{22}$$

p is the pressure, μ is the fluid viscosity and u the velocity. This equation must be solved in conjunction with the energy equation as the buoyancy term $\rho(T)g$ is temperature dependent.

The problem of combined conduction, convection and radiation heat transfer is extremely complex. All the issues of radiative transfer such as optical depth, spectral property variations and directional intensity variations are still involved. In addition all the

---

[3] The diffusion approximation is only valid within nearly isotropic medium but not at boundaries with a large anisotropy of intensity. To apply diffusion methods at such interfaces artificial boundary conditions are used that results in the correct heat flux

complexities of convective heat transfer may be involved such as viscous effects, inertial effects and turbulence [3].

THE TPV MODEL - IDEAS FOR REALISTIC SIMULATION

How can be the heat transfer equations be implemented for TPV-modeling? Fig.(5) schematically shows a TPV generator.

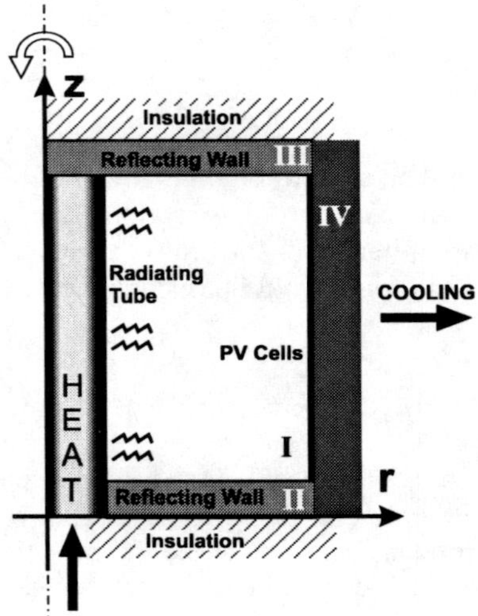

FIGURE 5. Simulated rotational symmetric TPV generator in cylindrical coordinates. The cross section is partitioned in four regions as indicated by the Latin numbers.

Region I evacuated - Radiation is the only transfer mode in Region I, whereas under the assumption of opaque surfaces in II+III+IV only conductive transfer occurs. The radiation is absorbed at the surface and the energy balance between radiation and conduction provides a boundary condition. For the conduction regions (II, III, IV) the equation

$$\nabla(k\nabla T) = 0 \tag{23}$$

can be solved. For T and $\partial T/\partial n$ at the wall surfaces the boundary condition

$$-k\frac{\partial T}{\partial n}|_{wall} = q_r, \tag{24}$$

is valid. q_r is calculated from radiative heat transfer for a given wall temperature. If the energy flux was given the $q_r(r,z)|_B$ value is determined on the basis of an assumed initial temperature distribution $T^{(0)}(r,z)|_B$. The resulting $q_r^{(0)}(r,z)|_B$ is substituted in Eq.(24) to give a new temperature distribution $T^{(1)}(r,z)|_B$ as solution of Eq.(23). This new $T^{(1)}(r,z)|_B$ is then used to find a new $q_r^{(1)}(r,z)|_B$ and the iteration process is repeated until convergence [5].

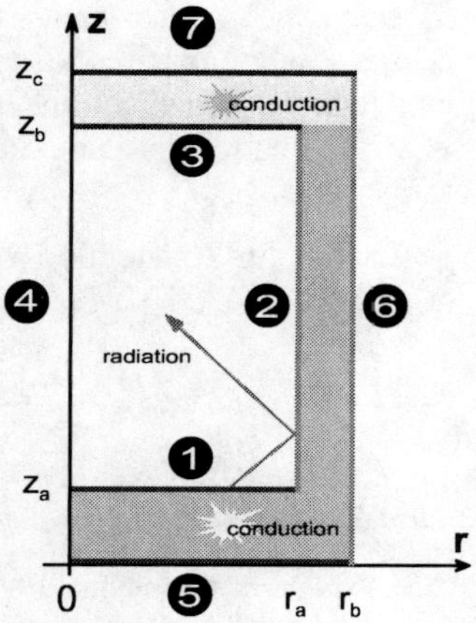

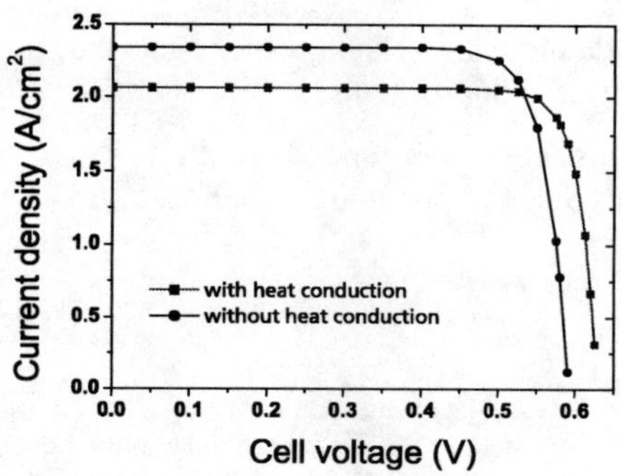

FIGURE 6. Left: Cross section of a TPV generator with radiator (1) on the bottom, PV cell (3) on the top and reflector (2). Conduction of heat is restricted to the grey shaded regions only. Right: Comparison of I-V curves for calculation with and without heat conduction.

Area I filled with transparent gas - Because of the non-participating gas in region I radiation, conduction and free convection are decoupled in region I. Again only conduction transfer occurs in regions II, III and IV. Although a transparent gas in area I is assumed the presence of free convection causes a coupling of the conservation equations. Therefore a simultaneous solution is required.

EXAMPLE CALCULATION

A first test calculation was done for a TPV converter consisting of one real GaSb photovoltaic cell ($l_g = 1.76\mu m$) cooled from the outside according to left side of Fig.(6) with $T_7 = 330K$, a selective tungsten emitter heated by a source of $T_5 = 1500K$, a perfect side reflector with a reflectivity of 100%. The spectrum was subdivided in 12 wavelength elements in the range from 0-10μm. The conductivity was assumed to be $k = 1W\,m^{-1}\,K^{-1}$ for both regions in on the left of Fig.(6). On the right side of Fig.(6) the I-V curves for the PV cell with and without heat conduction are compared. For further details see [5].

CONCLUSION

This paper is an introduction into the field of realistic TPV modeling. It points out different ways for solving the problem including radiation, conduction and convection. The Monte Carlo method is a generally applicable, expandable and suitable technique to treat all the different effects relating to radiative transfer. In the case of coupled radiation and conduction a simultaneous solution of the radiative source term with the energy equation is required. If the analysis is restricted to opaque surfaces and perfectly transparent media the multimode transfer interaction occurs as a boundary condition. For coupled radiation, conduction and convection additional conservation equations have to be included and the problem gets extremely complex. Simplifications are unevitable to solve such problems.

This work was funded by the German Funding Organization for Academic Research "Deutsche Forschungsgemeinschaft DFG" under Contract No. LU 204/12-1.

REFERENCES

1. Robert Siegel and John R. Howell. *Thermal Radiation Heat Transfer*. Hemisphere Publishing Corporation, Washington, Philadelphia, London, third edition, 1992.
2. Michael F. Modest. *Radiative Heat Transfer*. McGraw-Hill, Inc., New York, first edition, 1993.
3. M. Quinn Brewster. *Thermal Radiative Transfer and Properties*. John Wiley & Sons, Inc., New York, Chichester, Brisbane, Toronto, Singapore, first edition, 1992.
4. Jeffery T. Farmer and John R. Howell. Comparison of Monte Carlo Strategies for Radiative Transfer in Participating Media. *Advances in Heat Transfer*, 31:333 – 429, 1992.
5. Johannes Aschaber et al. Modelling of a Thermophotovoltaic System Including Radiation and Conduction Heat Transfer. In *17th European Photovoltaic Solar Energy Conference*, 2001.

500 Watt Diesel Fueled TPV Portable Power Supply

W. E. Horne, M. D. Morgan, V. S. Sundaram
EDTEK, Inc., Kent, Washington
T. Butcher
Brookhaven National Laboratory, Upton, New York

ABSTRACT

A test-bed 500 watt diesel fueled thermophotovoltaic (TPV) portable power supply is described. The goal of the design is a compact, rugged field portable unit weighing less than 15 pounds without fuel. The conversion efficiency goal is set at 15% fuel energy to electric energy delivered to an external load at 24 volts.

A burner/recuperator system has been developed to meet the objectives of high combustion air preheat temperatures with a compact heat exchanger, low excess air operation, and high convective heat transfer rates to the silicon carbide emitter surface. The burner incorporates a air blast atomizer with 100% of the combustion air passing through the nozzle. Designed firing rate of 2900 watts at 0.07 gallons of oil per hour. This incorporates a single air supply dc motor/fan set and avoids the need for a system air compressor. The recuperator consists of three annular, concentric laminar flow passages.

Heat from the combustion of the diesel fuel is both radiantly and convectively coupled to the inside wall of a cylindrical silicon carbide emitter. The outer wall of the emitter then radiates blackbody energy at the design temperature of 1400°C. The cylindrical emitter is enclosed in a quartz envelope that separates it from the photovoltaic (PV) cells. Spectral control is accomplished by a resonant mesh IR band-pass filter placed between the emitter and the PV array. The narrow band of energy transmitted by the filter is intercepted and converted to electricity by an array of GaSb PV cells. The array consists of 216 1-cm x 1-cm GaSb cells arranged into series and parallel arrays. An array of heat pipes couple the PV cell arrays to a heat exchanger which is cooled by forced air convection.

A brief status of the key TPV technologies is presented followed by data characterizing the performance of the 500 watt TPV system.

I INTRODUCTION

To fully realize the potential of efficiently converting infrared (IR) energy from a thermal source to electricity presents a substantial challenge. To meet this challenge, spectral control within the thermophotovoltaic (TPV) system must be implemented allowing the recovery of those portions of the emitted IR spectrum to which the PV cells do not respond. In addition to the requirement of close spectral control over the emitted IR energy, care must be exercised to minimize thermal losses from the TPV system. The potential for thermal loss exists in every aspect of the TPV generator and strict attention to the details of thermal isolation and recovery is critical for efficient thermophotovoltaic conversion. Our approach to realizing TPV energy conversion is predicated on three underlying technologies: 1) GaSb photovoltaic cells, 2) an IR band-pass filter, and 3) a liquid fueled burner with accompanying recuperator employed to recover burner exhaust-gas energy.

CP653, *Thermophotovoltaic Generation of Electricity: 5th Conference*
edited by T. J. Coutts, G. Guazzoni, and J. Luther

A short discussion of the status of the three key technologies will be presented below. Data will also be presented characterizing the performance of a 500 watt diesel fueled TPV system built in our laboratory.

II STATUS OF KEY TPV TECHNOLOGIES

GaSb has a great advantage for the fabrication of TPV cells in that the photo-junction can be produced by a relatively inexpensive diffusion process. Maintaining low-cost PV cell production has been, and continues to be, a primary focus of our cell development. To make TPV commercially viable, overall costs, and especially cell costs, must be minimized.

GaSb device fabrication began with a process originally developed by The Boeing Company and licensed to our laboratory on a non-exclusive basis. Subsequently, our laboratory has made significant improvements to the original process thereby improving PV cell performance and reproducibility. Through our continuous process improvement efforts, our GaSb PV cell fab line now enjoys a cell production yield of 90%. Efforts to reduce cell cost are ongoing.

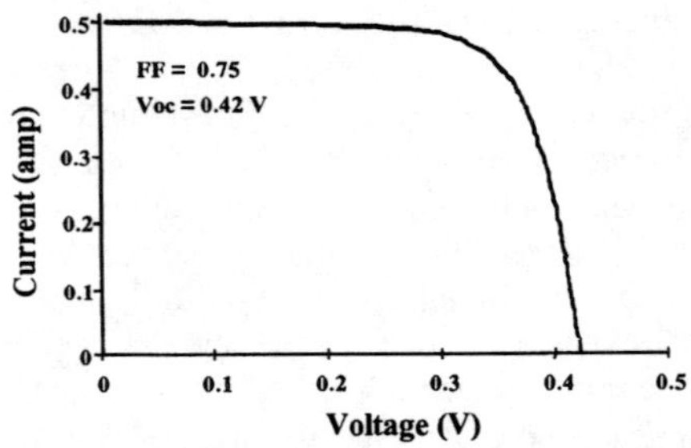

Figure 1. Typical I-V Curve of a GaSb Cell

GaSb cell electrical characteristics were measured using a tungsten lamp whose intensity was calibrated by a photodiode standard. The cells were water cooled and the lamp was operated in continuous mode. Under these conditions, individual 1 cm^2 cells yielded Voc ≥ 0.4 V, FF ≥ 75%, and short circuit current (Isc) 500 mA/cm^2. The I-V curve from a typical GaSb cell is shown in figure 1. The external quantum efficiency (QE), or spectral response, of such a device is shown in figure 2.

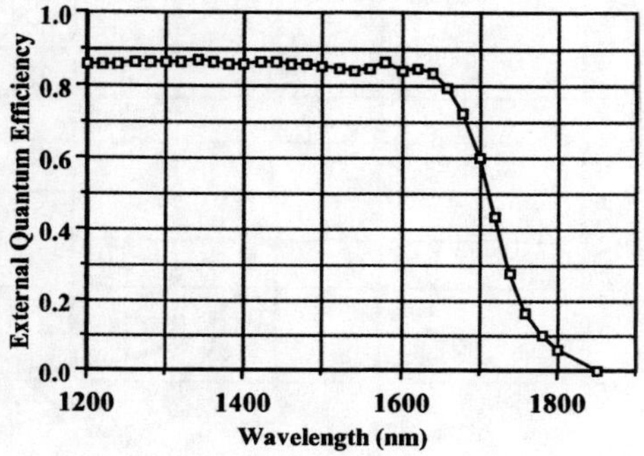

Figure 2. External QE of a GaSb Cell

The EDTEK IR band-pass filter is an array of antenna elements with dipole lengths near 450 nm and dipole line-widths of 80 nm and a resonant peak occurring at a wavelength of approximately 1.45 microns; near the peak response of the GaSb PV cell. At wavelengths greater than the resonant wavelength, the transmittance of the filter drops to zero as, concurrently, the reflectance increases until it matches that of a gold thin film (>98%). This reflectance remains constant out to the far infrared.

The challenge of these filters is in their fabrication[1)]. The feature size of the antenna elements of the filters require nano-fabrication techniques. We have found that the simplest and most robust process employs masked ion beam lithography. This process utilizes a collimated ion beam and a stencil mask to print the filter elements onto a glass substrate. Further processing results in ~4.2 x 10^8 antenna elements per cm^2 of filter. Each 500 watt TPV generator requires 460 cm^2 of filter area.

The filter performance depicted in figure 3 demonstrates filter transmittance and long wavelength reflectance. A vacuum deposited gold film was employed as a reflectance standard. Unfortunately, the typical peak transmittance performance of the filters fabricated for the 500 watt TPV generator was only about 50%. This lower-than-design throughput reduces the power density transmitted to the PV cells thereby limiting the generator's peak power output to a predicted 400 watts at 1400C emitter temperature.

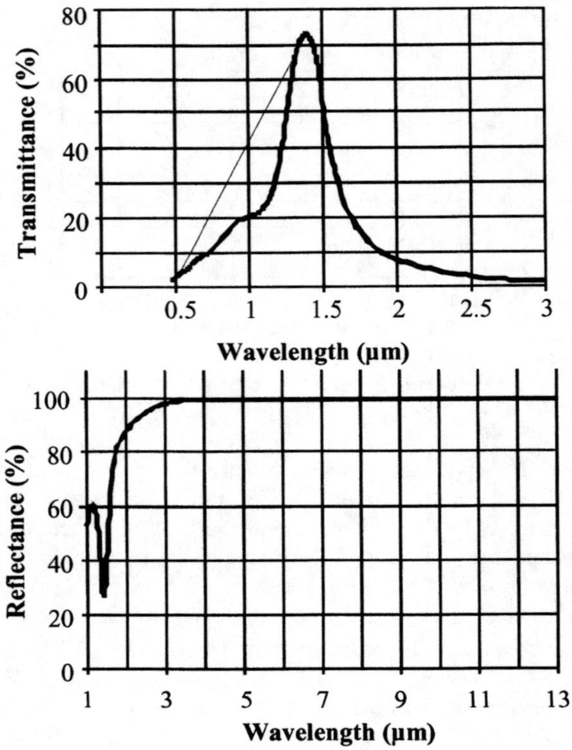

Figure 3. Transmittance and Reflectance Performance of IR Band-Pass Filter

Infrared filter improvements center on increasing in-band transmittance, reducing the transmittance of out-of-band photons and reducing the cost of filter production. Reduction of filter production cost is achievable through simplification of the fabrication process. A simpler process is possible due to recent improvements in electron-beam resists[2)]. Our investigations of these resists now concentrate on their applicability to ion beam lithography.

In our 500 watt TPV generator test bed, the burner design, developed by Brookhaven National Laboratories, employs an air blast atomization nozzle. Fuel is injected onto a conical surface and forms a uniform film as air flows over the inside and outside edges of this surface, the film is broken into a fine mist. The air and fuel circulation in the flame tube produce re-circulation zones which define the location of the flame front, the flame stability, and turndown characteristics. This nozzle offers the advantage of being able to operate at the required low firing rates without fouling.

The purpose of the flame tube is to define the flame zone and, in combination with the nozzle, ensure that the residence time of the fuel in the flame zone is sufficient for complete combustion. In this geometry, the combustion products turn at the end of the flame tube and flow back in the annular volume between the flame tube and the emitter. This provides a “convective zone” where the products transfer heat to the emitter.

The emitter cup defines the overall combustion area and directs the flow of exhaust gases along the channel between the flame tube and the inside surface of the emitter until they exit through the exhaust ports in the bulkhead. Ignition is achieved by means of a hot surface igniter.

III EXPERIMENTAL PROCEDURE

A 500 watt TPV generator test-bed was fabricated to assess the performance of the TPV heat source and electrical converter as a system. The burner/emitter/recuperator (BER) was fabricated as a separate unit that was then inserted into the TPV converter consisting of the PV cell arrays and output wiring. It is critical to thermally isolate the BER to minimize thermal energy out of the TPV system. To achieve this, the BER and the converter portion were separated by a double walled evacuated quartz tube. The BER was inserted into the inner tube of this double wall 'thermos bottle' enclosure to minimize thermal transfer to the converter except by radiative transfer from the emitter to the PV cell array. The converter PV cell array was wrapped around the outside of the 'thermos bottle'. The inside of the outer wall of the enclosure is coated everywhere with a gold layer to reflect thermal energy except the section between the emitter and the PV cell array which is coated with the IR band-pass filters. The filters pass only those photons that the GaSb cells can efficiently convert; other wavelengths are reflected back to the emitter.

The test-bed was designed to be immersed in a water bath calorimeter making it possible to accurately evaluate the heat transmitted out through the walls and PV cells of the TPV unit. The unit was also instrumented throughout to continuously monitor gas pressures, air flows, and fuel consumption. Thermocouples to monitor key temperatures were also provided.

The TPV cell array was separated into 36 series connected strings of six TPV cells each. These strings were wrapped around the heat source with four six-cell strings fitting on the circumference and nine rings of cells along the length of the emitter. A curve tracer was employed with a multiplexer to allow monitoring the I-V characteristics during testing. All the data were recorded on a computer and stored for analysis. Figure 4 is a photograph of the converter assembled, wired, and ready for test.

Figure 4. TPV Test-bed Unit Assembled for Calorimeter Test-bed Testing

Table 1 is a list of the parameters monitored during test-bed experiments and the instrumentation employed. This instrumentation allowed full evaluation of the performance of the system and its subsystems.

Table 1. List of Parameters Monitored During Test-Bed Experiments

Parameter	Monitoring Method/Instrumentation
Fuel Consumption	Volumetric Flowmeter
	Fuel Mass by Digital Scale
Combustion Air Volume	Intake Fan Vacuum
	Nozzle Air Pressure
	Combustion Chamber Pressure
Combustion	Fuel Pump Pressure
	Combustion Chamber Pressure
Quantity Excess Air	Oxygen Sensor in Exhaust Stream
PV Cell I-V Characteristic	Electronic Load Bank to Load PV Arrays
	Multiplexer to access 36 strings of 6 PV cells each
Temperature of Test-Bed	Thermocouples at 22 locations within test-bed
	1-2. Flame zone
	3-5. Three points along Recuperator exhaust channel (enter, middle, exit)
	6-8. Three points along Recuperator air channel (enter, middle, exit)
	9. Air entering nozzle
	10. Bulkhead (between recuperator and burner)
	11. Fuel Tank
	12. Outer Quartz Sleeve
	13. Calorimeter Coolant-Bath
	14. Calorimeter Coolant-Bath Input
	15. Calorimeter Coolant-Bath Output
	16-18. Three points along PV cell array
	19-20. Two locations on system Lid Plate
	21-22. Emitter Cup

The instrumentation outlined in table 1 allowed the monitoring and measurement of thermal and electrical mechanisms operating simultaneously in the TPV system. The thermal mechanisms are associated with the burner, the heat source, and parasitic thermal losses. Fuel input was measured to determine input power delivered to the system. The electrical conversion issues are related to PV cell performance, cell array performance and filter performance.

IV RESULTS AND DISCUSSION

The PV cells were arranged in 36 arrays or strings of six series-connected cells. During the test-bed evaluation tests, leads were brought out from each of the 36 strings so that they could be measured individually. Figure 5 presents a typical I-V characteristic for one of the six-cell strings during the testing. The cells were monitored as a function of emitter temperature. The power produced over the surface of the emitter was found to be uniform within 10%.

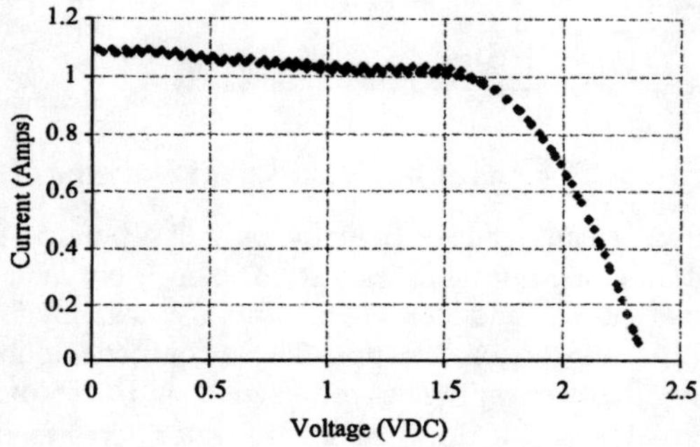

Figure 5. Typical I-V Characteristic for Six-Cell Series-Connected GaSb PV Cell String Under Test

Two important results from our system performance measurements were the power delivered to the PV cells and the thermal losses in the system. Under no-load conditions, these two factors encompass the waste heat in the PV cells, the heat absorbed in the filters, the thermal losses in the gold-film reflectors throughout the evacuated quartz sleeve, and other parasitic thermal losses from the system. There was no practical way to measure all these parameters discretely, however, we measured their aggregate sum via the water bath calorimeter. By measuring the temperature rise of the known mass of the water bath surrounding the generator as well as the temperature drop in the water bath heat exchanger at constant flow rate, it was possible to determine the power lost by the generator through its housing into the calorimeter. This heat, when no electrical load was applied to the PV cells, represented the aggregate sum of power transmitted into the PV cells, into the gold-film reflectors, and all the other parasitic thermal losses in the system. The heat expelled in the exhaust gas stream and that returned to the flame zone by the recuperator was determined by the measured gas mass flow rates, and by the temperatures of the exit exhaust, the pre-heated combustion air, combustion gases, and the ambient air.

A summary of the TPV thermal performance is outlined in figure 6 which shows a schematic of the burner and recuperator and summarizes system heat flows. The efficiency of the generator was determined by dividing the measured electrical power out by the total input power (fuel consumed) into the system. Figure 7 shows the measured power out of the system compared to the predicted performance curve.

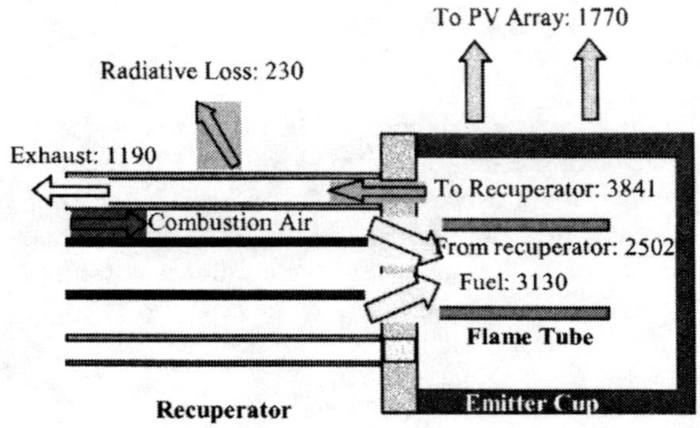

Measured System Heat Flow

Emitter Temperature:	**1200 C**
Heat Produced by Fuel:	3130watts
Heat from Flame Zone to Recuperator:	3841 watts
Heat From Recuperator:	2502 watts
Heat Lost to Exhaust:	1190 watts
Heat to Converter:	1770 watts
Radiative Loss from Recuperator:	230 watts
Electric Power Out:	185 watts

Recuperator Effectiveness: 61%

Figure 6. Schematic of TPV System Showing Heat Flows

The measured steady-state efficiency of the generator as a function of emitter operating temperature along with a calculated efficiency curve is shown in figure 8. Both the predicted output power and predicted system efficiency were based on measured performances of TPV system components. Those components included PV cell electrical characteristics, measured transmittance, reflectance, and absorptance of IR filters, burner output, system thermal mass, air flows, and recuperator performance.

To fully characterize the system, testing was started at low emitter temperatures and raised to the design temperature of 1400°C. Each of the data points in figures 7 and 8 represent a steady-state run of 2 to 3 hours in duration. The total time of the unit under test was about 50 hours. During the course of these repetitive tests, it was found the recuperator had developed air leaks between the spot welds in the heat exchanger seams. The air leaks degraded recuperator effectiveness and resulted in excess air being run through the recuperator exhaust plenum allowing a significant amount of parasitic heat loss out the exhaust. With these air leaks, which prevented running without excess air to the system, the total output electrical power was 185 watts at 1200°C, system efficiency was measured at 5.9% at a recuperator effectiveness of 61%. When the ~36 watts housekeeping power, required to operate system fuel pump and air fans, is included the overall system efficiency becomes 4.8%.

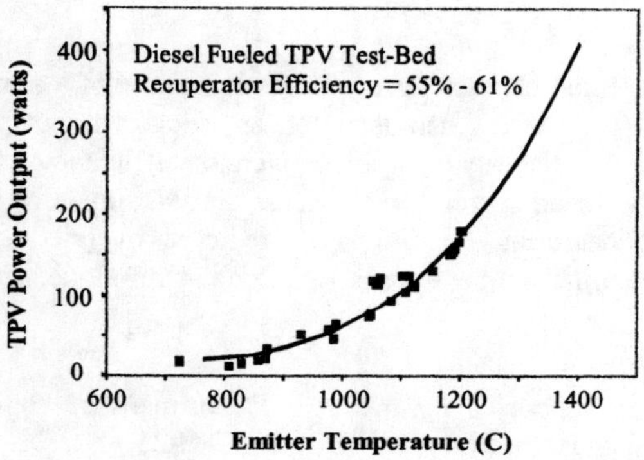

Figure 7. Electrical Power Output From TPV Test-Bed Shown with Curve of Calculated Output Power

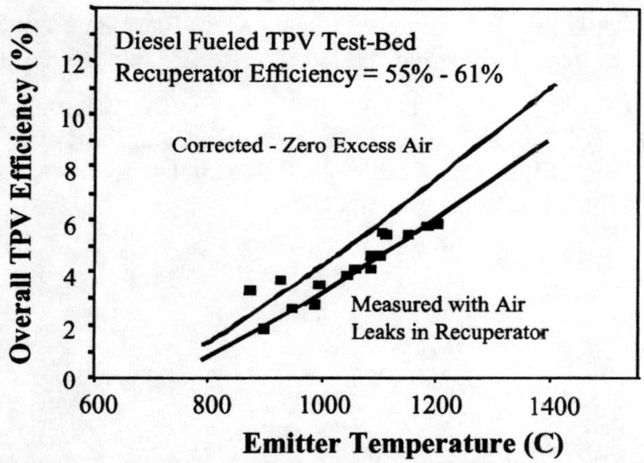

Figure 8. Measured System Efficiency Shown with Zero Excess Air Correction

Due to air leaks in the recuperator, the system could not be pushed to 1400°C with the installed combustion air fan. An improved recuperator was designed. Unfortunately, before the new design could be implemented an operator error destroyed the converter portion of the system preventing 1400°C testing. Using existing using air flow, temperature, and calorimeter measurements, the excess air was carefully quantified and then the system efficiencies were corrected for zero excess air, i.e., an air-tight recuperator. The corrected curve is shown in figure 8. The system efficiency at zero excess air was 7.5%, reduced to 6% with housekeeping power included. Extrapolation of the data along the predicted curve of figure 8 indicates an overall efficiency at 1400°C and zero excess air of 10% with housekeeping power included.

IV SUMMARY

Data were presented characterizing the performance of a nominal 500 watt diesel fueled TPV system. The system demonstrated an electric output of 185 watts with an overall fuel consumed to electrical output conversion efficiency of 7.5% at an emitter temperature of 1200°C when corrected for excess air. Recuperator efficiency was found to be 61% at these operating conditions. Predicted overall fuel to electric output efficiency at 1400°C emitter temperature is 10% for a stand-alone, self-powered system.

ACKNOWLEDGEMENTS

This work was supported by the U.S. Department of Defense Communication Electronics Command under contract DAAB07-97-C-6025.

REFERENCES

1) Morgan, M. D., et al., "Infrared Resonant Mesh Band-Pass Filter Fabricated with Masked Ion Beam Lithography", J. Vacuum Sci & Tech., 3903, 1996.

2) Falco, C. M., Van Delft, J. M., and Weterings, Jos P., "Hydrogen Silsesquioxiane/Novalak Bilayer Resist for High Aspect Ration Nanoscale electron-Beam Lithography", Philips Research Laboratories, Prof. Holstaan 4, 5656 AA Eindohoven, The Netherlands.

The Potential of Thermophotovoltaic Heat Recovery for the Glass Industry

T. Bauer, I. Forbes, R. Penlington and N. Pearsall

Northumbria Photovoltaics Applications Centre (NPAC), School of Engineering University of Northumbria, Newcastle upon Tyne, NE1 8ST, UK, t.bauer@unn.ac.uk

Abstract. This paper aims to provide an overview of heat recovery by thermophotovoltaics (TPV) from industrial high-temperature processes and uses the glass industry in the UK as an example. The work is part of a study of potential industrial applications of TPV in the UK being carried out by the Northumbria Photovoltaics Applications Centre. The paper reviews the relevant facts about TPV technology and the glass industry and identifies locations of use for TPV. These are assessed in terms of glass sector, furnace type, process temperature, impact on the existing process, power scale and development effort of TPV. Knowledge of these factors should contribute to the design of an optimum TPV system. The paper estimates possible energy savings and reductions of CO_2 emissions using TPV in the glass industry.

INTRODUCTION AND MOTIVATION

In modern industrial society the majority of energy is consumed in the sectors of transportation, building and industry, and fossil fuels are the primary sources of this energy. The use of fossil fuels has led to worldwide concerns about security of supply, increasing energy demand, limitation of resources and local and global environment impacts (e.g. acid rain, climate change). A consequence is increased interest in non-fossil fuel energy resources and the efficient use of fuel. Possible improvements in efficiency of use include technologies such as combined heat and power, energy management, improved insulation and heat recovery. Heat recovery is usually associated with heat upgrading systems such as heat exchangers or heat pumps. In practice, these systems are often limited to applications where heat recovery and demand coincide since heat can only be transported and stored to a limited extent. Heat recovery technologies that generate electricity, such as thermoelectrics or turbines combined with a generator, offer more flexibility since electricity can be transmitted with lower losses than for heat, the grid can act as storage and electricity can be converted to other energy forms at high conversion efficiencies. This paper considers thermophotovoltaics (TPV) as a heat recovery method that generates electricity in high-temperature industries.

Classified by end-use, high-temperature processes account for the largest part of energy consumption in UK industry (Fig. 1)[1]. The main energy uses in high-temperature processing are coke ovens, blast and other furnaces, kilns, metal smelters

[1] For this study worldwide data were not available, but it is thought that other industrial countries show similar distributions.

CP653, *Thermophotovoltaic Generation of Electricity: 5th Conference*
edited by T. J. Coutts, G. Guazzoni, and J. Luther

and glass tanks [1]. High temperature processes can be further classified by industry (Fig. 1). Most of these industries have energy-intensive processes operating at temperatures above 1000 °C and which are suitable for TPV operation [2].

Most high-temperature industries show a large diversity in terms of production volume, fuel type, furnace type, existing heat recovery method, specific process and process temperature. Although the glass industry is one of the smallest energy sectors, this industry was chosen as an example to examine TPV heat recovery systems because of its uniformity in terms of melting temperature (around 1500 °C), fuel[2] and type of furnace.

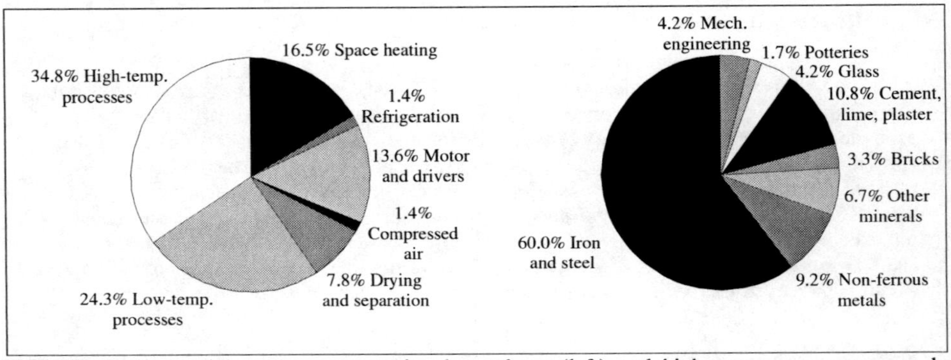

FIGURE 1. Industrial energy consumption by end-use (left) and high-temperature processes by industry (right) in 1992 for the UK [1]

THERMOPHOTOVOLTAICS

Thermophotovoltaics is the use of the photovoltaic effect to generate electricity from a high temperature thermal source [3]. Typically the temperature of a radiator (or emitter) ranges from 1300 to 2000 K, which leads to a theoretical hemispherical total radiation per unit area of about 16 to 91 W/cm^2 according to the Stefan-Boltzmann law. Photovoltaic cells convert part of the radiation below a certain wavelength into electricity and high wavelength radiation is usually reflected back to the heat source by some form of spectral control. The main TPV cell materials in use are Gallium Antimonide (GaSb), Silicon (Si) and Indium Gallium Arsenide (InGaAs), absorbing radiation below 1.7 μm, 1.1 μm and from 0.9 to 3.5 μm, depending on composition, respectively. Several combustion-based systems, mainly for combined heat and power or portable power applications, are in a prototype stage. Only a few publications [3-5] suggest the use of TPV for high-temperature industrial heat recovery.

[2] In 1998 for the UK natural gas has a share of 82 %, with oil at 6 % and electricity at 12 % [9]. In 1994 for the US a similar distribution is given: electricity (17.3 %), oil (1.6 %), natural gas (79.5 %) and others (0.8 %) [10].

Characterisation

Similar to combustion TPV systems, heat recovery TPV systems can be characterised by the output, total power (W) and power density (W/cm^2), the efficiency and cost. High power densities are required to improve the economics but can be associated with high losses. Although power densities as high as 2.5 W/cm^2 are reported [6], in this work a value of 1 W/cm^2 is seen as a reasonable compromise [3]. Unlike combustion TPV systems, TPV heat recovery systems should be seen as an interaction with a given high-temperature process, rather than a stand-alone system. The TPV heat recovery system will extract energy from the high-temperature process. For each specific process, the extent of process changes, such as process temperature, product quality or fuel consumption, and their acceptability should be considered.

Efficiency

Efficiency is usually defined as the ratio of the useful output to the total input. For TPV, the useful output is commonly defined as the generated power[3] but the total input can be defined in different ways. For example, the input for a combustion TPV system may be defined by the fuel entering the system, so the efficiency of the combustion can decrease the overall system efficiency. Other TPV systems exclude the heat source. For example, solar TPV systems are likely to be characterised by the total incident radiation onto the system. For TPV heat recovery systems, there are applications where the radiator may not be defined as an inherent part of the system, since the process itself, such as production of molten glass, may radiate with a suitable infrared spectrum[4]. The TPV heat recovery system efficiency η_{sys} will exclude the heat source and possibly the radiator, which should lead to an inherently higher value of η_{sys} compared to combustion TPV. Combustion based TPV systems under development are estimated to be able to reach an overall efficiency of the order of 15% [7]. Since TPV heat recovery excludes the combustion efficiency (of the order of 80%), a value of the order of $\eta_{sys} = 20\%$ is seen as realistically achievable.

A possible definition of η_{sys} is given by the following equation, where P_{el} is the electrical output power and P_{exheat} the total amount of heat extracted from the process. In practice, P_{exheat} is difficult to determine and can be replaced by the sum of all outputs of the TPV system: P_{el}, the cooling power P_{cool} and the heat losses not ascribed to the cooling power P_{loss} (see Fig. 2). Usually P_{cool} can be derived from the flow, the density, the heat capacity and the temperature change of the cooling fluid and, for a properly designed system, P_{loss} will be minimised and may be negligible.

$$\eta_{sys} = \frac{P_{el}}{P_{exheat}} = \frac{P_{el}}{P_{el} + P_{cool} + P_{loss}} \qquad (1)$$

---

[3] For combined heat and power (CHP) the output may also include the useful heat.

[4] For the exclusion of heat source and radiator, the efficiency may be defined as the ratio of the electricity output to the total incident radiation minus the backward reflected radiation to the heat source, where the efficiency becomes a function of the spectrum of the incident radiation.

THE GLASS INDUSTRY

Glass Industry Statistics

The major producers of glass are located in the US, Western Europe, Japan, China, Republic of Korea and the former USSR [8]. Glass production can be classified into six sectors [9]: container, flat, fibre, domestic, scientific and technical, and giftware (Table 1). Flat and container glass are the two dominating sectors by energy use. Their glass furnaces tend to be large and efficient and the industry is dominated by only a few large manufacturers. The other sectors tend to show smaller production volume and lower efficiency. On a glass site, the melting process in the furnace accounts for about 70% of the total energy use [9,10]. The electricity demand accounts for 10 - 30% of the total site energy in the US, depending on the sector, and a mean value of 14% is found for the UK [10].

TABLE 1. Overview of the glass industry in the UK in 1998 (data from [9])

Sector (share by energy)	Furnace Number	Typical Furnace Type and Melting process	Furnace Output (tonnes/ day)	Site Energy (GJ/ tonne[5])	Typical Method of Comb. Air Preheating	Typical Secondary Process
Container (55%)	14	Tank, cross or end fired, max. 1500 ºC	150 – 300	9.0	Regenerator	Forehearth
Flat (20%)	3	Tank, cross-fired, max. 1600 ºC	More than 300	9.2	Regenerator	Float bath
Fibre (10%)	8	Tank often with electrical boosting	20 – 300[6]	15-25	Regenerator, recuperator	Forehearth
Domestic (5.7%)	3	Tank often with electrical boosting	150 – 300	20-30	Regenerator, recuperator	Forehearth
Sc. & tech. (8.3%)	13	Tank, day tank	1 – 300[6]	20-30	Regenerator, recuperator	Forehearth
Giftware (1%)	More than 30	Pot, day tank, 900-1600 ºC	Less than 1	More than 50	Recuperator	None

Glass Production Process

The upper part of Fig. 2 illustrates a generic type of melting process. The batch consisting of the raw material, cullet (recycled glass) or a mixture of both is fed into the furnace where the primary melting, the refining and conditioning phases take place sequentially. For large production volumes, tank furnaces are used (Fig. 3). The batch is fed continuously into one side of the furnace (doghouse), the melting process takes place throughout the furnace and the molten glass is extracted at the other end (forehearth). Tank furnaces have distinct local process regions with constant temperature.

[5] The theoretical requirement to heat glass from room temperature to 1500 ºC is about 1.7 GJ per tonne. The energy required for the chemical reaction of the raw material is about 0.6 GJ per tonne [11]. These values can differ for different glass compositions [18]. A modern furnace with an efficiency of 40% operating without cullet has an energy consumption of 5.75 GJ/tonne.

[6] Value for the US market [11]

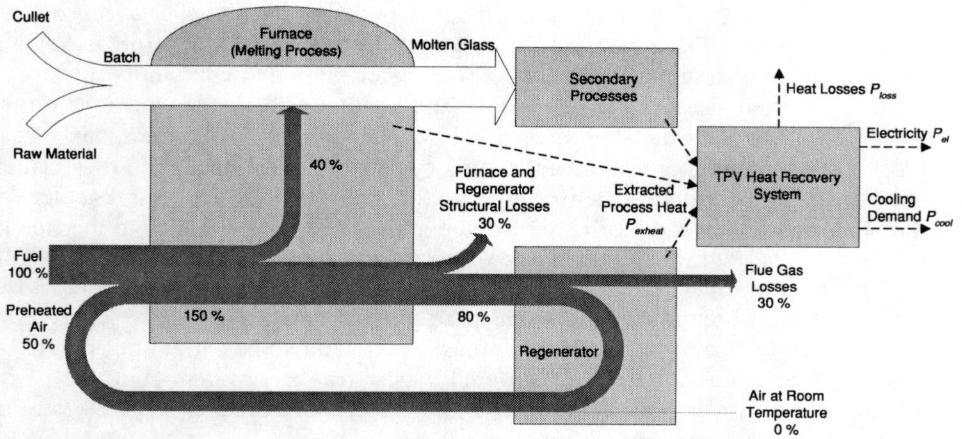

FIGURE 2. Sankey diagram of typical natural gas-powered regenerative tank furnace[5] and showing where TPV system could be sited for heat recovery[7].

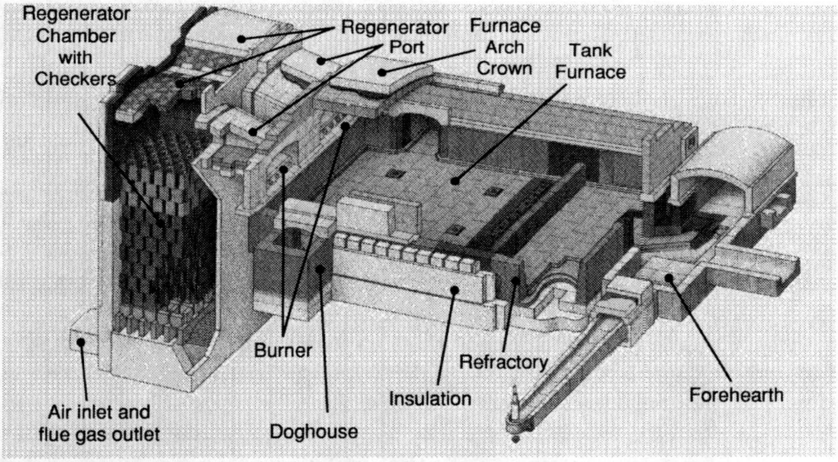

FIGURE 3. End-fired regenerative tank furnace with forehearth (adapted from [11])

For small production volumes, intermittently operating furnaces are used. These are likely to operate in a 24-hour cycle consisting of batch feeding, melting and extraction of the glass. Although there are some opportunities for TPV in regard to these furnaces and these are included in the overall study, this paper will concentrate on the continuous regenerative tank furnaces which represent the major part of the glass market.

Furnaces usually operate 24 hours per day. Sometimes electrical boosting is used and some furnaces are entirely electrically heated. For electrical heating, rods are usually introduced into the furnace and the glass acts as an electrical conductor. This achieves a high conversion efficiency from the electricity to the heat into the glass.

The most common furnace type for large production volume is a regenerative natural gas-powered tank furnace. The largest of these are float glass furnaces, with a daily production volume of up to 1000 tonnes of glass and containing up to 2500 tonnes of molten glass. Fig. 3 shows an example of an end-fired regenerative furnace and Fig. 2 gives the estimated[7] energy distribution of a typical modern tank furnace [10,12]. The regenerator uses an alternating cycle of combustion air and flue gas. The combustion air is heated in the first regenerator. Afterwards, natural gas together with the preheated combustion air is burned above the molten glass bulk and the flue gas leaves the furnace through the second regenerator. After 15-30 minutes the direction of firing is reversed and the second regenerator preheats the combustion air [13,14].

The energy content of the fuel is only partly transferred to the glass, with some lost in the structure (mainly through the furnace walls) and some carried out by the flue gas. There are three locations where a TPV heat recovery system could be placed, which are the furnace, the regenerator and the secondary processes. These locations are separately reviewed in the next three sections.

The Furnace

The tank furnace has refractory walls. These enclose three distinctive semitransparent areas, which are the molten glass bulk[8] on the bottom, the luminous flame above and the surrounding hot gas[9]. Heat transfer is dominated by radiative heat exchange resulting in a net radiation heat transfer from the luminous flame, the hot gas and the furnace walls (mainly the crown) to the glass [15]. For both the molten glass and the hot gas/flame region, the incident radiation heat flux at the inner furnace wall is high. Assuming grey-body radiation [16], the hemispherical total radiation of the molten glass bulk is estimated at about 50 W/cm². At the crown, the incident hemispherical total radiation depends on the firing regime and location. For a flat glass furnace, values of 32 - 71 W/cm² are reported [13] and the radiation is spectrally selective [15]. Most of the incident wall radiation is reflected or reradiated into the furnace and only a small fraction, 0.1 - 1.8 W/cm² [17,18], is lost through the walls. The lower value can be achieved by an insulation layer on the outside. However, insulation can increase the outer refractory temperature to above 1000 °C [17] leading to increased refractory corrosion. Thus, the insulation of a furnace is a compromise between energy losses through the wall and the furnace lifetime (typically 10 years [9]). Some regions of the furnace such as the crown[10], the sidewall at the glass line or the throat[11] are especially susceptible to refractory corrosion and sometimes cooling is applied instead of insulation. Interruption of cooling due to power failure can be critical [14,17]. Some regions are designed to have higher wall losses to reduce glass temperature, e.g. in the conditioning zone of the float glass furnace [17,19].

---

[7] The percentage values in Fig. 2 may vary about ±10% due to aspects like differences in optimisation and age of the furnace, the glass type, air temperature, the cullet to raw material ratio and the moisture content of the batch.

[8] The bulk of molten glass for practical furnace dimensions can be regarded as optically thick (the product of physical layer thickness and absorption coefficient is greater than 4) and a grey-body radiator with an emissivity of about 0.9 [16].

[9] The volume of gas for practical furnace dimensions can be regarded as optically thin, where some spectral regions are nearly transparent and others are strongly radiating and absorbing. The selectivity is mainly caused by vibration and rotation bands of CO_2 and H_2O [15].

[10] In the crown silica bricks are used, operating near their softening point at 1620 °C, to avoid contamination of the molten glass.

[11] Container and special glass furnaces are split into melting and working ends, connected by the throat whose primary function is to accommodate the relatively large temperature difference between these two sections [17].

The Regenerator

The most common method to improve the energy efficiency of a tank furnace is the use of a regenerator, which can partially transfer heat from the flue gas to the combustion air. In Fig. 2, typical values of reused (50%) and wasted (30%) energy are shown. The theoretical maximum limit of reused energy is about 58%, estimated from data provided in [12], but this is not achieved in practice, because of increased costs for large regenerators and increased production of NO_x due to high air preheating temperatures[12]. The fact that part of the energy in the regenerator is usually wasted is also reflected in the flue gas temperature on leaving the regenerator, which is typically 550-650 °C [10].

Secondary Processes

In continuously operating furnaces the molten glass is distributed via the forehearth for further processing such as blowing, pressing, casting or drawing. The exception is the float glass process, which does not require a forehearth. The forehearth also has the function of thermal homogenisation of the glass (heating and forced cooling is carried out) to reach a high glass quality. The glass temperature at the forehearth outlet (around 1100 °C [10]) and the wall losses (around 0.1 W/cm^2 [18]) are low compared to the furnace. The average forehearth energy consumption accounts for 5.6% of the total energy consumed on a typical UK glass site [9].

Today, nearly all of the flat glass is produced using the float glass process. The molten glass is poured continuously from the furnace onto a bath of molten tin, which is held on a temperature gradient of about 1000 ºC at the entrance and 600 ºC at the exit (where the glass has reached dimensional stability). To prevent oxidation of the tin bath, the process takes place in a chamber with a reducing atmosphere [14]. The energy use for heating the tin bath and the chamber above the glass accounts for a mean value of 12% of the total site energy in a float glass site in the US [10].

DISCUSSION OF A TPV SYSTEM IN THE GLASS INDUSTRY

Furnace

The furnace refractory walls offer large areas, where there may be two possibilities to incorporate a TPV heat recovery system. Firstly, one or more of the semitransparent areas could act as the radiator and the TPV system would ideally consist exclusively of a filter and a PV cell. This kind of operation may allow the use of Si PV cells, if a suitable filter can be designed. For all other locations discussed, GaSb and InGaAs are the more probable cell materials, because of the lower temperature range, which requires low bandgap PV cells. The corrosive atmosphere, possible contamination of the glass due to the TPV system and the high radiation levels inside the furnace as well as the structural importance of the refractories (especially the arch crown) may

[12] In the UK, stringent NO_x emission limits lead partly to lower air-preheating, which increases the waste heat of the regenerator.

cause difficulties. The second possibility is to incorporate the TPV system inside the wall. For an insulated wall, the temperature at the junction between the refractory and the insulation is still above 1000 °C, which would allow TPV operation.

For both options, the principal problem is seen as the small wall heat losses (commonly in the range of 0.1 to 0.3 W/cm^2 [17,18]) compared to the necessary heat flux to operate TPV (5 W/cm^2 for power density of 1 W/cm^2 and $\eta_{sys} = 20\%$). To overcome this, it may be possible to accept higher wall losses (resulting in higher fuel consumption), to accept lower power densities for TPV (resulting in higher TPV costs) or possibly to include a means to concentrate the heat flux locally whilst ideally keeping the average wall heat losses at the previous level. There are locations in the furnace with increased losses (higher heat flux) but of limited area or available energy for TPV use. These are the sidewall at the glass line, the throat, openings or parts of the conditioning zone of a float glass furnace. The redesign of burners to include a TPV device may also be considered.

The Regenerator

The conditions for TPV operation differ markedly between the furnace and the regenerator. In the former, the aim is to minimise wall losses, whereas in the regenerator a part of the available heat is used for air preheating and the other part is usually wasted. Thus, the efficiency of TPV may be less critical in the regenerator. The correct range of temperature for TPV operation is usually found in the region where the flue gas enters. TPV systems are likely to require a radiator, because of low and spectrally selective radiation and contamination effects from the flue gas. Possible difficulties can be seen in non-uniformity of temperature zones and the temperature cycling (with a period of 15-30 minutes). Storage may be required to bridge the thermal cycling. The extraction of heat by the TPV system may result in a lower working temperature of the regenerator, which could require it to be increased in size.

Secondary Processes

Forehearths are usually easily accessible and the waste heat could be recovered by TPV. However, the extent and arrangement of the cooling depends on the forming process to follow and may be varied within one furnace and for different furnaces. The area available limits the use of TPV to small-scale applications.

In terms of TPV use, float glass chambers can be considered in a similar way to furnaces. The disadvantages for TPV use are the low temperature, the limited amount of available energy and the limitation to one glass sector (flat glass).

TPV Contribution to Energy Efficiency on a Glass Site

About 60% of the melt energy is available as waste heat (Fig. 2), so, assuming around one third of this can be converted and $\eta_{sys} = 20\%$, the resulting generation would be equivalent to 4% of the melt energy. Furthermore, if the melt energy corresponds to 70% of the site energy, then the TPV output would provide about 3% of the total site energy. Assuming a TPV power density of 1 W/cm^2, an electricity generation of 3% of the site energy and a furnace of 100 tonnes/day, it is estimated that the TPV system would have a power output of about 270 kW and would require a high-temperature area of 27 m^2. Such areas are available both on the furnace walls and in the regenerator.

The potential for energy saving in the UK glass industry, from technologies such as batch preheating from the flue gas, optimised combustion systems, increased wall insulation and the use of cullet, has been estimated at 8.4% for the period 1999-2010 [9]. Thus, it can be seen that the predicted saving from TPV of 3% is similar to that from these other technologies and it is suggested that, in the long term, a heat recovery TPV system producing electricity equivalent to 3% of the site energy could be competitive with other energy saving technologies.

Usually all the electricity produced by TPV could be used on-site, because statistically at least 10% of the site energy is required as electricity [10]. The total energy related[13] CO_2 emissions of the UK glass industry are estimated at 1.87 million tonnes per year in 1998 (using data from [9]). Using a conversion factor of 122.2 $kgCO_2/GJ$ for electricity generation and assuming 3% of the total annual energy consumption of the UK glass industry (about 30 PJ), the possible CO_2 saving by using TPV is estimated at 0.11 million tonnes per year (approximately 6% of the CO_2 emissions from the glass industry). This is already valuable but there is also potential for higher savings due to TPV technology improvements in combination with developments in the glass industry such as oxygen-fuel firing or NO_x limitations. The savings could also be extended to other high-temperature industries and uptake will be affected by legislation on carbon and/or energy taxes.

RESULTS AND CONCLUSIONS

TABLE 2. Assessment of locations for TPV use in continuous tank furnaces

Location for TPV use	Small or large scale	Typical furnace and process	Impact on the existing process	Process temperature	R&D Time
Furnace walls	Large	All	High	Locally constant	Long
Regenerator	Large	Most	Low	Varying, 15-30 minute cycle	Medium
Float glass process	Small	Float glass process	High	Locally constant	Long
Conditioning zone wall	Small	Float glass furnace	High	Locally constant	Medium
Sidewalls at the glass line	Small	All	High	Locally constant	Medium
Throat	Small	Container and special glass tanks	High	Locally constant	Medium
Forehearth	Small	All but float glass	High	Locally constant	Medium
Openings	Small	Most	Low	Locally constant	Short
Redesign of the burner	Small	All combustion	Low	Varying, 15-30 minute cycle	Medium

[13] About 0.56 million tonnes of CO_2 emission comes from batch gas, which is not included here.

The potential of TPV heat recovery systems to improve energy efficiency in the glass industry has been investigated. Various locations within glass production with a suitable temperature range have been identified. Small-scale applications on the side walls of the glass line, the throat, the float glass conditioning zone and float glass chamber, furnace openings, redesigned burners and the forehearth could be used to test and launch TPV technology and to offer a reliable grid-independent power supply on the glass site. The large-scale applications on furnace walls and in the regenerator are possible locations to provide energy efficiency improvements on a glass site. Table 2 (above) summarises the findings. If the technical difficulties can be overcome, the large-scale use of TPV heat recovery in the UK glass industry could provide about 21% of the site electricity on average and reduce energy related CO_2 emissions by about 6%. Other high temperature industries can be assessed with a similar methodology and would be expected to show similar potential.

REFERENCES

1. "Energy Consumption in the United Kingdom (Energy Paper 66)", Department of Trade and Industry, UK, 1997, pp. 43-48
2. Brown, H. L., "Energy Analysis of 108 Industrial Processes", Fairmont Press, 1996
3. Coutts, T. J., "A review of progress in thermophotovoltaic generation of electricity" in *Renewable and Sustainable Energy Reviews* 3, 1998, pp. 77-184
4. Fraas, L., Avery, J., Daniels, W., Huang, H. X., Malfa, E., Testi, G., "TPV Tube Generators for Apartment Building and Industrial Furnace Applications", *17. European Photovoltaic Solar Energy Conference*, 2001
5. "Glass Project Fact Sheet - Thermophotovoltaic electric power generation using exhaust heat", U.S. Department of Energy, Office of Industrial Technologies, Energy Efficiency and Renewable Energy, 2001
6. Fraas, L., Groeneveld, M., Magendanz, G., Custard, P., "A single TPV cell power density and efficiency measurement technique", *4. NREL Conference on Thermophotovoltaic Generation of Electricity*, 1999, pp. 312-316
7. Fraas, L. M., Avery, J. E., Huang, H. X., "Thermophotovoltaics: heat and electric power from low bandgap solar cells around gas fired radiant tube burners", *29. IEEE Photovoltaic Specialists Conference*, 2002
8. Darnay, A. J., "Manufacturing worldwide: industry analyses, statistics, products, and leading companies and countries", Gale Group, 1999, pp. 452-453
9. "Technology Benchmarking in the UK Glass Sector (General Information Report 75)", ETSU, Energy Efficiency Best Practice programme (EEBPP), 2000
10. "Glass Industry of the future, Energy and Environmental Profile of the U.S. Glass Industry", Office of Industrial Technologies, Energy Efficiency and Renewable Energy, US Department of Energy, 2002
11. "Holloware Glass Furnaces", Fused Cast Products Division, Product Catalogue, SEPR, France
12. Beerkens, R., "Application of Energy Saving Technologies for Glass Furnaces: A comparative study", in *Improved technologies for the rational use of energy in the glass industry*, Fachinformationszentrum Karlsruhe, 1992, pp. 325-339
13. Hayes, R. R., Brewster, S., Webb, B. W., McQuay, M. Q., Huber, A. M., "Crown incident radiant heat flux measurements in an industrial regenerative, gas-fired, flat-glass furnace" in *Experimental Thermal and Fluid Science* 24, 2001, pp. 35-46
14. Tooley, F. V., "The Handbook of Glass Manufacture", Vol. 1 and 2, 3. Edition, Ashlee Publishing, 1984
15. Wieringa, J. A., Elich, J. J. Ph., Hoogendoorn, C. J., "Spectral effects of radiative heat-transfer in high-temperature furnaces burning natural gas", *Journal of the Institute of Energy*, Vol LXIII No 456 Sep., 1990, pp. 101-108
16. Gardon, R., "A Review of Radiant Heat Transfer in Glass", in *Journal of The American Ceramic Society*, Vol. 44, No. 7, 1961, pp. 305-312
17. Trier, W., "Glass Furnaces, Design construction and operation", translated by K. L. Loewenstein, Society of Glass Technology Sheffield, 1987
18. "The Use of Refractories for Container Glass Furnaces", SEPR Group, 2000
19. "The Use of Refractories for Flat Glass Furnaces", SEPR Group, 2000

Thermophotovoltaic Power Generation by Super-Adiabatic Combustion in Porous Quartz Glass

Katsunori Hanamura and Tomoyuki Kumano

Department of Environmental and Renewable Energy Systems, Gifu University, 1-1 Yanagido, Gifu 501-1193, Japan
Phone: (+81) 58-293-2536, Fax: (+81) 58-230-1892, e-mail: hanamura@cc.gifu-u.ac.jp

Abstract. A new thermophotovoltaic (TPV) power generation system has been proposed on the basis of reciprocating-flow super-adiabatic combustion in porous media and has been investigated through numerical simulation. The porous media consists of a thin foam-ceramic plate sandwitched by a pair of porous quartz glass plates packed perpendicular to the flow direction in an insulated flowing-duct. By energy recirculation through the reciprocating-flow combustion system, the foam-ceramic temperature becomes about two times higher than the theoretical (adiabatic) flame temperature, using only a small amount of combustion heat. Strong radiant energy is emitted from the foam-ceramic plate. Of those, long wavelength components are absorbed by the porous quartz glass and are regenerated to preheat working-gases. On the other hand, short wavelength components pass through the porous quartz glass. About 48% of the combustion heat is converted into the radiant energy in a short wavelength range which are useful for (GaSb) TPV power generation.

INTRODUCTION

Electricity is generated, in general, using huge energy conversion systems, such as gas turbine and diesel engine power plants, though a micro gas turbine and a fuel cell have, recently, begun to be used as a distributed power generation. On the other hand, a dry cell or a battery is very convenient for any handy electronic facilities and instruments. However, the energy densities of the dry cell and the battery are about one hundredth of those of fossil fuels. As a result, researches on milli-size gas turbines and micro-size thermoelectric power devices integrated with catalytic combustion become grown in relation to electric power supply for MEMSs (Micro Electro-Machine Systems) [1]. In these cases, thermal energy released by combustion of fuel is converted into electric power. Therefore, the total thermal efficiency becomes higher as the exhaust gas enthalpy is more effectively regenerated into the enthalpy increase of a premixture.

In the past decade, on the other hand, an electric power density and a conversion

CP653, *Thermophotovoltaic Generation of Electricity: 5th Conference*
edited by T. J. Coutts, G. Guazzoni, and J. Luther

efficiency of GaSb thermophotovoltaic cells become higher up to, respectively, 4~5W/cm^2 and 25~30% [2]. As a result, if all combustion heat could be converted into radiant energy of an active wavelength-range for GaSb TPV cells, an electric power generation system which has a thermal efficiency of 30% will be realized without huge movable elements. Therefore, research and development on emitters, filters and heat recovery burners are continued to reduce long wavelength radiation components and to increase total thermal efficiency [3].

The authors are conducting a research on reciprocating-flow super-adiabatic combustion in porous media through numerical simulation [4] and experiments [5, 6]. In this system, the maximum temperature in the porous medium becomes eleven times higher than the theoretical (adiabatic) flame temperature of a low calorific gas introduced. At the same time, a strong radiant energy field is established in the porous medium, using only a small amount of combustion heat.

In the current study, a new energy conversion system has been proposed on the basis of the reciprocating-flow super-adiabatic combustion in a porous quartz glass and the conversion efficiency from the combustion heat to radiant energy has been investigated through numerical simulation.

CONCEPT OF SUPERADIABATIC COMBUSTION GENERATOR

Figure 1 shows a schematic diagram of the concept of an electric power generation system proposed here. A thin porous medium (foam ceramics) made of alumina or zirconia is installed at the middle in an insulated flowing-duct. At the both sides, a porous quartz glass is installed with a small combustion space between them. Air is introduced into the porous media; where, the flow direction reverses regularly. When air flows in the direction from left to right, fuel is supplied only into the left-hand side

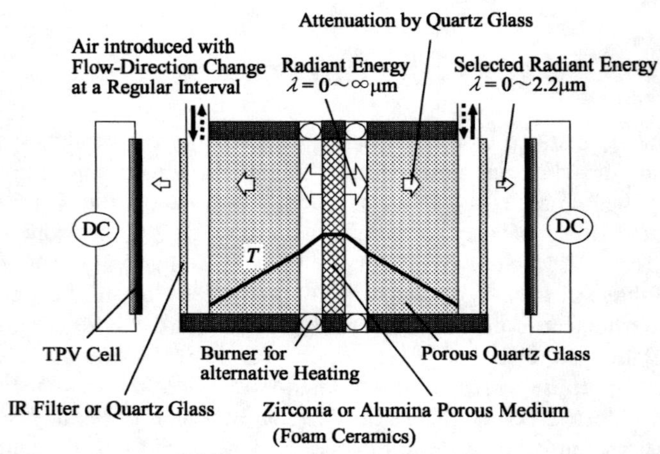

FIGURE 1. Schematic diagram of a concept of recyprocating-flow super-adiabatic combustion thermophotovoltaic power generation system

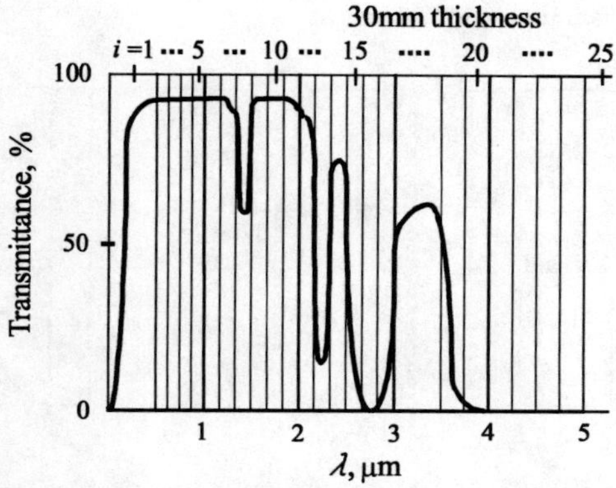

FIGURE 2. Transmittance of quartz glass plate with a thickness of 30mm

combustion space, and vice versa. The foam ceramics is heated by the combustion gas. The combustion gas enthalpy is stored by the downstream side porous quartz glass. When the flow direction is changed, air is preheated by the stored energy since the porous quartz glass provides heat storage. By the energy recirculation through reciprocating-flow system, the maximum foam ceramic temperature becomes higher than the theoretical (adiabatic) flame temperature. Radiation emitted from the foam ceramics travels through the porous quartz glass which has a transmittance with respect to the wavelength of an incident radiation, as shown in Fig.2. As a result, most of radiation of having a wavelength longer than 2 microns is absorbed by the porous quartz glass and is converted into the enthalpy increase of air. On the other hand, radiation of having a wavelength shorter than 2 microns passes mainly through and is incident on a TPV cell installed at the outside of the combustion system.

NUMERICAL SIMULATION

A one-dimensional physical model based on the abovementioned concept of the energy conversion system is shown in Fig.3. Air flows uniformly in the entire region with a velocity u. The inlet gas temperature is T_o. When air flows in the direction from left to right, there is uniform heating by combustion in the region of $x_{-q} < x < x_{-c}$, and vice versa. The amount of fuel injected for combustion is negligibly small, i.e., a few percents of air. It can be assumed that the velocity is maintained at constant and the flowing-gas is non-radiating. Both ends of the system are faced with TPV cell surfaces which provide radiosities R_{-ei} and R_{ei}.

Using these assumptions, the energy equations for both gas and solid phases are, respectively, as follows.
In the regions of $x_{-c} < x < x_c$, $x_{-qo} < x < x_{-q}$ and $x_q < x < x_{qo}$;

$$\rho\, c_p \frac{\partial T}{\partial t} + \rho\, c_p u \frac{\partial T}{\partial x} = \lambda \frac{\partial^2 T}{\partial x^2} - K\left(T - T_0\right) - \alpha_{c,\,q} A_{c,\,q}\left(T - T_{c,\,q}\right) \tag{1}$$

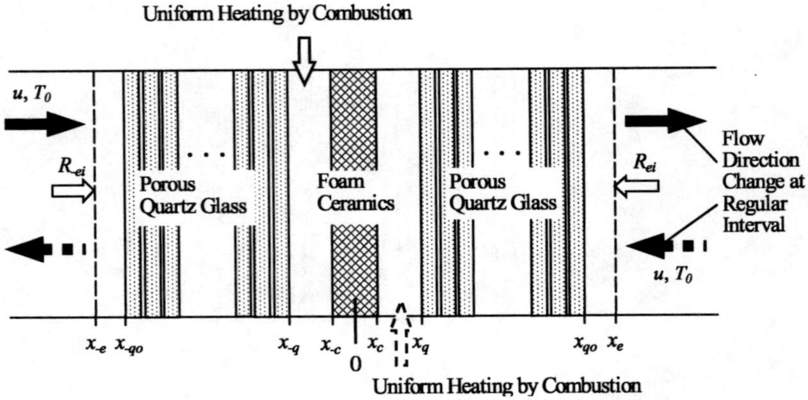

FIGURE 3. A one-dimensional physical model for numerical simulation

$$(\rho\, c_P)_{c,\,q}\frac{\partial T_{c,\,q}}{\partial t}=\lambda_{c,\,q}\frac{\partial^2 T_{c,\,q}}{\partial x^2}-\sum\frac{\partial q_{rci,\ rqi}}{\partial x}+\alpha_{c,\,q}A_{c,\,q}\left(T-T_{c,\,q}\right) \tag{2}$$

In the regions of $x_{-q} < x < x_{-c}$, $x_c < x < x_q$, $x_{-e} < x < x_{-qo}$ and $x_{qo} < x < x_e$;

$$\rho\, c_p\frac{\partial T}{\partial t}+\rho\, c_p u\frac{\partial T}{\partial x}=\lambda\frac{\partial^2 T}{\partial x^2}-K\left(T-T_0\right)+Q \text{ (Only in combustion space)} \tag{3}$$

Here, ρ, c_p, T and λ are, respectively, the density, the specific heat, the temperature and the thermal conductivity of the gas phase, and those with subscript of c or q mean quantities of the foam ceramics and the porous quartz glass. Furthermore, Q and K mean, respectively, the uniform heating term and a characteristic over-all heat transmission coefficient for heat loss to the periphery. The third terms on the right hand sides in Eqs.(1) and (2) express the heat transfer terms between gas and solid phases; where, α and A are, respectively, a heat transfer coefficient and a specific surface area.

The second term on the right hand side in Eq.(2) shows summation of divergence of radiation fluxes. For calculation, the radiation intensity is divided into i = 25 wavelength-bands corresponding to the spectral distribution of transmittance, as shown in Fig.2. For the foam ceramics, assuming isotropic scattering, the divergence of the radiation flux for ith wavelength-band is expressed as follows.

$$\frac{\partial q_{rci}}{\partial \tau_i}=-\frac{4\pi\left(1-\omega_i\right)}{\omega_i}\left(S(\tau_i)-F_i I_b\right) \tag{4}$$

$$S(\tau_i)=(1-\omega_i)F_i I_b+\frac{\omega_i}{2}\left[G_{oi}E_2(\tau_i-\tau_{-ci})+G_{ei}E_2(\tau_{ci}-\tau_i)\right]+\frac{\omega_i}{2}\int_{\tau_{-ci}}^{\tau_{ci}}S(\tau'_i)E_1\left(\left|\tau_i-\tau_i'\right|\right)d\tau_i' \tag{5}$$

Here, τ, ω, $S(\tau)$, F, I_b, G, $E_1(\tau)$ and $E_2(\tau)$ are, respectively, the optical thickness, the scattering albedo, the source function, the blackbody fraction, the radiation intensity of the blackbody, the irradiation from the adjacent porous quartz glass plate, the 1st and 2nd order exponential integral functions. On the other hand, it is assumed that the porous quartz glass consists of thin quartz glass plates (1.5, 3, 4.5 or 6mm) with an open-slit ratio of $a_R = 0.3$, and those are arranged in parallel with a small clearance between them, as shown in Fig.3 and 4. As a result, multi-reflection as well as absorption are taken into account in estimating the radiosities and irradiation for each porous quartz glass plate. Finally, the divergence of the radiation flux for *i*th wavelength-band is expressed as follows.

$$\frac{\partial q_{rqi}}{\partial \tau_i} = -2\pi(1-a_R)[\{-R_i H_{oi} + (1-R_i)G_{qoi}\}E_2(\tau_i) + \{-R_i H_{ei} + (1-R_i)G_{qei}\}E_2(\tau_{si}-\tau_i) + \int_0^{\tau_{si}} F_i I_b E_1(|\tau_i - \tau_i'|)d\tau_i'] + 4\pi(1-a_R)F_i I_b \quad (6)$$

Here, *R* and *H* are, respectively, the surface reflectance and radiation flux coming out from the glass. Furthermore, it is assumed that the radiation passing through the open-slit arrives directly at the next glass plate.

Basic equations were transformed into dimensionless forms and calculated using a finite difference method. The principal dimensionless variables and parameters adopted here are

$$X = x/2x_c, \quad \theta = T/T_o, \quad \beta_{hc} = ut_{hc}/2x_c, \quad Pe = 2x_c\rho c_p u/\lambda, \quad r_q = \lambda_q/\lambda, \quad r_c = \lambda_c/\lambda,$$
$$\Gamma_c = (\rho c_p)_c/\rho c_p, \quad \Gamma_q = (\rho c_p)_q/\rho c_p, \quad H_o = (T_{th} - T_o)/T_o = Q(x_q - x_c)/\rho c_p u T_o,$$
$$M_c = 4x_c^2 \alpha_c A_c/\lambda, \quad M_q = 4x_c^2 \alpha_q A_q/\lambda. \quad (7)$$

Considering a spongy porous ceramics made of alumina, the standard values of r_c, Γ_c and M_c were assumed to be, respectively, unity, 300 and 210, while those of r_q, Γ_q and M_q, respectively, unity, 600 and 2500 for the quartz glass. The geometrical and

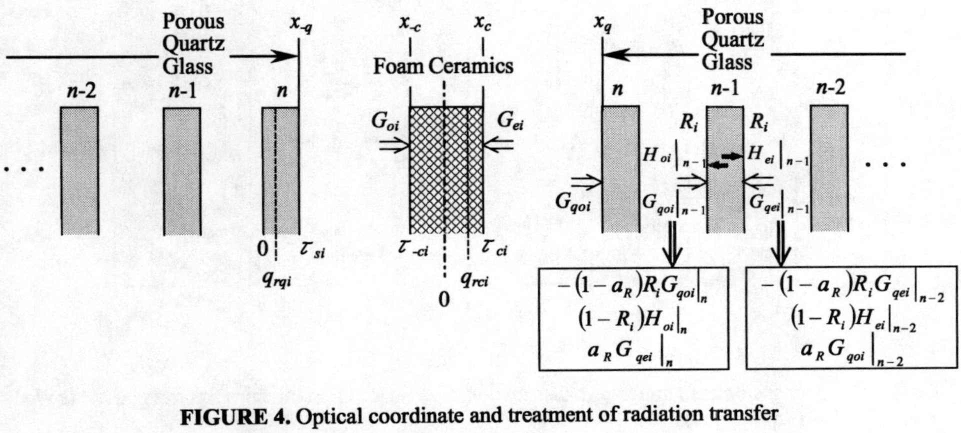

FIGURE 4. Optical coordinate and treatment of radiation transfer

optical thicknesses of the foam ceramics, $2x_c$ and $2\tau_c$, are, respectively, 10mm and unity; it was assumed to be gray in most cases of calculation. The scattering albedo, ω, was assumed to be 0.5. The inlet temperature of air, T_o, is 300K.

RESULTS AND DISCUSSION

Temperature Distributions in Porous Media

Figures 5(a) and (b) show distributions of dimensionless temperatures of gas and solid phases. The Peclet number, Pe, of 70 is equivalent to the flow velocity of 0.7m/s. The dimensionless half-cycle, β_{hc}, for the change of the flow direction is 15 which corresponds to the real time of 0.21 seconds. The dimensionless heating value, H_o, of 2 is equivalent to the dimensionless temperature increase of θ_{th} - θ_o = θ_{th} - 1 = 2; as a result, the dimensionless theoretical flame temperature is 3, as shown in Fig.5. The temperature distributions are varied with time in a cycle; therefore, those at $\beta = \beta_{hc}$ are shown; that is, just before the flow direction changes. Air is preheated by the porous quartz glass, into which it flows from the right hand side. It is further heated by uniform heating in the combustion chamber; where, both premixed and diffusion combustions give the same temperature rise. The combustion gas enthalpy is effectively converted into the radiant energy emitted from the foam ceramics, in which the gas temperature decreases drastically. Furthermore, the residual enthalpy is stored by the downstream side porous quartz glass, in which the gas temperature decreases to that close to the inlet temperature. The stored energy is regenerated into the enthalpy increase of air when the flow direction is changed. By the energy recirculation, the maximum gas and solid temperatures become, respectively, 3 and 2.2 times higher than

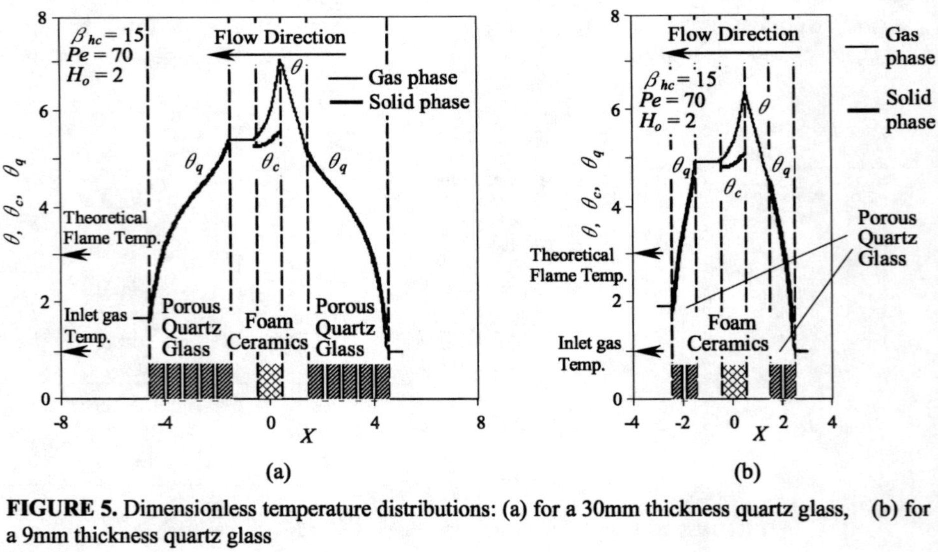

FIGURE 5. Dimensionless temperature distributions: (a) for a 30mm thickness quartz glass, (b) for a 9mm thickness quartz glass

the theoretical (adiabatic) flame temperature. From a comparison between Figs.5(a) and (b), the maximum temperature increases with the thickness of the porous quartz glass since the more exhaust gas enthalpy is regenerated.

Radiation Flux Distributions in Porous Quartz Glass

Figures 6(a) and (b) show distributions of dimensionless radiation fluxes for unit wavelength for each wavelength band. The radiation emitted from the foam ceramics is decreased by absorption and/or surface reflection as traveling through porous quartz glass plates in both positive and negative directions. Especially, the radiation fluxes of having long wavelengths are attenuated drastically since the quartz glass has large absorption coefficient as described in the spectral distribution of the transmittance of the quartz glass (Fig.2). However, large amount of fluxes for wavelength bands of 2.5 and 3.5 microns also come out since there are transmitting windows as shown in Fig.2. On the other hand, the attenuation of the radiation flux of having wavelengths shorter than 2.2 microns is only due to the surface reflection of the glasses; because the absorption coefficient is negligibly small. Even in the case of thin porous quartz glass (Fig.6(b)), attenuation for the long wavelength components are enough, except for the transmitting window bands, though the total radiation flux becomes smaller.

Total Energy Balance in Generation System

Figures 7(a), (b), (c) and (d) show variations of total energy balance. Here, Q_{rad}, Q_{out} and Q_{loss} present the ratios of, respectively, the radiant energy emitted from the system, the out-flowing enthalpy of the exhaust gas and the heat loss to the periphery to the

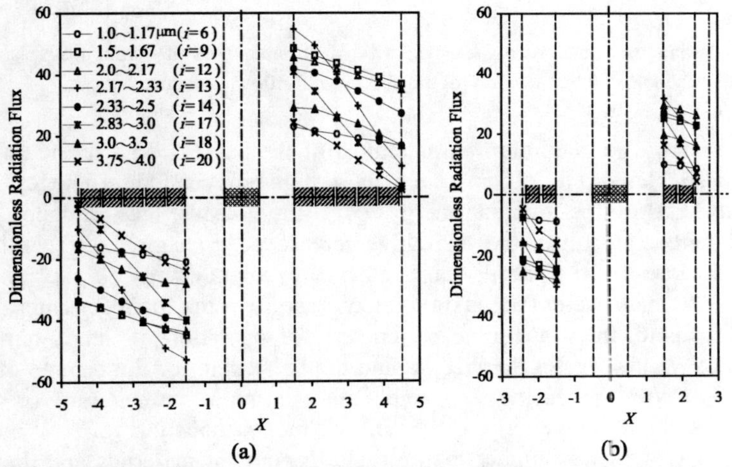

FIGURE 6. Distributions of dimensionless radiation fluxes for unit wavelength for each wavelength band: (a) for a 30mm thickness quartz glass, (b) for a 9mm thickness quartz glass

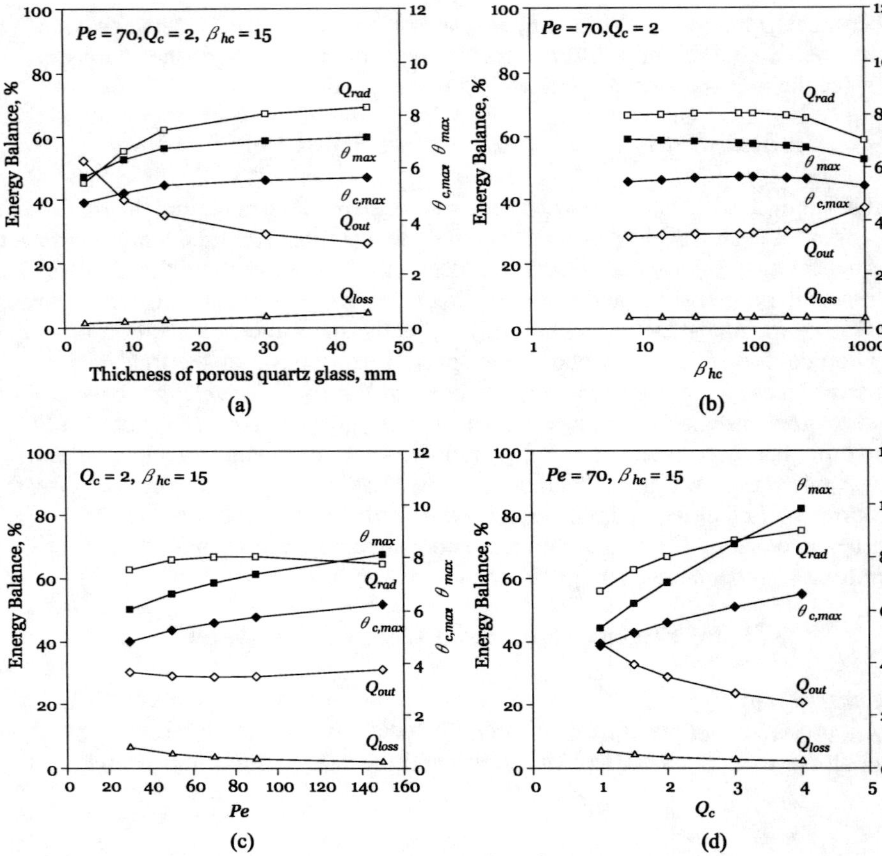

FIGURE 7. Variation of total energy balance with (a) the thickness of porous quartz glass, (b) the heating value, (c) the Peclet number and (d) the half cycle for flow direction change

combustion heat. The maximum temperatures of the gas θ_{max} and the foam ceramics $\theta_{c,max}$ are also shown. In order to obtain a high conversion efficiency from the combustion heat to the radiant energy Q_{rad}, the exhaust gas enthalpy must be effectively regenerated into the enthalpy increase of air supplied. Therefore, with increasing thickness of the porous quartz glass, the radiant energy Q_{rad} becomes higher, according to the increase in the maximum temperature of the foam ceramics (Fig.7(a)). On the other hand, there are three parameters for operation of the system, i.e., the heating value H_o, the Peclet number Pe and the half cycle for the change of the flow direction β_{hc}. With increasing H_o, since the maximum temperature of the foam ceramics rises, the radiant energy Q_{rad} increases monotonously (Fig.7(b)); but it is not realistic for a temperature higher than a melting point of materials and that yielding high emission of nitrogen oxides. Similarly, the maximum temperature increases with Pe, monotonously, as shown in Fig.7(c). However, the radiant energy Q_{rad}, first, increases, but then, decreases with increasing *Pe*. Because the combustion heat is proportional to *Pe*, while the conversion efficiency from the flowing gas enthalpy to radiant energy is not; as a result, the ratio between them Q_{rad} decreases in a high *Pe* number range. On the other hand, the exit temperature is prevented from rising by the high frequency change of the flow direction in this system. As a result, as shown in Fig.7(d), the radiant energy increases with decreasing β_{hc}. However, in the case of an extremely small β_{hc}, the foam ceramics is not heated enough by the combustion heat; that is, before the foam ceramic temperature rises enough, the flow direction changes. Therefore, there is the maximum radiant energy around $\beta_{hc} = 70$. As shown in Fig.7, about 70% of the combustion heat is converted into the radiant energy emitted from the system.

Spectral Distributions of Radiant Energy

Figures 8(a) and (b) show spectral distributions of the radiant energy emitted from the system; where, the effect of the porous-quartz-glass thickness is shown in Fig.8(a), while a comparison between a gray and a nongray foam-ceramics is made in Fig.8(b). In the gray model, according to the optical properties of the quartz glass, large amount of radiant energy in a wavelength range shorter than 2.2 microns is obtained without strong absorption, while the radiant energy of long wavelength components is small except for those for 2.5 and 3.5 micron bands. As shown in Fig.8(a), with increasing thickness of the porous quartz glass, the shorter wavelength radiation becomes increased since the foam ceramic temperature rises. Even in the case of a thin quartz glass, the long wavelength radiation is attenuated enough, though the total radiant energy decreases. However, it is not easy to remove the radiant energy in the transmitting window bands (2.5 and 3.5 micron bands), using only a quartz glass. As a result, a nongray foam ceramics was considered here; where, the scattering albedo ω

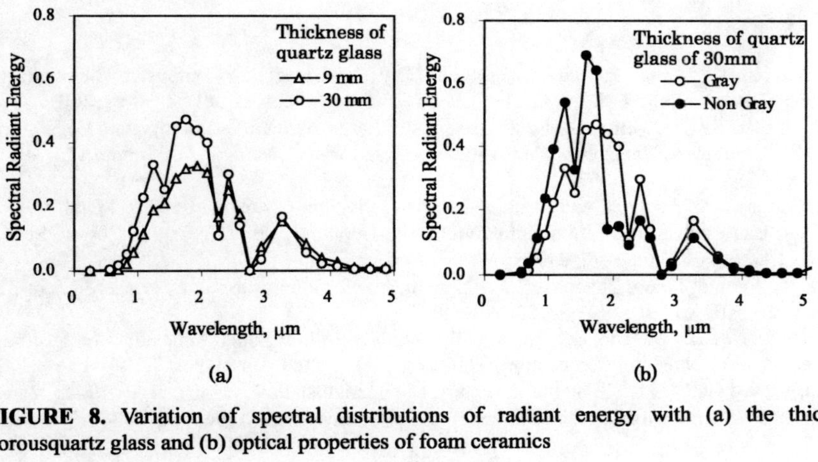

FIGURE 8. Variation of spectral distributions of radiant energy with (a) the thickness of porousquartz glass and (b) optical properties of foam ceramics

was assumed to be 0.5 in the wavelength range shorter than 1.8 microns, while 0.9 in the range longer than that. As shown in Fig.8(b), much amount of radiant energy, i.e., about 70%, is converged in the range of the wavelength shorter than that. In this case, about 48% of the combustion heat is converted into the radiant energy in a wavelength range shorter than 1.8 microns which is useful for electric power generation using GaSb TPV cells.

CONCLUSIONS

A thermophotovoltaic power generation system based on super-adiabatic combustion in a ceramic foam with a porous quartz glass heat storage has been investigated through numerical simulation. By heat recirculation, the maximum foam ceramic temperature becomes about two times higher than the theoretical (adiabatic) flame temperature. Therefore, about 70% of the combustion heat was converted into the radiant energy emitted from the foam ceramics. Much amount of the long wavelength components is absorbed by the porous quartz glass; that is, about 60% of the radiant energy exists in the wavelengh range from visible region to 1.8 microns, using a gray foam ceramics. A nongray ceramic foam, e.g., with anti-reflection coating, might be useful to prevent from emitting radiation for window bands of the quartz glass. In this case, about 48% of the combustion heat was converted into the radiant energy in a wavelength range which is useful for GaSb TPV cells. The total thermal efficiency will be expected to be 14.4%.

ACKNOWLEDGEMENT

The authors would like to express appreciation for the financial support from a Grant-in-Aid for Scientific Research (C) of the Ministry of Education and Science of Japan (No.13650217).

REFERENCES

1. Zhang, C., Najafi, K., Bernal, P.L. and Washabaugh, D.P., "An integrated Combustor-Thermoelectric Micro Power Generator", the 11th Int. Conference on Solid-State Sensors and Actuators, 2001.
2. Bett, A.W., Keser, S., Stollwerck, G. and Sulima, O.V., "Large-Area GaSb Photovoltaic Cells", Proc. of 3th NREL Conference on Thermophotovoltaic Generation of Electricity, Colorado, May 18-21, 1997, p.41 ff.
3. Fraas, L., Samaras, J., Avery, J. and Minkin, L., "Antireflection Coated Refractory Metal Matched Emitters for Use with GaSb Thermophotovoltaic Generators", Proc. of 39th Power Sources Conference, Cherry Hill, June 14th, 2000.
4. Hanamura, K., Echigo, R. and Zhdanok, S.A., "Superadiabatic Combustion in a Porous Medium", Int. J. Heat Mass Transfer, Vol.36, 1993, pp.3201-3209.
5. Hanamura, K., Adachi, M. and Echigo, R., "Reciprocating-Flow Super-Adiabatic Combustion in Porous Media", Environmental Combustion Technology, Vol.1, 2000, pp.81-94.
6. Hanamura, K. and Echigo, R., "Thermal Structure of Superadiabatic Combustion in Porous Media", Proc. of the 3rd KSME-JSME Thermal Engineering Conference, Vol.2, 1996, pp.339-342.

SESSION 3

Emitters/Filters

Chair: Christopher Hebling

An Overview Of TPV Emitter Technologies

A. Gombert
Fraunhofer Institute for Solar Energy Systems ISE, Oltmannsstr. 5, 79100 Freiburg, Germany
phone: +49 761 4016683, fax: +49 761 4016681, e-mail: andreas.gombert@ise.fhg.de

Abstract. Radiating surfaces at typical temperatures in the range of 1200 – 2000 K are the source of energy which is incident on the PV cells of TPV systems. The radiators or emitters may have very pronounced or weak spectrally selective properties on which TPV system efficiencies depend very much. This paper gives an overview of existing emitter technologies. Two different approaches to realize spectrally selective emitters are discussed in detail: rare-earth oxides and engineered tungsten emitters.

1. INTRODUCTION

In a TPV system a body which is heated to typical temperatures between 1200 K and 2000 K emits radiation which is partly converted to electric energy by using PV cells (Fig. 1). In order to increase the efficiency the spectral emissivity of the heated body should be matched to the bandgap of the PV cell. Thus the wanted properties of the emitter are: durability at high temperatures, low emissivity for photons with energies lower than the bandgap energy of the PV cell and high emissivity for photons with higher energy. Specifications which are more precise depend on the TPV system. It should be mentioned that it is already a difficult task to fit the very fundamental requirements.

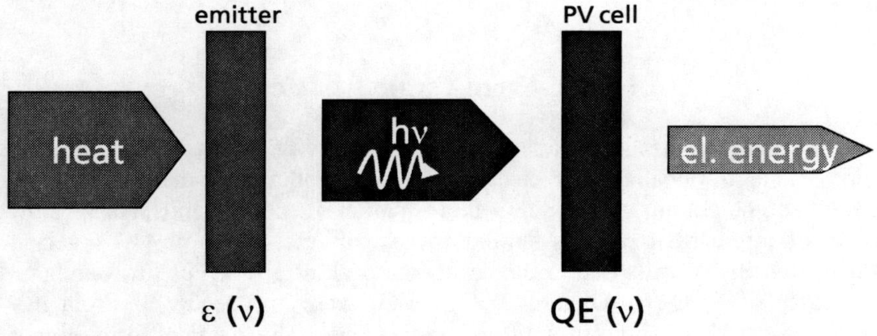

FIGURE 1. Sketch of a TPV system.

CP653, *Thermophotovoltaic Generation of Electricity: 5th Conference*
edited by T. J. Coutts, G. Guazzoni, and J. Luther

The modification of thermal radiation can be described through the definition of spectral emissivity $\varepsilon(\lambda)$ with $I_{BB}(\lambda,T)$ being the thermal radiation from a blackbody as given by the Planck's law.

$$I(\lambda,T) = \varepsilon(\lambda) \cdot I_{BB}(\lambda,T) \qquad (1)$$

The spectral emissivity is often thought to be a property of the radiating material. From quantum optics we know that the interaction of radiation and material is in fact a function of the photonic density of states which is the number of optical modes [1]. Having a cavity in mind, it is obvious that the number of optical modes can be influenced by physically structuring materials. Thus, there are in general two possibilities to tailor the spectral emissivity:

- choice of suitable materials.
- engineering, i.e. by physical structuring.

It should be kept in mind that the spectral emissivity equals the spectral absorptivity according to Kirchhoff's law. Thus, we can not only make use of absorptivity measurements for emitter characterization but also have an indirect method for the design of emitters.

2. CERAMIC EMITTERS

Ceramic emitters were the first materials which were taken into account because they can be operated at high temperatures in an oxidizing atmosphere, e.g. air. As broadband emitter silicon carbide (SiC) is often used. It has a very high emissivity and can have quite different geometries adapted to the TPV system. As material with a moderate spectral selectivity cobalt doped refractory oxides such as Al_2O_3, MgO or Al_2MgO_4 were proposed [2]. The disadvantage of all ceramic emitters is the high emissivity in the mid infrared due to lattice vibrations.

2.1 Rare-earth Oxide Emitters

Rare-earth oxide emitters possess a selective emissivity over a very small spectral bandwidth. Due to localized 4f electrons of the lanthanides there is almost no interaction with neighbour ions and thus no formation of energy bands. As a result, the emissivity of rare-earth oxides is similar to gaseous materials. Yb_2O_3 is very well suited to match Si PV cells due to the emittance peak at 1 μm. For low bandgap PV cells mixtures of Er_2O_3/Ho_2O_3 and Er_2O_3/Yb_2O_3 were proposed [3]. Erbia has an emissivity peak 1.55 μm, holmia at 1.2 μm and at 2 μm. The classical form of an rare-earth oxide emitter is the "Welsbach" mantle. A mantle made out of organic fibres is soaked in a solution containing the rare-earths. After burning off the organic fibres a rare-earth oxide fibre mantle remains. Due to the open structure of the mantle the emittance of the combustion flame is also present in emittance spectra of rare-earth oxide mantles (Fig.2) [4]. It can also be seen that the emissivity of the ytterbia mantle is close to 1 for wavelengths > 8 μm as the emittance is close to the blackbody emittance. The mechanical stability of mantles limits the size of the emitters. Therefore other ways to achieve extended rare-earth oxide emitters were investigated: fiber bundle plate emitters [5], thin film emitters [6], thick film emitters [7] and also small particle emitters [8].

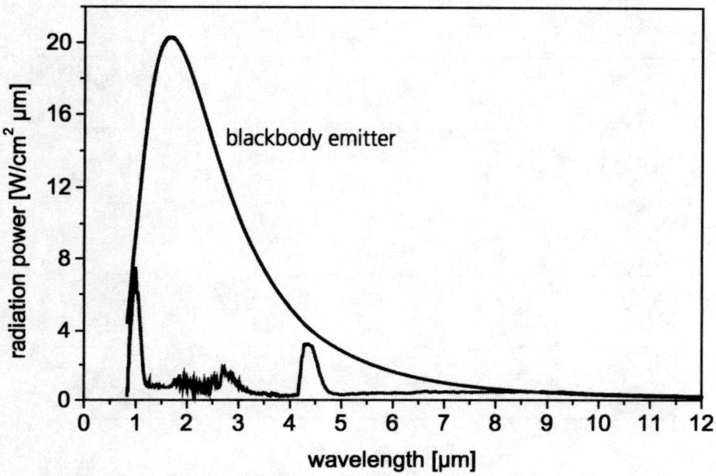

FIGURE 2. Emittance spectrum of an Yb_2O_3 emitter in form of a Welsbach mantle at a temperature of 1735 K. The peak at 1μm is the emittance of the Yb_2O_3, the other peaks are related to the flame (courtesy: B. Bitnar, Paul-Scherrer-Institute, Switzerland [4]).

3. TUNGSTEN EMITTERS

Metals were considered as emitters for TPV systems because of their low emissivity in the mid and far infrared. At shorter wavelengths metals have an increased emissivity due to electronic transitions and interband and intraband absorption. In the MIR and FIR their optical properties are dominated by the free electrons which means that the emissivity of a flat surface is very low. This natural selectivity makes metals emitters suitable for TPV systems. Of course, metal emitters can only be used in vacuum or in inert gas atmosphere. Tungsten as emitter was proposed in combination with a GaSb PV cell. In this case tungsten is very suitable because its interband absorption coincides almost with the bandgap of GaSb (see also

Fig. 3). Furthermore, tungsten is a material with a very low evaporation rate in vacuum at temperatures of 1200 K to 2000 K.
The low emissivity in the MIR and FIR makes tungsten very suitable as a base material which is coated or physically structured in order to increase the emissivity at convertible wavelengths. Three different approches to engineer tungsten emitters were investigated so far:

- AR coatings
- 1D and 2D surface-relief structures
- 3D photonic crystals

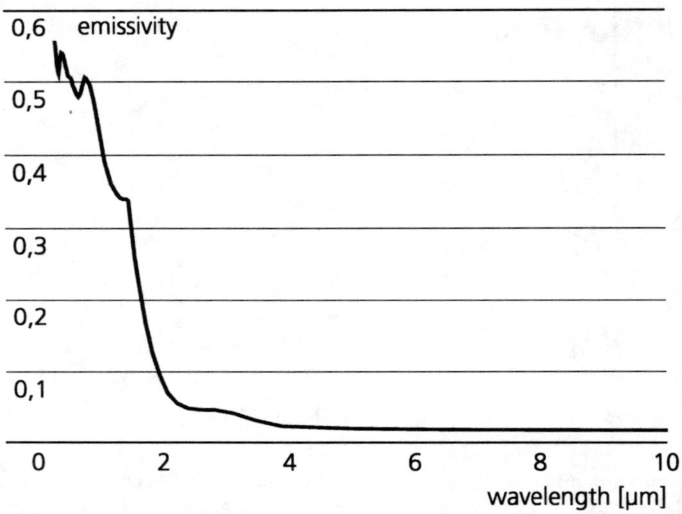

FIGURE 3. Emissivity of a flat tungsten surface (calculated from the optical constants at 293 K).

3.1 AR Coated Tungsten Emitters

A very efficient way to increase the emissivity of tungsten is the application of an antireflective (AR) coating. On metals, dielectric AR coatings are only effective at wavelengths at which the absorptivity of the metals is not too low. Thus, the natural selectivity of the metal can be enhanced very efficiently. In order to achieve highest emissivity of AR coated tungsten at wavelengths in the range of 1.0 – 1.7 μm the AR coating should be slightly absorbing. Emissivities of up to 95% can be obtained theoretically in this wavelength range with an AR coating with refractive index of

$\tilde{n} = 2.0 + 0.15i$ (Fig. 4). Due to the high refractive index of the tungsten such an AR coated emitter exhibits a low angle dependency.
AR coated tungsten emitters were already used in TPV systems [9]. The advantage of the coating is that it can by applied relatively easily also on cylindrical substrates, e.g. on tungsten foils. Many TPV systems are cylindrical.

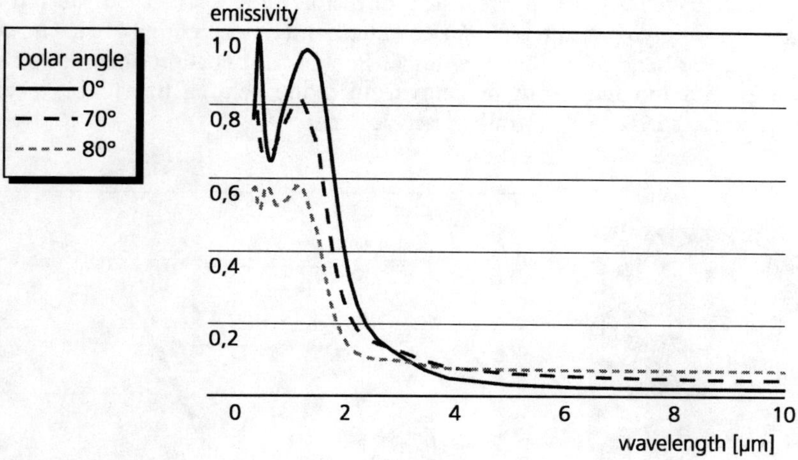

FIGURE 4. Angle dependence of the emissivity of an AR coated tungsten surface (calculated for a coating with n = 2.0 + 0.15i and a thickness of 138nm).

3.2 Tungsten Surface-Relief Gratings

Absorptance peaks in metallic gratings are known as "Wood anomalies" since the beginning of the 20th century [10]. This observation lead to the concept of periodically structured metal emitters in which thermally exited surface plasmons couple to the emitted radiation [11, 12]. If the surface-relief grating is modulated only in one direction, the coupling of surface plasmons is restricted to TM polarized radiation. Thus, gratings which are modulated in two dimensions were mainly investigated. Very sharp emissivity peaks can be obtained by this method. The main disadvantage is the strong angle dependence of the surface plasmon resonance. Therefore, other phenonema than the classical surface plasmon resonance were taken into consideration: vertical standing waves in deep cavities [13] and antireflective subwavelength gratings.
In Fig. 5a the near field intensity distribution of a subwavelength grating was calculated for TM polarization using rigorous diffraction theory. A plane wave is incident perpendicular to the surface. The grating has a period of 250nm, a depth of 200nm and a rectangular cross section. The lamella have the same width as the grooves. The high intensity in the grooves indicates a strong absorption. Because of

the small period of the grating compared to the wavelengths of interest, the decrease of the reflectance and thus the increase of the emissivity is relatively broadband and shows a low angle dependence (Fig. 5b). Similar to AR coatings, an increase of emissivity is possible especially in the spectral region of the interband absorption.
The durability of tungsten surface-relief gratings is limited by surface diffusion. Tungsten surface-relief gratings are almost planar after few hours at temperatures > 1500 K. It was found that the surface diffusion process can be slowed down significantly by applying a thin coating, i.e. a hafnium oxide coating [14]. The coating can be only a few tens of nanometers thick. Such a thin coating does not effect the optical properties too much but in general an oxide coating has to be taken into account when the surface-relief gratings are designed.

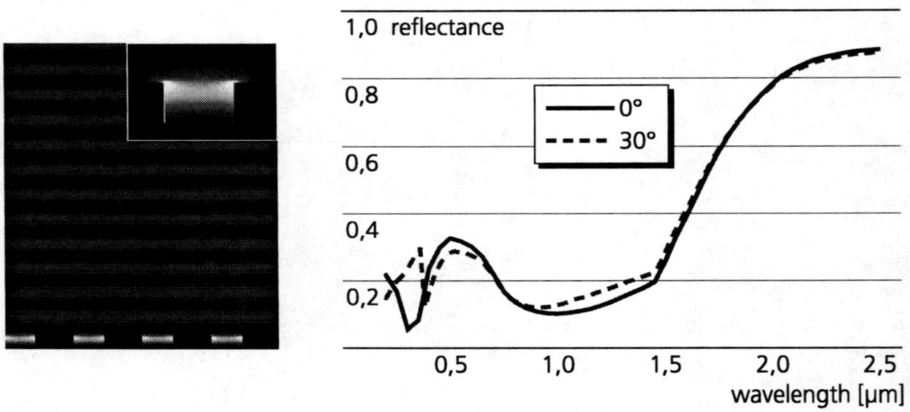

a) b)

FIGURE 5. Near field intensity distribution of a one-dimensional tungsten subwavelength grating for a TM polarized incident wave with a wavelength of 1000nm (a) and reflectance spectra of this grating for the angles of incidence of 0° and 30° and TM polarization (b). All shown data were calculated by using rigorous diffraction theory.

3.3 Tungsten Photonic Crystals

Media in which the refractive index varies periodically can be described in analogy to the periodic potential in a crystal as "photonic crystals". Of course, in photonic crystals the band structures depend on the polarization of the optical modes. Therefore, complete bandgaps for photons with certain energies exist only where the bandgaps for each of the polarizations overlap. In incomplete bandgaps photons with certain energies cannot propagate in certain directions. In complete bandgaps photons of certain energies cannot exist. Then, the photonic density of states and thus the emissivity is zero. At the band edge or at localized modes, the photonic density of states can be very high.

Complete bandgaps can only be formed in three-dimensional (3D) photonic crystals. Fleming *et al.* proposed a 3D tungsten photonic crystal as selective emitter for TPV applications [15]. They manufactured a 3D tungsten photonic crystal by selectively removing Si from already fabricated polysilicon/SiO_2 structures and refilling the mould with tungsten by using a CVD process. The resulting 3D tungsten photonic crystal with a diamond lattice structure is shown in Fig. 6 [16]. By using Fourier transform spectrometry they characterized the optical properties of the 3D tungsten photonic crystal and compared the measurements with theoretical results obtained by a transfer-matrix calculation. The 3D tungsten photonic crystal with a lattice constant of 4.2 μm has a very large bandgap for infrared wavelengths ranging from 8 μm to 20 μm. At the band edge, the reflectance of the 3D tungsten photonic crystal decreases and the transmittance increases. At the same time a significantly increased absorptance is found at the band edge as predicted theoretically (Fig. 7). As a result, one can say that 3D photonic crystals represent a very elegant description which gives insight to emissivity as a function of the photonic density of states.

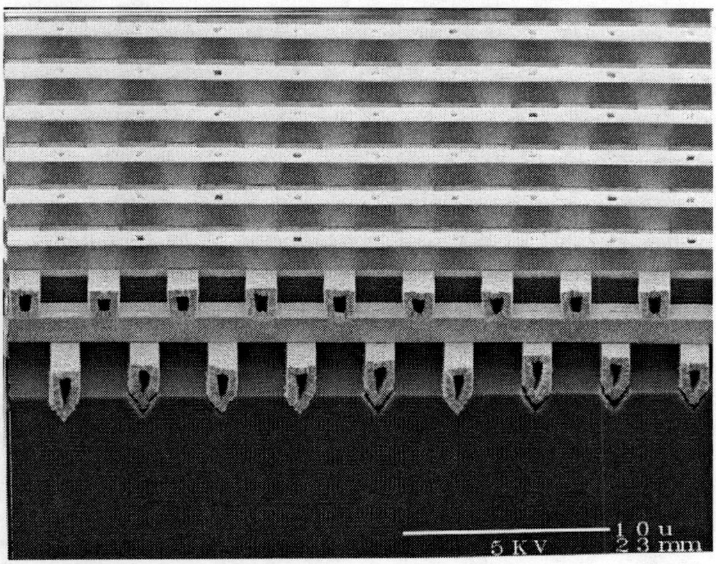

FIGURE 6. SEM picture of a four-layer 3D tungsten photonic crystal with a lattice constant of 4.2 μm (courtesy: J.M. Gee, Sandia National Laboratories [16]).

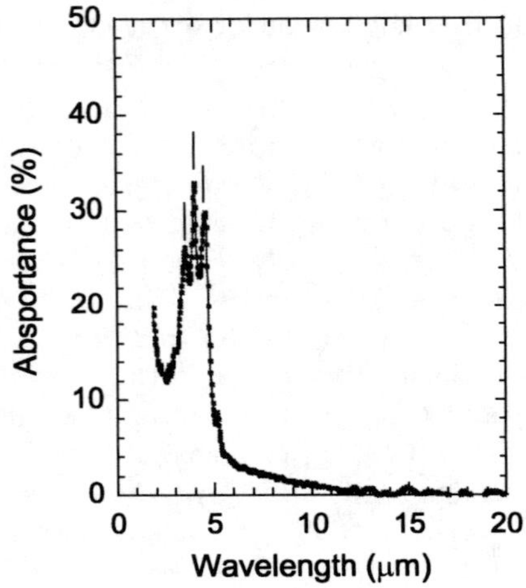

FIGURE 7. Measured absorptance of a four-layer 3D tungsten photonic crystal with a lattice constant of 4.2 μm (courtesy: J.M. Gee, Sandia National Laboratories [16]. The absorptance peak is situated at the band edge.

4. CONCLUSIONS

A variety of emitter concepts is available for different TPV systems. Ceramic emitters can be used at high temperatures in an oxidizing atmosphere, tungsten emitters need an inert gas atmosphere or vacuum. Recent work mainly addresses the possibility to engineer emissivity e.g. by physical structuring materials. An important issue is still the durability of the emitter. For many of the proposed TPV emitters thermal and mechanical stability are not sufficiently tested so far.

ACKNOWLEDGEMENTS

The author thanks B. Bitnar from the Paul-Scherrer-Institute, L.M. Fraas from JXCrystals, C. Schlemmer from Fraunhofer ISE and J.M. Gee from Sandia National Laboratories for providing results of their appreciated work and for very fruitful discussions.

REFERENCES

1. Yablonovitch, E., *Phys. Rev. Lett.* **20**, 2059-2062 (1987).
2. Ferguson, L. and Fraas, L., "Matched Infrared Emitters for Use with GaSb TPV Cells", in *The Third NREL Conference on Thermophotovoltaic Generation of Electricity*, edited by T. J. Coutts et al., AIP Conference Proceedings 401, New York: American Institute of Physics, 1997, pp. 169-179.

3. Wong, E.M., et al., "Multiband Spectral Emitters Matched to MBE Grown Photovoltaic Cells", in *The Second NREL Conference on Thermophotovoltaic Generation of Electricity*, edited by J. P. Benner et al., AIP Conference Proceedings 358, New York: American Institute of Physics, 1995, pp. 278-289.
4. Bitnar, B., Paul-Scherrer-Institute, Switzerland, private communication.
5. Nelson, R.E., "Thermophotovoltaic Emitter Development" in *The First NREL Conference on Thermophotovoltaic Generation of Electricity*, edited by T. J. Coutts and J. P. Benner, AIP Conference Proceedings 321, New York: American Institute of Physics, 1994, pp. 80-96.
6. Chubb, D.L. and Lowe R.A., "Emittance Theory for Thin Film Selective Emitter" in *The First NREL Conference on Thermophotovoltaic Generation of Electricity*, edited by T. J. Coutts and J. P. Benner, AIP Conference Proceedings 321, New York: American Institute of Physics, 1994, pp. 229-244.
7. Chubb, D.L., et al., "Effect on Temperature Gradient on Thick Film Selective Emitter Emittance" in *The Third NREL Conference on Thermophotovoltaic Generation of Electricity*, edited by T. J. Coutts and J. P. Benner, AIP Conference Proceedings 401, New York: American Institute of Physics, 1997, pp. 293-313.
8. Chubb, D.L. and Lowe, R.A., "A Small Particle Selective Emitter for Thermophotovoltaic Energy" in *Thermophotovoltaic Generation of Electricity: Second NREL Conference*, edited by J. P. Benner et al., AIP Conference Proceedings 358, New York: American Institute of Physics, 1995, pp. 263-277.
9. www.jxcrystals.com and private communication with Lewis M. Fraas, JXCrystals.
10. Wood, R.W., *Proc. Phys. Soc.* **18**, p. 396 (1902).
11. Heinzel, A., et al., "Surface-relief Grating Structures in Tungsten for Use as a Selective Emitter in Thermophotovoltaic Systems", in *European Optical Society Topical Meetings Digests Series*, Volume 12, Savonlinna: European Optical Society, 1997, pp. 250 – 251.
12. Kreiter, M., et al., *Opt. Comm.* **168**, pp. 117-122 (1999).
13. Heinzel, A., et al., *J. Mod. Optics* **47**, pp. 2399-2419 (2000).
14. Schlemmer, C., et al., to be published in this volume.
15. Fleming, J.G., et al., *Nature* **417**, pp. 52-55 (2002).
16. Gee, J.M., Sandia National Laboratories, Albuquerque, USA, private communication.

Selective Emitters for High Efficiency TPV Conversion: Materials Preparation and Characterisation

D. Diso[a], A. Licciulli[b], A. Bianco[b], G. Leo[a], G. Torsello[a], S. Tundo[a], A. De Risi[b], M. Mazzer[a]

[a] *CNR-IMM Campus Universitario, Via Arnesano, 73100 Lecce (I)*
[b] *Dip. di Ingegneria dell'Innovazione Univ. di Lecce - Campus Universitario, Via Arnesano - 73100 Lecce (I)*

Abstract. Optimising the spectral emissivity of the IR radiation source in a TPV generator is one of the crucial steps towards high efficiency TPV conversion. In this paper we present different approaches to the preparation of selective emitters to be coupled to high efficiency photovoltaic cells. The emitters are designed to work at a temperature of about 1500K and they have been prepared to be used either as external coatings for the burner or as a structural material for the burner itself. Composite ceramics containing rare earth cations, prepared by slip-casting, with various concentration of rare earths were prepared by Slip Casting and Slurry Coating. Rare earth oxides have been incorporated into different oxide matrices, namely Silica, Alumina, Zirconia and their combination. The final aim was to find the material that exhibits the best performance in terms of both high selective power emission, good efficiency along with acceptable thermo-structural properties (high temperature thermal shock resistance, good strength, no creep). The power density emitted by samples as function of the temperature has been tested in the range 1000nm-5000nm. The high temperature emission measurements and the structural tests indicate that a good compromise between the functional and the thermo-structural properties may be reached. The results of the tests on the emitter coatings carried out in a TPV generator at the operating conditions are also presented in this paper.

1. INTRODUCTION

It is well known that one of the major limitations to the conversion efficiency of a Thermophotovoltaic (TPV) system operating at moderate temperatures (T < 1300 °C) is the fact that the largest part of the electromagnetic radiation, emitted by a blackbody- or a greybody-like source at those temperatures is wasted because of the mismatch with the spectral response of any of the photovoltaic cells proposed so far. Radiant energy which is neither converted nor recycled can result in a destructive heat

CP653, *Thermophotovoltaic Generation of Electricity: 5th Conference*
edited by T. J. Coutts, G. Guazzoni, and J. Luther

load on the generator components, including the temperature-sensitive photovoltaic cells and will drastically lower the conversion efficiency of the system.

One of the solutions to increase the overall TPV efficiency is the use of selective emitters, radiators whose spectral emissivity is high in a narrow wavelength range (namely in the absorption band of the photovoltaic cell) and low elsewhere. Actually, the presence of a selective emitter does not increase the power density available for the PV conversion with respect to a high emissivity material such as SiC, but it reduces losses due to low energy photons and carrier thermalisation. The effectiveness of Rare-Earth (RE) oxides as spectrally selective emitters was evidenced in 1972 by Guazzoni et al.[1] who published a detailed study on high-temperature spectral emittances of Erbium, Samarium, Neodymium and Ytterbium oxides. They prepared the emitters by slip casting using slurries of rare earth oxide powders. These oxide ceramics in bulk monolithic form, were severely limited by thermal shock. To overcome this limit, Nelson [2] prepared the rare earth oxides in fibrous form, obtaining good thermal resistant emitters which, thanks to small thickness, exhibited also limited off-band radiation as well as a superior resistance to thermal cycling. Being the emissivity a surface effect, the efficiency enhancement observed in those fibrous emitters was mostly related to the large surface roughness compared to the same material with a flat or polished surface. Although not mechanically durable, gas lighting mantles are excellent examples of thermally stress-tolerant structure because the 10-μm-diameter of the fibres is small enough to survive thermal stresses.

The problem of fragility of the mantle-type emitters can be overcome by using a composite-based emitter, incorporating the desired rare earth oxides into a structural material consisting of quartz or Nextel 440 fibres as proposed by Adair and Rose [3]. Recently, L.G.Ferguson et. al. [4] fabricated a new "band-gap matched" emitter, consisting of a composite Co- or Ni-doped MgO, which exhibited high mechanical integrity, thermal shock resistance, excellent heat transfer characteristics, near-ideal spectral efficiency and peak emissive power of 0.1 MW/m^2 μm. Some interesting experimental results have been also found for Erbium, Thulium and Erbium-Holmium aluminium garnet thin films (0.25-0.40 mm) by Lowe R.A., Chubb et al. [5].

Oxide based materials are generally transparent to infrared radiation since the valence electrons are tightly joined to the atoms through ionic and covalent bonding. Their optical absorption from the visible to the infrared range is very low since no electronic transitions are allowed. At wavelengths larger than 4-5 microns, lattice and molecular vibration are resonant with long wavelength photons and therefore they are responsible for both high absorbivity and high emissivity. The low absorption and emissivity that characterise oxide ceramics in the range of 0.5-5 microns [6] suggest that they can be good host for Rare Earth emitter. In particular Silica, Zirconia, Alumina and Aluminium based garnet ceramics are also very important thermo-structural ceramics. This is the reason why we have decided to investigate the above mentioned oxides to prepare an optimised emitter.

We have produced samples incorporating rare earth oxides into different oxide matrices, namely Silica, Alumina, Zirconia and their combination in the form of bulk ceramic and porous thick coatings. Er was selected as the preferred rare earth element to match the spectral response of photovoltaic cells with the maximum conversion efficiency at about 1500 nm.

2. EXPERIMENTAL

Composite Preparation

Silica glass was obtained from a modified sol-gel process previously reported [7]. Bulk ceramic and porous coatings were prepared from ceramic slips containing a reactive colloidal phase.

The ceramic powders contained in the slip were prepared in house by a wet chemical route: rare earth salt and Alumina sol where mixed dried and calcined. Garnet powders could be obtained, provided that the calcining temperature was above 1050°C. Garnet powders, bulk and films were prepared with composition $Er_xY_{3-x}Al_5O_{10}$, with x ranging from 0 to 3. For x = 0 and for x = 3 we will use the abbreviations YAG and ErAG respectively. Short fibre reinforced composites have been prepared from ceramic slurries containing short Alumina or Zirconia fibres. Details on the preparation of the composites are presented elsewhere[8], whereas details on the preparation of garnet films and bulk structures are proprietary and patents are pending. The reactive slips were cast in plastic molds, gelled dried and sintered. Disks of various size and thickness where prepared for emissivity testing. Porous coatings on greybody combustors e.g. SiC and Alumina where obtained with good adhesion and good thermal shock resistance. The thickness of the coatings could be controlled from 50 to 500 microns.

Experimental setup for high-temperature optical characterisation

The high-temperature optical characterisation was carried out by recording the emission spectra of the analysed samples together with the emission spectrum of a Silicon Carbide reference sample in the same experimental conditions. Samples were heated up to temperatures as high as 1800K by a gpl-oxygen flame both in back-heating and front heating configuration with respect to the acquisition equipment.

For the measurement of the surface temperature two optical pyrometers with spectral response at 3.9 μm and 1.5 μm, and an S-thermocouple were used. Samples faced a multimode fluoride glass optical fibre which funnels the radiation into an F-matcher connected to a 74100-M Lot-Oriel double grating monochromator.

Finally, the radiation is detected by a InGaAs sensor and a InSb detector. To prevent the fibre from heating, a molybdenum shield with a 1 mm diameter hole was positioned between the fibre and the sample. A preliminary calibration was performed with the aim of verifying emittance values of the SiC specimen at the wavelength of analysis of pyrometers. SiC emissivity is equal to 90% in the working range of the near infrared pyrometer (1000 to 1900 nm), and to 75% in the spectral window of the infrared pyrometer [9].

Since the size of the emitting area was selected by the viewing angle of the ensemble fibre-Molybdenum shield, the calibration curves included all the geometrical factors. The power density measurements were performed at different temperatures of the emitters, and for a wide range of RE concentrations.

3. RESULTS AND DISCUSSION

The measured high-temperature power density is reported in Fig.1 as a function of wavelength at a temperature of 1250°C for specimens hosting the same Er concentration by weight into different matrices. The spectra refer to the wavelength region of Erbium emission. The monocrystalline YAG spectrum has the lowest off band contribution. The peak at 1500 nm is attributed to the Erbium ion transition $^4I_{13/2} \rightarrow {}^4I_{15/2}$ and the peak at 1000 nm to $^4I_{11/2} \rightarrow {}^4I_{15/2}$. The 1.5μm-band structure appears to be the convolution of the transitions between the Stark levels resulting from the 2J+1 degeneration removal of the ground and the excite levels of Er^{3+} by the electric host field. More details on the excitation and de-excitation mechanisms in rare earth-doped selective emitters are described in the paper by Torsello et al. presented at this conference [10]

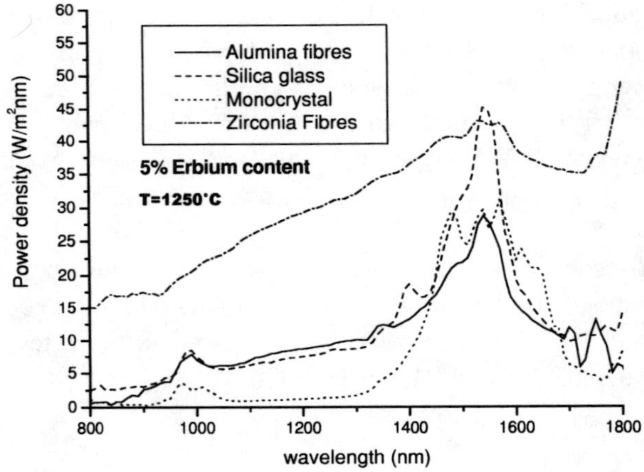

FIGURE 1. Power density emitted by samples in which the same Erbium ions concentration was inserted into different host matrix, (solid line) Alumina fibres, (dashed lines) Silica glass, (dot line) YAG monocrystal and (dash dot line) Zirconia fibres.

The two composite samples contain the rare earth oxide providing the functional performances and special fibres to provide good mechanical properties (strength, creep resistance) at high temperature (1500°C). Although they differ only for the type of fibre the high temperature emission behaviour is quite different. Fig.1 evidences that the specimen containing Zirconia fibres does not exhibit any selective emission, whereas the sample reinforced with Alumina fibres exhibit a lower off-band emission and a higher peak to background ratio at 1.5 μm. Torsello et al. [10] have shown that in composites containing Zirconia fibres, efficient non-radiative channels prevent Erbium ions from decaying radiatively and therefore the in-band emission is strongly reduced compared to the Alumina-based composites where the Erbium ions are bounded to Aluminium and oxygen into ternary compounds. This structure is shown to inhibit nonradiative decay thus increasing the probability of radiative deexcitation [10]. As a result we can state that a strong absorption is a necessary but not a sufficient condition for an efficient emission at high temperature. In the Zirconia-based composite it is, in fact, clearly evident that non-radiative phenomena are able to quench and even suppress the Er^{3+} emission around 1500nm.

Bulk Silica exhibits the highest in-band intensity in the wavelength range shown in Fig.1, but it is limited to a maximum operating temperature of about 1250 °C and it has a very high emissivity (about 90%) at wavelengths larger than 3.5μm [6].

A series of ceramic Aluminium Garnet samples with different Erbium/Yttrium ratio were also prepared. Contrary to YAG monocrystals, ceramic YAG is opaque and its surface is characterised by a higher roughness. In Fig. 2 the peak power density at 1480nm is plotted versus the Erbium concentration at about 1200°C for both monocrystalline and ceramic YAG. For Erbium fractions from 0 to 40% the emitted power increases with increasing Erbium content but, between 40% and 100% the emission does not change significantly. This indicates a trade-off between the increase of radiating centres and the efficiency of the de-excitation energy transfer processes. The observed behaviour, reported in Fig.2, is quite similar for ceramic and single-crystal material even if the saturation effect in the monocrystal matrix seems to occur at a lower concentration than for the ceramic garnet. The experimental errors have been estimated at 10%.

This result, along with the high peak emission found in transparent amorphous Silica shown in Fig.1, suggests that in ceramic materials emissivity is a bulk property influenced by factors such as the scattering at grain boundaries and at the pore surfaces, or the dimensions of grains and pores. The lower power density that is found in ceramic samples at low Er concentration is in fact attributed to scattering phenomena which prevent the deeper Er cation to contribute to the surface emissivity.

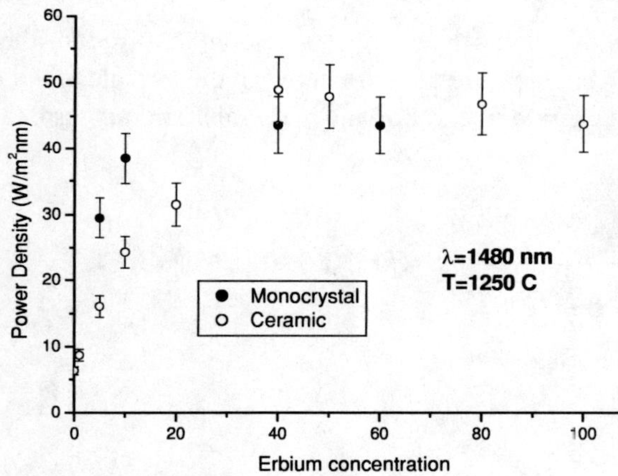

FIGURE 2. Power density emitted by ceramic and monocrystal YAG versus the percentage of substitutional Erbium ions at 1480nm and T=1250°C.

In fig.3 the spectral emission of Erbium oxide is compared to ceramic Erbium Aluminium Garnet (ErAG). The efficiency of a selective emitter is usually defined as the ratio of the power emitted in the wavelength spectrum of absorption of the photovoltaic converters to the total power irradiated at a given temperature.

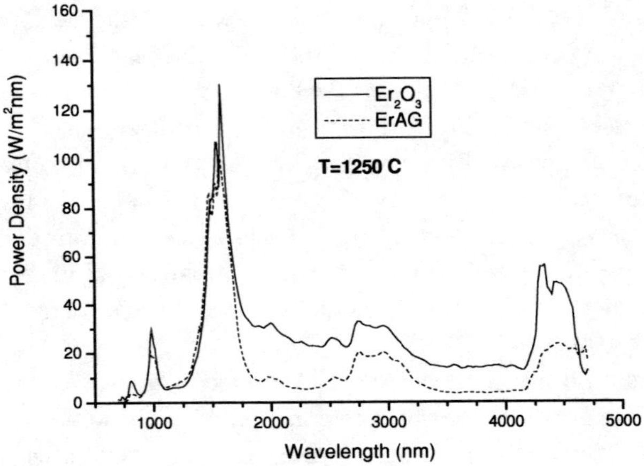

FIGURE 3. Spectral power density emitted by Erbium Oxide and Erbium Aluminium Garnet samples at 1250°C.

From the experimental point of view it is convenient to introduce the notion of partial efficiency (1) (Fig.4) to measure the ratio between the in-band and the off-band power emission in the wavelength interval of analysis which, in our case, ranges from 800 to 4700 nm.

$$Part\ Eff = \frac{\int_{1360nm}^{2000nm} PowDen_i[\lambda]d\lambda}{\int_{1360nm}^{4700nm} PowDen_i[\lambda]d\lambda} \qquad i = Er_2O_3, ErAG \qquad (1)$$

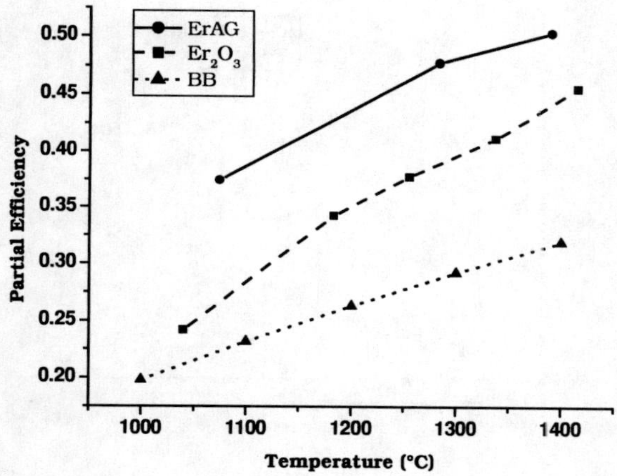

FIGURE 4. Erbium Aluminium Garnet, Erbium Oxide and blackbody partial efficiencies versus temperature.

Erbium Aluminium Garnet and Erbium Oxide exhibits a comparable peak emission but the garnet has a much higher partial efficiency.

These experimental results are in agreement with the theory proposed by Torsello [10] which predicts that Erbium Aluminium Garnet should be more efficient as a selective emitter than Erbium Oxide where non radiative deexcitation channels dominate over radiative deexcitation.

A slurry technique was used to apply an ErAG coating on the external surface of a SiC cylinder to be used in a TPV burner (Fig.5). The thickness of the coating was about 200 microns and its porosity was quite high (~50%).

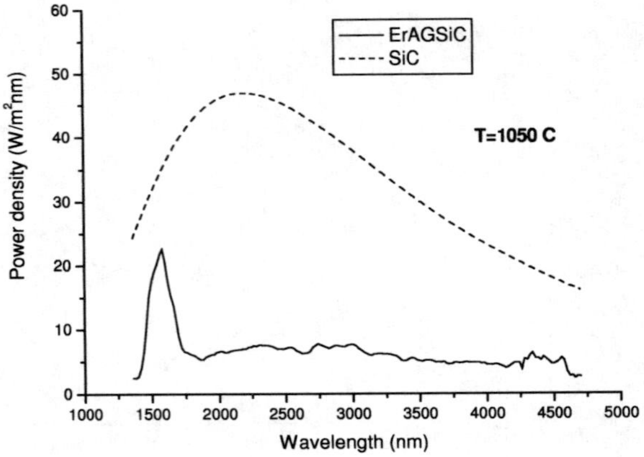

FIGURE 5. Spectra of SiC and SiC covered by a ErAG coating at 1050°C.

Due to its porosity the coating could match the difference in thermal expansion coefficient with the substrate without the development of cracks or delamination. Fig.5 shows that the use of the coating allows to drastically change the emissivity properties of the substrate. In particular the power emitted at wavelengths outside the region of optimum spectral response of the photovoltaic cells was strongly reduced.

Even if the coatings do not absorb the radiation from the substrate, the high porosity and the thickness of the ErAG layer enhances the radiation scattering which suppresses almost entirely the contribution of the SiC substrate to the surface emissivity.

CONCLUSIONS

We have studied the selective emission properties of a variety of mixed oxide ceramics containing rare earths. Amorphous Silica exhibits a large in-band emissivity even at low RE doping concentration, Alumina based garnets described by the general formula $Er_xY_{3-x}Al_5O_{10}$, $(0 < x < 3)$ are characterised by a very low emissivity in a very broad range outside the Er absorption band at 1500nm. We have also investigated bulk and surface phenomena by comparing polycrystalline porous ceramics with transparent amorphous Silica and single crystals with similar chemical composition and crystalline structure. Depending on the physical properties of the host matrix i.e. porosity and crystalline structure, a minimum amount of Er incorporated in a YAG structure is necessary for the in-band power density to outbalance the off-band emission.

The thermo-structural properties of bulk garnet ceramics up to 1600°C are quite good in terms of refractoriness, creep resistance and mechanical resistance, but the high coefficient of thermal expansion and low thermal conductivity limit the resistance to thermal shocks.

As a practical and effective alternative to the use of bulk oxide ceramics we have showed that thick porous films can be used as coating of bulk SiC combustors. Bulk SiC, due to the high thermal conductivity and lower coefficient of thermal expansion has excellent thermal shock resistance and unlimited lifetime. The combined system, porous coating of ErAG on SiC combustor, looks therefore very efficient, practical and economical solution.

ACKNOWLEDGMENTS

This research was funded by the European Commission through the project THE REV of the 5th Framework Programme of Research and Development, and by the Italian Space Agency.

REFERENCES

1. Guazzoni G.E., Appl. Spect., 1972, 26, (1), pp.60-65.
2. Nelson R.E.: AIP Conference Proceedings, 1995, 321, pp.80-95.
3. Adair P.L. and Rose M.F.: AIP Conference Proceeding, 1995, 321, pp.245-262.
4. L.G.Ferguson and F.Dogan: J.of Mat. Sci., 2001, 36, pp.137-146.
5. Lowe R.A., Chubb D.L., Serene C.F. and Good B.S.: Appl. Phys. Lett., 1994, 64, (26), 3551-3.
6. Sova R.M., Linevsky M.J., Thomas M.E., Mark F.F., Infrared Physics & Technology 39, 1998, pp.251-261.
7. G. De, A. Licciulli, M. Nacucchi, J. of Non-Crystalline Solids 201(1996) 153-158
8. A. Licciulli, A. Maffezzoli, G. Torsello, D. Diso, S. Tundo, M. Rella, M. Mazzer, Journal of Sol-Gel Science and Technology, 26, (2003)
9. Neuer G. and Jaroma-Weiland G., Paper presented at the Thirteenth symposium on thermophysical properties, June 22-27, Boulder, Colorado, USA, 1997.
10 G. Torsello, M. Lomascolo, A. Bianco, A. Licciulli, S. Tundo, D. Diso, M. Mazzer, This conference

Theoretical Comparison of Erbium, Holmium and Thulium Aluminum Garnet Selective Emitters

Brian S. Good and Donald L. Chubb

NASA Glenn Research Center, 21000 Brookpark Road, Cleveland, OH 44135

Abstract. Spectral control through the use of selective emitters is an important means of improving the efficiency of thermophotovoltaic (TPV) systems. The availability of PV cells having a wide range of bandgap energies, along with the development of selective emitters having a range of emission band energies, gives the developer of TPV power systems a wide range of options. The rare earth aluminum garnet selective emitters developed in our laboratory offer a number of advantages when used in such systems [1-3]. In this paper, we present results of a detailed computational study of the effects of a number of design and operating parameters, including emitter film thickness, scattering within the film, and the temperature gradient across the film thickness, on the performance of three rare-earth aluminum garnet selective emitters. Specifically, we present emitter efficiency, net emittance, wavelength-dependent emittance, total output power, and the useful output power, for $Er_3Al_5O_{12}$, $Ho_3Al_5O_{12}$, and $Tm_3Al_5O_{12}$ emitters, as functions of the above parameters.

INTRODUCTION

It has long been recognized that spectral control through the use of selective emitters is an effective means of improving the performance of thermophotovoltaic power systems [4]. The rare-earth aluminum garnet selective emitters developed in our laboratory offer a number of advantages when used in such systems [5]. One of the keys to obtaining optimal performance in a selective emitter TPV system is the matching of the energy of the strong, narrow emission band of the selective emitter with the bandgap energy of the photovoltaic cells used in the system. While it is possible to specify the cell bandgap energy over a wide range by varying the cell structure and the composition of the cell material, the development of a variety of selective emitters having a range of emission band energies offers the system designer additional flexibility.

In this paper, we present the results of a detailed computational study of the effects of a number of design and operating parameters, including emitter film thickness, scattering within the film, and the temperature gradient across the film thickness, on the performance of rare-earth aluminum garnet ($Er_3Al_5O_{12}$, $Ho_3Al_5O_{12}$, or $Tm_3Al_5O_{12}$) selective emitters. It is found that the behavior of the emitters can be understood by considering the effects of the study's parameters on the temperature profiles across the emitter thickness.

CP653, *Thermophotovoltaic Generation of Electricity: 5th Conference*
edited by T. J. Coutts, G. Guazzoni, and J. Luther
2003 American Institute of Physics 0-7354-0113-6

FILM EMITTANCE MODEL

The film geometry and fundamental quantities used in our model are displayed in Fig. 1. A thermal flux Q_{in}, consisting of conductive and radiative components, passes from the fixed-temperature substrate into the film. The internal fluxes, also incorporating radiative and conductive components, are characterized by the positive and negative-going intensities i^{+} and i^{-}. The detailed form of the output flux Q_{out} at the emitting surface depends on whether the vacuum interface or lossy boundary condition (discussed below) is used. It is assumed that the problem is one-dimensional, that is, the emitter film thickness is small compared to its lateral dimensions, and edge effects are ignored. Scattering within the film, where included, is assumed to be isotropic. Reflection at both boundaries is taken to be diffuse. The film's index of refraction and the extinction coefficient are assumed to be independent of position within the film, but dependent on wavelength. The film's environment, i.e. the thermal cavity in a TPV system, is ignored; no radiative flux impinges on the emitter from the environment.

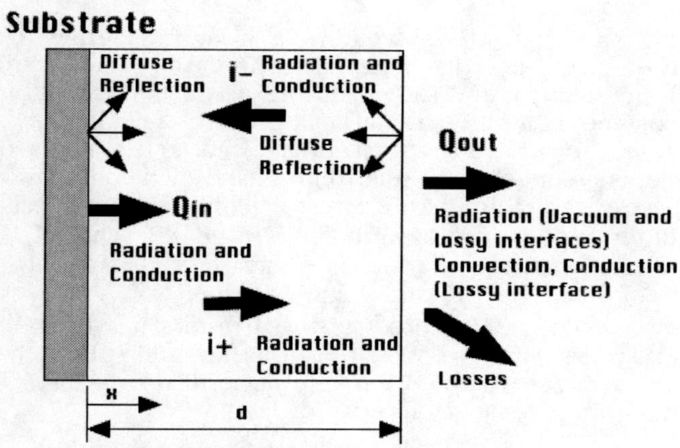

FIGURE 1. Model system and parameters

THEORETICAL DEVELOPMENT

Equations for the net heat flux, incorporating radiative and conductive contributions, have been derived and presented elsewhere [6]; only a brief outline will be presented here. General expressions for the positive- and negative-going radiative intensities (as shown in Fig. 1) are obtained following Siegel and Howell [7]. These fluxes are expressed in terms of a radiative source function. The flux expressions can be evaluated at the film boundaries to yield $i^{+}(d)$ and $i^{-}(0)$, and the boundary conditions applied to relate $i^{-}(d)$ and $i^{+}(0)$. These expressions can be simplified by assuming that the boundaries are diffuse, and that the source function is angle-independent. The wavelength-dependent fluxes are obtained by integrating the intensities over all angles (at the film-substrate interface), or over a range of angles consistent with Snell's law (at the film-vacuum interface). The total wavelength-dependent radiative flux at the radiating surface is the signed sum of the positive- and negative-going fluxes. Energy conservation within the film requires that, in the steady state, the sum of the wavelength-integrated radiative flux and the conductive flux at each point within the film is a constant, equal to the output power at the outer film surface plus losses, if any.

Input data for the model include the extinction coefficient and film index of refraction, which are obtained from experimental measurements performed in our laboratory. The substrate index of refraction is obtained from the literature, using data for platinum. The default film thickness is 0.05 cm, and the film thermal conductivity is taken to be 0.01 W/cm/K for all three materials. The wavelength corresponding to the PV cell bandgap energy is assumed to be 1650 nm for the $Er_3Al_5O_{12}$ emitter, 1900 nm for the $Tm_3Al_5O_{12}$ emitter, and 2150 nm for the $Ho_3Al_5O_{12}$ film. Scattering is incorporated by assuming that the scattering coefficient is a constant fraction of the extinction coefficient, and substrate temperatures of 1200 K and 2000 K are used.

One of two boundary conditions is used. Under vacuum interface conditions, the temperature at the radiating surface is computed self-consistently, incorporating the requirement that the conductive component of the flux (and hence the temperature gradient) be zero at the surface, which is assumed to be in contact with a vacuum.

On the other hand, when lossy conditions exist, the surface temperature is specified in advance, and held fixed. In such cases, the temperature gradient at the surface will be nonzero, and the conditions at the surface will not be consistent with the existence of a purely radiative flux there. The difference between the total flux reaching the surface from within the film, and the flux radiated away from the surface, is assumed to be due to an unspecified combination of conductive and convective losses. In the following calculations, we specify the surface temperatures in terms of the average gradient $(T_s - T_r)/T_s$, where T_s and T_r are the temperatures at the substrate-film interface and the radiating face, respectively. Gradient values of 0.05, 0.1, 0.15 and 0.2 are used.

The computer code used to produce these results has been described elsewhere [3]. Because the optical properties of the film and substrate materials are wavelength-dependent, all computations are performed over a wavelength range of 1-10000 nm, with a wavelength interval of 1 nm.

RESULTS AND DISCUSSION

We have investigated the effects on emitter performance of the choice of emitter film material, the temperature gradient across the film thickness, the emitter film thickness, and the degree of scattering within the film.

Emitter performance is characterized by the wavelength-dependent spectral emittance, the total output power, the "useful" power (that portion of the emitter's total output power radiated at energies above the bandgap energy of the PV cell, i.e. the portion of the output power that is available for conversion by the PV cell), the efficiency, defined as the ratio of the useful to total power, and the wavelength-independent total emittance.

Because the intensity of the radiation emitted by the film is proportional to T^4, even relatively small variations in temperature across the film thickness may give rise to large differences in selective emitter performance.

The detailed behavior of the computed properties can be understood by considering the fact that the radiation emerging from the emitter can be broken down into three spectral regions: (1) the emission band, i.e. the narrow region of the spectrum over which the spectral emittance of the film is large; (2) the short-wavelength region, i.e. the region of the spectrum having wavelength smaller than the emission band, where (with the exception of some secondary narrow emission bands) the spectral emittance of the film is small, and the dominant emission is that of the substrate; (3) the long-wavelength region, where the spectral emittance may be large, but the blackbody function is small. The dominant temperature dependence arises from the emission band; while the power output in all three regions goes as T^4, the power from the short-wavelength region is dominated by the contribution from the substrate, which is held at a fixed temperature. In the long wavelength region, the temperature dependence is strong, but the absolute value of the power from that region of the spectrum is small.

Temperature Profiles

Temperature profiles for the three materials are displayed in Figs. 2a-c, for substrate temperatures of 1200 K and 2000 K, and for gradients of 0.05-0.2, along with vacuum interface conditions. It should be noted that the vacuum interface

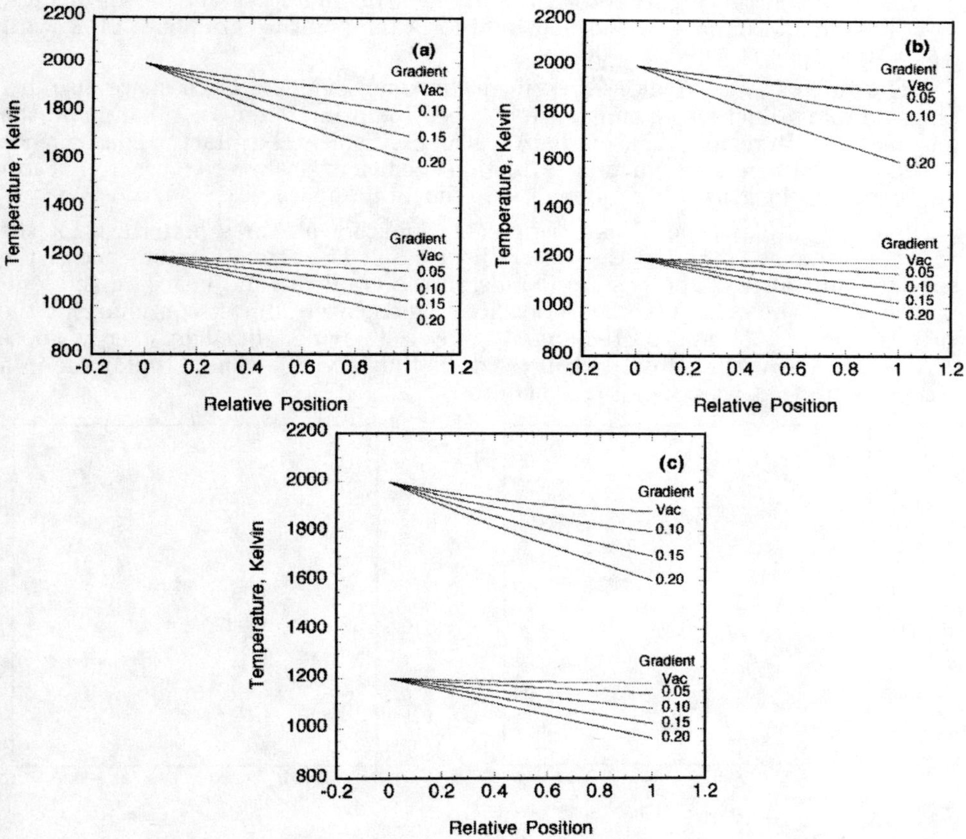

FIGURE 2. Emitter Temperature Profiles. (a) $Er_3Al_5O_{12}$ (b) $Ho_3Al_5O_{12}$ (c) $Tm_3Al_5O_{12}$

condition gives rise to the minimum possible average temperature gradient, so that in those cases where the vacuum interface condition gives rise to a gradient greater than 0.05, no results for that value of the gradient are reported.

It can be seen that in all cases, the temperature profiles are close to linear, with the degree of curvature being an increasing function of substrate temperature, a result of the competition between radiative and conductive heat transfer within the film. For lower substrate temperatures, internal conductive transfer completely dominates, while at larger substrate temperatures, the effects of radiative transfer are evident. For T_s = 1200 K, all temperature profiles are very close to linear, while for T_s = 2000 K, the profiles having small gradients exhibit a slight curvature. Given the T^4 temperature dependence of the radiative component of the flux, even a slight deviation from linearity may have an appreciable effect on the properties and on the performance of the emitter; modeling performed using these profiles may yield results quantitatively different from earlier modeling work that assumed a linear temperature profile [2].

Spectral Emittance

Spectral emittance curves for the three emitter materials are shown in Figs. 3a-c, for vacuum interface conditions and for a temperature gradient of 0.2, with a substrate temperature of 2000 K.

The three spectral regions described above are evident. Each curve also displays a rapidly fluctuating emittance between the primary emission band and the long-wavelength region. This oscillation is an experimental artifact, a consequence of the fact that the experimental extinction coefficient data were measured at low and high wavelengths using different experimental apparatus.

Fig. 3d displays the primary emission bands for all three materials, for the vacuum interface case, and the 0.2 gradient case. The strong dependence of the spectral emittance in the primary bands on the gradient (and hence on the temperature) can be clearly seen. These three bands cover, almost continuously, the spectrum from 1350 nm to 2150 nm. Any PV cell having a bandgap energy whose wavelength lies within this spectral region is thus a viable candidate for use in a TPV system that uses one of these emitters.

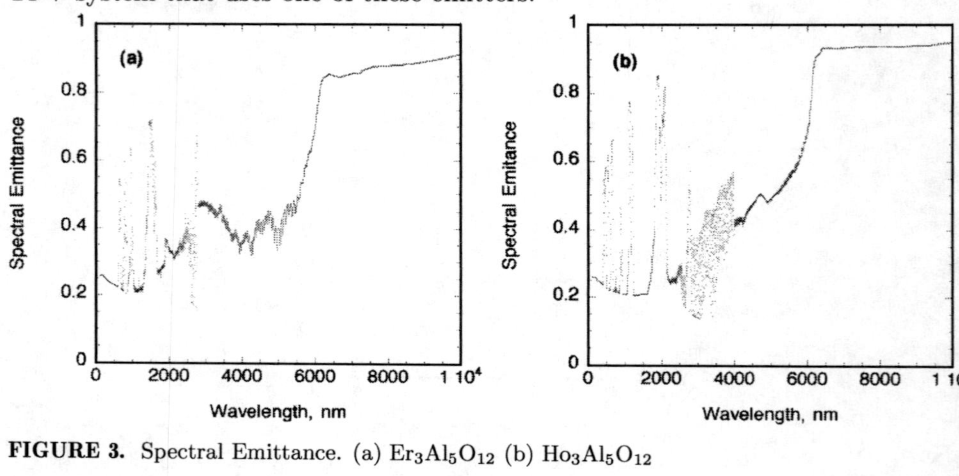

FIGURE 3. Spectral Emittance. (a) $Er_3Al_5O_{12}$ (b) $Ho_3Al_5O_{12}$

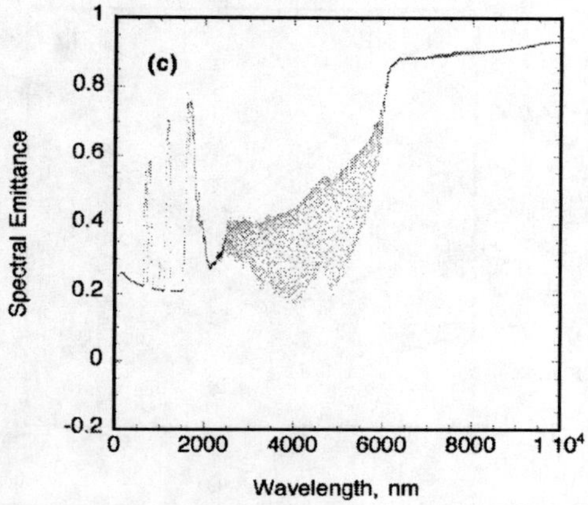

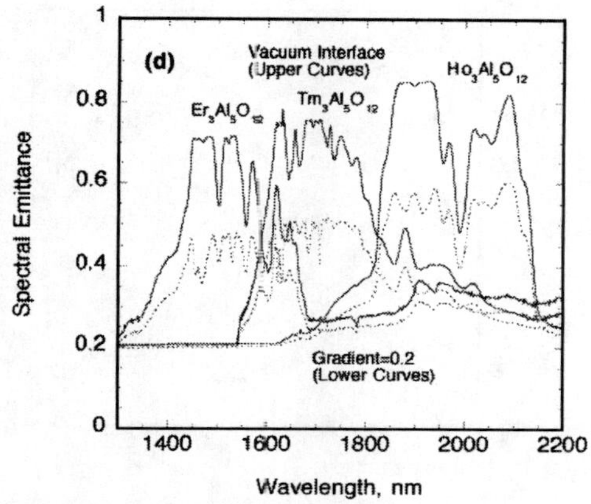

FIGURE 3. Spectral Emittance. (c) $Tm_3Al_5O_{12}$ (d) Primary emission bands

Temperature Gradient Effects

We next consider the effects of the average temperature gradient across the film thickness on emitter performance. The efficiency (Fig. 4) at each temperature is highest for the $Ho_3Al_5O_{12}$ emitter and lowest for the $Er_3Al_5O_{12}$ film. The efficiency decreases slightly with increasing average gradient for all three materials, which can be understood by considering separately the effects of the temperature gradient on the useful (Fig. 5) and total (Fig. 6) powers.

As discussed above, the useful power (Fig. 5) consists primarily of a strongly temperature- and gradient-dependent contribution from the narrow primary emission band of the film. On the other hand, the total power (Fig. 6) includes, in addition, short- and long-wavelength contributions that are either less temperature sensitive or small in magnitude. Because the efficiency is the ratio of useful to total power, the temperature dependence of the efficiency is dominated by that of the useful power. The temperature dependence of both the useful and total powers is greater for the T_s=2000 K substrate temperature.

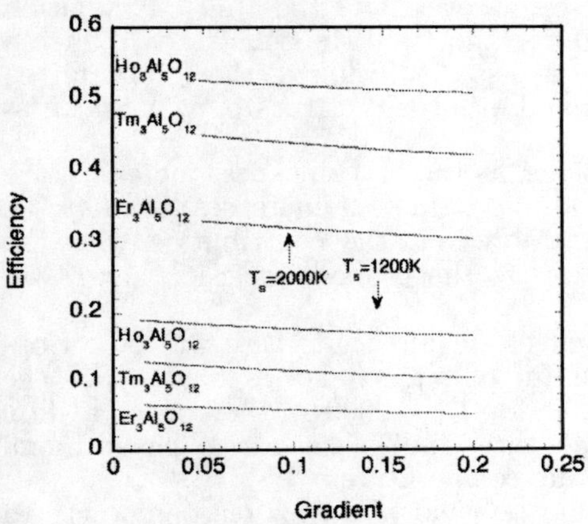

FIGURE 4. Emitter efficiency versus temperature gradient.

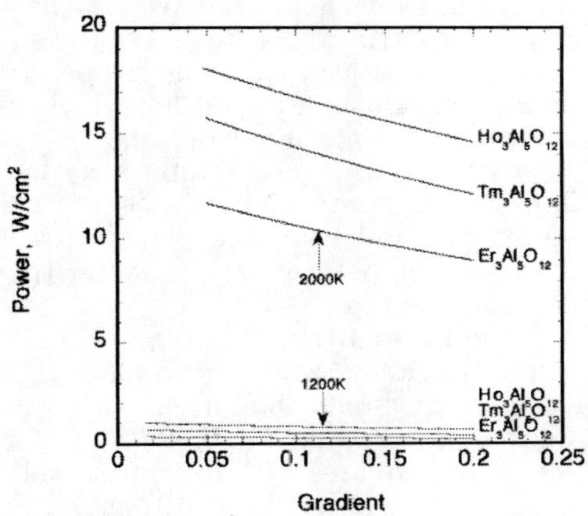

FIGURE 5. Useful power versus temperature gradient.

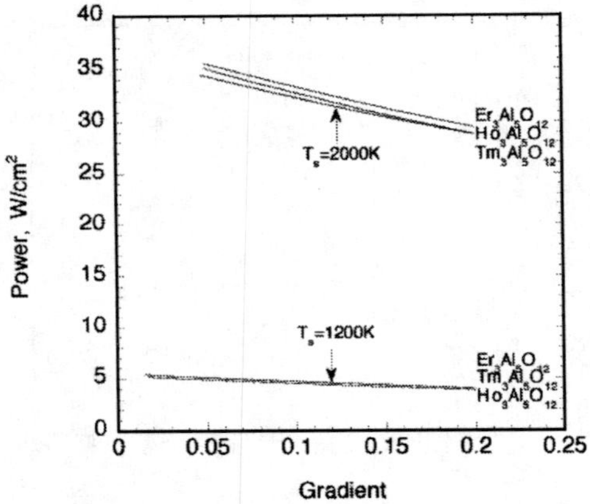

FIGURE 6. Total power versus temperature gradient.

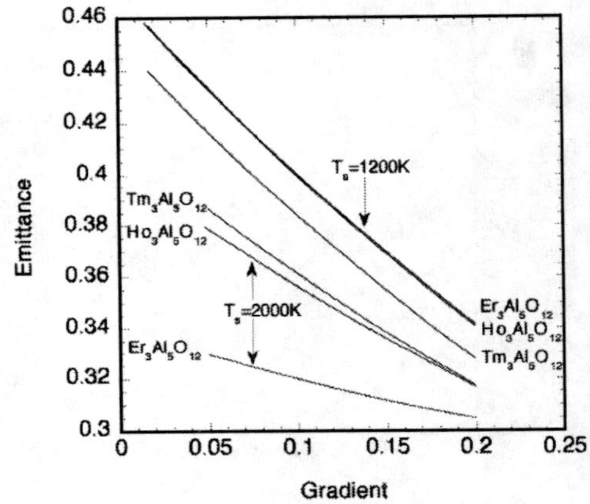

FIGURE 7. Total emittance versus temperature gradient.

The wavelength-independent total emittance (Fig. 7) shows a substantial decrease with increasing temperature gradient. The emittance is defined as the ratio of the total output power to σT_s^4. As the gradient increases, the temperature of an increasing portion of the film is substantially lower in temperature than the fixed-temperature substrate, so the emittance decreases.

Effects of Film Thickness

Vacuum interface temperature profiles, normalized to film thickness, are presented for each emitter material as a function of film thickness (in cm) in Figs. 8a-f. For a given substrate temperature, as the film thickness increases, the surface temperature decreases. For each film thickness, the surface temperature is highest for $Ho_3Al_5O_{12}$, and lowest for $Tm_3Al_5O_{12}$.

If the emitter film were opaque, the radiation emitted would be strictly a surface phenomenon. This is not the case, however, and the emitted radiation includes contributions from the film's interior and from the substrate. If such a film were held at a constant temperature across its thickness, the total emitted power would increase with film thickness and eventually saturate, at a thickness determined by the film's opacity.

However, for vacuum interface conditions, as the film thickness increases, the film surface temperature drops and the average gradient increases, so that the power radiated from the film decreases. Moreover, the contribution with the strongest temperature dependence, that due to the primary emission band, will decrease the most.

The competition between these two effects means that there will be an optimum thickness which maximizes the useful power, as shown in Fig. 9. The maxima are broad, but it can be seen that for T_s = 2000 K they occur at film thicknesses of around 0.03-0.07 cm for all three film materials. The optimum thickness increases with decreasing substrate temperature.

The total power (Fig. 10) shows similar behavior, but, for reasons described above, its temperature sensitivity is less than that of the useful power. The optimum in each case occurs at a greater thickness than for the useful power.

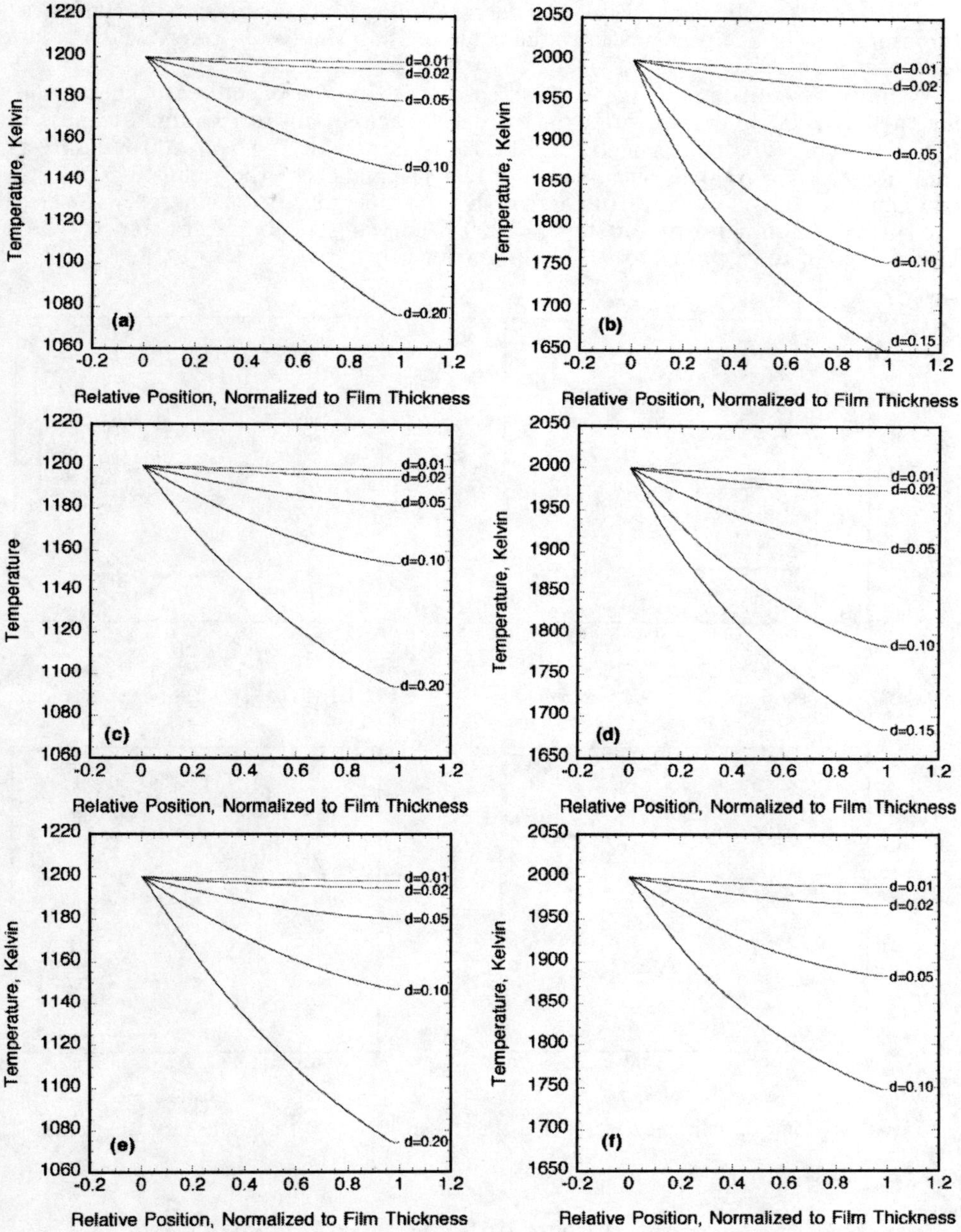

FIGURE 8. Temperature versus film thickness. (a) $Er_3Al_5O_{12}, T_s = 1200K$ (b) $Er_3Al_5O_{12}, T_s = 2000K$ (c) $Ho_3Al_5O_{12}, T_s = 1200K$ (d) $Ho_3Al_5O_{12}, T_s = 2000K$ (e) $Tm_3Al_5O_{12}, T_s = 1200K$ (f) $Tm_3Al_5O_{12}, T_s = 2000K$

The emitter efficiency (Fig. 11) decreases with increasing film thickness, as the useful power decreases with gradient faster than the total power, which is less temperature sensitive.

The total emittance (Fig. 12) increases with increasing film thickness, and, for the T_s=2000 K curves, appears to reach a maximum for a film thickness of about 0.1 cm. The emittance is substantially larger for the T_s=1200 K emitters than for the T_s=2000 K ones. This is a consequence of the definition of total emittance used; the temperature across the film thickness is considerably closer to the substrate temperature for the T_s=1200 K curves, so that the emittance, which is the ratio of total power to σT_s^4, is larger for these cases.

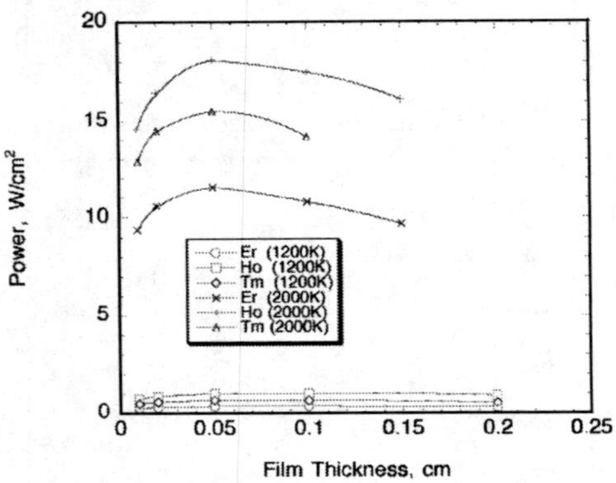

FIGURE 9. Useful power versus film thickness.

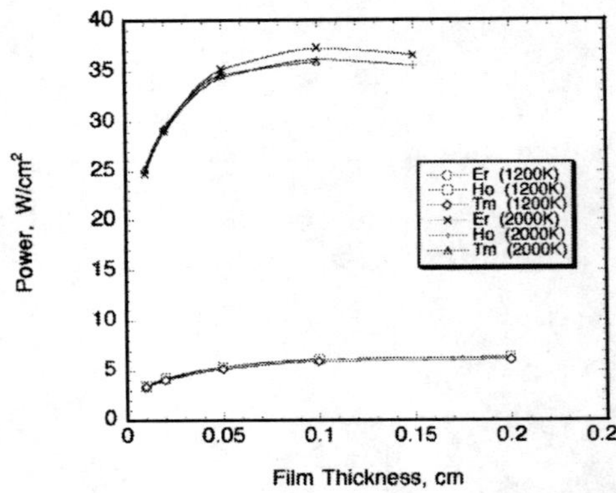

FIGURE 10. Total power versus film thickness.

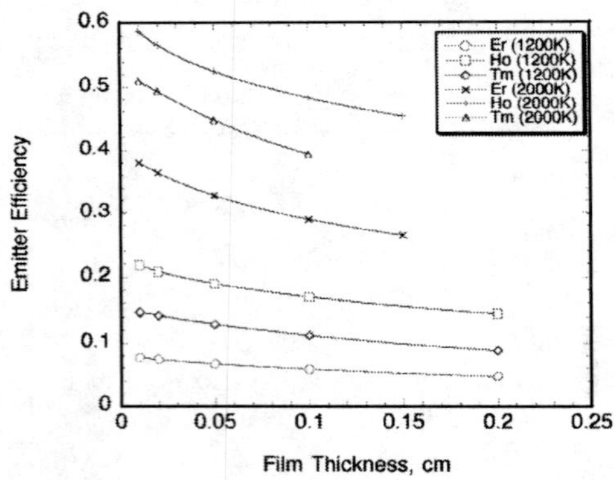

FIGURE 11. Emitter efficiency versus film thickness.

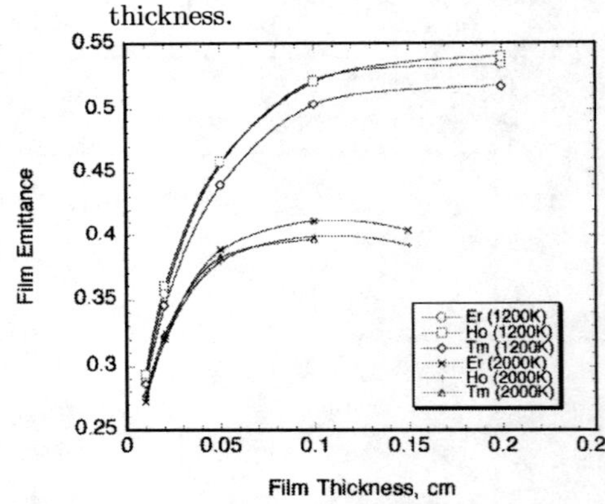

FIGURE 12. Emittance versus film thickness.

Effects of Scattering

Finally, we consider the effects of changes in the scattering coefficient on emitter performance. We consider values of the scattering coefficient (expressed as a fraction of the extinction coefficient) of 0.0 to 0.9, using the vacuum interface boundary condition. For a given substrate temperature, increasing the scattering

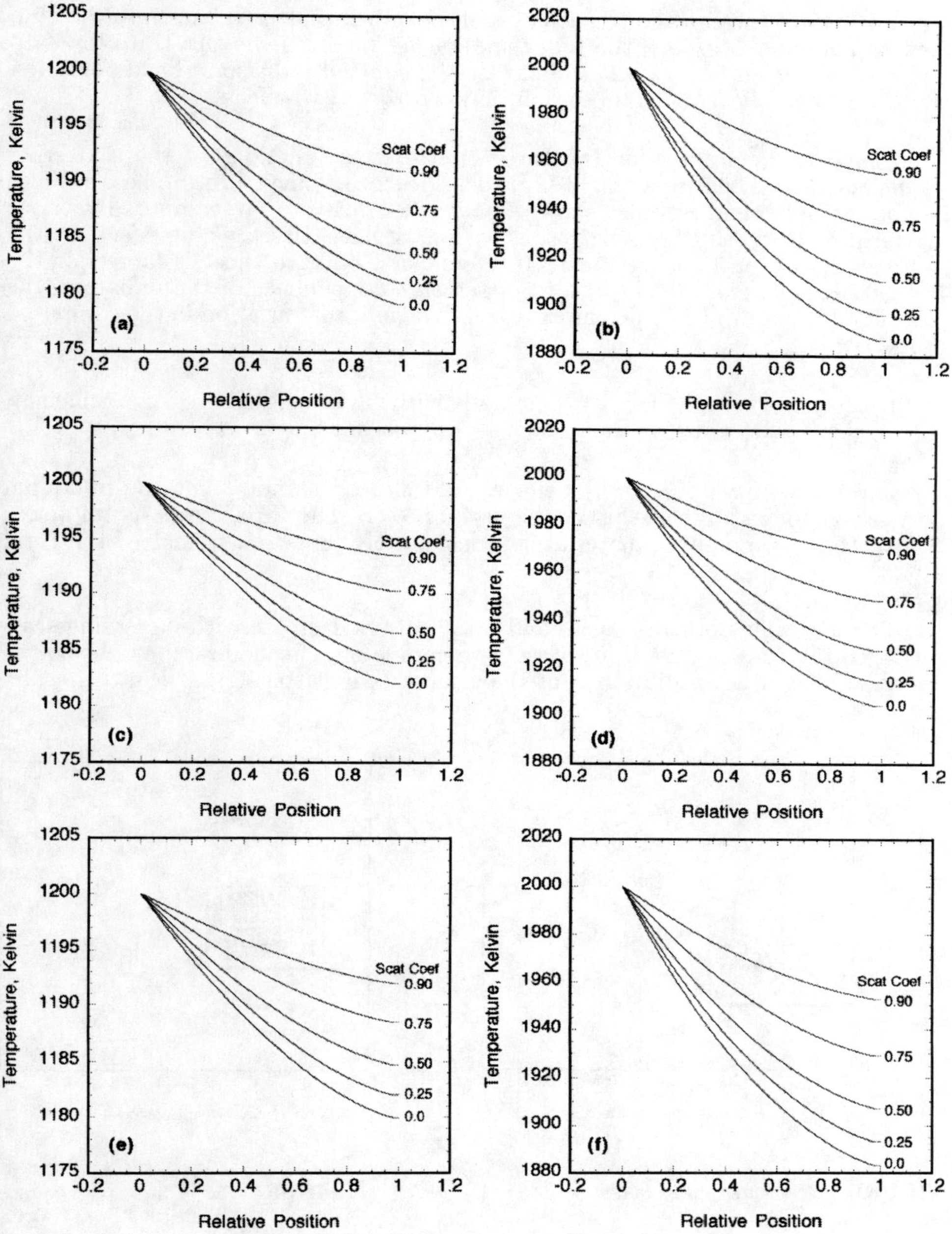

FIGURE 13. Temperature versus scattering. (a) $Er_3Al_5O_{12}, T_s = 1200K$ (b) $Er_3Al_5O_{12}$, $T_s = 2000K$ (c) $Ho_3Al_5O_{12}, T_s = 1200K$ (d) $Ho_3Al_5O_{12}, T_s = 2000K$ (e) $Tm_3Al_5O_{12}$, $T_s = 1200K$ (f) $Tm_3Al_5O_{12}$, $T_s = 2000K$

tends to reduce the emitted radiation, as more of it is scattered back into the film. As this happens, however, the film temperature far from the substrate tends to increase, which, in turn, tends to increase the emitted radiation. The actual flux emitted is a result of the competition between these two effects.

Temperature profiles for the three emitters, as functions of the scattering coefficient, are shown in Figs. 13a-f. The degree of curvature in these profiles is more evident than in those shown in Figs 2a-c, which were computed without scattering. Increasing the scattering coefficient has only a slight effect on the surface temperature for the T_s=1200 K emitters, but the effect is larger for the T_s=2000 K emitters. Given the T^4 temperature dependence of the radiation, the variation in scattering coefficient can cause a significant variation in the computed properties.

The useful power (Fig. 14) decreases with increasing scattering coefficient, with a sharper decrease as the coefficient approaches unity.

The total power (Fig. 15) behaves in a similar manner, but the fractional decrease is not as large as that of the useful power. The variation in useful power among the three emitter materials is considerably larger than that of the total power.

The fact that both the useful and total powers decrease with increasing scattering coefficient suggests that, of the two competing effects described above, the decrease in emitted radiation caused by increased scattering back into the film dominates.

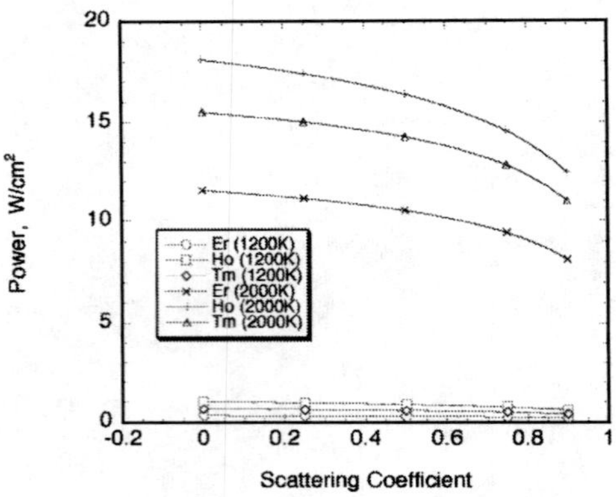

FIGURE 14. Useful power versus scattering.

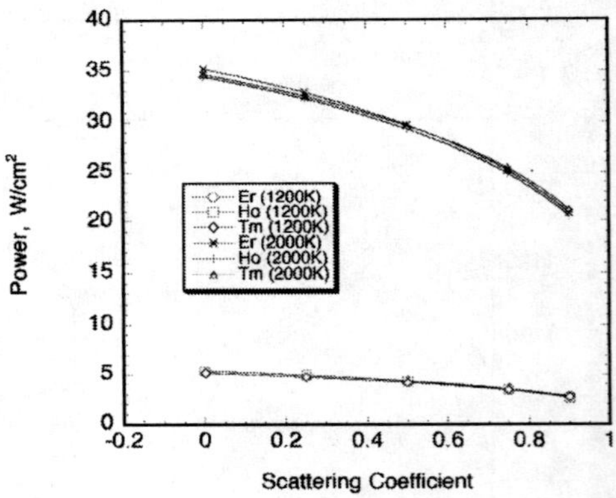

FIGURE 15. Total power versus scattering.

The efficiency (Fig. 16) increases slightly with increasing scattering for all three materials. This is a manifestation of the relative decreases in useful and total power. The efficiency is highest for $Ho_3Al_5O_{12}$ and lowest for $Er_3Al_5O_{12}$.

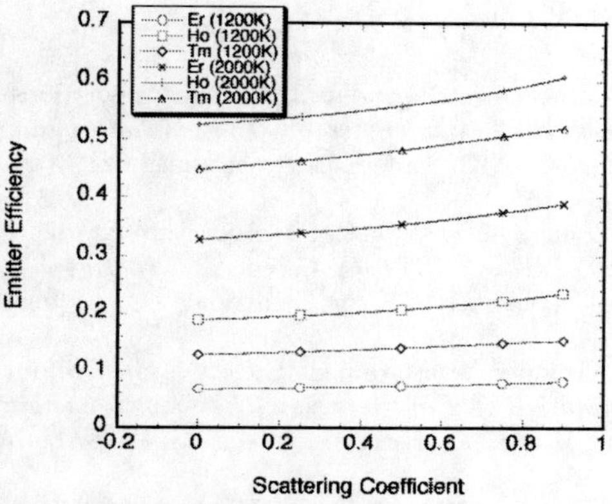

FIGURE 16. Emitter efficiency versus scattering.

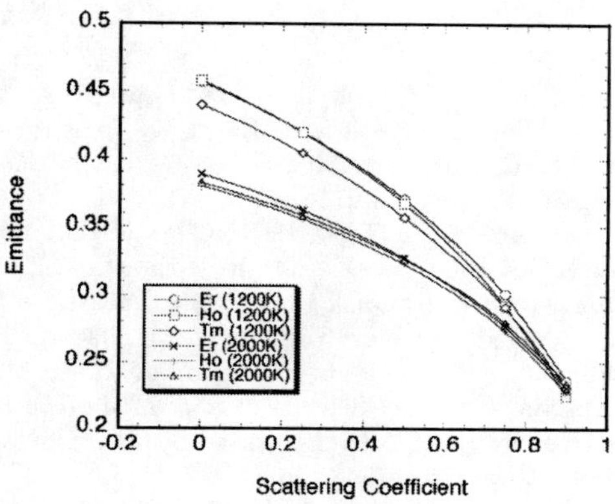

FIGURE 17. Emittance versus scattering.

The total emittance (Fig. 17) decreases sharply with increasing scattering coefficient. This behavior is a consequence of the dominance of the effects of backscattering over those of the corresponding temperature increase across the film thickness. The emittances of the T_s=1200K films are larger than those of the T_s=2000K films, as, again, the temperature across the 1200K films are closer to the substrate temperature than are those across the 2000K films. This difference essentially disappears at large values of the scattering coefficient, where the temperature profiles are close to uniform.

CONCLUSIONS

We have presented results of a study of the calculated performance of $Er_3Al_5O_{12}$, $Tm_3Al_5O_{12}$ and $Ho_3Al_5O_{12}$ selective emitters, as functions of several material and film properties.

For the range of properties considered, we have obtained both linear temperature profiles and profiles that show appreciable curvature. Simple performance models based on the assumption of a linear profile are a reasonable first approximation; however, because the useful power output depends strongly on the temperature, if specific and detailed information about emitter performance is required, more rigorous calculations are appropriate.

The useful power is a decreasing function of temperature gradient and scattering coefficient, and exhibits a maximum at a value of film thickness that depends on substrate temperature and film material. The total power exhibits similar behavior, but its dependence on temperature and on parameters that affect the temperature (e.g. film thickness) is weaker than that of the useful power. The efficiency is a decreasing function of film thickness and gradient, and increases slightly with increasing scattering coefficient. The emittance is a decreasing function of substrate temperature, temperature gradient and scattering coefficient, and has a dependence on film thickness that is qualitatively similar to that of the total power.

REFERENCES

1. Good, B. S., Chubb, D. L., and Lowe, R. A., "Temperature-Dependent Efficiency Calculations for a Thin-Film Selective Emitter," in *The First NREL Conference on Thermophotovoltaic Generation of Electricity,* edited by T. J. Coutts et al., AIP Conference Proceedings 321, New York, 1994, pp 263-275.
2. Chubb, D. L., Good, B. S., Clark, E. B., and Zheng Chen, "Effect of Temperature Gradient on Thick Film Selective Emitter Emittance," in *The Third NREL Conference on Thermophotovoltaic Generation of Electricity,* edited by T. J. Coutts et al., AIP Conference Proceedings 401, Woodbury, New York, 1997, pp 293-313.
3. Good, B. S., Chubb, D. L. and Pal, A. M., "Temperature Gradient Effects in an Erbium Aluminum Garnet Selective Emitter," in *The Fourth NREL Conference on Thermophotovoltaic Generation of Electricity*, edited by T. J. Coutts et al., AIP Conference Proceedings 460, Woodbury, New York, 1998, pp 214-223.
4. Guazzioni, G. E. and Kittl, E.,"Cylindrical Erbium Oxide Structures for Thermophotovoltaic Generators," in *Research and Development Technical Report ECOM 4249*, U.S. Army Electronics Command, 1974.
5. Chubb, D. L. and Lowe, R. A., *J. Appl. Phys.* **74**, 5687 (1993).
6. Chubb, D. L., NASA Technical Memorandum, in press.
7. Siegel, R., and Howell, J. R., *Thermal Radiation Heat Transfer,* Hemisphere Publishing, Washington, 1981, Ch. 14.

Spectrally Selective Emitters with Deep Rectangular Cavities Fabricated with Fast Atom Beam Etching

Hitoshi Sai, Yoshiaki Kanamori, and Hiroo Yugami

Graduate school of Engineering, Tohoku University
Aoba01, Aramaki, Aoba-ku, Sendai, 980-8579, Japan

Abstract. Spectral emittance and thermal stability of two-dimensional W surface gratings are investigated to obtain high-temperature resistive selective emitters for TPV applications. Numerical calculations based on rigorous coupled-wave analysis are performed to determine the structural profile of surface gratings. According to the determined parameters, W gratings composed of rectangular microcavities with the period of 1.0 μm are fabricated on single crystalline and polycrystalline W substrates by means of fast atom beam etching technique. The grating shows a strong emission peak which can be explained by the confined modes inside the cavities. The grating made from a single crystalline W shows very high thermal stability up to 1400K, while the polycrystalline grating is deformed at a high temperature because of the grain growth.

INTRODUCTION

With recent developments in micromachining techniques, many researchers have been actively studying on thermal radiation control by means of periodic microstructures, for example, one-dimensional (1D) [1,2] and 2D surface gratings [3-6], and 3D photonic crystals [7,8], to realize ideal spectral emittance or very efficient heat transfer. Greffet et al. have reported on thermal emittance from 1D surface gratings on a SiC substrate, and showed sharp emittance peaks due to the surface phonon polariton modulated by the grating in the infrared (IR) region [1,2]. Hienzel et al. [3] and Sai et al. [4] have fabricated 2D metallic grating to develop selective emitters, and demonstrated strong emittance peaks originated from surface plasmon polariton supported by the gratings in the near IR (NIR) region. The research of Heinzel et al. has been developed in relatively deep gratings by Ferber et al [5]. As another report on 2D gratings, Maruyama et al. [6] has investigated thermal radiation from 2D gratings composed of deep microcavities in the IR wavelengths and discussed the effect of confined modes inside the cavities. On the other hand, Lin et al. [7] and Fleming et al. [8] have reported on thermal emission from 3D photonic crystals with woodpile structure and observed emittance decrease attributed to the photonic bandgap.

Control of thermal radiation by means of periodic microstructures has several advantages such as adjustability of design, applicability to various materials, and so on. All above-mentioned reports have demonstrated that spectral emittance can be

CP653, *Thermophotovoltaic Generation of Electricity: 5th Conference*
edited by T. J. Coutts, G. Guazzoni, and J. Luther

controlled by varying the profile of microstructures and materials. Naturally, anyone expects to apply this method to high-temperature applications where thermal radiation mostly dominates heat transfer phenomena. Thermophotovoltaics (TPV) [9] is a typical application for this technique. Actually, several groups have researched surface gratings intending to develop selective emitters for TPV applications [3-5].

However, thermal stability of such microstructures is supposed to be lower than that of bulk materials, and therefore it is important to investigate the thermal stability of microstructures at high temperatures. Some researchers have fabricated microstructured radiators with refractory metals aiming at high thermal resistibility [3,5,8], but the detailed feature has not been given. In the past publications, the maximum temperature at which thermal emittance is measured has been limited at 1200K [3] as far as authors' knowledge. Since emitters are operated at over 1300K (1000°C) in ordinary TPV systems, further investigation and development are needed for practical utilization.

The objective of this research is to develop spectrally selective emitters with high thermal stability enough to be used in TPV applications. We have conducted numerical calculations based on rigorous coupled-wave analysis (RCWA) [10] to obtain a reasonable emitter design. Based on the calculated results, we have fabricated 2D surface gratings on single and polycrystalline W substrates by means of electron beam (EB) lithography and fast atom beam (FAB) etching technique [11]. Spectral emittance of the samples has been measured at high temperatures, and evaluated their thermal resistivity.

NUMERICAL CALCULATIONS

Surface Design of Selective Emitters

In TPV systems, it is necessary that spectral feature of radiators is matched with spectral response of photovoltaic cells. For example, the sensitive spectral band of GaSb cells is about 0.8 to 1.8 μm, and this requires emitters to have a high emittance only within that band and low outside it. To design the grating profile which materializes such spectral features, we have performed RCWA calculations on W gratings varying structural parameters and the state of incident wave. Figure 1 is a schematic drawing of the calculation model of 2D W gratings with rectangular microcavities. All gratings are symmetric to the *x*- and *y*- axes. Parameters are defined as periodicity Λ, aperture size a, depth d, incident angle θ, azimuthal angle ϕ, polarization angle ψ, and wavelength λ. With use of optical constants of W at room temperature reported in the literature [12], we have simulated spectral absorptance of W gratings, and regarded it equal to emittance according to Kirchhoff's law.

Figure 2 (a) shows the calculated normal emittance spectra of 2D W gratings with rectangular cavities with three different a/Λ. Λ and d/a are fixed at 1.0 μm and 1.0, respectively. Spectral emittance of the W gratings increases drastically with increasing a/Λ in λ of 0.3 to 2.0 μm. In addition, high emittance region broadens

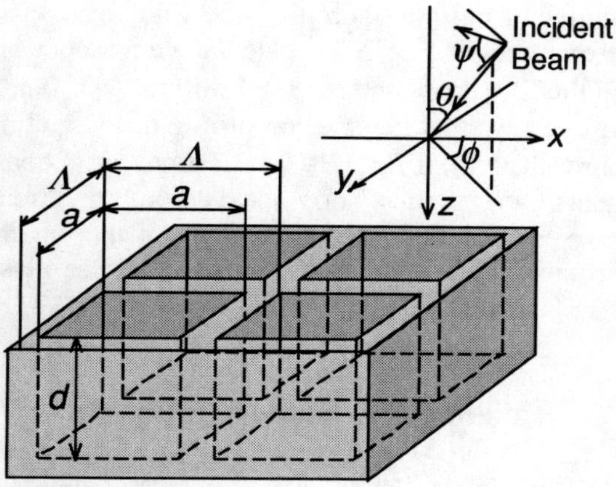

FIGURE 1. Schematic diagram of the calculation model of 2D surface gratings composed of rectangular parallelepiped microcavities.

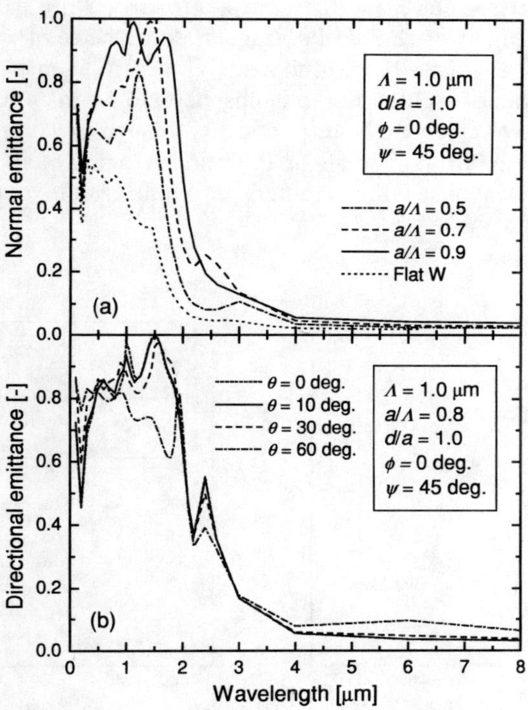

FIGURE 2. Calculated spectral emittance of 2D W gratings with rectangular microcavities. In all calculations, $\Lambda = 1.0\ \mu m$, $d/a = 1.0$, $\phi = 0°$ and $\psi = 45°$. (a), Normal emittance of W gratings with the different value of a. (b), incident angle dependence of spectral emittance of W grating with $a/\Lambda = 0.8$ and $d/a = 1.0$.

simultaneously. It is derived from the figure that a/Λ value of 0.8 to 0.9 is suitable to realize good selective emitters. In Fig. 2 (b), calculated emittance spectra are plotted as a function of θ with the fixed parameters, $\Lambda = 1.0$ μm, $a/\Lambda = 0.8$, $d/a = 1.0$, $\phi = 0°$, and $\psi = 45°$. This figure shows that the spectral profile of the grating doesn't depend on θ very much. This result is ideal for TPV applications. It is considered that these emittance enhancements are mainly originated from the resonance between electromagnetic waves and standing wave modes inside the cavities [5] instead of surface plasmon polaritons, because W cannot support surface plasmons strongly in these wavelengths.

Efficiency and Radiative Power

Based on the calculated spectra, we have estimated selective emission efficiency η_{se} and relative radiative power of microstructured W emitters. η_{se} is defined as the ratio of the radiative power emitted in the PV cell's sensitive wavelengths to the total radiative power. Relative radiative power is also defined as the ratio of the radiative power of selective emitters to that of an ideal blackbody emitter in that region. Here, the region is set to $\lambda = 0.8$ to 1.8 μm, which corresponds to the sensitive wavelengths of GaSb cells. Angular dependence of emittance is also a significant problem to evaluate them. In this study, we use hemispherical emittance, which is the mean value of emittance in a hemisphere, to estimate reasonable indices in TPV applications. We calculate hemispherical emittance by computing directional emittance from $\theta = 0°$ to 90° at intervals of 15°, and averaging them in solid angle. In all calculations, it is supposed that emitters' spectral feature doesn't depend on

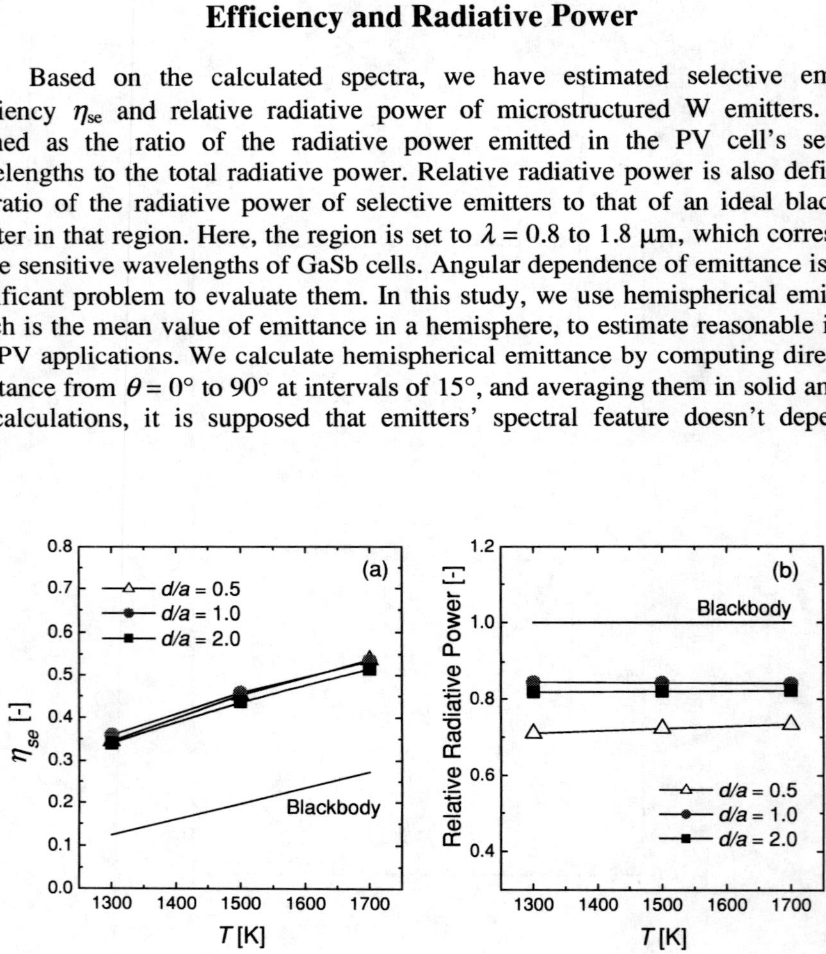

FIGURE 3. η_{se} (a) and relative radiation energy (b) of 2D W gratings with rectangular cavities as functions of temperature. All plotted data are calculated with the hemispherical emittance simulated under the condition of $\Lambda = 1.0$ μm and $a/\Lambda = 0.8$.

temperature.

Figure 3 (a) shows the η_{se} of microstructured W emitters for GaSb PV cells as a function of temperature at three different aspect ratio (*d*/*a*). Other structural parameters are fixed as $\Lambda = 1.0$ μm and $a/\Lambda = 0.8$. W emitters possess higher η_{se} than the blackbody, and it increases from 0.35 to 0.55 with temperature rising. It can be said from the figure that η_{se} does not depend on *d*/*a* significantly. Relative radiative power of the W emitters is plotted in Fig. 3 (b) in the same manner as Fig.3 (a). Relatively high energy density up to 0.85 is achieved. It is almost independent from temperature, but improves from 0.71 to 0.85 by increasing *d*/*a* from 0.5 to 1.0. However, further increasing of *d*/*a* is not effective in raising radiative power.

Although the spectral feature of W emitters has not been optimized for GaSb cells yet, we have chosen $\Lambda = 1.0$ μm, $a/\Lambda = 0.8$, and $d/a = 1.0$, which are the best solution obtained in these calculations, as parameters to fabricate samples.

EXPERIMENTAL

Sample Preparation

2D surface grating have been processed on W substrates by the following procedure. First, a thick resist layer, an Al thin film and an EB resist layer are stacked one after another on a mechanically polished W substrate. The top EB resist is

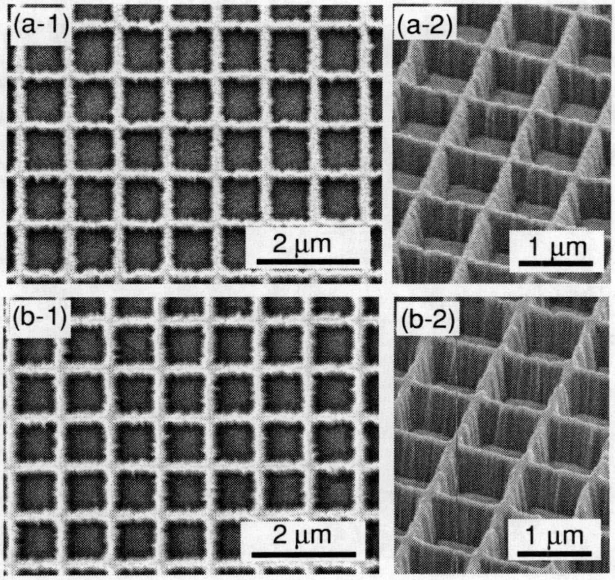

FIGURE 4. SEM images of the 2D W gratings: (a) single crystalline W grating (1: top view, 2: oblique view), (b) polycrystalline W grating (1: top view, 2: oblique view).

exposed by EB lithography system (ELIONICS, ELS-3700) and a grating pattern is drawn. After the development of the pattern, FAB etching with SF_6 gas is carried out with a FAB system (EBARA, FAB-60ML) to replicate it on the Al film. Since FAB is electrically neutral atomic or molecular beam, it is possible to obtain fine patterns with nanometer order without deformation of etching shape due to accumulated charge on samples. Next, FAB etching with O_2 gas is successively performed to replicate the pattern on the thick resist under the Al film. With this double layer method, we can obtain a deep resist mask. Finally, FAB etching with SF_6 is performed again and the grating pattern is transferred on the W substrate consuming the resist mask. The same procedure is used to fabricate mono- and polycrystalline W gratings.

Figures 4 (a) and (b) show scanning electron microscope (SEM) images of 2D W gratings with 1.0 μm periodicity processed on mono- and poly crystalline W substrates. It can be confirmed that 2D array of rectangular microcavities are fabricated on each substrate. a and d are estimated as 0.8 μm and 0.75 μm, respectively, through SEM observations. It is observed that the cross section of a cavity is not a perfect rectangular but a trapezoid whose base length of the bottom is a little bit shorter than that of the top. It is also observed that the walls of each cavity are uneven; however, the roughness is sufficiently smaller than Λ.

Experimental Setup

We have measured thermal emittance spectra of the samples with the emission measurement system equipped with an electric heater, a Fourier transform NIR spectrometer with an emission port (Perkin Elmer, GX2000), and some optical elements. The schematic drawing of the experimental setup is shown in Fig. 5. Thermal emission from a heated sample passes though a sapphire window and is focused on a pinhole, which shuts out stray lights, by a CaF_2 lens. After that, the emission beam is collimated by another CaF_2 lens, and analyzed by the FT-IR. The measured spectra are calibrated and given units by comparing with the emissive spectra of the materials whose emittance has been already known, for example, SiC.

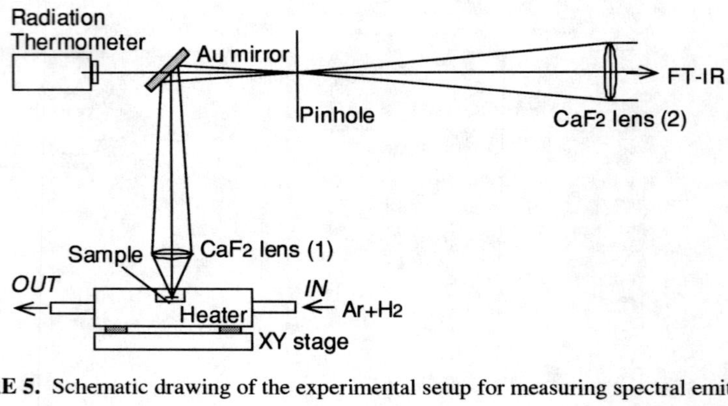

FIGURE 5. Schematic drawing of the experimental setup for measuring spectral emittance.

During heating samples, 5%H_2-Ar gas is flown inside the heater casing to prevent samples from oxidation. Surface temperature of samples is determined with a radiation thermometer by measuring the temperature of a flat W heated simultaneously with microstructured samples.

RESULTS AND DISCUSSION

Figure 6 shows the normal emittance spectra of the single crystalline W grating shown in Fig. 4 (a) at high temperatures. In the figure, measured emittance and literature data [12] of flat W are added as references. The emittance of the grating increases drastically in λ < 2.0 μm. In addition, a significant emission peak is appeared at about 1.3 μm, which can be explained by the standing wave modes inside cavities [5]. The spectral selectivity of the measured data is somewhat worse than that of the calculated shown in Fig.1. This may be caused by the difference of the cavity shape: smaller *d* and the tapered walls. Emittance measurement has been carried out varying temperature up to 1400K. The spectral feature of the single crystalline grating has slightly changed with increasing temperature, and this will be explained by temperature dependence of optical constants. On the other hand, that of the polycrystalline grating has been clearly aggravated in the point of spectral selectivity at over 1300K.

After a heating at 1400K for 1h, we have measured reflectance spectra on both samples. The results are plotted in Fig.7 with reflectance spectra measured before the heating. Before the heating, each sample has similar reflectance spectrum as shown in the figure, and shows strong absorption at about λ = 1.3 μm, which corresponds to the emission peak in Fig.6. After the heating, however, the reflectance of both samples is

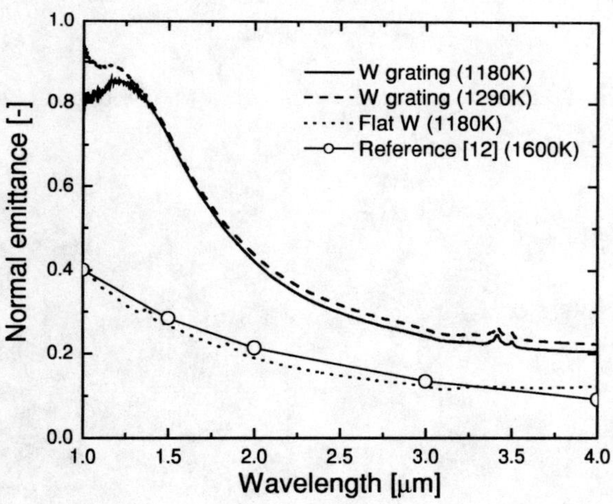

FIGURE 6. Measured spectral emittance of the 2D grating fabricated on a single crystalline W substrate shown in Fig. 4 (a) and a flat W measured at 1180K in a reductive atmosphere.

quite different; the single crystalline sample maintains its spectral feature, while that of the other becomes duller and no clear absorption peak is observed in the NIR region.

SEM observation has been conducted to check surface conditions of the samples after the heating. Figure 8 shows the SEM images of the W gratings after a heating at 1400K for 1h in 5%H_2-Ar atmosphere. The grating made from single crystalline W keeps its microstructure almost perfectly. However, a severe deformation and creation of voids are observed on the polycrystalline W grating, and this causes the degradation of spectral feature. Significant grain growth is also observed on the sample. Considering these results and the very high melting point of W, we regard the deformation of the microstructures on a polycrystalline W is mainly caused by the grain growth. In other words, high-temperature resistive microstructures can be produced by using single crystalline materials.

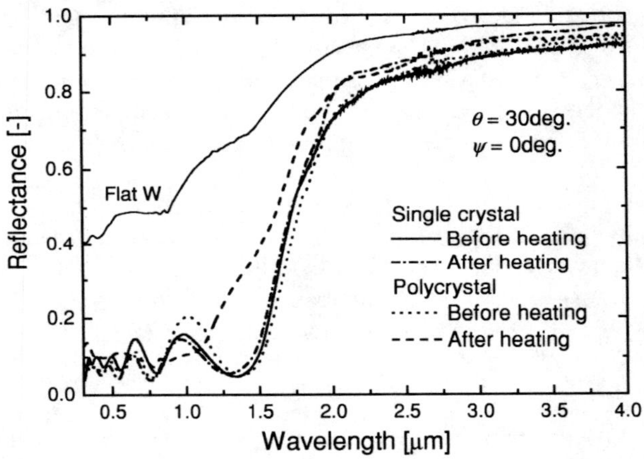

FIGURE 7. Measured reflectance spectra of the W gratings after a heating at 1400K for 1 hour in a reductive atmosphere.

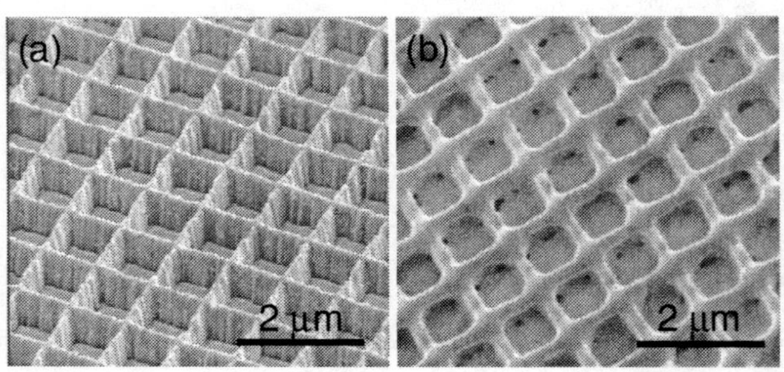

FIGURE 8. SEM images of the W grating made from (a) a single crystalline W and (b) polycrystalline W after a heating at 1400K for 1h in a reductive atmosphere.

CONCLUSIONS

To obtain high-performance selective emitters for TPV generation, surface design of 2D W surface gratings are investigated with numerical calculation based on RCWA. Although optimization has not been achieved, it is showed that microstructured W emitters can realize good spectral selectivity and high energy density at the same time. According to the calculated results, 2D surface grating with 1.0 μm periodicity composed of rectangular microcavities have been fabricated on both of mono- and polycrystalline W substrates with the combination of EB lithography and FAB etching techniques. The W gratings show a strong emittance peak, which can be explained by the resonance between electromagnetic waves and confined modes inside the cavities. The grating structure on a polycrystalline W is deformed and damaged by a heating at 1400K in a reductive atmosphere. It is speculated that this phenomenon is mainly due to grain growth. On the other hand, the grating structure on a single crystalline W is stable over 1400K. From these results, we conclude that the thermal resistivity of microstructures can be drastically improved by using single crystalline substrates.

ACKNOWLEDGEMENTS

The authors appreciate the monetary supports by the Ministry of Education, Science, Sports and Culture, Japan, Japan Space Forum, and the Asahi Glass Foundation. The authors also thank the workers at the Venture Business Laboratory in Tohoku University for their technical supports in preparing the samples.

REFERENCES

1. Le Gall, J., Oliver, M., and Greffet, J.-J., *Phys. Rev. B*, **55**, 10105-10114 (1997).
2. Greffet, J.-J, Carminati, R., Joulain, K., Mulet, J-P, Mainguy, S., and Chen, Y., *Nature*, **416**, 61-64 (2002).
3. Heinzel, A., Boerner, V., Gombert, A., Bläsi, B., Wittwer, V., and Luther, J., *J. Mod. Opt.*, **47**, 2399-2419 (2000).
4. Sai, H., Yugami, H., Akiyama, Y., Kanamori, Y., and Hane, K., *J. Opt. Soc. Am. A*, **18**, 1471-1476 (2001).
5. Maruyama, S., Kashiwa, T., Yugami, H., and Esashi, E., *Appl. Phys. Lett.*, **79**, 1393-1395 (2001).
6. Ferber, J., Aschaber, J., Hebling, C., Heinzel, A., Wiehle, R., Zenker, M., Luther, J., in *16th European Photovoltaic Solar Energy Conference and Exhibition, Glasgow 2000*.
7. Lin, S.Y., Fleming, J.G., Chow, E., Bur, J., Choi, K.K., and Goldburg, A., *Phys. Rev. B*, **62**, R2243-2246 (2000).
8. Fleming, J.G., Lin, S.Y., El-Kady, I., Biswas, B., and Ho, K.M., *Nature*, **417**, 52-55 (2002).
9. Coutts, T.J., *Renew. Sustain. Energy Rev.* **3**, 77-184 (1999).
10. Moharam, M.G., in *Holographic Optics: Design and Applications*, edited by Cindrich, I., Proceedings of SPIE 883, 1988, pp. 8 - 11.
11. Toma, Y., Hatakeyama, M., Ichiki, K., Huang, J., Yamaguchi, K., Watanabe, K., and Kato, T., *Jpn. J. Appl. Phys.*, **36**, 7655-7659 (1997).
12. Lynch, D.W., and Hunter, W.R., in *Handbook of Optical Constants of Solids I*, edited by E.D. Palik, New York: Academic Press, 1985, pp.334-341.

Thermal stability of micro-structured selective tungsten emitters

C. Schlemmer[1], J.Aschaber[2], V. Boerner[1], J. Luther[1]

[1] *Fraunhofer Institute for Solar Energy Systems ISE, Heidenhofstrasse 2, 79110 Freiburg, Germany*
phone: +49 761 40166 89, fax: +49 761 40166 81,
e-mail: schlemme@ise.fhg.de
[2] *Freiburger Materialforschungszentrum, Stefan-Meier-Strasse 21, 79104 Freiburg, Germany*

Abstract. The microstructuring of a tungsten emitter allows a remarkable increase of the emittance in a defined spectral range with low mid IR emission at the same time.
The optical properties of the selective emitter are determined by the grating period and the shape of the grating. Furthermore for the spectral angle dependancy the structure depth plays a significant role. As theoretical investigations using Rigorous Coupled Wave Analysis (RCWA) have shown the spectral angle dependency decreases in case of deep or high spatial-frequency structures.
At usual TPV emitter temperatures (typically above 1400K) microstructured selective tungsten emitters show both surface diffusion and the so-called recrystallisation effect. In some cases a modification and destruction of the surface structure due to this phenomena have been observed. At first, the technique of producing selective tungsten emitters is described. In addition, the result of our investigations concerning thermal stability of both shallow and deep gratings is presented and some attempts to handle thermal deterioration are discussed.

Keywords: Tungsten emitter - 1, Thermal stability - 2, Recrystallization – 3, Surface diffusion – 4, Hafnium oxide - 5

1. Introduction

Beside of rare earth oxides [1] and anti-reflection coated emitters [9], micro-structured tungsten is a promising alternative as selective emitter in TPV as first investigations have shown [2]. Due to its selective emittance a tungsten emitter is naturally well adapted to a TPV system in which GaSb cells are used. Below the bandgap of GaSb (0.72eV resp. 1.73μm) i.e. in the mid IR a plane tungsten emitter

CP653, *Thermophotovoltaic Generation of Electricity: 5th Conference*
edited by T. J. Coutts, G. Guazzoni, and J. Luther

shows an emittance of less than 20%[1] [3]. Furthermore, the low evaporation rate is an important advantage. Even at 2000°C it is negligible [4]. To operate microstructured tungsten as TPV emitter, protective gas atmosphere or vacuum is necessary to avoid oxidation.

A great increase of the emittance is achieved by microstructuring the emitter surface. Surface plasmon resonance (SPR) does occur which encourage the emittance in a favoured spectral wavelength range. In addition, in the case of high spatial-frequency gratings, the microstructure acts as anti-reflecting surface with higher emittance. But not only the grating period and the grating depth are important. Also the shape of a real grating and the natural surface roughness do influence the emittance spectrum. To work out suitable grating parameters, we use the Rigorous Coupled Wave Analysis (RCWA). Theoratically any arbitrary periodic structure can be simulated. In addition, a simulation of gratings consisting of several different emitter materials is possible.

2. RCWA

The RCWA method allows to simulate the emission spectrum of arbitrary periodic structures. In case of selective tungsten emitters we use this possibility to find appropriate grating parameters to adapt the emitter spectrum to GaSb cell sensitivity. The algorithm of RCWA works as follows [5,6]. The em-field outside of the grating region is expanded in a Rayleigh series and the grating profile (i.e. the dielectric function) is also expanded in a Fourier series. In case of a non-rectangular grating profile the dielectric function is represented by a multi-layer stack of binary shifts with the desired appropriate conformity. By inserting both expansions into the Maxwell equations one gets the coupled wave equations. The grating problem is reduced to a matrix eigenvalue problem [7]. Finally, the solution is received by matching the solution of the coupled wave equations with the initial boundary

[1] In effect the reflectance is greater than 80% at this wavelength. However since a tungsten emitter is opaque at IR Kirchhoff law in the form of emittance=1-reflectance can be applied

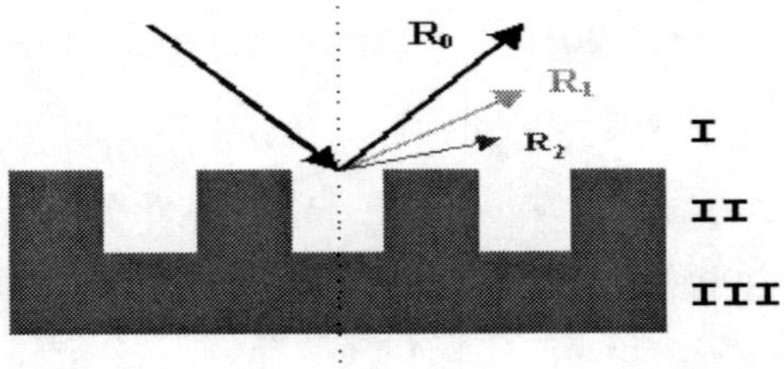

Fig.2.1: Side view of a linear grating. In the case of metallic gratings many evanescent orders of diffraction R_n do exist

conditions of the different regions I-III of the diffraction grating (Fig.2.1). In order to receive accurate results the number of orders of diffraction which has to be taken into account is rather high in the case of metallic gratings. In the following examples up to 49 orders were calculated. An incident wave can couple with the metallic grating by excitation of surface plasmons. Inversely in the case of a selective emitter radiation decouples from the surface grating. Unfortunately SPR shows a strong angle dependance.

In the first example the spectral emittance of a grating with a 1400nm period, a cavity width of 700nm and a structure depth of 1000nm at different polar angles was calculated. At normal incidence a broad emittance maximum at about 1650nm can be observed. As for SPR typical, it shifts significantly to the far IR region when the polar angle is increased (Fig. 2.2).

The prevention of the angle dependance is therefore of upmost interest in emitter modelling. For gratings with either narrow and deep cavities or small periods another kind of emittance enhancement, the anti-reflection effect occurs. It is nearly angle independent (see Fig.2.3). In this example a period of 600nm was chosen. The cavity width is 300nm and its depth 200nm. In the anti-reflection affected region the emittance reaches more than 60% with a fast decrease in the ambit of the 1.73µm GaSb bandgap. While the still existing SPR is represented by the small peak which moves towards far IR at 30°polar incidence, the large emittance maximum is mainly caused by broad-band anti-reflection. As predicted in effective medium theory, RCWA has shown that such a resonance depends less on the grating period but strongly on the cavity width and its depth.

Due to the weak angle dependent emittance of anti-reflection sub-wavelength micro-structures, we prefer the manufacture of such grating types. A grating with

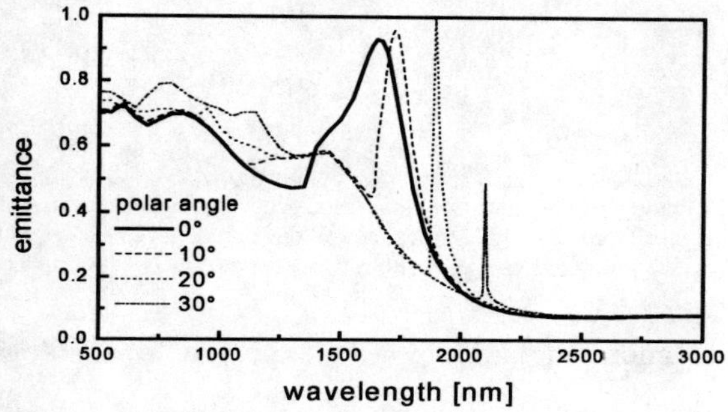

Fig.2.2: Typical angle dependance of SPR at TM polarisation

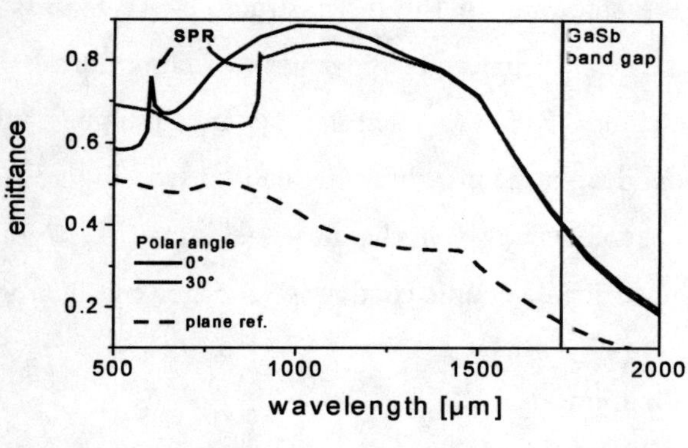

Fig.2.3: Weak angle dependent emittance spectrum of a linear grating at TM polarisation

these parameters was not yet produced but we are in process to realize it with the present holographic process very soon.

3. The manufacture of deep gratings

In order to receive a one or two dimensional grating structure a standard holographic process is used. The polished tungsten emitter sheet is coated with a 500-1400nm

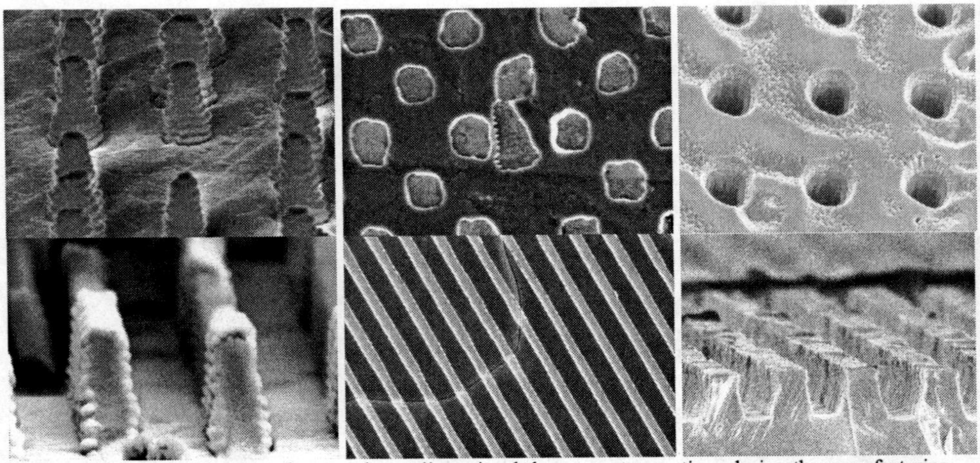

Figures 3.1-3.6: SEM images of two and one dimensional deep tungsten gratings during the manufacturing process. From left to right: Photoresist after exposure and development (Cr deposition, Fig.3.4) / Cr etch mask after Lift Off/ Final surface grating after RIE. In both cases the period is 1400nm. The depth of the gratings is about 500-800nm resp. 900nm.

photoresist layer and structured by UV exposure in a holographic device. After developing a thin layer (50-100nm) of chromium is evaporated onto the resist structure. Grace of the great anisotropy of the evaporation there rests only little chromium at the sidewalls of the resist structure. Acetone is used to lift off the remaining resist and a negative etching mask of chromium is left. We use SF_6 for reactive ion etching (RIE). Its etch rate is much higher for tungsten than for chromium. Thus deep structures may be realised (see Figures 3.3 and 3.6). Finally the Cr mask is removed by wet chemical etching resp. Argon sputter deposition. Furthermore, an optional ceramic coating is deposited by electron beam evaporation.

4. Recrystallization

A general problem in micro-structuring metallic emitters in TPV is the phenomenon of primary recrystallzation (PR). As a result of the manufacture of tungsten in sheet form, thermal stress is produced by rolling. Furthermore especially at the surface a non-negligible deformation zone is always present after the indispensable polishing and lapping in order to receive good results for the micro-structuring process. As in TPV service provided the emitter is annealed up to temperatures where PR takes place (Fig.4.2). In this region of temperature the main part of internal strain and tension is reduced, internal tension and the density of dislocations decreases rapidly [8]. For tungsten PR starts at app. 1180°C [4]. Total PR can be reached at 1400°C after one hour. A destruction of several micro-structured tungsten emitters after annealing has yet been observed (Fig. 4.4). A possibility to avoid this effect is to anneal the emitter before microstructuring. Furthermore by annealing the emitter for a longer time at higher temperatures the remaining grain structure changes and grain growth takes place (secondary recrystallization) which should have a positive effect due to bigger plane areas on the emitter surface.

5. Surface diffusion

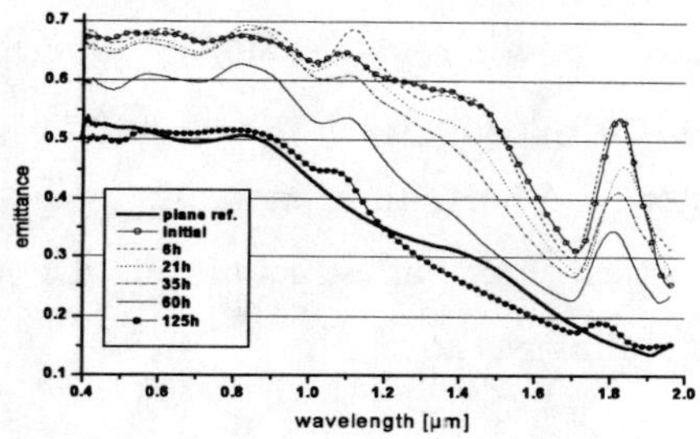

Fig.5.1 Decreasing emittance of the grating in figure 5.1 with a period of 1400nm and conical cavities in TM polarisation and 15°/20° polar/azimuth angle during annealing at 1000°C.

Surface diffusion represents a second important effect which is responsible for the destruction of micro-structures of metallic emitters during thermal treatment. Even for materials with a very high melting point (for tungsten: T_m=3400°C) surface diffusion starts already at rather low temperatures. Figure 5.2 shows the thermal deterioration of a micro-structured tungsten emitter before and after annealing for 125 hours at 1000°C with a continuous decreasing emittance (figure 5.1).

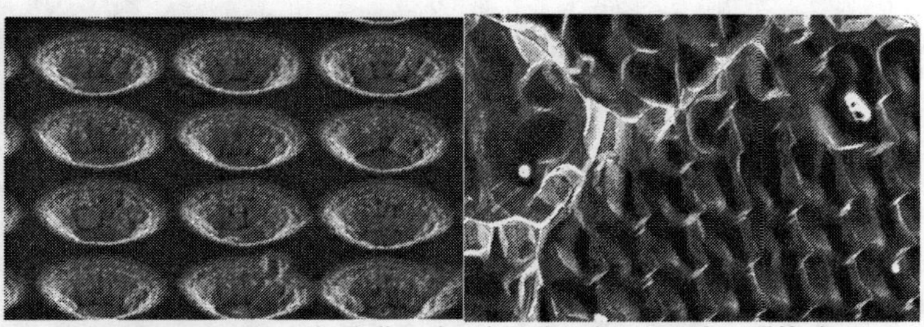

Fig.5.2 Destruction of a 2-dimesional grating due to surface diffusion

6. Surface diffusion control

In order to keep the idea of a stable micro-structured surface as TPV emitter, surface diffusion has to be controlled. The only promising approach was the coating of the emitter structure by a thin ceramic layer. After etching with SF_6, a 20-300nm thin coating with a high temperature stable material should slow down surface diffusion. First of all, flat tungsten emitters were coated by different materials to find out the thermal properties of such thin coatings. Finally, hafnium oxide (HfO_2) has shown the most suitable properties of all investigated materials. An example for a coated microstructure is shown in figure 6.1. After a short time annealing, a characteristical granular structure of the hafnium coating has arisen.

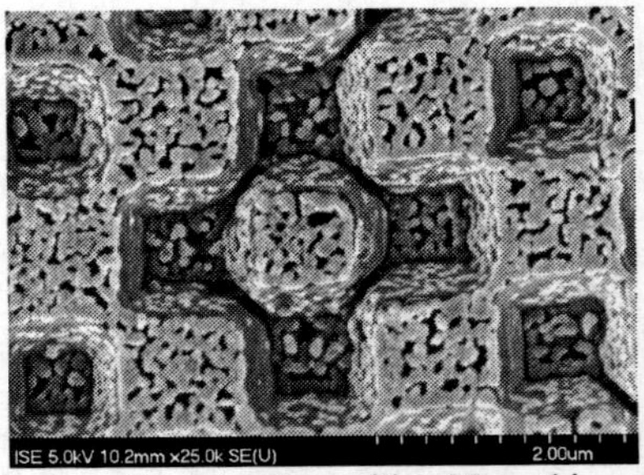

Fig.6.1: Microstructure with a 25nm thin layer of HfO_2

7. Results

The thermal stability of both linear and 2-dimensional coated gratings has been investigated. As already in chapter 5 presented the emittance spectrum was determined by fourier transform spectroscopy at room temperature. With our set-up at Fraunhofer ISE it is possible to measure spectral hemispherical reflectance with a variation of both polarization and polar angle. For an accurate measurement of the emittance quality a polar angle of 10° and an azimuth angle of 20° was chosen. Thus there is less intensity lost by reflected orders leaving the set-up in the incident direction (Littrow condition, see Fig.7.1).

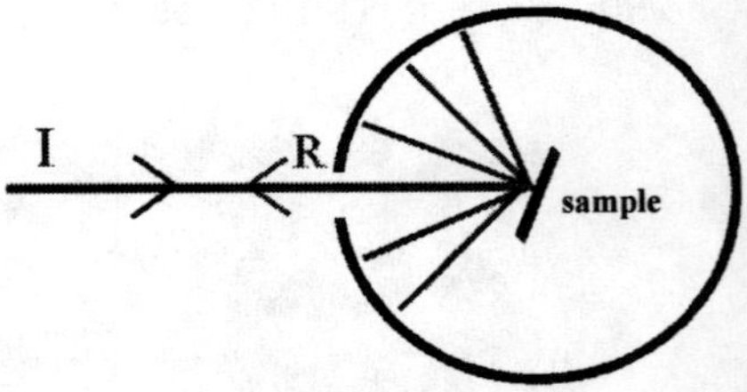

Fig. 7.1: Littrow Condition: The reflected order R is leaving the hemispheric set-up in the direction of the incident light beam I

First, the thermal stability of a shallow non-recrystallized 1-dimensional microstructured hafnium oxide coated emitter is presented. The structure period was 1400nm, the structure depth was about 200nm (see fig. 7.2, left image). In order to show the increased thermal stability of a coated emitter surface, only a half of the structured area has been coated with hafnium dioxide.

The initial emittance spectra are shown in figure 7.3 (left side). Under a polar angle of 20°, the SPR has divided into two emission peaks for both the coated and the uncoated half of the emitter. Due to a rather thin coating of about 50nm, the two spectra are very similar to each other. After the optical characterization, a long term thermal treatment followed. The annealing temperature was about 1200°C in a high vacuum environment. After 99 hours annealing time the emittance was remeasured. The result was noticeable (fig. 7.3, right side). Although the emittance of the hafnium coated area decreased for 0.2 at 1700nm, it remained above (about 0.3) the value of the uncoated areas.

The SEM shows the modification of the surface topology of both areas after the

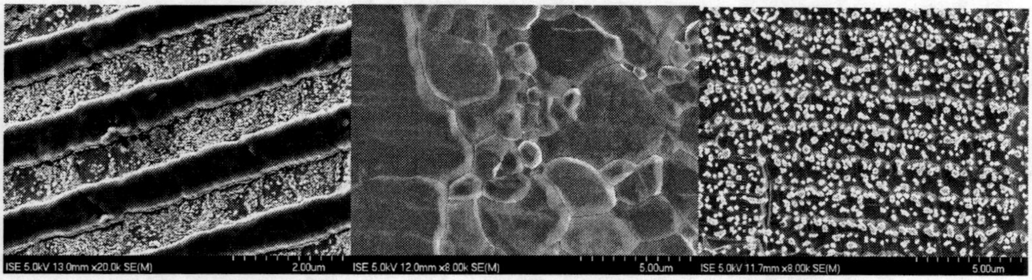

Fig. 7.2: Surface topology and stability of a 1.4μm linear and shallow microstructure. After 99 hours annealing time the initial surface profile (left) has been destroyed (centre). A hafnium oxide coated area has been conserved partially (right).

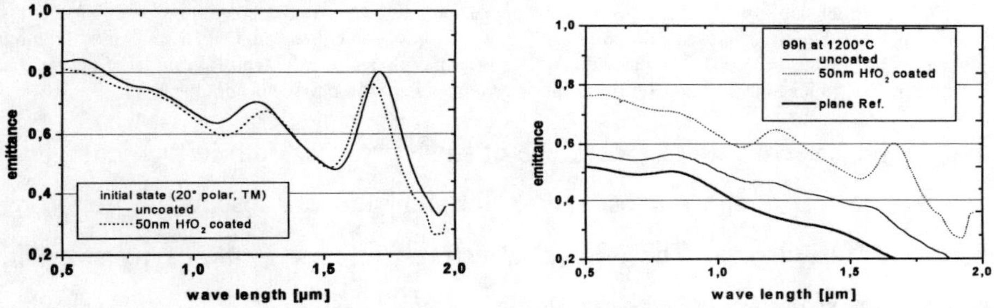

Fig. 7.3: Initial (left) and final (right) spectra of an uncoated/HfO_2 coated linear tungsten emitter. The annealing time was 99hours in a high vacuum environment.

annealing procedure. In both cases a modification due to recrystallization was expected. In the case of the uncoated surface, the microstructure has been already destroyed the lamella structure changed to a near flat tiled surface. The initial homogeneously coated surface has been modified, too. But in difference to the first example, the microstructure has been conserved at least partially under the rearranged hafnium dioxide coating.

Furthermore, the thermal stability of a deep 2-dimensional micro-structured hafnium oxide coated emitter is presented (see fig. 6.1). In this case, the surface was fully recrystallized before the microstructuring process. The initial reflectance is a result of conventional SPR. Below a wavelength of 2.0μm it decreases rapidly and it lies both for TM and for TE polarisation lower than 0.4 in the band below 1.7μm (fig. 7.2). The annealing temperature was the same than in the first example (1200°C). After 24 and 176 hours annealing time the reflectance was remeasured. The result was more than convincing (fig. 7.3). Even after 176 hours of thermal treatment the reflectance remained consistent and very close to the initial spectrum for both polarisations. The surface topology was examined again by SEM. The thin hafnium

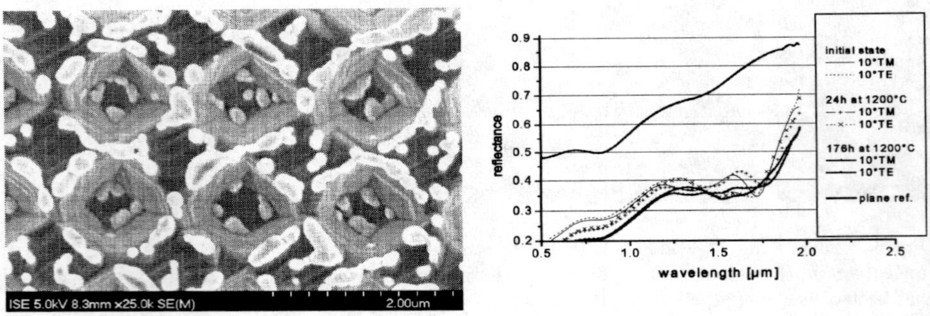

Fig. 7.4: Thermal stability of a 2-dimensional, 1.4μm period hafnium coated deep emitter. After 176 hours annealing time, the initial surface structure (see. Figure 6.2) has been conserved. Only a small modifications of the cavity profile has occurred. The hafnium dioxide layer has changed to the typical granular structure. The emittance spectra for both TM and TE polarisation remained very close to the original spectra.

oxide layer has changed to a coverage of small crystals. Both in the grating cavities and on the top of the microstructure these crystals are responsible for very low surface diffusion rates. The original structure profile, e.g. the cavity size and the depth of the structure, have been hold up (fig. 7.4, left).

8. Conclusion

With RCWA it is possible to generate grating parameters which lead to a suitable angle independent emission spectrum of micro-structured tungsten emitter to be used in combination with GaSb cell devices. In order to receive deep gratings with less angle dependent emittance a chromium etch mask is used. Concerning the thermal stability of the microstructures the result is not very promising without a ceramic coating even at temperatures at 1000°C. Due to thermal mobility surface atoms rearrange which leads to a destruction of the structure. In order to keep the idea of micro-structured emitters for TPV the emitter has to be coated with a thin ceramic layer or semi conductor layer of high thermal stability. In this case surface diffusion is extremely reduced. In the frame of our investigations a 1.4µm periodic two-dimensional emitter structure with a thin, 25nm coating of HfO_2 remained stable. After 176h annealing time at approx. 1200°C the initial reflectance spectrum has been hold. Further long time annealing will be obligatory to work out the overall lifetime of such ceramic-coated micro-structures.

REFERENCES

1. E.M.Wong et al.: Multiband Spectral Emitters Matched to MBE Grown Photovoltaic Cells, 1995, Proceedings of the 2nd NREL Conference on Thermophotovoltaic Generation of Electricity
2. A.Heinzel et al.: Radiation filters and emitters for the NIR based on periodically structured metal surfaces, J. of Mod. Opt., 2000, Vol.47, 13, 2399-2419
3. E.D. Palik: Handbook of optical constants of solids, Academic Press (1985)
4. E. Lassner, W.-D. Schubert: Tungsten, Kluwer Academic (1990)
5. M.G. Moharam, T.K. Gaylord: Rigorous coupled wave analysis of metallic surface relief gratings, J.Opt.Soc.Am. A3, 11, 1780-1787 (1986)
6. T.K. Gaylord: Planar dielectric grating diffraction theories, Appl. Phys. B28, 894-937 (1985)
7. S.Peng, G.M.Morris: Efficient implementation of rigorous-coupled wave analysis for surface-relief gratings, J.Opt. Soc. Am. A12, 5, 1087-1096 (1995)
8. F.J. Humphreys: Recrystallization and Related Annealing Phenomena, Pergamon (1996)
9. US Pat. 6,271,461

Semiconductor Silicon as a Selective Emitter

Donald L. Chubb,* David S. Wolford,* Andrew Meulenberg,† and Robert S. DiMatteo†

*NASA Glenn Research Center, Cleveland, Ohio 44135, U.S.A.
†The Charles Draper Laboratory, Cambridge, Massachusetts, 02139–3563, U.S.A.

Abstract. Silicon operating in a vacuum is a good candidate thermal emitter since it has a high melting point (1680 K). The semiconductor bandgap, which can provide selective emission, adds to the potential for high operating temperature and, therefore, high radiated power. We present the detailed emitter theory, along with both theoretical and experimental results for spectral emittance of thin (~1 μm) silicon films on sapphire substrates with a platinum backing. These results show the importance of temperature and film thickness in determining the selective spectral emittance and, with the proper material parameters, can be readily extended to other materials and systems.

INTRODUCTION

Thermphotovoltaic (TPV) research has mainly been directed at combustion-driven TPV systems that operate in the atmosphere. As a result, selective emitter research has concentrated on ceramic materials, which are mainly oxides, that are stable in the atmosphere at high temperatures. However, operation of a radioisotope thermphotovoltaic (RTPV) system, that would be used in a space power system, and the new microgap thermophotovoltaic [1] (MTPV) system are at vacuum conditions. Therefore, emitter materials other than oxides can be used in these systems.

Semiconductors are a logical choice for a selective emitter. They have large absorptance (and thus large emittance) for photon energies, E, greater than their bandgap energy, E_g, and low absorptance (and thus low emittance) for photon energies less than the bandgap energy. Thus, using the same semiconductor for the emitter and photovoltaic (PV) array might make an ideal emitter-PV array match for a TPV system. Unfortunately, high temperatures reduce the bandgap of semiconductors. This spoils the emitter-PV match. Furthermore, only a limited number of semiconductors are suitable for high temperature operation. Silicon (Si) has a high melting point (1680 K) and its band-narrowed spectral emittance would be suitable for InGaAs PV arrays, which would be cooled.

The emissive properties of Si are the subject of this paper. For dielectric selective emitters such as the rare-earths [2], spontaneous electron transitions of the valence electrons from the lowest excited states to the ground state accounts for the emittance in most regions of the spectrum. However, at long wavelengths (λ >5 μm) vibrations of the crystal lattice (phonons) result in $\varepsilon(\lambda) \approx 1$ regardless of the temperature. Both of these processes are nearly independent of temperature. As a result, $\varepsilon(\lambda)$ is nearly independent of temperature if the temperature change across the emitter is negligible [2]. However, the lattice structure of

CP653, *Thermophotovoltaic Generation of Electricity: 5th Conference*
edited by T. J. Coutts, G. Guazzoni, and J. Luther

semiconductors results in a strong temperature dependence for $\varepsilon(\lambda)$. For photon energies, $E > E_g$ or $\lambda < \lambda_g = hc_o/E_g$ the emittance results from interband transitions of electrons and holes. For $E < E_g$ or $\lambda > \lambda_g$ the emittance results from transitions of electrons or holes between states within the same energy band. This latter emittance is the inverse of free carrier absorption and is strongly temperature dependent. Similar to the dielectric selective emitters, for $\lambda >> \lambda_g$, $\varepsilon(\lambda) \approx 1$ regardless of the temperature as a result of the vibrational modes of the crystal lattice.

The processes described above, that determine $\varepsilon(\lambda)$, are quantitatively expressed by the extinction coefficient, $K(\lambda)$, which is the sum of the absorption coefficient, $\alpha(\lambda)$, and the scattering coefficient, $\sigma(\lambda)$.

$$K(\lambda) = \alpha(\lambda) + \sigma(\lambda), \quad \text{cm}^{-1} \tag{1}$$

It is the product of $K(\lambda)$ and the emitter thickness, d, which is called the optical depth, $\kappa(\lambda)$, that is the main parameter for determining $\varepsilon(\lambda)$.

$$\kappa(\lambda) = K(\lambda)d \tag{2}$$

Thus, emitter thickness is the key variable over a large range of values for controlling the emittance. As will be shown later, for $\kappa(\lambda) >> 1$ the emittance is independent of κ. For $\lambda < \lambda_g$ the extinction coefficient is very large so that $\kappa(\lambda)$ is also large and as a result $\varepsilon(\lambda)$ will be nearly independent of d above a certain thickness. For $\lambda > \lambda_g$, where free carrier absorptance (emittance) is the dominant process, K is much smaller than K for $\lambda < \lambda_g$. As a result, thickness becomes important in determining $\varepsilon(\lambda)$ for free carriers in a range where thickness has little effect on $\varepsilon(\lambda)$ for $\lambda < \lambda_g$. Thus, by varying the emitter thickness the ratio of the emittance for $\lambda < \lambda_g$ to the emittance for $\lambda > \lambda_g$ can be varied.

Several authors [3–5] have measured the emittance of silicon at elevated temperatures for thick samples (>100 μm). Those authors were not interested in using silicon as a selective emitter. Therefore, they did not investigate the effect of thickness on the emittance.

In the next section the model used to calculate the emittance will be discussed. Then the experimental results for the emittance of single crystal Si on a sapphire substrate will be presented, and compared with the theoretical results.

EMITTANCE THEORY

Figure 1 is a schematic drawing of the emitter structure. A semiconductor emitting film of thickness, d_f, is deposited on a dielectric substrate of thickness, d_s. On the backside of the substrate is deposited a highly reflective metallic film. The metallic film is necessary to block radiation from the thermal source that heats the emitter. For the experimental emitter, the semiconductor is Si, the substrate is sapphire and the metallic film is platinum, Pt.

The large extinction coefficient, characteristic of semiconductors, for photon energies above the bandedge translates into large emittance above the bandgap energy, even for very thin layers of the semiconductor. The thin layers, possible because of the high emittance in the useable regions, reduce the free-carrier radiation of the unusable long-wavelength light

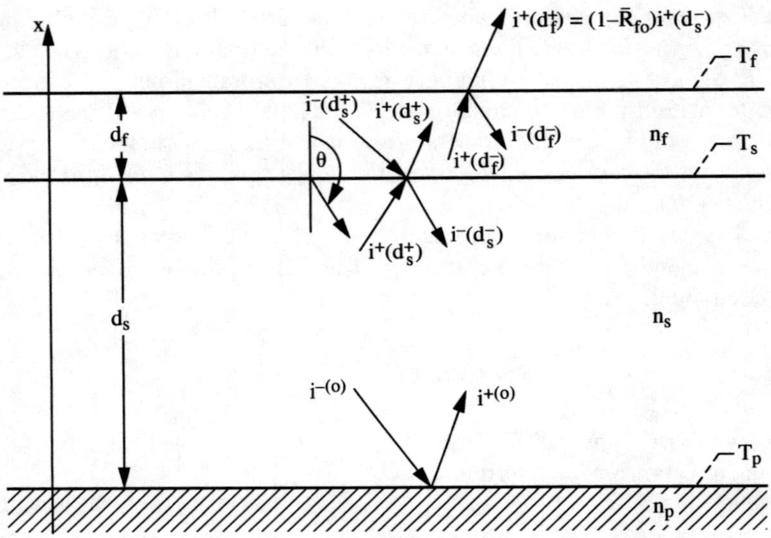

Figure 1. Schematic drawing of one dimensional emittance model.

(where K(λ) is orders of magnitude lower). As a consequence, if the semiconductor layer is thin enough, the long-wavelength light emission will be controlled by the metal reflector and the substrate, not by the semiconductor layer. This could reduce critical regions of the "background" (or out-of-band) emittance by an order-of-magnitude (e.g., ~0.03 versus ~0.25) relative to that of present selective emitters which must be thick.

The conceptual model below provides a simplified view of the detailed model in Appendix A (which includes temperature gradients in the emitter substrate and emitting film.

For regions of the spectrum where the optical depth, $\kappa_f(\lambda)$, is large and there is negligible temperature change across the emitting medium, the emittance from the film *f* into medium o is the following simple result (Eq. (A–58)).

$$\varepsilon(\lambda) \approx 1 - \overline{R}_{fo} \tag{3}$$

Where $\overline{R}_{fo}$ is the reflectivity at the interface between media *f* and *o*. Equation (3) applies to a semiconductor emitter for $\lambda < \lambda_g$. If the indices of refraction of the two media are nearly the same, $\overline{R}_{fo} \approx 0$ and $\varepsilon(\lambda) \approx 1$.

For regions of the spectrum where $K_f(\lambda) \approx 0$ ($\lambda > \lambda_g$) the film is transparent and the emittance is determined by the optical properties of the substrate and metallic film behind the substrate (Eq. (A–61)). This emittance can be reduced if the reflectivity at the substrate-film interface, $\overline{R}_{fs}$, is large.

Since K_f (in the free-carrier region) increases dramatically with increasing temperature (e.g., to >100/cm^{-1} at 1000 K), emitter-film thickness is critical and must be kept in the d_f << 0.01 cm (i.e., the 100 μm) range. With $K_f \sim 10^4$/cm^{-1} (in the direct-bandgap region), semiconductor films on the order of 1 μm are useable for efficient emitters. Such thin films make them highly-selective emitters, if their free-carrier concentrations can be kept low enough at the operating temperature.

Theoretical Emittance Results

Using the extinction coefficient and index of refraction data presented in Appendix B, ε(λ) (Eq. (A–54)) was calculated as a function of λ. The source function integrals (Eqs. (A–43) to (A–48)) were numerically integrated. The iterative process described in Appendix A was used to obtain the temperature changes ΔT_s and ΔT_f.

Figure 2 shows ε(λ) for several values of T_p with the silicon film thickness, d_f = 1 μm, and the sapphire substrate thickness, d_s = 525 μm, which are representative of experimental values, since d_f is small the temperature change, $\Delta T_f < 10^{-5}$. Also, $\Delta T_s < 0.01$. As a result, only for small λ, where $K_f \rightarrow \infty$, does ε(λ) show any effect of the temperature changes ΔT_f and ΔT_s (see Eq. (A–58)).

For $\lambda < \lambda_g$ the extinction coefficient is determined by band-to-band transitions ($K = K_b$, Eq. (A–64)) and $K_b >> 1$. As a result, $\kappa_f > 1$ and ε(λ) can be approximated by Eq. (A–58). Since $\Delta T_f \approx 0$, Eq. (A–58) yields the result, $\varepsilon(\lambda) \approx 1 - \overline{R}_{fo}$, which is independent of T_p. Figure 2 agrees with this result. For $\lambda > \lambda_g$ free carrier intraband transitions determined K (Eq. (A–63), $K = K_{fr}$) and κ_f. As Fig. 2 shows, ε(λ) increases rapidly with T_p in this wavelength

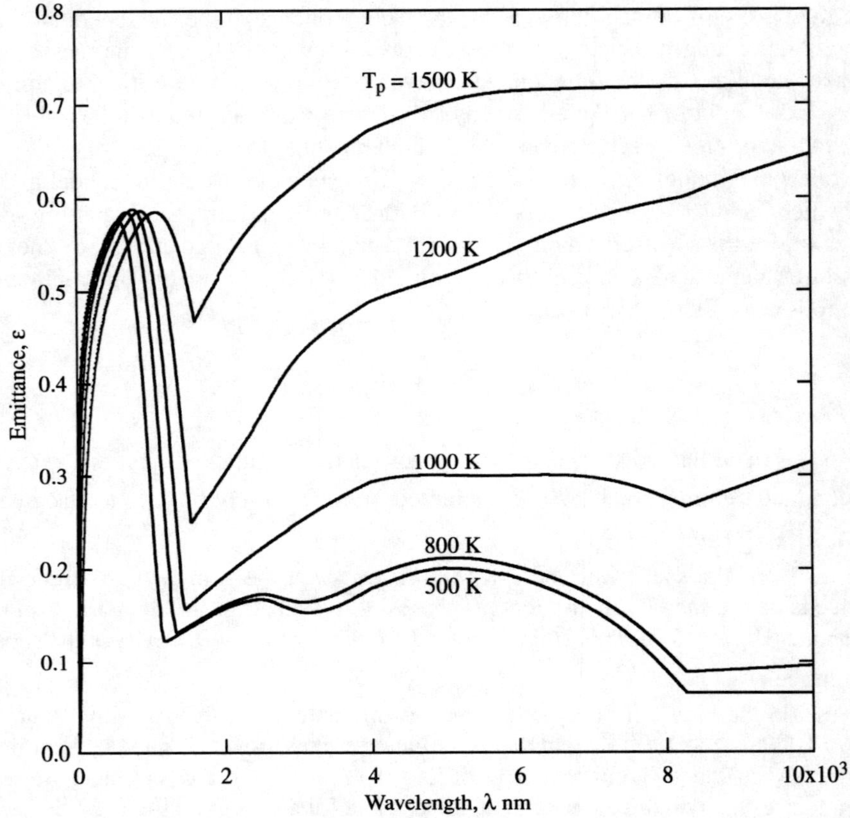

Figure 2. Calculated spectral emittance of a pure silicon film of thickness, $d_f = 1$ μm, grown on a sapphire substrate of thickness, $d_s = 525$ μm, with a platinum film behind sapphire substrate at several platinum temperatures, T_p.

region since K_{fr} is a sensitive function of T. For $T_p > 1000$ K, $\varepsilon(\lambda)$ is undesirably large. The sharp drop in $\varepsilon(\lambda)$ for $\lambda < \lambda_g$ results because of the large value of the index of refraction, n_{fr}, and thus large value for $\overline{R}_{fo}$. In this wavelength region the emittance could be increased by the addition of an antireflective film [13,14].

The effect on $\varepsilon(\lambda)$ of silicon film thickness, d_f, is illustrated in Fig. 3 for $T_P = 1200$ K. The silicon film thickness effect on $\varepsilon(\lambda)$ is similar to the T_P effect. For $\lambda > \lambda_g$, where $\kappa_f = K_{fr} d_f$ the emittance increases rapidly with increasing d_f. To keep $\varepsilon(\lambda)$ small in this wavelength region d_f must be less than 0.1 μm. For $\lambda < \lambda_g$, comparing Figs. 2 and 3, we see that d_f has a larger effect on $\varepsilon(\lambda)$ than T_p. The reason being that $\varepsilon(\lambda)$ is an increasing function of $\kappa_f = K_b d_f$ in Fig. 3 whereas κ_f is a constant in Fig. 2.

EXPERIMENTAL RESULTS

Silicon Samples and Emittance Apparatus

The spectral emittance of several silicon films on a sapphire substrate with a platinum backing was measured at vacuum conditions. Figure 4 is a schematic drawing of the experimental apparatus. The n-type silicon films were epitaxially grown on $d_s = 525$ μm thick sapphire substrates by Lawrence Semiconductor Research Laboratory.[1] In order to have the lowest possible number of free carriers the purest silicon possible was grown. The sample resistivities were >1000 Ω-cm. To insure adhesion of the platinum, a thin layer of chromium (<20 nm) was first sputter deposited on the sapphire. The sputtered platinum film thickness was ≈ 60 nm. Transmittance and reflectance measurements were made of each sample before and after a sample was tested. Transmittance of all the samples before testing was less than 0.01, indicating that the platinum would block nearly all radiation from the tungsten heater. After heating the transmittance increased to approximately 0.01 for the $d_f = 1$ μm sample for $\lambda > 1200$ nm, and to approximately 0.02 for the $d_f = 2$ μm sample.

A blackbody was used to calibrate the system. As a result, the spectral emittance was calculated using the following expression.

$$\varepsilon\left(\lambda, T_p\right) = \frac{i\left(\lambda, T_p\right)}{i_b\left(\lambda, T_b\right)} \frac{e_b\left(\lambda, T_b\right)}{e_b\left(\lambda, T_p\right)} \tag{4}$$

Where T_p is the sample temperature, T_b is the blackbody temperature, e_b is the blackbody emissive power (Eq. (A–5)), $i\,(\lambda, T_p)$ is the intensity measured by the monochromator detector when viewing the sample and $i_b(\lambda, T_b)$ is the intensity measured by the monochromator detector when viewing the blackbody.

Because of the strong temperature dependence of $e_b\,(\lambda, T)$, the sample temperature, T_p, is the critical parameter in Eq. (4). Since silicon is a conductor and the tungsten heater current has an alternating current component there was an induced voltage on the silicon samples that prevented the use of a thermocouple to measure T_p. As a result, T was determined from

[1]Lawrence Semiconductor Research Laboratory, Inc., 2300 W. Huntington Dr., Tempe, Arizona 85282, U.S.A., phone 602–438–2300.

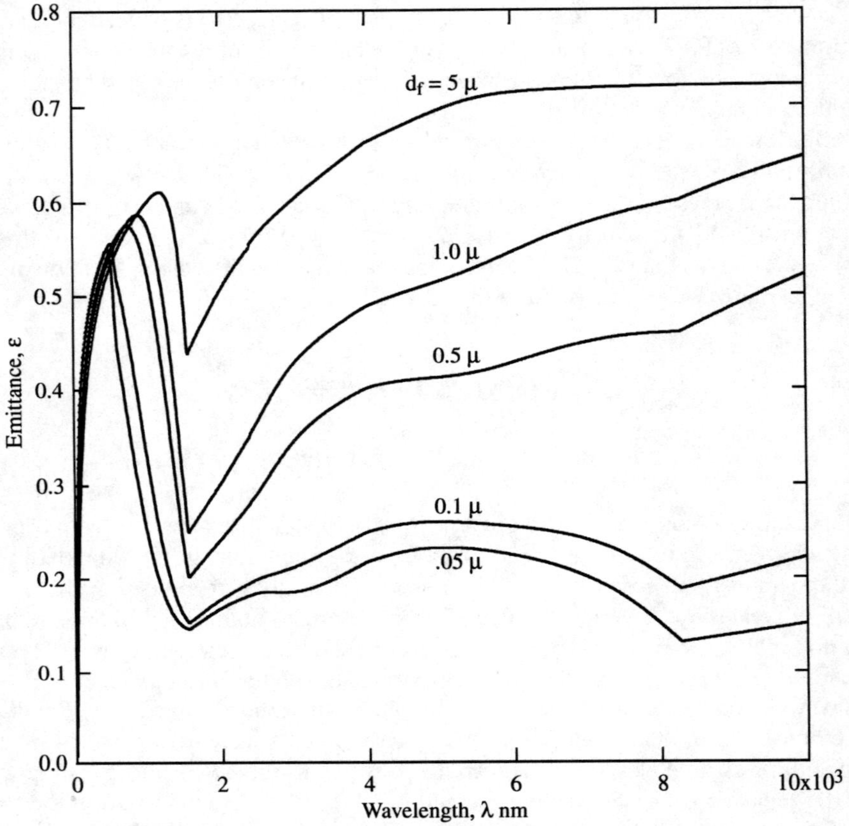

Figure 3. Calculated spectral emittance of a pure silicon film on a sapphire substrate of thickness, $d_s = 525$ μm, with platinum film behind substrate at temperature, $T_p = 1200$ K, for several silicon film thicknesses, d_f.

the measured intensities at two different wavelengths that have the same emittance. If the emittance at λ_1, is the same as the emittance at λ_2, then the following expression results.

$$\frac{e_b(\lambda_1, T_p)}{e_b(\lambda_2, T_p)} = \frac{i(\lambda_1, T_p)}{i(\lambda_2, T_p)} \tag{5}$$

Since $u = C_2/\lambda T_p > 8.5$ the approximation $e^u - 1 \approx e^u$ can be made. As a result, using Eq. (A–5) for e_b, Eq. (5) yields the following.

$$T_p = \frac{c_2\left(\frac{1}{\lambda_2} - \frac{1}{\lambda_1}\right)}{\ln\left[\left(\frac{\lambda_1}{\lambda_2}\right)^5 \frac{i(\lambda_1, T_p)}{i(\lambda_2, T_p)}\right]} \tag{6}$$

The area on the sample viewed by the monochromator was controlled by the two pinhole apertures. They were adjusted such that the diameter of the viewed spot in the center of the 2.54 cm diameter samples was approximately 1.27 cm. Only radiation that left the sample surface within this 1.27 cm diameter spot could be viewed by the monochromator. Therefore, the only way radiation from any source, other than the sample, could enter the monochromator was by reflection off the 1.27 cm diameter spot. To eliminate any other source of radiation the walls of the vacuum chamber and viewing window were water cooled.

Three photovoltaic detectors were used to measure the intensity. For $600 \le \lambda \le 1130$ nm a silicon detector was used. For $1130 \le \lambda \le 1800$ nm a germanium detector was used. And for $1800 \le \lambda \le 5600$ nm an indium antimonide detector was used.

EMITTANCE RESULTS

Figure 5 compares the calculated and measured spectral emittance for a silicon film thickness, $d_f = 1$ µm, on a sapphire substrate, $d_s = 525$ µm, with a platinum backing. Using the intensities measured at $\lambda_1 = 1550$ nm and $\lambda_2 = 2061$ nm and Eq. (6) the temperature was 831 K. Since λ_1 and λ_2 are in a region where $\kappa_f = \kappa_s \approx 0$ the silicon film and substrate are transparent and the emitted radiation is coming from the platinum, (see Eq. (A–61)). Thus, the measured temperature is the platinum temperature, $T_p = 831$ K. From the theoretical model, $\Delta T_s < 0.005$ and $\Delta T_f < 10^{-5}$. Therefore, assuming $T_p = T_s = T_f$ is probably a good approximation.

Since the silicon film thickness, d_f, is the same order as the wavelength interference effects produce a sinusoidal oscillation in the experimental $\varepsilon(\lambda)$. The theoretical model does not include interference effects. Even so, there is good agreement between the theoretical and experimental $\varepsilon(\lambda)$. For $\lambda < 800$ nm the measured intensity was too small to determine $\varepsilon(\lambda)$. However, for higher temperatures where the intensity is detectable for $\lambda < 800$ nm, the emittance exceeds the theoretical value. For $\lambda < 800$ nm, $\kappa_f \rightarrow \infty$. As a result, for $\Delta T_f = 0$, $\varepsilon(\lambda) = 1 - \overline{R}_{fo}$ (Eq. (A–58)). This implies the value used for n_{fR} (Eq. (A–67)) to calculate

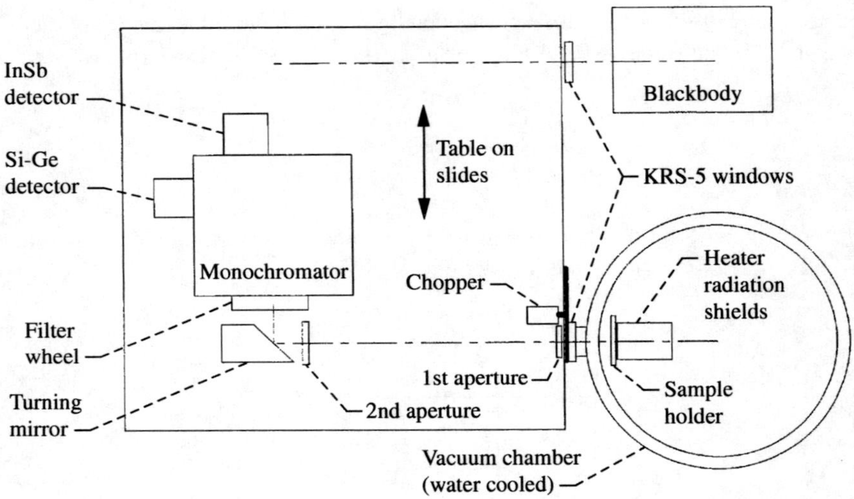

Figure 4. Schematic drawing of apparatus for measuring spectral emittance at vacuum conditions.

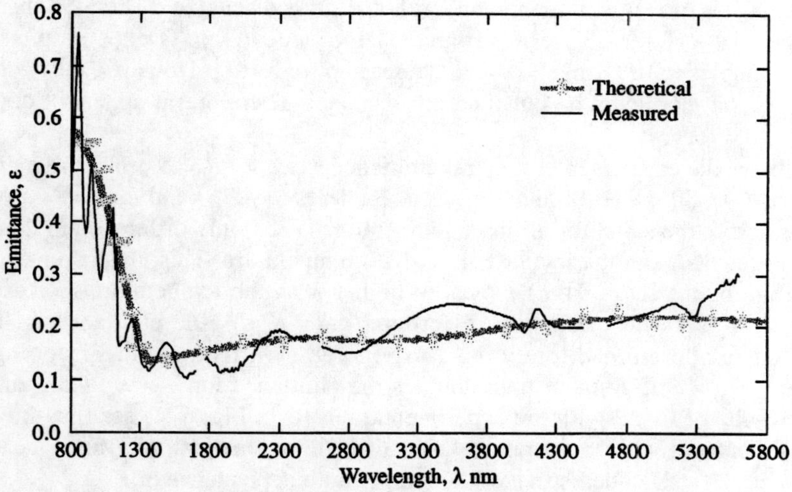

Figure 5. Comparison of experimental and theoretical spectral, $\varepsilon(\lambda)$, for pure silicon film of thickness, $d_f = 1$ μm, on a sapphire substrate of thickness, $d_s = 525$ μm, with a platinum backing film. The measured temperature is $T_p = 831$ K.

$\overline{R}_{fo}$ (Eq. (A–20)) is too large in the region $\lambda < 800$ nm. As discussed earlier, the sharp drop in the theoretical $\varepsilon(\lambda)$ for $\lambda < \lambda_g$ results from the large value of n_{fR}. The $\varepsilon(\lambda)$ results of Sato [3] for thick samples (>100 μm) show $\varepsilon(\lambda)$ decreasing for $\lambda < \lambda_g$. However, the decrease is not as large as that predicted by our theoretical model. Therefore, it appears Eq. (A–67) overestimates n_{fR} for $\lambda < \lambda_g$.

Figure 6 shows the experimental spectral emittance, $\varepsilon(\lambda)$, for $d_f = 1$ μm at three different temperatures, T_p = 805, 831, 905K. For $\lambda < \lambda_g$, where $\kappa_f \rightarrow \infty$ and $\varepsilon(\lambda) \approx 1 - \overline{R}_{fo}$, the emittance is nearly the same for all three temperatures. The only difference in this wavelength region results from the increase in λ_g with temperature. This is apparent in Fig. 6 where the minimum value of $\varepsilon(\lambda)$ moves to higher λ as the temperature increases. For $\lambda > \lambda_g$, where free carrier transitions determine κ_f (= $K_{fr}d_f$), the emittance is strongly dependent on temperature. Merely a 100 K increase in temperature in going from T_p = 805 K to T_p = 905 K more than doubles the emittance for $\lambda > \lambda_g$. Comparing the theoretical results of Fig. 2 with the experimental results in Fig. 6 we see that the experimental $\varepsilon(\lambda)$ increases with temperature faster than the theoretical $\varepsilon(\lambda)$ for $\lambda > \lambda_g$. This implies that Eq. (A–63) underestimates the temperature dependence of K_{fr}.

The effect of silicon film thickness, d_f, on $\varepsilon(\lambda)$ is illustrated in Fig. 7 where $\varepsilon(\lambda)$ is shown for two different values of d_f = 1 and 2 μm at T_p = 831 K. Similar to the temperature effect, the emittance increases with film thickness in the wavelength region $\lambda > \lambda_g$, where free carrier intraband transitions determine the extinction coefficient, K_{fr}. For $\lambda < \lambda_g$, where interband transitions determine the extinction coefficient, K_b, the emittance increases with d_f. Also, the maximum value of $\varepsilon(\lambda)$ in this λ region moves to larger λ as d_f increases. This same result is predicted by the theoretical model (Fig. 3).

CONCLUSION

This work has explored silicon emitters because of their potential for high-temperature operation. The theoretical modeling provides a realistic basis for investigating options for optimizing a system. The experimental effort, supported by the extensive data available for silicon, has verified the theoretical model and indicates where additional experimental and optimization work is needed. More importantly, it opens up a field of endeavor that could identify some of the emitter materials and structures required for high-efficiency energy conversion. Near-term studies of this type should include emitters of germanium (as a high-temperature, narrower-bandgap option) and InGaAs (as a lower-temperature, but direct-bandgap, material with higher extinction coefficients near its bandedge). As the optical-constant database grows for materials at these higher temperatures, designer materials (e.g., silicon-germanium alloys) and structures (e.g., quantum wells with exotic materials) or combinations will be considered. The potential for TPV to compete with mature solar-PV technologies as a cost-effective technology, for both space and terrestrial applications, appears real.

ACKNOWLEDGMENT

The authors thank and acknowledge Mark Smith of the Ohio Aerospace Institute for sputter depositing the platinum films on the backside of the sapphire substrates of the silicon films.

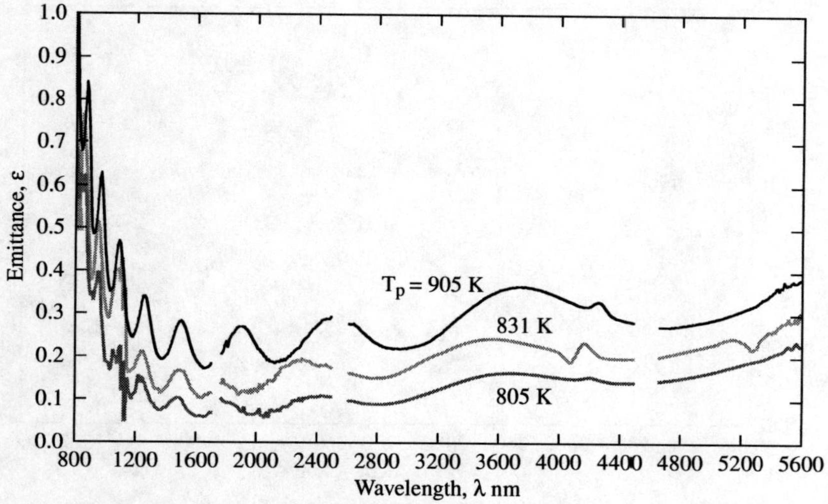

Figure 6. Experimental spectral emittance, $\varepsilon(\lambda)$, for pure silicon film of thickness, $d_f = 1$ μm, on a sapphire substrate of thickness, $d_s = 525$ μm, with platinum backing at temperatures, $T_p = 805$, 831 and 905 K.

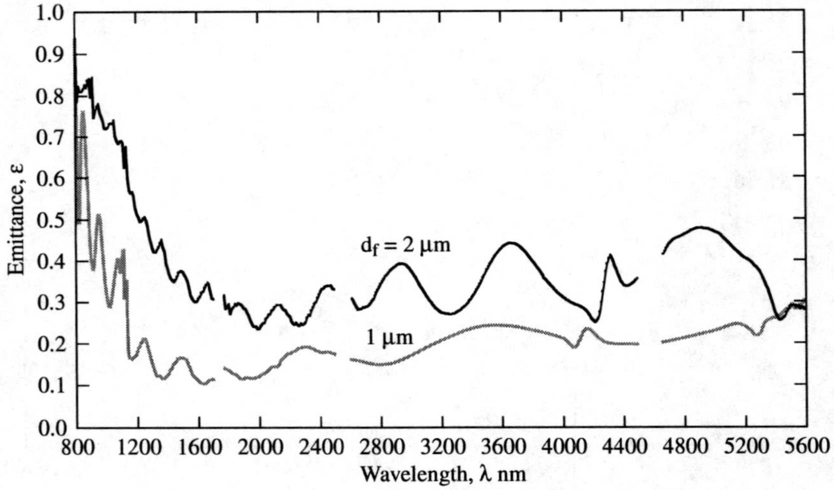

Figure 7. Experimental spectral emittance, $\varepsilon(\lambda)$, for pure silicon films at a temperature, $T_p = 831$ K, on a sapphire substrate of thickness, $d_s = 525$ μm, with platinum backing at two silicon film thicknesses, $d_f = 1$ and 2 μm.

APPENDIX A—EMITTANCE MODEL

A radiation transfer analysis similar to that in Ref. 2 was used to calculate $\varepsilon(\lambda)$. Referring to figure 1, the radiation intensity, i^+ in the +x direction and the intensity, i^-, in the –x direction are given by the following equations [6] for a one-dimensional geometry with boundaries at $x = 0$ and $x = d$.

$$i^+(\kappa,\mu) = i^+(0,\mu)\exp\left(-\frac{\kappa}{\mu}\right) + \int_o^{\kappa} S(\kappa^*,\mu)\exp\left[-\frac{(\kappa-\kappa^*)}{\mu}\right]\frac{d\kappa^*}{\mu} \qquad 0 \le \mu \le 1 \qquad (A-1)$$

$$i^-(\kappa,\mu) = i^-(\kappa_d,\mu)\exp\left(\frac{\kappa_d-\kappa}{\mu}\right) - \int_{\kappa}^{\kappa_d} S(\kappa^*,\mu)\exp\left[\frac{\kappa^*-\kappa}{\mu}\right]\frac{d\kappa^*}{\mu} \qquad -1 \le \mu \le 0 \qquad (A-2)$$

Where $\mu = \cos\theta$ and θ is the angle between the $+x$-axis and the intensity direction. $S(\kappa,\mu)$ is the so-called source function, $i^+(0, \mu)$ is the intensity in the +direction at $x = 0$ ($\kappa = 0$) and $i^-(\kappa_d, \mu)$ is the intensity in the –direction at $x = d$ ($\kappa = \kappa_d$). The optical depth, κ is the following, assuming the extinction coefficient K is independent of x.

$$\kappa = Kx \qquad (A-3)$$

For single crystal materials with few crystal defects, scattering is small compared to absorptance. As a result the source function is the following [6].

$$S(\kappa,\lambda) = n_f^2 i_b(\lambda,T) = \frac{n_f^2}{\pi} e_b(\lambda,T) \qquad (A-4)$$

Where $i_b(\lambda,T)$ is the blackbody intensity and $e_b(\lambda,T)$ is the blackbody emissive power.

$$e_b(\lambda,T)d\lambda = \pi i_b(\lambda,T)d\lambda = \frac{2\pi c_1\, d\lambda}{\lambda^5\left[\exp(c_2/\lambda T) - 1\right]} \qquad (A-5)$$

$$c_1 = hc_o^2 = 0.595\times10^{16}\ W\cdot\mathrm{nm}^4/\mathrm{cm}^2 \qquad (A-6)$$

$$c_2 = \frac{hc_o}{k} = 14.388\times10^6\ \mathrm{nmK} \qquad (A-7)$$

As Eq. (A–4) indicates, the source function for negligible scattering depends only on the temperature, T, and the wavelength, λ. Thus, if T is constant across the film the source function will be a constant. However, since S is exponentially dependent on T, even small changes in temperature have a significant effect on the emittance [2,7,8].

The hemispherical spectral emittance, $\varepsilon(\lambda)$, for emission into a dielectric with index of refraction, n_o, is defined as follows.

$$\varepsilon(\lambda) \equiv \frac{q_{out}(\lambda)}{n_o^2 e_b(\lambda,T)} \tag{A-8}$$

Where $q_{out}(\lambda)$ is the radiation flux leaving the emitter surface.

$$q_{out} = 2\pi\int_o^{\pi/2} i^+\left(d_f^+\right)\cos\theta\sin\theta d\theta = 2\pi\int_o^1 i^+\left(d_f^+\right)\mu\, d\mu \tag{A-9}$$

And $i^+(d_f^+)$ is the intensity in the $+x$-direction on the $+x$-side of the vacuum-film interface at $x = d_s + d_f$. To determine q_{out} the radiation transfer equations must be solved with appropriate boundary conditions at the three interfaces ($x = 0$, $x = d_s$, and $x = d_f + d_s$).

The boundary conditions are the following. At $x = 0$ the intensity moving in the $+x$-direction, $i^+(0)$, is the sum of the radiation emitted by the metal into the substrate and the reflected intensity.

$$i^+(0) = n_s^2\varepsilon_{ps} i_b\left(\lambda, T_s\right) + R'_{ps} i^-(0) \tag{A-10}$$

Where ε_{ps} is the directional hemispherical spectral emittance of the metal into the dielectric substrate and R'_{ps} is the directional reflectivity at the metal-substrate interface. The reflectivity can be approximated by the value for normal incidence, $\overline{R}_{ps}$, since only for $\theta \approx \pi/2$ does R' change significantly [9]. In that case.

$$R'_{ps} \approx \overline{R}_{ps} = \frac{\left(n_s - n_{pR}\right)^2 + n_{pI}^2}{\left(n_s + n_{pR}\right)^2 + n_{pI}^2} \tag{A-11}$$

Where n_s is the index of refraction of the dielectric substrate and n_{sR} and n_{sI} are the real and imaginary parts of the index of refraction of the metal. Also, from Kirchhoff's law [6] for an opaque metal.

$$\varepsilon_{ps} = 1 - R'_{ps} \approx 1 - \overline{R}_{ps} \tag{A-12}$$

In the results for $\varepsilon(\lambda)$ to be presented later, Eqs. (A–11) and (A–12) are used to calculate ε_{ps} and R'_{ps}.

At $x = d_s$ the boundary conditions are the following. The intensity in the $+x$ direction on the $+x$-side of the film-substrate interface, $i^+(d_s^+)$, is the sum of the intensity transmitted from the $-x$-side of the interface and the intensity reflected from the $+x$ side of the interface.

$$i^+\left(d_s^+\right) = \left(1 - R'_{fs}\right)i^+\left(d_s^-\right) + R'_{fs} i^-\left(d_s^+\right) \tag{A-13}$$

Similarly, for the intensity in the $-x$ direction on the $-x$ side of the interface, $i^-(d_s^-)$, the following condition applies.

$$i^-\left(d_s^-\right) = \left(1 - R'_{fs}\right)i^-\left(d_s^+\right) + R'_{fs}i^+\left(d_s^-\right) \quad \text{(A-14)}$$

At this interface Snell's law [9] must be satisfied. Also, since the film is a semiconductor, its index of refraction will have real and imaginary parts, $n_f = n_{fR} - jn_{fI}$. In addition, $n_{fR} > n_s$ for silicon and sapphire. As a result, for $i^-(d_s^+)$ there is a critical incidence angle, θ_{Mfs}, for total reflection. That is for $\theta > \theta_{Mfs}$, $R'_{fs} = 1$ for $i^-(d_s^+)$. For $i^+(d_s^-)$ when $n_{fR} > n_s$, $R'_{fs} < 1$ for all angles of incidence. In general the critical angle, θ_M, is a function of n_s, n_{fR} and n_{fI}. However, since $n_{fR} \approx 4$ for silicon and $n_s \approx 1.9$ for sapphire the approximation $n_{fR}^2 >> n_s^2$ can be made so that the critical angle is the same as that for the interface between two dielectrics.

$$\cos^2\theta_{Mfs} = \mu_{fs}^2 = 1 - \left(\frac{n_s}{n_{fR}}\right)^2 \quad \text{(A-15)}$$

We assume $n_{fR} >> n_{fI}$ and approximate R'_{fs} by the reflectively for normal incidence, $\overline{R}_{fs}$.

$$R'_{fs} \approx \overline{R}_{fs} = \left(\frac{n_{fR} - n_s}{n_{fR} + n_s}\right)^2 \quad \text{(A-16)}$$

Equation (A–16) applies to $i^+(d_s^-)$ for all angles of incidence. For $i^-(d_s^+)$ Eq. (A–16) applies for $\theta_{Mfs} + \pi/2 < \theta < \pi$ and $R'_{fs} = 1$ for $\pi/2 \le \theta \le \theta_{Mfs} + \pi/2$.

At $x = d_s + d_f$ the intensity in the $-x$-direction, $i^-(d_f^-)$, is just the reflected intensity.

$$i^-\left(d_f^-\right) = R'_{fo}i^+\left(d_f^-\right) \quad \text{(A-17)}$$

Again, since $n_{fR} > n_o$, Snell's law must be satisfied. Therefore,

$$R'_{fo} = 1 \qquad \text{for} \qquad \theta_{Mfo} \le \theta \le \pi/2 \quad \text{(A-18)}$$

and

$$\cos^2\theta_{Mfo} = \mu_{fo}^2 = 1 - \left(\frac{n_o}{n_{fR}}\right)^2 \quad \text{(A-19)}$$

Also, approximate R'_{fo} by the reflectivity for normal incidence, $\overline{R}_{fo}$, when $0 \le \theta \le \theta_{Mfo}$.

$$R'_{fo} \approx \overline{R}_{fo} = \left(\frac{n_{fR} - n_o}{n_{fR} + n_o}\right)^2 \quad \text{for} \quad 0 \le \theta < \theta_{Mfo} \tag{A-20}$$

The intensity leaving the film, $i^+(d_f^+)$, is just the transmitted intensity.

$$i^+\left(d_f^+\right) = \left(1 - R'_{fo}\right) i^+\left(d_f^-\right) \tag{A-21}$$

Using Eq. (A–21) in Eq. (A–9) yields the following result.

$$q_{out}(\lambda) = 2\pi \int_o^1 \left(1 - R'_{fo}\right) i^+\left(d_f^-\right) \mu \; d\mu \tag{A-22}$$

The intensity $i^+(d_f^-) = i^+(\kappa_f, \mu)$ is given by Eq. (A–1), where $i^+(0,\mu) = i^+(d_s^+)$, $S(\kappa^*, \mu)$ is given by Eq. (A–4) and the optical depth, κ_f, is the following.

$$\kappa_f = K_f d_f \tag{A-23}$$

Where K_f is the extinction coefficient of the semiconductor film. We assume the intensity $i^+(d_s^+)$ is isotropic (independent of angle). As a result, using Eq. (A–1) for $i^+(d_f^-)$ in Eq. (A–22) with R'_{fo} given by Eqs. (A–18) and (A–20) yields the following result.

$$q_{out}(\lambda) = \left(1 - \overline{R}_{fo}\right) \left\{ 2 \left[E_3\left(\kappa_f\right) - \mu_{fo}^2 E_3\left(\frac{\kappa_f}{\mu_{fo}}\right) \right] q^+\left(d_s^+\right) + \phi_+\left(\kappa_f\right) - \phi_{M+}\left(\kappa_f\right) \right\} \tag{A-24}$$

Where $E_3(x)$ is the third order exponential integral function, $q^+(d_s^+)$ the radiation flux in the $+x$-direction on the +side of $x = d_s$ and $\phi_{M+}(\kappa_f)$ and $\phi_{M+}(\kappa_f)$ are source function integrals defined below.

$$E_n(x) = \int_o^1 u^{n-2} e^{-\frac{x}{u}} du \tag{A-25}$$

$$q^+\left(d_s^+\right) = 2\pi \int_o^1 i^+\left(d_s^+\right) \mu d\mu = \pi i\left(d_s^+\right) \tag{A-26}$$

$$\phi_+\left(\kappa_f\right) = 2 n_f^2 \int_o^{\kappa_f} e_b(\lambda, T) E_2\left(\kappa_f - \kappa\right) d\kappa \tag{A-27}$$

$$\phi_{M+}(\kappa_f) = 2\mu_{fo} n_f^2 \int_o^{\kappa_f} e_b(\lambda,T) E_2\left(\frac{\kappa_f - \kappa}{\mu_{fo}}\right) d\kappa \qquad (A-28)$$

As Eq. (A–24) shows, q_{out} can be determined once $q^+(d_s^+)$, $\phi_+(\kappa_f)$ and $\phi_{M+}(\kappa_f)$ are known. The solution for $q^+(d_s^+)$ is obtained by applying the boundary conditions described earlier and the radiation transfer equations (Eqs. (A–1) and (A–2)). The source function integrals can be evaluated once $T(x)$ is known. For very thin materials (≤10 μm) the temperature will be nearly constant through the film. In that case the source function integrals an be easily evaluated. However, for thicker materials temperature changes will occur that significantly affect the source function integrals. Earlier studies [8] with dielectric low thermal conductivity selective emitters have shown that a linear temperature variation is a good approximation for films less than 1 mm thickness. In other words, thermal conduction dominates radiation energy transfer for thin dielectrics. Since Si has a higher thermal conductivity than dielectrics, the linear temperature variation will certainly be applicable for the Si film. A linear temperature variation will also be assumed for the sapphire substrate. With the assumed linear temperature variation the source function integrals can be evaluated.

We now proceed to outline the method for obtaining $q^+(d_s^+)$. Application of boundary conditions and the use of the radiation transfer equations produces a set of 6 linear algebraic equations for the radiation fluxes at the boundaries, $q^+(0)$, $q^+(d_s^+)$, $q^-(d_s^+)$, $q^+(d_s^-)$, $q^-(d_s^-)$ and $q^-(d_f^-)$. These are then solved for $q^+(d_s^+)$. Boundary conditions on the intensities are given by Eqs. (A–10), (A–13), (A–14), and (A–17). In addition, at the film-substrate interface ($x = d_s$) the total radiation flux must be continuous so that the following boundary condition on the radiation flux must be satisfied.

$$q^+(d_s^-) - q^-(d_s^-) = q^+(d_s^+) - q^-(d_s^+) \qquad (A-29)$$

The radiation fluxes $q^+(d_s^-)$ and $q^-(d_s^+)$ are obtained by using Eqs. (A–1) and (A–2).

$$q^+(d_s^-) = 2\pi \int_o^1 i^+(d_s^-)\mu d\mu = 2\pi \int_o^1 i^+(0) e^{-\frac{\kappa_s}{\mu}} \mu d\mu + 2\int_o^{\kappa_s} e_b(\lambda,T) d\kappa \int_o^1 \exp\left[\frac{\kappa_s - \kappa}{\mu}\right] d\mu$$

$$q^+(d_s^-) = 2q^+(0) E_3(\kappa_s) + \phi_+(\kappa_s) \qquad (A-30)$$

$$q^-(d_s^+) = -2\pi \int_{-1}^o i^-(d_s^+)\mu d\mu = -2\pi \int_{-1}^o i^-(d_f^-) e^{-\frac{\kappa_f}{\mu}} \mu d\mu - \int_o^{\kappa_f} e_b(\lambda,T) d\kappa \int_{-1}^o \exp\left[\frac{\kappa}{\mu}\right] d\mu$$

$$q^-(d_s^+) = 2q^-(d_f^-) E_3(\kappa_f) + \phi_-(\kappa_f) \qquad (A-31)$$

Where, $\kappa_s = K_s d_s$ and

$$\phi_+(\kappa_s) = 2n_s^2 \int_o^{\kappa_s} e_b(\lambda, T) E_2(\kappa_s - \kappa) d\kappa \quad \text{(A – 32)}$$

$$\phi_-(\kappa_f) = 2n_f^2 \int_o^{\kappa_f} e_b(\lambda, T) E_2(\kappa) d\kappa \quad \text{(A – 33)}$$

and we have assumed $i^+(0)$ and $i^-(d_f^-)$ are isotropic so that $q^+(0) = \pi i^+(0)$ and $q^-(d_f^-) = \pi i^-(d_f^-)$.

Using the boundary condition at $x = 0$ (Eq. (A–10)) and Eq. (A–2) for $i^-(0)$ and assuming $i^-(d_s^-)$ is isotropic $q^-(d_s^-) = \pi i^-(d_s^-)$ yields the following result.

$$q^+(0) = 2\pi \int_o^1 i^+(0)\mu d\mu = 2\pi \left[n_s^2 \varepsilon_{ps} \int_o^1 i_b(\lambda, T_p)\mu d\mu - \bar{R}_{ps} \int_{-1}^o i^-(0)\mu d\mu \right]$$

$$q^+(0) = n_s^2 \varepsilon_{ps} e_b(\lambda, T_p) + \bar{R}_{sp}\left[2q^-(d_s^-) E_3(\kappa_s) + \phi_-(\kappa_s) \right] \quad \text{(A – 34)}$$

Where

$$\phi_-(\kappa_s) = 2n_s^2 \int_o^{\kappa_s} e_b(\lambda, T) E_2(\kappa) d\kappa \quad \text{(A – 35)}$$

Using Eq. (A–1) for $i^+(d_s^-)$ and Eq. (A–2) for $i^-(d_s^+)$ in Eq. (A–13) yields the following result.

$$q^+(d_s^+) = 2\pi(1 - \bar{R}_{fs}) \int_o^1 i^+(d_s^-)\mu d\mu - 2\pi \left[\bar{R}_{fs} \int_{-1}^{-\mu_{fs}} i^-(d_s^+)\mu d\mu + \int_{-\mu_{fs}}^0 i^-(d_s^+)\mu d\mu \right]$$

$$q^+(d_s^+) = (1 - \bar{R}_{fs})\left[2q^+(0) E_3(\kappa_s) + \phi_+(\kappa_s) \right] + 2q^-(d_f^-)$$

$$\times \left[\bar{R}_{fs} E_3(\kappa_f) + \mu_{fs}^2 (1 - \bar{R}_{fs}) E_3\left(\frac{\kappa_s}{\mu_{fs}} \right) \right] + (1 - \bar{R}_{fs})\phi_{M-}(\kappa_f) + \bar{R}_{fs}\phi_-(\kappa_f) \quad \text{(A – 36)}$$

Where we have assumed $i^+(0)$ and $i^-(d_f^-)$ are isotropic so that $q^+(0) = \pi i^+(0)$ and $q^-(d_f^-) = \pi i^-(d_f^-)$, $\phi_+(\kappa_f)$ is given by Eq. (A–32), $\phi_-(\kappa_s)$ is given by Eq. (A–33) and $\phi_{M-}(\kappa_f)$ is the following

$$\phi_{M-}(\kappa_f) = 2\mu_{fs} n_f^2 \int_o^{\kappa_f} e_b(\lambda, T) E_2\left(\frac{\kappa}{\mu_{fs}} \right) d\kappa \quad \text{(A – 37)}$$

Now apply the boundary condition at $x = d_s + d_f$ (Eq. (A–17)). Using Eq. (A–1) for $i^+(d_f^-)$ yields the following result.

$$
\begin{aligned}
q^-\left(d_f^-\right) &= 2\pi\int_o^{\mu_{fo}} i^+\left(d_f^-\right)\mu d\mu + 2\pi\,\overline{R}_{fo}\int_{\mu_{fo}}^1 i^+\left(d_f^-\right)\mu d\mu\; q^-\left(d_f^-\right) \\
&= 2q^+\left(d_s^+\right)\left[\overline{R}_{fo}E_3\left(\kappa_f\right) + \mu_{fo}^2\left(1-\overline{R}_{fo}\right)E_3\left(\frac{\kappa_f}{\mu_{fo}}\right)\right] \\
&\quad + \overline{R}_{fo}\phi_+\left(\kappa_f\right) + \left(1-\overline{R}_{fo}\right)\phi_{M+}\left(\kappa_f\right) \qquad \text{(A – 38)}
\end{aligned}
$$

Where we have assumed $i^+(d_s^+)$ is isotropic so that $q^+(d_s^+) = \pi i^+(d_s^+)$ and $\phi_+(\kappa_f)$ and $\phi_{M+}(\kappa_f)$ are given by Eqs. (A–27) and (A–28).

Equations (A–29) to (A–31), (A–34), (A–36), and (A–38) make up a set of 6 linear algebraic equations for $q^-(d_s^-)$, $q^+(d_s^-)$, $q^-(d_s^+)$, $q^+(0)$, $q^+(d_s^+)$ and $q^-(d_f^-)$. After a great deal of algebra they can be solved for $q^+(d_s^+)$ to obtain the following result.

$$
\begin{aligned}
\frac{q^+\left(d_s^+\right)}{n_o^2 e_b\left(\lambda, T_p\right)} &= \frac{2}{D}\left(\frac{n_f}{n_o}\right)^2\Bigg\{\left(1-\overline{R}_{fs}\right)\Bigg[\frac{n_s^2}{n_f^2}\left(\varepsilon_{ps} + \overline{R}_{ps}\overline{\phi}_-\left(\kappa_s\right)\right)E_3\left(\kappa_s\right) \\
&\quad + \frac{n_s^2}{n_f^2}\overline{\phi}_+\left(\kappa_s\right) + h_m\overline{\phi}_{M-}\left(\kappa_f\right)\Bigg] + b\overline{\phi}_-\left(\kappa_f\right) + 2\left[bE_3\left(\kappa_f\right) + \mu_{fs}^2\left(1-\overline{R}_{fs}\right)h_M E_3\left(\frac{\kappa_f}{\mu_{fs}}\right)\right] \\
&\quad \times\left[\overline{R}_{fo}\overline{\phi}_+\left(\kappa_f\right) + \left(1-\overline{R}_{fo}\right)\overline{\phi}_{M+}\left(\kappa_f\right)\right]\Bigg\} n_f \ge n_s, n_f \ge n_o \qquad \text{(A – 39)}
\end{aligned}
$$

Where terms appearing in Eq. (A–39) are defined below.

$$
h_M = 1 - 4\overline{R}_{ps}E_3^2\left(\kappa_s\right) \qquad \text{(A – 40)}
$$

$$
b = \overline{R}_{fs} + 4\overline{R}_{ps}\left(1-2\overline{R}_{fs}\right)E_3^2\left(\kappa_s\right) \qquad \text{(A – 41)}
$$

$$
\begin{aligned}
D = 1 - 4\overline{R}_{ps}\overline{R}_{fs}E_3^2\left(\kappa_s\right) - 4\left[\overline{R}_{fo}E_3\left(\kappa_f\right) + \mu_{fo}^2\left(1-\overline{R}_{fo}\right)E_3\left(\frac{\kappa_f}{\mu_{fo}}\right)\right] \\
\times\left[bE_3\left(\kappa_f\right) + \mu_{fs}^2\left(1-\overline{R}_{fs}\right)h_M E_3\left(\frac{\kappa_f}{\mu_{fs}}\right)\right] \qquad \text{(A – 42)}
\end{aligned}
$$

$$\bar{\phi}_+(\kappa_f) \equiv \frac{\phi_+(\kappa_f)}{2n_f^2 e_b(\lambda, T_p)} = \kappa_f\left[e^{u_p} - 1\right]\int_o^1 \frac{E_2\left[\kappa_f(1-\bar{x})\right]}{\exp\left[\frac{u_p}{(1-\Delta T_s)(1-\bar{x}\Delta T_f)}\right]-1} d\bar{x} \qquad (A-43)$$

$$\bar{\phi}_+(\kappa_s) \equiv \frac{\phi_+(\kappa_s)}{2n_s^2 e_b(\lambda, T_p)} = \kappa_s\left[e^{u_p} - 1\right]\int_o^1 \frac{E_2\left[\kappa_s(1-\bar{x})\right]}{\exp\left[\frac{u_p}{1-\bar{x}\Delta T_s}\right]-1} d\bar{x} \qquad (A-44)$$

$$\bar{\phi}_-(\kappa_f) \equiv \frac{\phi_-(\kappa_f)}{2n_f^2 e_b(\lambda, T_p)} = \kappa_f\left[e^{u_p} - 1\right]\int_o^1 \frac{E_2\left[\kappa_s\bar{x}\right]}{\exp\left[\frac{u_p}{1-(\Delta T_s)(1-\bar{x}\Delta T_f)}\right]-1} d\bar{x} \qquad (A-45)$$

$$\bar{\phi}_-(\kappa_s) \equiv \frac{\phi_-(\kappa_s)}{2n_s^2 e_b(\lambda, T_p)} = \kappa_s\left[e^{u_p} - 1\right]\int_o^1 \frac{E_2\left[\kappa_s\bar{x}\right]}{\exp\left[\frac{u_p}{1-\bar{x}\Delta T_s}\right]-1} d\bar{x} \qquad (A-46)$$

$$\bar{\phi}_{M+}(\kappa_f) \equiv \frac{\phi_{M+}(\kappa_f)}{2n_f^2 e_b(\lambda, T_p)} = \mu_{fo}\kappa_f\left[e^{u_p} - 1\right]\int_o^1 \frac{E_2\left[\frac{\kappa_f}{\mu_{fo}}(1-\bar{x})\right]}{\exp\left[\frac{u_p}{(1-\Delta T_s)(1-\bar{x}\Delta T_f)}\right]-1} d\bar{x} \qquad (A-47)$$

$$\bar{\phi}_{M-}(\kappa_f) \equiv \frac{\phi_{M-}(\kappa_f)}{2n_f^2 e_b(\lambda, T_p)} = \mu_{fs}\kappa_f\left[e^{u_p} - 1\right]\int_0^1 \frac{E_2\left[\frac{\kappa_f}{\mu_{fs}}\bar{x}\right]}{\exp\left[\frac{u_p}{(1-\Delta T_s)(1-\bar{x}\Delta T_f)}\right]-1} d\bar{x} \qquad (A-48)$$

$$u_p \equiv \frac{c_2}{\lambda T_p} \qquad (A-49)$$

In obtaining the dimensionless source function integrals (Eqs. (A–43) to (A–48)) we have assumed linear temperature variations through the emitting film and substrate. The temperature changes ΔT_s and ΔT_f are defined as follows.

$$\Delta T_s \equiv \frac{T_p - T_s}{T_p} \qquad (A-50a)$$

$$\Delta T_f \equiv \frac{T_s - T_f}{T_s} \qquad (A-50b)$$

Also, remember that the results for the radiation fluxes apply when $n_f \geq n_s$ and $n_f \geq n_o$.

Determination of Temperature Changes

To determine ΔT_s and ΔT_p we apply an overall energy balance, again assuming that thermal conduction dominates radiation transfer through the film and substrate. As a result,

$$k_{ths}\frac{T_p - T_s}{d_s} = k_{ths}\frac{T_p - \Delta T_s}{d_s} = k_{thf}\frac{T_s - T_f}{d_f} = k_{thf}\frac{T_p(1-\Delta T_s)\Delta T_f}{d_f} = Q_{out} \qquad (A-51)$$

Where k_{ths} and k_{thf} are the thermal conductivities of the film and substrate. In a vacuum, all energy leaving the film is by radiation. As a result,

$$Q_{out} = \int_o^\infty n_o^2 \varepsilon\left(\lambda, T_p, \Delta T_s, \Delta T_f\right) e_b\left(\lambda, T_p\right) d\lambda \qquad (A-52)$$

Where $\varepsilon(\lambda, T_p, \Delta T_s, \Delta T_s)$ is the spectral emittance defined in terms of the temperature of the substrate-metal interface, T_p. Equations (A–51) and (A–52) can be solved for ΔT_s and ΔT_f in terms of T_p by an iteration process once ε is known.

An inconsistency exists at the emitting film-vacuum interface in assuming that Eq. (A–51) applies there. At this interface $-k_{th}\, dT/dx = 0$, whereas Eq. (A–51) applies for a linear temperature variation, $-\frac{dT}{dx} = \frac{T_p(1-\Delta T_s)T_f}{d_f}$. To satisfy $dT/dx = 0$ at the film-vacuum interface, the temperature change through the film will be larger than our assumed linear variation. As a result, the emittance calculated using the linear temperature variation will be larger than the actual emittance.

The emittance is defined in terms of T_p.

$$\varepsilon\left(\lambda,\kappa_f,\kappa_s,T_p,\Delta T_s,\Delta T_f\right)=\frac{q_{out}\left(\lambda,\kappa_f,\kappa_s,T_p,\Delta T_s,\Delta T_p\right)}{n_o^2 e_b\left(\lambda,T_p\right)} \qquad (A-53)$$

Where q_{out} is obtained from Eq. (A–24) so that ε is the following

$$\varepsilon\left(\lambda,\kappa_s,\kappa_f,T_p,\Delta T_s,\Delta T_p\right)=2\left(1-\bar{R}_{fo}\right)$$
$$\times\left\{\left[E_3\left(\kappa_f\right)-\mu_o^2 E_3\left(\frac{\kappa_f}{\mu_{fo}}\right)\right]\frac{q_+\left(d_s^+\right)}{n_o^2 e_b\left(\lambda,T_p\right)}+\left(\frac{n_f}{n_o}\right)^2\left[\bar{\phi}_+\left(\kappa_f\right)-\bar{\phi}_{M+}\left(\kappa_f\right)\right]\right\} \qquad (A-54)$$

The radiation flux $q^+(d_s^+)$ is given by Eq. (A–39).

The iterative procedure for solving Eqs. (A–51) and (A–52) for ΔT_s and ΔT_f in terms of T_p is the following. Initial guesses for ΔT_s and ΔT_f are used in Eq. (A–54) to calculate ε. Then Q_{out} is calculated using ε in Eq. (A–52). With this value of Q_{out}, new values of ΔT_s and ΔT_f are calculated using Eq. (A–51). With the new values for ΔT_s and ΔT_f the procedure is repeated until convergence occurs.

Limiting Solutions for Emittance

As Eqs. (A–39) and (A–54) indicate, ε is a rather complicated function of κ_f, κ_s, ΔT_s, ΔT_p, T_p, n_s, n_f and n_p. However, for $\kappa_f \to \infty$ and $\kappa_f = 0$ the emittance is greatly simplified. For a semiconductor, κ_f is large for photon energies, E, greater than the bandgap energy, E_g, ($\lambda < \lambda_g = hc_o/E_g$). For $\kappa_f \to \infty$, $E_3(\kappa_f) = E_3(\kappa_f/\mu_{fs}) = 0$. As a result Eq. (A–54) becomes the following.

$$\lim_{\kappa_f\to\infty}\varepsilon\left(\lambda,T_p,\Delta T_s,\Delta T_f\right)=\lim_{\kappa_f\to\infty}\left\{2\left(1-\bar{R}_{fo}\right)\left(\frac{n_f}{n_o}\right)^2\left[\bar{\phi}_+\left(\kappa_f\right)-\bar{\phi}_{M+}\left(\kappa_f\right)\right]\right\} \qquad (A-55)$$

If $u_p/(1-\Delta T_s) >> 1$ and $u_p\Delta T_f/(1-\Delta T_s)\kappa_f << 1$ then Eqs. (A–43) and (A–47) can be expanded to obtain the following.

$$\lim_{\kappa_f\to\infty}\bar{\phi}_+\left(\kappa_f\right)=\frac{1}{2}\exp\left[-\frac{u_p\Delta T_f}{\left(1-\Delta T_s\right)}\right]=\frac{1}{2}\exp\left[-\frac{c_2\Delta T_f}{\lambda T_p\left(1-\Delta T_s\right)}\right] \qquad (A-56)$$

$$\lim_{\kappa_f\to\infty}\bar{\phi}_{M+}\left(\kappa_f\right)=\frac{1}{2}\mu_{fo}^2\exp\left[-\frac{u_p\Delta T_f}{\left(1-\Delta T_s\right)}\right]=\frac{\mu_{fo}^2}{2}\exp\left[-\frac{c_2\Delta T_f}{\lambda T_p\left(1-\Delta T_s\right)}\right] \qquad (A-57)$$

For Si and T_p < 1500 K, u_p > 8.5 so that the approximation $u_p/(1 - \Delta T_s) >> 1$ applies. Therefore, Eqs. (A–56) and (A–57) are applicable. Using (A–56) and (A–57) in (A–55) yields the following result.

$$\lim_{\kappa_f \to \infty} \varepsilon(\lambda, T_p, \Delta T_s, \Delta T_f) = (1 - \overline{R}_{fo}) \exp\left[-\frac{c_2 \Delta T_f}{\lambda T_p (1 - \Delta T_s)} \right] \qquad c_2/\lambda T_p >> 1 \quad (A-58)$$

Thus $\varepsilon(\lambda)$ has an exponential dependence on the temperature change across the emitting film. This dependence on temperature change has been observed for rare earth selective emitters [2,7,8]. Notice that if $\Delta T_f = 0$, $\varepsilon = 1 - \overline{R}_{fo}$, which is the usual result for an opaque material [6].

Now consider $\varepsilon(\lambda)$ for $\kappa_f = 0$, which approximates the wavelength region $\lambda_g < \lambda < \lambda_c$, where λ_c is the wavelength region where the semiconductor lattice vibrations result in large κ_f. If $\kappa_f = 0$, Eqs. (A–43) and (A–47) show that $\overline{\phi}_+(\kappa_f) = \overline{\phi}_{M+}(\kappa_f) = 0$. Also, $E_3(0) = 1/2$ and as a result from Eq. (A–54) $\varepsilon(\lambda)$ becomes the following.

$$\varepsilon(\lambda, \kappa_f = 0) = (1 - \overline{R}_{fo}) \frac{q^+(d_s^+)}{n_o^2 e_b(\lambda, T_p)} \qquad (A-59)$$

Using Eq. (A–39) for $\kappa_f = 0$ results in the following.

$$\varepsilon(\lambda, \kappa_f = 0) = \frac{2}{D} \left(\frac{n_s}{n_o} \right)^2 (1 - \overline{R}_{fo})(1 - \overline{R}_{fs}) \left[(\varepsilon_{ps} + \overline{R}_{ps} \overline{\phi}_-(\kappa_s)) E_3(\kappa_s) + \overline{\phi}_+(\kappa_s) \right] \qquad (A-60)$$

For sapphire the extinction coefficient is small for $\lambda < 5$ μm. Therefore, we can also make the approximation $\kappa_s \approx 0$. In that case Eq. (A–60) becomes the following.

$$\varepsilon(\lambda, \kappa_f = \kappa_s = 0) = \frac{1}{D_o} \left(\frac{n_s}{n_o} \right)^2 (1 - \overline{R}_{fo})(1 - \overline{R}_{fs}) \, \varepsilon_{ps} \qquad (A-61)$$

Where,

$$D_o = 1 - \overline{R}_{fs} \overline{R}_{ps} - \left[\overline{R}_{fo} + \mu_{fo}^2 (1 - \overline{R}_{fo}) \right] \times \left[\overline{R}_{fs} + \overline{R}_{ps} (1 - 2\overline{R}_{fs}) + \mu_{fs}^2 (1 - \overline{R}_{fs}) (1 - \overline{R}_{ps}) \right] \qquad (A-62)$$

As Eq. (A–61) shows, even through $\kappa_f = \kappa_s = 0$, small emittance can not be achieved if the metal emittance into the substrate, ε_{ps}, is large and $\overline{R}_{fo}$ and $\overline{R}_{fs}$ are small.

APPENDIX B—OPTICAL PROPERTIES OF SILICON, SAPPHIRE AND PLATINUM

Before the emittance can be calculated the extinction coefficients of the emitting semiconductor film and substrate, as well as, the indices of refraction for the film, substrate and metal backing must be known. As noted earlier, the extinction coefficient for a semiconductor will be temperature dependent, as well as, wavelength dependent.

In the wavelength region where the photon energy is less than the bandgap energy, $\lambda > \lambda_g$, we used the free carrier absorption coefficient of Vandenabeele and Maer [4] to approximate the extinction coefficient of silicon.

$$K_{fr} = \alpha_{fr} = 1.23 \times 10^{-9} \lambda^{1.51} T^{2.95} \exp\left[-\frac{7000}{T}\right] \text{ cm}^{-1} \qquad (B-1)$$

Where λ is in nm and T in kelvin. As Eq. (A–63) indicates, α_{fr} is a rapidly increasing function of temperature. Thus, the emittance for $E < E_g$ ($\lambda > \lambda_g$) will also increase with increasing T; an undesirable characteristics for a selective emitter.

For $E > E_g$ where interband transitions account for absorption (or emittance) we approximated the extinction coefficient by the absorption coefficient given in Ref. [10], which includes the variation of E_g with temperature.

$$K_b = \sum_{\substack{i=1,2 \\ j=1,2}} C_i A_j \left[\frac{\left(\frac{1240}{\lambda} - E_{gj}(T) + E_{pi}\right)^2}{\exp\left(\frac{E_{pi}}{kT}\right) - 1} + \frac{\left(\frac{1240}{\lambda} - E_{gj}(T) - E_{pi}\right)^2}{1 - \exp\left(-\frac{E_{pi}}{kT}\right)} \right] + A_d \left[\frac{1240}{\lambda} - E_{gd}\right]^{1/2} \text{ cm}^{-1} \qquad (B-2)$$

Where, $E_{g1} = 1.557$eV, $E_{g2} = 2.5$eV, $E_{gd} = 2.5$eV, $E_{p1} = 1.827 \times 10^{-2}$eV, $E_{p2} = 5.773 \times 10^{-2}$eV, $C_1 = 5.5$, $C_2 = 4.0$, $A_1 = 3.231 \times 10^2$ cm^{-1}eV^{-2}, $A_2 = 7.237 \times 10^3$ cm^{-1}eV^{-2}, $A_d = 1.052 \times 10^6$cm^{-1}eV^{-2} and λ is nm and energies are in eV. In the numerical calculation of ε versus λ, $K = K_b$ for $0 \le \lambda \le \lambda_{fr}$ where λ_{fr} is the wavelength where $K_b = K_{fr}$. For $\lambda > \lambda_{fr}$, $K = K_{fr}$ was used.

For the sapphire substrate, which is transparent for $\lambda < 3.0$ µm, we used a curve fit to the absorption coefficient at $T = 1000$ C supplied on the website of Saphikon (www.saphikon.com), a manufacturer of sapphire.

$$K_s = 0.2 \text{ cm}^{-1} \qquad \text{for} \qquad \lambda < 3000 \text{ nm}$$

$$(B-3)$$

$$K_s = 5.76 \times 10^{-4} \exp\left[1.80 \times 10^{-3} \lambda\right] \text{cm}^{-1} \qquad \text{for} \qquad \lambda > 3000 \text{ nm}$$

Where λ is in nm.

The index of refraction for sapphire used in the calculation of $\varepsilon(\lambda)$ was the following curve fit to the data supplied by Saphikon.

$$n_s = 1.874 - 1.865\times10^{-4}\lambda + 1.284\times^{-7}\lambda^2 - 4.057\times^{-11}\lambda^3 + 5.563\times10^{-15}\lambda^4 - 2.891\times10^{-19}\lambda^5 \qquad (B-4)$$

Where λ is in nm.

For the index of refraction of silicon we used the emittance data of Sato [3] at $T = 1000$ K in the wavelength region $\lambda < \lambda_g$ where ε is nearly constant and $\kappa_f \rightarrow \infty$ and Eq. (A–58) with $\Delta T_f = 0$ to obtain $\bar{R}_{fo}$. And then Eq. (A–20) to obtain n_{fR}. This data was then fit with the following expressions.

$$n_{fR} = 16.044\lambda^{-0.194} \qquad 0 \leq \lambda \leq 4000 \text{ nm}$$

$$n_{fR} = 3.20 \qquad 4000 \leq \lambda \leq \infty \qquad (B-5)$$

Where again λ is in nm. This data applies for $T = 1000$ K. However, since variation in n_{fR} with T is small we used Eq. (A–67) for all temperatures. For the real and imaginary parts of the index of refraction of the platinum backing on the sapphire substrate the tabular data in Ref. 11 was used.

Calculation of the temperature changes ΔT_f and ΔT_s requires the thermal conductivities of silicon, k_{thf}, and sapphire, k_{ths}. For sapphire we used $k_{ths} = 0.1$ W/cmK which was obtained from Saphikon. This value applies for $T \geq 1200$ K, but we used it for $T < 1200$ K. Since $k_{ths} > 0.1$ for $T < 1200$K, our estimate of ΔT_s will be larger than the actual value. For silicon, we used the result that applies for a semiconductor at high temperature [12], $k_{th} \sim 1/T$. Using the value [12] $k_{thf} = 1.3$ W/cmK at $T = 300$ K we obtain the following.

$$k_{thf} = \frac{390}{T} \text{W / cmK} \qquad (B-6)$$

REFERENCES

1. DiMatteo, R.S., Grieff, P., Finberg, S.L., Young-Waithe, K.A., Choy, H.K.H., Masaki, M.M., and Fonstad, C.G., *Appl. Phys. Lett.*, 79, 1894 (2001).
2. Chubb, D.L., Pal, A.T., Patton, M.O., and Jenkins, P.P., *Journal of the European Ceramic Soc.*, 19, 1999, pp. 2551–2562.
3. Sato, T., *Japanese J. Appl. Phys.* 6, 339 (1967).
4. Vandenabeele, P., and Maex, K., *J. Appl. Phys.* 72, 5867 (1992).
5. Timans, P.J., *J. Appl. Phys.* 74, 6353 (1993).
6. Siegel, R. and Howell, J.R., *Thermal Radiation Heat Transfer*, 2nd edition, Washington DC, Hemisphere, 1981, Chs. 4, 14.
7. Chubb, D.L., Good, B.S., Clark, E.B., and Chen, Z., "Effect of Temperature gradient on Thick Film Selective Emitter Emittance," *Proceedings of 3rd NREL Conference on Thermophotovoltaic Generation of Electricity*, AIP Conference Proceedings 401, 1997, pp. 293–313.
8. Good, B.S., Chubb, D.L., and Pal, A., "Temperature Gradient Effects in an Erbirum Aluminum Granet Selective Emitter," *4th NREL Conference on Thermphotovoltaic Generation of Electricity,* AIP Conference Proceedings 460, 1999, pp. 214–223.
9. Born, M., and Wolf, E., *Principles of Optics, Electromagnetic Theory of Propagation Interference and Diffraction of Light*, Pergamon Press, Fourth edition, 1970, Chapter 1.
10. Rajkanan, K., Singh, R., and Shewchun, *J., Solid State Electronics*, 22, 792 (1979).
11. Lide, D.R. (Editor), *CRC Handbook of Chemistry and Physics*, 71st ed. CRC Press, 1990.
12. Wolf, H.F., *Semiconductors*, Wiley-Interscience, New York, 1971.
13. Höfler, H., Würfel, P. and Ruppel, W., *Solar Cells*, 10, 257 (1983).
14. Fraas, L., Samaras, J., Avery, J., and Groeneweld, M., "Antireflection Coated Refractory Metal Matched Emitters for use With Thermophotovoltaic Generators," *16th European Photovoltaic Solar Energy Conference and Exhibition*, Glasgow, U.K., May 1–5, 2000.

A STRUCTURAL STUDY OF InGaAs/InGaAs STRAIN-BALANCED MQW FOR TPV APPLICATIONS

S. Tundo[a,b,*], M. Mazzer[b], L. Nasi[c], L. Lazzarini[c], C. Ferrari[c], G. Salviati[c], A. Passaseo[d], G. Torsello[b], D. Diso[b], A. Licciulli[a], K. Barnham[e], N. Ekins Daukes[e], C. Rohr[e], P. Abbott[e], G. Clarke[f]

[a] *Dip. Ingegneria dell'Innovazione, Università di Lecce, via Monteroni, 73100– Lecce, Italy*
[b] *CNR-IMM, Sez. di Lecce, Via Arnesano, I-73100 Lecce, Italy*
[c]*CNR-MASPEC, Parco Area delle Scienze 37/A, I-43010 Fontanini- Parma, Italy*
[d] *NNL-INFM via Arnesano, I-73100 Lecce, Italy*
[e]*Blackett Laboratory, Imperial College, London SW7 2BW, UK*
[f]*IQE Ltd, Cardiff, UK*

Abstract. Multi-quantum well photovoltaic cells offer a number of advantages over conventional "single-gap" cells for thermophotovoltaic applications, first of all because they can reach a higher open circuit voltage under the same radiation source and with the same absorption edge. Material quality issues and the constraints imposed by the commercial available substrates indicate that $In_xGa_{1-x}As/In_yGa_{1-y}As$ /InP strain-balanced heterostructures are suitable to obtain good quality multi-quantum wells with an absorption edge just below 2.0 μm. Structural stability in the presence of a high density of elastic energy such as in the case of a strain-balanced multi-layer is a very important issue to be addressed by optimising key parameters like composition, thickness of wells and barriers and number of periods. In this paper we present and discuss the mechanisms of plastic relaxation of these structures with a particular attention to the impact of the extended defects generated by the local breakdown of the crystal lattice to the electrical properties of the devices. Then, after the presentation of the optimum structure with an absorption edge at 1.96 μm, we discuss the issue of a further extension of the absorption edge through the use of a so-called virtual substrate, that is a buffer structure between the substrate and the device designed to relax to a given extent with a minimum number of dislocations propagating towards the active region. On the basis of a recipe based on the experimental results on InGaAs single and multi-layers grown on GaAs, we have designed a series of step-graded buffer structures providing good virtual substrates with a lattice parameter larger than GaAs. Strain-balanced multi-quantum wells have been grown on $In_xGa_{1-x}As$ virtual substrates with $0.14 < x < 0.35$ with a residual density of threading dislocations of about 10^5 cm^{-2}. Work is in progress to remove the residual morphological undulation (cross hatch) induced by the misfit dislocations confined in the buffer structure and to extend this approach to InP.

*** corresponding author stefania.tundo@ime.le.cnr.it**

CP653, *Thermophotovoltaic Generation of Electricity: 5th Conference*
edited by T. J. Coutts, G. Guazzoni, and J. Luther

1. INTRODUCTION

Multi-quantum well photovoltaic cells has been proved to be highly advantageous for TPV energy conversion thanks to the enhancement of the open circuit voltage over a single-gap cell under the same radiation source and with the same absorption edge [1]. In order to extend the absorption edge to match long wavelength selective emitters, $In_xGa_{1-x}As/In_yGa_{1-y}As$ strain balanced multi-quantum wells (InGaAs SBMQW) are one of the best system both from the point of view of its electro-optical and its structural properties. For a given substrate, the absorption edge of an InGaAs SBMQW can be increased by increasing the compressive strain in the wells and the tensile strain in the barriers under the constraint that the balancing condition is satisfied. Although the balancing condition ensures that the multi-quantum well as a whole does not relax, as it happens in the case of unbalanced systems, the strain differential between wells and barriers cannot be pushed beyond a critical limit where the structure becomes unstable and starts to breakdown. We will briefly discuss the mechanisms of this structural breakdown and show how the structural parameters on which the critical strain-differential depends, can be adjusted in order to maximise the absorption edge.

In order to push the absorption edge beyond this limit it is necessary to increase the lattice parameter of the substrate. The only commercially available substrate whose lattice parameter is between GaAs and InAs is InP, therefore to balance the structure around a different value, one needs to develop a so-called virtual substrate, that is a partially relaxed buffer layer or multilayer designed to reach the desired lattice parameter.

A number of different models and recipes have been proposed so far to design virtual substrates [2] with the aim to confine the dislocations well away from the active regions of the devices. Most of these models are based on general laws extrapolated from the experiments [3] or on generalisations of the original Matthews and Blakeslee model [4].

In this paper we report a structural study of novel structures based on $In_yGa_{1-y}As/In_xGa_{1-x}As$ SB MQWs grown on InP substrates and on InGaAs virtual substrates designed to be grown on GaAs. The nature of the defects and the mechanisms of their formation are analysed and discussed in term of the optimisation of the photovoltaic cells.

2. EXPERIMENTAL DETAILS

The first series of samples investigated in this work are based on $In_yGa_{1-y}As$ (10 nm)/$In_xGa_{1-x}As$(10 nm) strain balanced (SB) multi quantum wells (MQWs) placed in the intrinsic region of an InGaAs p-i-n junction grown on (100) InP substrates at 670°C by MOVPE (Samples A). Compositions and thickness of wells and barriers were designed in order to extend the absorption edge between 1800 and 2000 nm. The sample details in terms of the MQWs strain arrangement are reported in TABLE 1.

TABLE 1. MQW structural details of samples A					
Sample	**Well strain**	**Barrier strain**	**Net strain**	N_w	**Defect clusters**
#1	0.61 %	-0.48 %	0.04 %	27	yes
#3	0.67 %	-0.48 %	0.07 %	none	no
#4	0.89 %	-1.35 %	-0.44 %	9	no
#5	0.99 %	-0.92 %	0.02 %	10	yes
#6	1.09 %	-1.35 %	-0.26 %	6	yes
#7	1.18 %	-1.54 %	-0.36 %	7	yes
#8	1.18 %	-1.58 %	-0.39 %	7	yes
#9	1.33 %	-1.12 %	-0.20 %	9	no

TABLE 1. Summary of the MQW main structural details of samples A obtained by the XRD and TEM data. N_{tot} is the entire number of periods in the MQWs and N_w the layer number at which the thickness modulations are observed to begin.

The second series of samples (Samples B) described in this paper were specifically designed to test the effect of virtual substrates on GaAs- based devices. Virtual substrates with predetermined lattice parameter at the top of the buffer layer and MD profiles were designed according to a semi-empirical recipe based on the strain relaxation model in InGaAs/GaAs multi-layer buffers proposed by Romanato et al.[2]. That model is based on the assumption that, once the equilibrium critical thickness is exceeded, the residual strain, ε, of a mismatched epitaxial layer decreases with the inverse of the square of the layer thickness [3]. The details of the recipe used to design the step-graded buffer layers for the virtual substrates of Samples B, will be reported in a forthcoming paper.

InGaAs p-i-n junctions containing strain balanced MQWs in the intrinsic region were deposited on virtual substrates corresponding to the equilibrium lattice parameter of $In_xGa_{1-x}As$ with $0.14 < x < 0.35$. Compositions and thickness of the 10-period InGaAs/InGaAs MQWs were chosen to meet the strain balance condition with respect to virtual substrate lattice parameter according to the design proposed by Ekins-Daukes et al.[5]. The samples were grown in a horizontal LP/MOCVD system (AIXTRON 200 AIX) on (001) exactly oriented Si doped GaAs substrates at the growth temperature of 650°C and growth rate of 4 μm/h. Only the InGaAs well were grown at the rate of 2 μm/h.

The TEM study was performed on a JEOL JEM 2000FX microscope operated at 200 kV. Observations were made on (100) plan view and (-110) cross section specimens with Bragg reflections g = 220 and g = 200, which are known to be highly sensitive to strain field and chemical composition, respectively. The X-ray characterisation was performed by the methods of Double Crystal X-Ray Topography (XRT) and High Resolution X-Ray Diffraction (HRXRD) in the CuKα 004 symmetrical and the 335 asymmetrical geometries using a Philips high resolution X-ray diffractometer.

The EBIC observations were performed at room temperature in a Philips XL20 scanning electron microscope equipped with a variable gain current amplifier (Keithley 428). The electron beam current was measured by means of a Faraday cage located on the stage near the sample before and after each EBIC measurement. The CL investigations were performed on a

Cambridge Stereoscan 350 SEM equipped with an Oxford Instruments mono-CL system at a sample temperature of 77K with the specimen in plan view (p layers on top).

3. RESULTS AND DISCUSSION

Strain-balanced multi-quantum wells grown on InP

The X-ray topography experimental results rule out the presence of misfit dislocations in all the type-A samples. The superlattice peaks of the X-ray rocking curves turn out to be broadening while a high density of randomly distributed defects were revealed by XRT, when the strain in the W/B layers becomes higher.

EBIC analysis, performed at several values of accelerating voltage, confirm XRT results and reveal that no plastic relaxation has occurred at the interface between the strain–balanced MQW and n region, that is all the samples are free of misfit dislocations. Taking into account the Monte Carlo simulations [6], the (electron-hole pair) generation volume already includes the n-region at 15 keV and no dislocations affect the collection efficiency even if we used higher values of accelerating voltage.

Nevertheless there is a clear correlation between the defects observed at the surface and the non radiative recombination centres giving rise to a strong EBIC contrast in the whole range of accelerating voltage. EBIC analysis performed at low accelerating voltage show that these defects have a high electrical activity and appears like dark dots (Fig.1a). Comparing with secondary electrons (SE) images (not reported here) these dark spots seem correlated with the centre of the hillock. Increasing the accelerating voltage these defects have a different electric response. When the interaction volume reaches the intrinsic region (electron beam voltage >15 keV) the defects appear like dark spots with a very strong non-radiative recombination activity, surrounded by a brighter halo (Fig.1b).

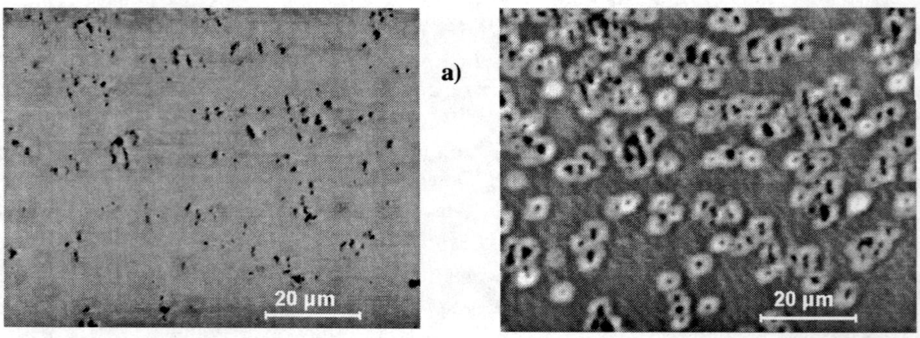

a) b)

FIGURE 1. EBIC micrographs of the same sample area at different beam energies E_b: $E_b = 5$ keV (a), $E_b = 20$ keV (b).

Despite these defects are electrically active, they have a limited effect on the dark current of these cells, which is comparable to the best control cells (same structure but with no quantum

wells). At the same time, however, the photocurrent drops dramatically at a relatively low forward bias voltage. The impact of these defects on the electric response of the photovoltaic cells in term of the microscopic properties of the defects is discussed elsewhere [7]
The TEM cross section analysis revealed that the generation of these defects is correlated to the onset of the thickness modulation in the MQWs. As a typical example, Fig. 2 shows the development of wavy growth in one of the structures investigated (W/B mismatch=0.95%).

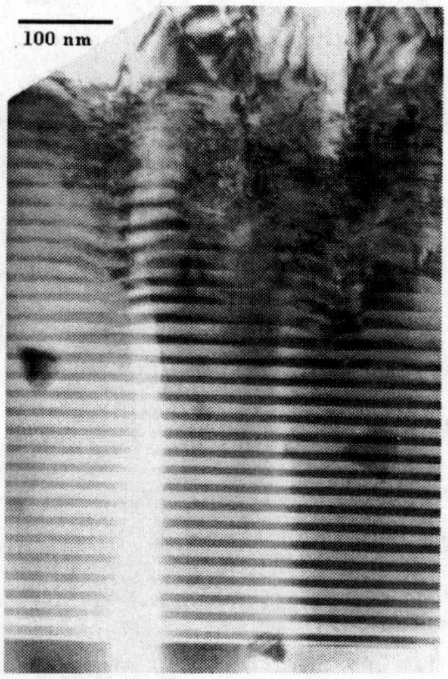

FIGURE 2. Bright field (g=002) TEM cross section showing the development of wavy growth in one of the structures investigated (W/B mismatch=0.95%). Wells and barriers correspond to darker and brighter layers respectively.

It was observed that the epitaxial growth front in the MQWs is flat up to a certain thickness where an observable undulation starts to develop. In agreement with the findings of other authors [8], these undulations always originate in the tensile barrier layers and propagate vertically with increasing amplitude through the MQW structure. It is also observed that wells and barriers are modulated in opposite phase and eventually columns of highly stressed material are developed. These highly stressed regions are the favourite sites for the local occurrence of a catastrophic plastic relaxation whose probability increases with increasing misfit between wells and barriers. The local collapse of one of either the tensile-strained or compressive strained layer brakes the balance in favour of the lattice parameter of one of the two materials. As a result the threshold for the onset of a 3D growth regime can be exceeded for the complementary layer and a cluster of extended defects develops rapidly and forms a conical structure threading up to the free surface where it gives rise to morphological depressions. The cluster density observed in cross section is about 1 in 5-6 μm. Away from the dislocated areas the MQW has a good structural quality with flat layers and sharp well and barrier interfaces.
Keeping the growth temperature constant, the thickness modulations are observed to originate after a number of periods, N_w, which decreases by increasing the W/B mismatch. The

predominant parameter controlling this process is found to be the elastic energy density accumulated in one period E_p:

$E_p = (A_w \varepsilon_w^2 t_w + A_b \varepsilon_b^2 t_b)$

being $\varepsilon_{w,b}$ and $t_{w,b}$ the strain and the thickness of wells and barriers calculated by using the XRD data while $A_{w,b}$ are function of the elastic stiffness coefficients of each layer. It was found that the maximum number of layers that can be grown without modulations is a function of E_p^{-1}. This result is consistent with the existence of a critical elastic energy density for the flat-to-rough morphology transition in the MQW (see Nasi et al [9] for a detailed explanation).

It is worth noting that, with the well\barrier thickness values of samples A, it has not been possible to approach the 2 μm emission wavelength with a defect free structure. Taking into account that in the structures investigated the tensile barriers seem responsible of the wavy growth of the MQW and that the quantity N_w increases by decreasing the elastic energy density per period, we designed a structure with the same well thickness of Samples A, but with thicker and Indium-richer barriers(details will be published elsewhere). Cross section TEM analysis, reported in Fig.3, shows that a defect free MQW with 40 repetitions, emitting at 1.96 μm at room temperature, was successfully grown in this way.

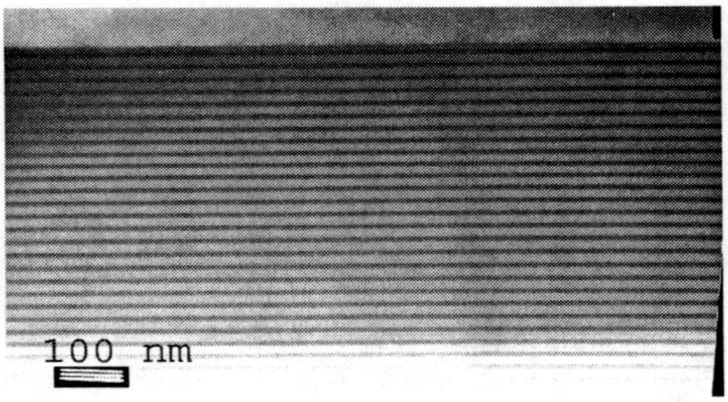

FIGURE 3. TEM (g=200) cross section of the top part of the MQW grown with thicker and Indium richer barriers with respect of type-A samples.

Defect-free devices grown on virtual substrates

XRD and TEM investigations [10] performed on InGaAs SBMQW grown on a virtual substrate (Samples B) reveal that misfit dislocations are strictly confined at the buffer/GaAs substrate interface, in a few nm thick layer, well outside the active region of the device. (Fig.4)

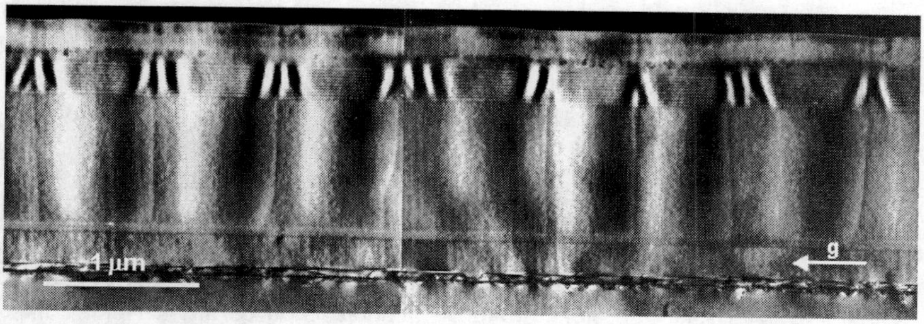

FIGURE 4. DF 220 cross sectional micrograph of the whole structure. MDs are strictly confined at the buffer/substrate interface.

Plan view TEM analysis of a total area of about 0.1 mm^2 showed that the density of threading dislocation reaching the free surface is less than 10^5 cm^{-2}. EBIC analysis performed at several beam energies confirm that there are not MDs in the intrinsic region and that the density of threading dislocations is not higher than 10^5 cm^{-2} [11]. No other electrical active defects were observed. However the stress fields associated with the underlying MDs network cause a roughening of the epilayer surface resulting in a cross hatched (CH) surface with a periodicity of about 1 μm. The CH morphological features are spatially correlated to the presence of both large- and fine-scale lateral contrast modulations forming a typical strain-induced contrast pattern with regular bright and dark columnar regions. The coarse modulation extending over the whole structure has the same wavelength as the CH. This evidences that the CH pattern is accompanied with stress concentration regions that elastically relax when thin foils for TEM observations are made.

Actually, CH consists of nearly flat topmost surface separated by V-shaped valley [10], where an higher density of step is present; the geometrical distribution of the valley being the same of the fine strain modulation observed in TEM plan view (not reported here). The ribbon-like morphology is then developed by the preferential incorporation of the growing species at these steps. The fine strain modulation extends only in the MQW region in correspondence of the CH valleys and it must be ascribed to thickness fluctuations of the MQW layers (Fig.5).

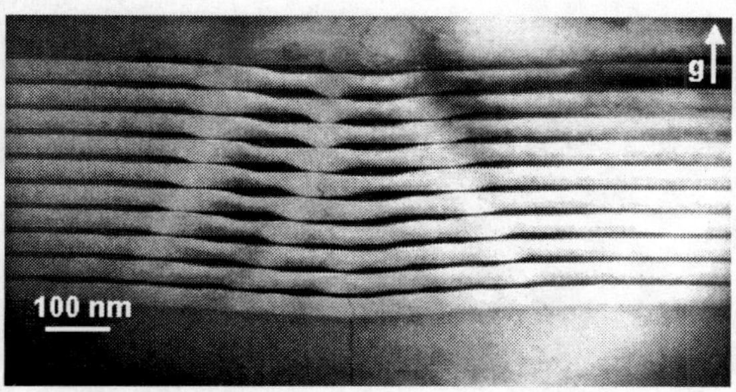

FIGURE 5. DF 220 cross sectional image showing the MQW thickness modulation at the CH valleys.

The step bunching phenomena can be considered responsible for MQW lateral thickness modulation. In this case the strain balance growth mechanisms can not be considered responsible of the MQW thickness fluctuations.

We observed that the described morphological instability affects more severely samples grown on virtual substrates having higher equilibrium lattice parameter. In the samples grown with an equilibrium lattice parameter corresponding to the In content $x = 0.35$ the MDs are still efficiently confined at the buffer-substrate interface, therefore the intensity of the strain induced contrast is significantly stronger respect to the sample reported in Fig.4 grown on virtual substrates with equilibrium lattice parameter corresponding to the In content $x = 0.15$. The stronger CH-induced lateral strain distribution found on this sample results in a larger thickness fluctuations of the MQW layers which thickness locally overcome the critical layer thickness, inducing the generation of new dislocation [12].

Concerning the spectral response of these structure, CL results showed that the emission edge has been moved to higher wavelengths with respect to GaAs, giving a value of the MQW emission peak of about 1230 nm at 77K. [12]

4. CONCLUSIONS

Strain-balancing and the use of virtual substrates makes it possible to significantly extend the absorption edge of InGaAs multi-quantum well photovoltaic cells towards higher wavelengths compared to the case of unstrained MQWs grown on commercially available substrates like InP or GaAs. By studying the structural problems related to the stability of strain-balanced multi-quantum wells with a large strain differential we could explain the origin of the local structural breakdown and its impact on the electrical properties of the devices. Good quality PV cells with an absorption edge at 1.96 μm were obtained in this way. The use of virtual substrates to red-shift the absorption edge of the PV cells was also investigated starting from a series of test samples on GaAs substrates. Strain-balanced InGaAs MQWs were successfully grown on virtual $In_xGa_{1-x}As$ substrates with $0.14 < x < 0.35$, the residual density of threading dislocations in the active region being lower that 10^5 cm^{-2} .

ACKNOWLEDGMENTS

This research was funded by the European Commission through the project THE REV of the 5th Framework Programme of Research and Development, and by the Italian Space Agency.

REFERENCES

1. P.Griffin, I.Ballard, K.Barnham, J.Nelson, A.Zachariou, J.Epler, G.Hill, C.Button and M.Pate, "The Application of Quantum Well Solar Cells to Thermophotovoltaics", *Solar Energy Materials and Solar Cells* **50**, 213 (1998).

2. Romanato F., Napoletani E., Carnera A.,Drigo A. V.,Lazzarini L., Salviati G.,Ferrari C., Cosacchi A and Franchi S., *J. Appl. Phys*. **86**, 4748 (1999)

3. Drigo A.V.,Aydinli A., Carnera A., Genova F., Rigo C., Ferrari C., Franzosi P. and Salviati G., *J. Appl. Phys*., **66**, 1975 (1989)

4. J.W. Matthews and A.E. Blakeslee, J. Crystal Growth, **32**, 265 (1976)

5. Ekins-Daukes N J, Zhang J, Bushnell D B, Barnham K W J, Mazzer M and Roberts J S 2000 Proc. 28th IEEE Photovoltaic Specialists Conference, Anchorage, Alaska, USA.

6. Napchan E. *Microscopy Anal*., 9 (1992)

7. S. Tundo, M. Mazzer, L. Lazzarini, L. Nasi, G. Torsello, D. Diso, G. Clarke, C. Rohr, K. Barnham "EBIC studies of strain balanced InGaAs/InGaAs MQWs" accepted by Materials Science and Technology, IOM

8. Ponchet A, Rocher A, Ougazzaden A and Mircea A , *J. Appl. Phys*. **75**, 7881 (1994)

9. Nasi L, Ferrari C, Lazzarini L, Salviati G, Tundo S, Mazzer M, Clarke G and Rohr C, 2002, EDS2002 Conference, 2-6 june 2002, Bologna. In press on *Journal of Physiscs: Condensed matter*

10. Lazzarini L, Nasi L, Ferrari C, Salviati G, Mazzer M, Passaseo A and Barnham KW *Inst. Phys. Conf. Ser*. No 169, 109 (2001)

11. S. Tundo, M. Mazzer, A. Passaseo, L. Nasi, C. Ferrari, L. Lazzarini, G. Salviati *Inst. Phys. Conf. Ser*. No 169, 567 (2001)

12. A Passaseo, R. Cingolani, M. Mazzer, M. Lomascolo, S. Tundo, L. Lazzarini, L. Nasi, G. Salviati, K.W. Barnham, 2002 Proc. MRS Spring Meeting, April 1-5 2002 San Francisco, California

SESSION 4

Emitters/Innovative Approaches

Chair: Robert Nelson

Characterisation Of Strain-Compensated InGaAs/InGaAs Quantum Well Cells For TPV Applications

Paul Abbott[a], Carsten Rohr[a], James P. Connolly[a], Ian Ballard[a], Keith W.J. Barnham[a], Ravin Ginige[b], Graham Clarke[c], Lucia Nasi[d], Massimo Mazzer[e]

[a] *Experimental Solid State Physics, Blackett Laboratory, Imperial College London, SW7 2BW, U.K.*
[b] *NMRC, University College, Cork, Eire*
[c] *IQE Europe Ltd., Cardiff CF3 0EG, Wales, U.K.*
[d] *CNR-IMEM, Fontanini-Parma, Italy.*
[e] *CNR-IMM, University of Lecce, 73100 Lecce, Italy*

Abstract. Thermophotovoltaic (TPV) generators can reduce pollution by lowering their operating temperature, but the choice of semiconductor materials for this purpose is limited. We present results on an InGaAs p-n cell lattice-matched to InP which is optimised for the Erbia emission spectrum peak at a wavelength of 1.5μm. However, for lower temperature TPV applications at longer wavelengths one is constrained by the lack of lattice-matched materials. In order to extend the absorption towards 1900 nm for a selective emitter based on Thulium strain-compensated InGaAs/InGaAs quantum well cells (QWCs) on InP have been designed and characterised. We present data showing that strain-compensated QWCs extend the spectral response (SR) to longer wavelengths and can show a lower dark current density than the bulk InGaAs p-n cell despite the QWC having a lower band-gap. We have developed a model for the SR of strained multi-quantum well (MQW) systems in InGaAsP, including quantum effects as well as strain-induced changes. SR modelling of strain-compensated structures is compared with experimental data, and efficiencies, for a Thulia spectrum, are predicted. Our study also shows that back surface reflection must be taken into account in these devices.

INTRODUCTION

For thermophotovoltaic (TPV) systems powered by the combustion of a conventional fuel, such as methane or diesel, minimising pollution is one of the primary objectives. For nitrogen oxides in particular this requires a reduction in combustion temperature [1]. The system under consideration in this paper includes a selective emitter, which absorbs the heat from the combustion and emits in a fixed, narrow wavelength range. As the temperature of the system is reduced the most efficient emitters become those which radiate at longer wavelengths. In order to maintain high system efficiency, the cells must be well matched to the incident spectrum.

An erbium oxide based emitter operating at about 1900K produces an emission spectrum centred at around 1500nm. At this wavelength InGaAs p-n cells lattice-

CP653, *Thermophotovoltaic Generation of Electricity: 5th Conference*
edited by T. J. Coutts, G. Guazzoni, and J. Luther

matched to InP have been shown to perform well [2], and such a cell is characterised and reported here. At temperatures around 1700K one of the more suitable emitters is based on thulium oxide and this produces an emission peak at around 1700nm. Unfortunately, this radiation cannot be absorbed by any material lattice-matched to InP within the InGaAsP system. Therefore, a method is needed to incorporate lattice-mismatched materials without the formation of dislocations and their subsequent destructive effect on device performance [3].

In quantum well cells (QWCs), thin layers of lower band-gap material are inserted into the i-region of a conventional p-i-n structure [4]. The increase in short-circuit current can be larger than the decrease in open-circuit voltage, hence increasing the overall efficiency above that of a p-i-n cell containing the barrier material [5]. Another favourable aspect of QWCs regarding TPV is their superior temperature characteristic [6].

Placing lattice-mismatched quantum wells in the structure appears, at first glance, to favour the formation of dislocations, but QWCs also enable the use of the strain compensation technique: the compressive strain experienced by one of the materials (here the quantum wells) is balanced by the inclusion of another material under tension (the barriers). The total magnitudes of the strains are designed to cancel each other, which means that, providing each layer does not exceed its individual critical thickness, there is minimal total strain within the structure and a much larger number of wells can be incorporated. In this way, the effective lattice constant of the quantum well region can be tuned to match the substrate [7]. A number of such QWCs, designed to extend absorption to longer wavelengths than lattice-matched InGaAs, are characterised in this paper.

EXPERIMENTAL

All cells reported were grown by Metal Organic Vapour Phase Epitaxy (MOVPE) and were processed into 1mm photodiode devices with 600μm optical windows. Only the best cell from each structure is included here in order to give a better indication of the performance of the cell rather than variations in processing.

The first cell, E004, is a p-n cell containing only lattice-matched InGaAs, designed to match an Erbia emitter at 1500nm (see table 1). It contains InP minority carrier window layers on either side of the active region and an InGaAs lateral conduction layer. This conduction layer was to be used in conjunction with semi-insulating substrates in Monolithic Interconnected Modules (MIMs) [2]. The uppermost InGaAs capping layer is removed from the optical window by etching as it is for contacting purposes only.

TABLE 1. Structure Of Bulk InGaAs Cell E004.

Repeats	Material	Thickness (μm)	Doping	Concentration (cm^{-3})
1	InGaAs	0.1	p	1×10^{19}
1	InP	0.1	p	2×10^{18}
1	InGaAs	0.3	p	1×10^{19}
1	InGaAs	2.0	n	5×10^{17}
1	InP	0.1	n	1×10^{18}
1	InGaAs	1.0	n	2×10^{19}
1	InP	0.2	n	2×10^{19}
	InP	Substrate	n	

The two strain-compensated QWCs were designed to extend the absorption of the cell beyond that of lattice-matched InGaAs. The first of these, E012 (see table 2), has lattice matched quaternary p and n regions with high band-gaps (1100nm) to reduce the dark current and InP window layers. InGaAs conduction and contact layers are retained and there is an undoped quaternary layer to reduce zinc diffusion from the p-region into the undoped quantum well region. This region itself begins and ends with a half barrier.

TABLE 2. Structure Of Strain-Compensated Cell E012.

Repeats	Material	Thickness (μm)	Doping	Concentration (cm^{-3})
1	InGaAs	0.1	p	1×10^{19}
1	InP	0.1	p	2×10^{18}
1	InGaAsP (1.1μm)	0.2	p	1×10^{18}
1	InGaAsP (1.1μm)	0.25	-	-
30	InGaAs (62% In)	0.012	-	-
30	InGaAs (47% In)	0.012	-	-
1	InGaAsP (1.1μm)	0.5	n	5×10^{17}
1	InP	0.1	n	1×10^{18}
1	InGaAs	2.0	n	2×10^{19}
1	InP	0.2	n	2×10^{19}
	InP	Substrate	n	

The second strain-compensated QWC, structure E1055 (see table 3), was designed to demonstrate absorption matching the thulia emission. The structure is simply an InP p-i-n incorporating 40 quantum wells. The E1055 device characterised in this report included a 214nm silicon nitride anti-reflection coating with an expected minimum in reflection at around 1700nm. Again, the quantum well region begins and ends with a half barrier.

TABLE 3. Structure Of Strain-Compensated Cell E1055.

Repeats	Material	Thickness (μm)	Doping	Concentration (cm^{-3})
1	InGaAs	300	p	1×10^{19}
1	InP	300	p	2×10^{18}
1	InP	200	-	-
40	InGaAs (45% In)	8	-	-
40	InGaAs (75% In)	16	-	-
1	InP	20	-	-
1	InP	500	n	1×10^{18}
	InP	Substrate	n	

RESULTS

The two strain-compensated QWCs demonstrate an extension in absorption beyond that of the lattice matched structure (see Figure 1). Both QWCs exhibit excitonic features at the absorption edge, which is 1790nm for E012 and 1930nm for E1055. The measured spectrum of a Thulia emitter at about 1600K is included in figure 1 for reference [8]. Clearly, as the absorption edge moves to longer wavelengths, more of the incident spectrum can be absorbed giving greater potential efficiency.

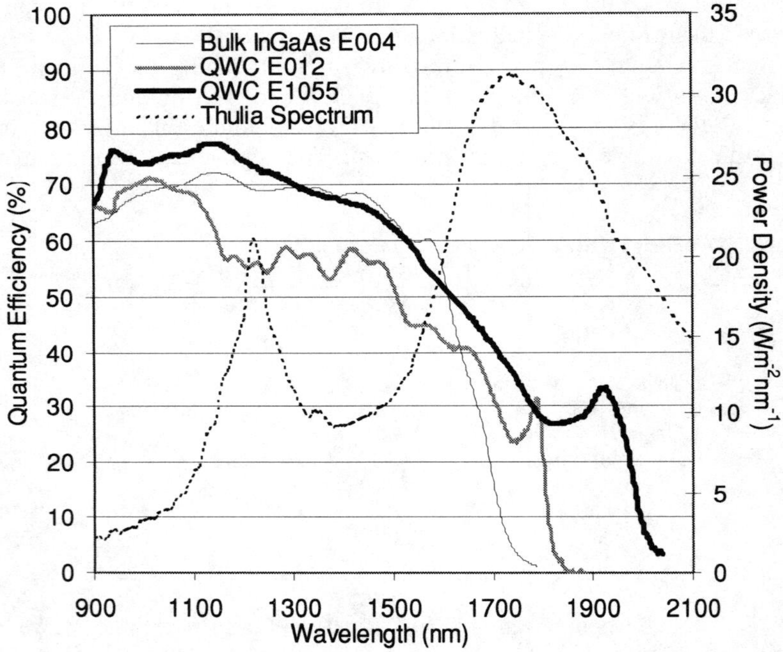

FIGURE 1. Zero bias spectral response comparison of the structures E004, E012 and E1055. The right hand axis corresponds to the thulia spectrum.

Compared with E012 the structure E1055 contains 10 more quantum well periods, with every individual layer suffering a greater strain than its equivalent in E012. Any residual strain produced by one period of well and barrier will therefore build up to a greater overall value in the structure. This would bring it closer to the critical limit at which dislocations form and two-dimensional growth can cease to be possible.

Figure 2 is a Transmission Electron Microscope (TEM) image of the quantum well region of structure E1055. It can be seen that the quantum wells are not affected by structural defects and that the interfaces between the wells and the barriers are sharp. This indicates that the strains have been well compensated and the whole quantum well region has remained below the critical limit for dislocation formation.

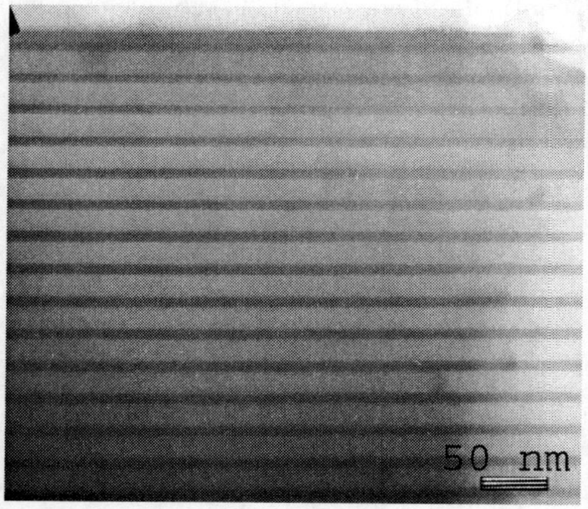

FIGURE 2. Transmission Electron Microscope (TEM) picture of a section of the quantum well region of structure E1055.

The dark currents of the three structures are shown in figure 3. Although the quantum wells in E1055 are structurally good, there is a poor surface morphology associated with the growth of the top layers. This may in some part explain why its dark current is relatively high. However, the QWC E012 demonstrates that dark currents below that of lattice matched InGaAs cells can be achieved, despite the presence of lower band-gap material.

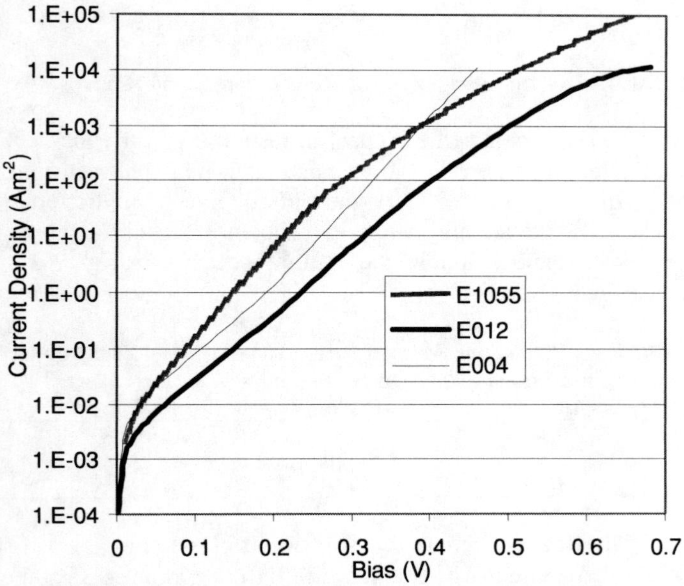

FIGURE 3. Dark current comparison of the structures E004, E012 and E1055.

Figure 4 shows the normalised bias dependence of the photocurrent in the structure E1055 at the monochromatic wavelengths 1200nm and 1800nm. The 1800nm measurement is of particular interest as it excites solely into the quantum wells. The constant level strongly suggests that there is full carrier collection in reverse bias and this extends into forward bias. This is encouraging as a demonstration that the carriers are able to successfully escape such deep quantum wells. However, this measurement was conducted at low intensity and the collection efficiency in forward bias begins to fall as the intensity is increased. This again may be connected to the problems with the overgrowth.

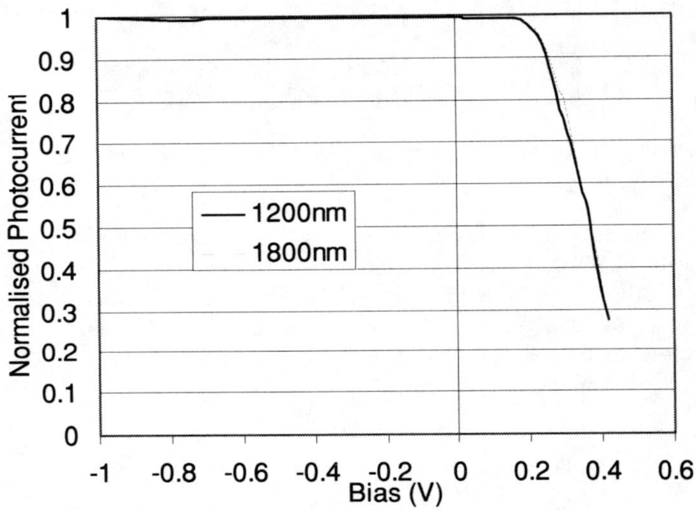

FIGURE 4. Bias dependent photocurrent of structure E1055 at various wavelengths.

By integrating the product of the spectral response and the incident spectrum, the amount of collected photocurrent can be calculated. This photocurrent can be used in conjunction with the measured dark current to give a prediction of the light-IV characteristics of the device and hence its efficiency. Table 4 shows the results for each of the three structures under the Thulia spectrum in figure 1. These figures assume no shading loss.

TABLE 4. Efficiency Predictions For Structures E004, E012 And E1055 Under Thulia Emission.

Structure	Current Produced ($\times 10^3$ A/m^2)	Output Power (kW/m^2)	Incident Power (kW/m^2)	Efficiency (%)
E004	6.48	2.14	19.58	11.0
E012	6.85	2.61	19.58	13.3
E1055	10.16	3.54	19.58	18.1

The beneficial effect of increased absorption is clearly evident as E012 produces a 22% increase in efficiency over E004, while E1055 produces 35% more power than E012 and 65% more than E004.

Modelling Spectral Response

The experimental results have been compared with a model developed at Imperial College [9]. The model covers the quantum nature of QWCs, but has also been refined to include the effects of strain on factors such as band-gap. Figure 5 shows the modelled spectral response of structure E012, which is compared with the experimental data.

The model suggests that a single pass of the light through the device is insufficient to produce all the photocurrent that is measured. There is a contribution from reflection at the rear surface, an annealed Au/Ge contact. By neglecting absorption within the substrate a 65% reflecting rear surface appears to enhance the response sufficiently to give good agreement with the data at the absorption edge. As expected, the enhancement is minimal above the absorption edge of InP, where most light is absorbed in the first pass.

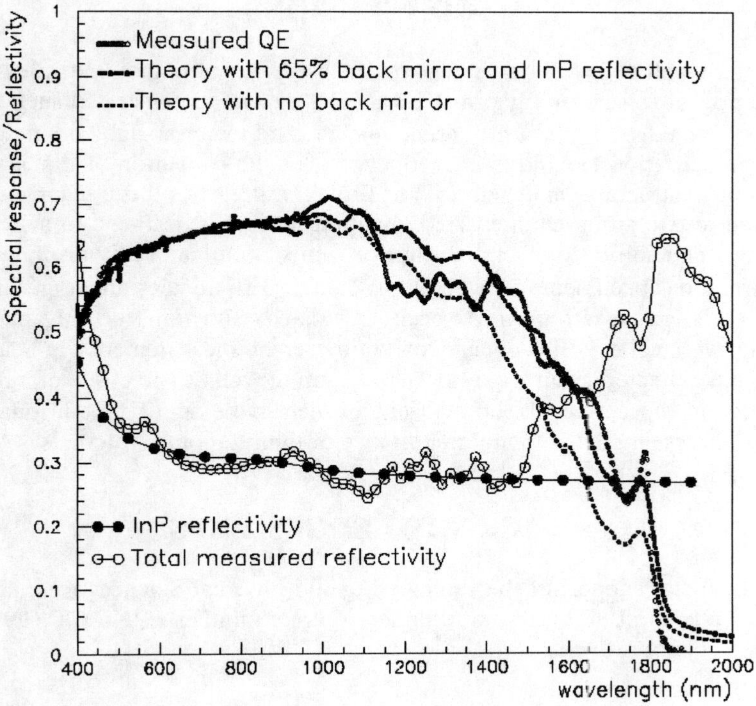

FIGURE 5. Modelled spectral response of strain-compensated structure E012, with and without a 65% rear surface reflection. The oscillations in the data are due to interference within the structure, which is not modelled. Also, the measured reflectivity is compared with the expected reflection at the front surface.

The figure of 65% is consistent with that obtained by another method previously reported [10], where the situation is treated as an absorbing material bounded by reflecting surfaces. This second method uses tabulated material data and experimentally determined parameters such as device reflectivity. It also requires the free carrier absorption of an InP substrate, which has been measured and found to be equivalent to a loss of less than 2.5% for a single pass through a 350μm substrate [10].

The presence of a rear surface reflection is also consistent with the reflectivity measurement of this structure. Figure 5 compares the measured reflectivity with the reflection expected from the front surface alone, an air/InP interface. Due to the mirroring effect the measurement increases sharply beyond the absorption edge to over double that which reflects from the front surface.

A significant reflection at the rear of the device coupled with low absorption losses in the substrate suggest that the inclusion of light trapping structures would be beneficial. This would be a further step towards increasing the absorption in the quantum wells and thus improving the efficiency of such devices.

CONCLUSIONS

Quantum well cells offer an extension in absorption of photovoltaic devices by the introduction of lower band-gap material, and we have demonstrated such a cell with an absorption edge of 1.93μm. Lattice mismatched materials may be included if the strain compensation technique is used to prevent the formation of dislocations. An InGaAs p-n structure matched to an Erbia emitter has been characterised and compared to two strain-compensated quantum well cells designed to match a Thulia spectrum. The ability to extend absorption whilst simultaneously displaying a lower dark current has been demonstrated out to 1.8μm. Efficiencies under an experimental Thulia emission spectrum have been calculated, showing the advantage of the quantum well cells. Bias-dependent photocurrent measurements indicate that all carriers are collected, even from the deep quantum wells. The effect of a rear surface reflection on the collected photocurrent is seen to be significant in quantum well samples, increasing the potential importance of light trapping structures.

ACKNOWLEDGEMENTS

The financial support of the Engineering and Physical Sciences Research Council (EPSRC) and the European Union under contract number ERK6-CT-1999-00019 is gratefully acknowledged.

REFERENCES

1. *Combustion Generated Pollution,* Science Research Council, London, 1976, p. 78.
2. Wilt, D. M., Fatemi, N. S., Jenkins, P. P., Weizer, V. G., Hoffman, R. W., Jain, R. K., Murray, C. S., and Riley, D. R., "Electrical and Optical Performance Characteristics of 0.74eV p/n InGaAs Monolithic Interconnected Modules" *in Thermophotovoltaic Generation of Electricity: Third NREL Conference,* edited by T. Coutts et al., American Institute of Physics, 1997, vol. 401 of AIP Conference Proceedings, pp. 237-247.

3. Griffin, P. R., Barnes, J., Barnham, K. W. J., Mazzer, M., Zanotti-Fregonara, C., Olson, C., Rohr, C., Haarpaintner, G., David, J. P. R., Roberts, J. S., Grey, R., and Pate, M. A., *J. Appl. Phys.*, **80**, 5815 (1996).
4. Barnham, K. W. J., and Duggan, G., *J. Appl. Phys.*, **67**, 3490-3493 (1990).
5. Barnham, K. W. J., Ballard, I., Barnes, J., Connolly, J., Griffin, P., Kluftinger, B., Nelson, J., Tsui, E., Zachariou, A., *Appl. Surface Science*, **113/114**, 722-733 (1997).
6. Ballard, I., Barnham, K. W. J., Nelson, J., Connolly, J. P., Roberts, C., Roberts, J. S., and Pate, M. A., "The Effect of Temperature on the Efficiency of Multi-Quantum Well Solar Cells" *in Proc. 2nd World Photovoltaics Conference*, Vienna, 1996, pp. 3624-3626.
7. Rohr, C., Connolly, J. P., Barnham, K. W. J., Mazzer, M., Button, C. C., and Clark, J., "Strain-balanced In0.62Ga0.38As/In0.47Ga0.53As (InP) Quantum Well Cell for Thermophotovoltaics" *in Proc. 28th IEEE Photovoltaic Specialists Conference* (Anchorage, USA), IEEE, USA, 2000, pp. 1234-1237.
8. Diso, D., Licciulli, A., Bianco, A., Leo, G., Torsello, G., Tundo, S., Sinisi, M., Larizza, P., Mazzer, M., "Selective Emitters for High Efficiency TPV Conversion: Materials Preparation and Characterisation" *in The 5th Conference on Thermophotovoltaic Generation of Electricity*, 2002 this volume.
9. Connolly, J. P., Nelson, J., Ballard, I., Barnham, K. W. J., Rohr, C., Button, C., Roberts, J., Foxon, T., "Modelling Multi Quantum Well Solar Cell Efficiency" *in Proc. 17th European Photovoltaic Solar Energy Conference*, Munich, 2001.
10. Abbott, P., Rohr, C., Connolly, J. P., Ballard, I., Barnham, K. W. J., Ginige, R., Corbett, B., Clarke, G., Bland, S. W., Mazzer, M., "A Comparative Study of Bulk InGaAs and InGaAs/InGaAs Strain-Compensated Quantum Well Cells for Thermophotovoltaic Applications" *in Proc. 29th IEEE Photovoltaic Specialists Conference*, (New Orleans, USA), IEEE, USA, 2002.

Non-imaging optics in a thermophotovoltaic generator

E. Lindberg and L. Broman
Solar Energy Research Center SERC, Dalarna University College
SE 781 88 Borlange, SWEDEN

Abstract. In order to achieve high efficiency in a TPV generator, it is important that a high fraction of emitted photons with energies below the TPV cell bandgap are reflected back to the emitter. This can be accomplished in several ways. We present the idea of an internally reflecting egg-shaped double cone with the emitter at one end, an edge filter at the wide center, and the TPV array at the other end. So far, we have studied this geometry by means of both ray tracing analysis in a computer program and by means of measurements with an emulated emitter. A sharp switchover from transmission to reflection in a multiple layer edge filter can be achieved only if the angles of incident rays are confined to a fairly narrow angular interval. The two used methods both show that the studied optics can lower the angular spread of rays incident onto the filter and that some 96% of the emitted rays (in the ideal case) reach their goal without passing the filter or being reflected by the filter more than once. A suggestion of a whole wood powder fuelled TPV system with this egg-shaped double cone and edge filter implemented is also given, as well as an animation tool for modelling the optical part of the TPV system.

INTRODUCTION

A thermophotovoltaic generator consists basically of an emitter of infrared radiation and an array of photovoltaic cells. The emitter is commonly heated by means of combustion of a fuel and has a typical temperature between 1000 K and 2000 K. A non-selective emitter radiates over a wide wavelength interval while a PV cell can only utilize energy from photons with energies over the bandgap E_g of the cell. As a high efficiency generator is desired it is therefore necessary either to decrease the irradiation of low-energy (high-wavelength) photons by a selective emitter or to reflect a large fraction of those photons back to the emitter.

This work concentrates on the possibility of using a special geometry in combination with interference filters in order to increasing optical efficiency. Interference filters (edge filters) can be designed to get a very good spectral selectivity, but they are sensitive to the incident angles, the edge is shifted substantially when the angle of incidence is changed, and the edge gets less steep (see Figure 3) when the range of incidence angles increases [1]. Therefore, a geometry that limits the angular distribution of the radiation incident on the filter should be designed, a geometry that, at the same time, does not (unnecessarily) increase the cell area over that of the emitter. In our previous research [2], it has been shown that decreasing the angular distribution to half may increase the overall generator efficiency by up to 15%.

In order to facilitate calculations of TPV generator optics efficiencies, an Excel® document has been constructed which displays the case when a selective edge filter is placed between the emitter and the PV cells. The previously presented [1] animation program has been further improved giving six factors that can be varied independently, see Figure 1. The factors are; temperature T for the normalized blackbody spectrum; bandgap E_g of the TPV cell; low reflectance ρ_1 of the edge filter and cut off wavelength λ_1 for the low reflectance; high reflectance ρ_2 of the edge filter and cut off wavelength λ_2. Output

CP653, *Thermophotovoltaic Generation of Electricity: 5th Conference*
edited by T. J. Coutts, G. Guazzoni, and J. Luther

from the animator is the optical availability η_A, standing for the ratio of the part of the emitted radiation which is useful for conversion over the emitted radiation that is transmitted through the filter. The animator is available at http://www.du.se/tpv.

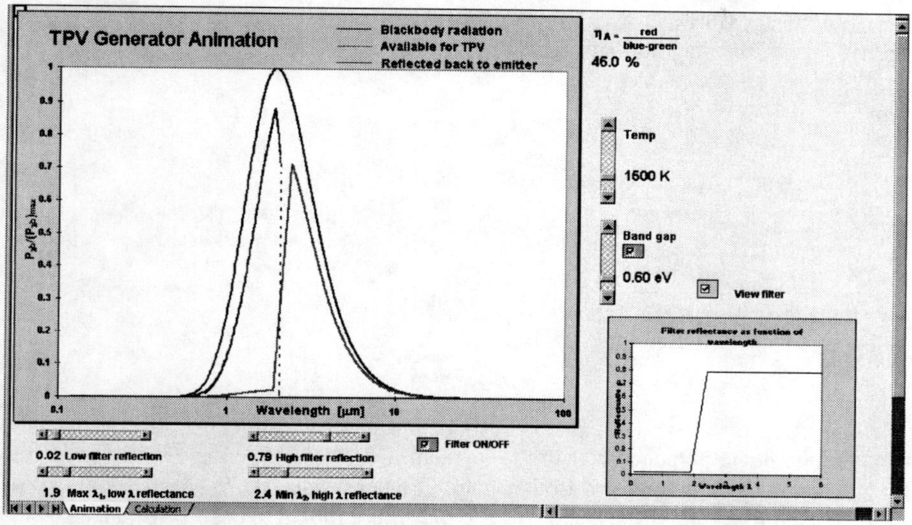

FIGURE 1. The full screen of TPV Generator Animator.

METHODS

The computer program TFCalc has been used for simulating edge filters. The filter consists of a stack of layers and the user needs knowledge to design these. The input to TFCalc is the proposal of such a stack, and the output is a diagram showing the edge between transmission and reflection as a function of radiation wavelength. The program can optimise the stack for better performance depending on special input parameters.

The simulation program for ray tracing analyses in different cones was constructed by the author and colleagues in the LabWindows® environment, programming in C++. A Lambertian emitter was simulated and each ray (of 100 000) was traced when passing the cones, and the direction of each ray when exiting the cones was monitored, as well as the coordinates on the exit surface of each ray. The intensities of the rays in small angular intervals were summarized as the angular distribution of power, shown in Figs. 6 and 9-11.

An apparatus for measuring radiation from a model emitter has been constructed to compare simulations with measurements. Three types of cones were investigated with this radiant transfer meter, one of each of the square cornets, circular cones, and Fabergé cones. Angular distribution of power was measured using a visible monochromatic beam of light of appropriate shape and size which was reflected onto a 5.0 cm diameter Al plate covered with sintered $BaSO_4$, emulating an emitter of Lambertian radiation [3], see Figure 2. The intensity of the light reflected by the emitter was first measured from the emitter alone, then from the emitter with a single cone, and finally with the double cone placed close to the emitter. Light intensities in different directions were measured with the help of an Si detector mounted on a manually rotated arm [1]. The signal output from the Si detector was monitored, together with the angle data, by a computer program developed in the LabWindows® environment. The data files, *i.e.* two columns of angle values and corresponding signal values of light intensity, were then transferred to the Excel® program in order to create the diagrams shown in Figure 13.

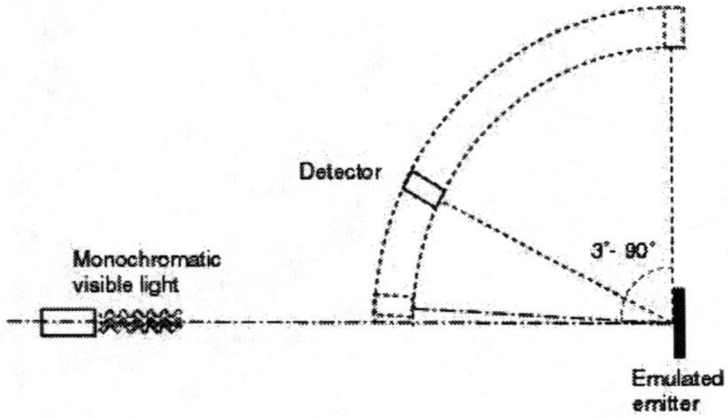

FIGURE 2. The radiant tranfer meter for measurements of angular distribution of power.

The direction of a ray is determined by the exit angle θ (3° to 90°) and the azimuth angle ϕ in the plane perpendicular to the incident light beam (the plane where the emitter is). The intensity of light radiated from a small Lambertian surface is proportional to $\cos\theta$. Power radiated from this surface at a specific θ is proportional to $\cos\theta \sin\theta$ since power at a specific exit angle θ equals intensity integrated over the interval $0° \leq \phi \leq 360°$ [2]. One important consequence of this for TPV applications is that the maximum radiative power from a Lambertian emitter is at $\theta = 45°$.

THE TPV – FILTER SYSTEM

The importance of limiting the range of incident angles is clearly demonstrated by Figure 3. This figure shows the reflectance curves for an edge filter with the edge at 2.07 μm (corresponding to photon energy $E_{ph} = 0.60$ eV) at an incident angle of 20° (bold curve).

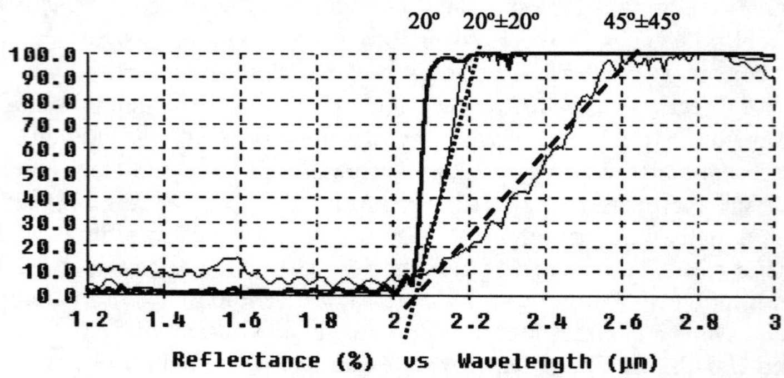

FIGURE 3. Calculated reflectance vs wavelength curves for three edge filters designed for incident angles of 20°, 20°±20°, and 45°±45°.

Also shown are the curves with incident angles 20°±20° and 45°±45°. As can be seen, the edge is shifted substantially when the angle of incidence is changed, and the slope gets less steep when the range of incident angles increases. Since the radiation onto the filter is not always of parallel rays, but of intervals of incident ray angles, these two intervals will show the situation of well collimated rays in a cone and of none-collimated rays, respectively.

The geometries considered in this study assume that the emitter is a square or circular slab (e.g. made of a ceramic material). Instead of placing the filter and the TPV cell array close to the side of the emitter, these are separated, but optically connected, by a square or circular double cone (or similar geometry) where the inside walls consist of highly reflecting material. The small opening of the primary cone is thus placed close to the emitter, and the other end of the first cone is intended to be covered by the filter, which is thus larger than the emitter. The primary cone acts as a collimator which reduce the range of incident angles of the radiation onto the filter. The second cone, similar to the first one, acts as a concentrator which directs radiation towards the TPV cell array, which has the same size as the emitter. Figure 4 shows the principle of the setup.

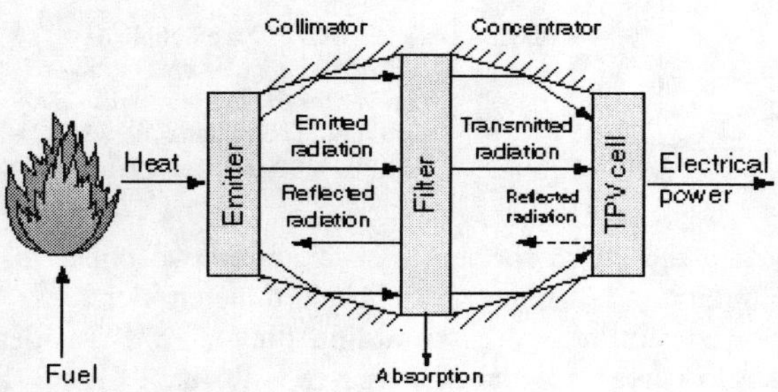

FIGURE 4. A basic TPV system including a flame-heated emitter, a non-imaging collimator, a selective edge filter, a non-imaging concentrator, and TPV cells [2].

INVESTGATIONS OF SQUARE CORNETS AND CIRCULAR CONES

Square cornets and truncated circular cones have been investigated (see Figure 5a-b) using ray tracing analyses and measurements. These types of geometries were chosen since they have previously been investigated as examples of non-imaging concentrators for moderate concen-tration of sunlight onto PV cells [4]

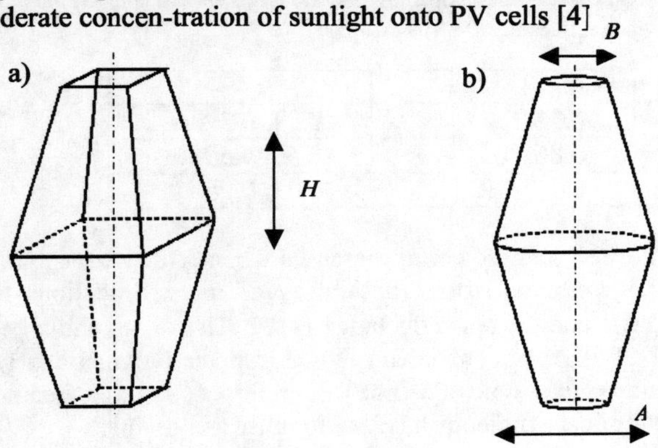

FIGURE 5. a) The double square cornet geometry and b) the double circular cone geometry.

Geometric concentration C is the ratio between the small opening area over the large opening area, A^2/B^2, and H is the height for a single cornet and cone in cm. $B = 5$ cm for all geometries. The simulations included $C = 2, 3, 4, 6, 9$, and $H = 13, 15, 20, 30, 40, 50$ cm. The measured square cornet and circular cone were both of $C = 4$ and $H = 13$.

Figure 6 shows an example of simulated angular distribution of power for the emitter, for single and double cones with C = 4 and H = 30 cm. The reflectance of the inner walls is assumed $\rho = 1$. θ is the incident angle of radiation on the large area joining the cones as compared to the cone's axis. As can be seen, the range of incident angles θ is around 18°±18° with a peak at 26° for the single cone, compared to 45°±45° for the emitter. The power P from the emitter which passes the cones has been calculated and is shown as the area under the curves. Parts of the emitted radiation will not reach the exit from the double cone (since the area under the curve for the double cone is smaller than the area under the curve for the emitter), parts of the rays will be reflected back.

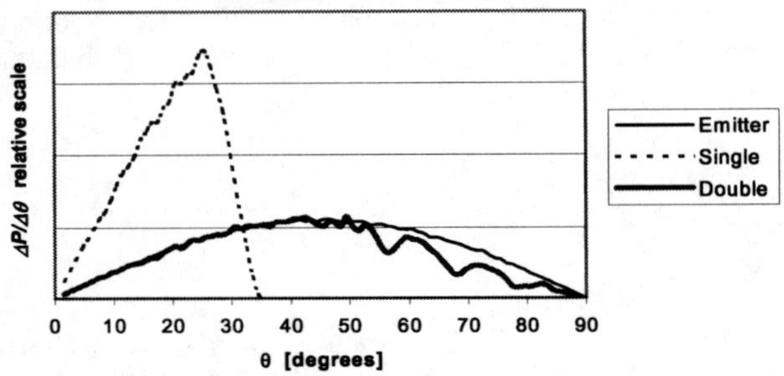

FIGURE 6. An example of simulated angular distribution of power for the emitter, the single and double circular cone with $C = 4$ and $H = 30$ cm. Reflectance $\rho = 1$.

Since the double cone is symmetric, the curve for the intensity after two cones also gives the intensity of long wavelength radiation that is reflected back to the emitter.

The two types of geometries showed similar results regarding the range of incident angles onto the second cone. The range decreases when H increases with fixed C, and the range decreases slightly when C increases with fixed H. Optimum regarding smallest range is when $C = 9$ and $H = 50$.

Results from calculations to find the optical efficiency, *i.e.* the part of emitted power which passes the double cornets and cones (the latter in italics), are summarized in Table 1.

TABLE 1. Part of emitted power that passes the exit opening of double cornets and cones (in italics) when reflectance $\rho = 1$ is assumed.

	$C = 4$	$C = 9$
$H = 15$	0.72, *0.76*	0.23, *0.29*
$H = 30$	0.86, *0.87*	0.56, *0.60*
$H = 50$	0.91, *0.93*	0.73, *0.76*

The optical efficiency of the cornets and the cones increases when H increases, but, on the other hand, long cornets and cones will offer possibility of more reflections than shorter ones, which will reduce efficiency when reflectance $\rho < 1$. The cones show better optical efficiency than the cornets, but are not sufficiently good, especially when a large C (and thus large decrease in angular spread onto the filter) is preferred. Another geometry must be developed in order to get optical efficiency as close to unity as possible.

THE FABERGÉ CONE

In order to limit the region of incident angles on the filter, which will make the filter act more efficiently, a special geometry of the internally reflecting cones has been developed; here called the Fabergé cone, see Figure 7. Equation 1 is the equation of the ellipse, where the constants *a* and *b* are the long and short half axes of the ellipse, and ξ and η are the first and second axis, respectively.

$$\frac{\xi^2}{a^2}+\frac{\eta^2}{b^2}=1 \qquad (1)$$

The curve intersects the ξ-axis in $(a, 0)$ and $(-a, 0)$. The curve intersects the η-axis in $(0, b)$ and $(0, -b)$. The coordinates of the focal points are $(c, 0)$ and $(-c, 0)$, where $c=\sqrt{a^2-b^2}$.

The ellipse is cut at the lines $\xi = -c$ and $\xi = c$. When the upper part of the truncated ellipse is reflected in the line $\eta = b^2/2a$, giving the desired curve in Figure 8, an ideal 2D non-imaging optical device, a CEC or Compound Elliptic Concentrator, with $C = 1$ is created, since every ray leaving the left opening (traveling right) will reach the right opening if the two sides are ideal reflectors. When the upper part of the truncated ellipse is rotated around the line $\eta = b^2/2a$, the two openings will be circular and the rotational body resembles somewhat a Fabergé egg[1]. Such an egg with $\rho = 1$ surfaces will probably be a near-perfect 3D CEC, a possibility that will be explored in this paper.

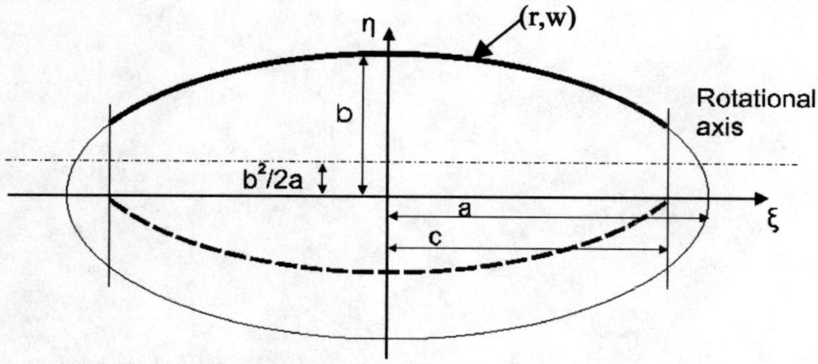

FIGURE 7. Principal construction of the Fabergé cones

A point (r,w) on the curve at the distance *r* from the midpoint on the rotational axis and at distance *w* from this axis can be established through Equation 2:

$$\frac{r^2}{a^2}+\frac{\left(w+\dfrac{b^2}{2a}\right)^2}{b^2}=1 \qquad (2)$$

When the coordinate system is changed to a 3D-system, where the rotational axis becomes the *z*-axis, the former η-axis becomes the *x*-axis, and a *y*-axis is added, Equation 2

[1] Carl Fabergé (1846-1920), Russian goldsmith who made beautiful small sculptures, some resembling hollow eggs, for, among others, the Russian Tsar. (Encyclopaedia Britannica)

in the new coordi-nate system becomes Equation 3, which is then rearranged to Eq. 4:

$$\frac{z^2}{a^2}+\frac{\left(\sqrt{x^2+y^2}+\frac{b^2}{2a}\right)^2}{b^2}=1 \quad (3)$$

$$z=\pm\sqrt{a^2-\frac{a^2}{b^2}\left(\sqrt{x^2+y^2}+\frac{b^2}{2a}\right)^2}\ ;\ |z|\le\sqrt{a^2-b^2} \quad (4)$$

The positive z-values refer to the part of the body above the x-y-plane with the small opening upwards, and negative z-values mean the body below the x-y-plane with the large opening upwards. Figure 8 shows an example where $C = 4$, $A = 10$, and $B = 5$. The height $H = c$ and depends on a and b. Here $H = 8.4$.

$b=\frac{A}{2}+\frac{B}{2}$ and $a=\frac{b^2}{B}$, A and B are inputs to the computer ray tracing program.

Two hypotheses: **i)** Nearly all beams which leave one of the opening surfaces (even if the surface is Lambertian) will reach the other opening surface (if the inner wall has ideal reflectance). **ii)** When the rays pass the plane $z = 0$ the spread of angles between the rays and the rotational axis will be low.

If these hypotheses are correct this geometry can be ideally suited to being placed between the emitter and the TPV cell array, with an edge filter in the plane $z = 0$.

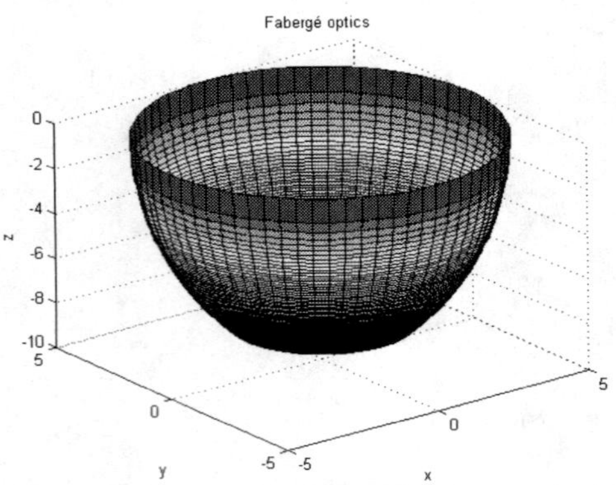

FIGURE 8. An example of a Fabergé cone (half an egg shell) with $C = 4$.

Simulations of Fabergé cones

Five Fabergé geometries with $C = 2$, 3, 4, 6, and 9 have been investigated with 100 000 rays leaving the emitter at the entrance aperture; the results are shown in Figure 9. The curves are named by the geometric concentration C. θ is the angle between the ray and the axis when the rays leave the exit apertures, and power is at a relative scale, the same for all curves in a figure. As can be seen, in the Fabergé cones the rays are much closer to the axis and the angular spread is thus lower, the $C = 9$ geometry being the most efficient in this respect.

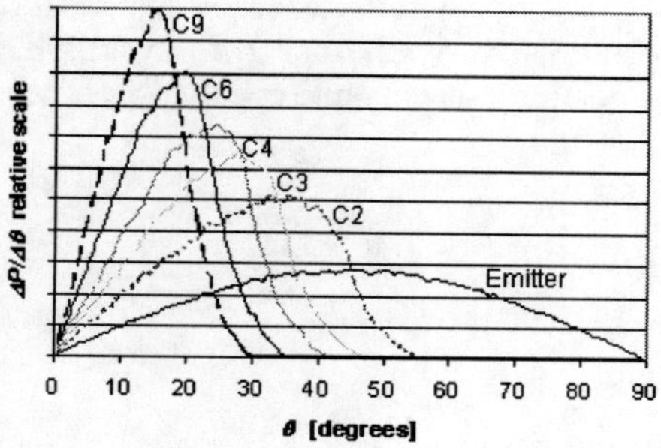

FIGURE 9. Simulated angular disrtribution of power in single Fabergé cones with different C values.

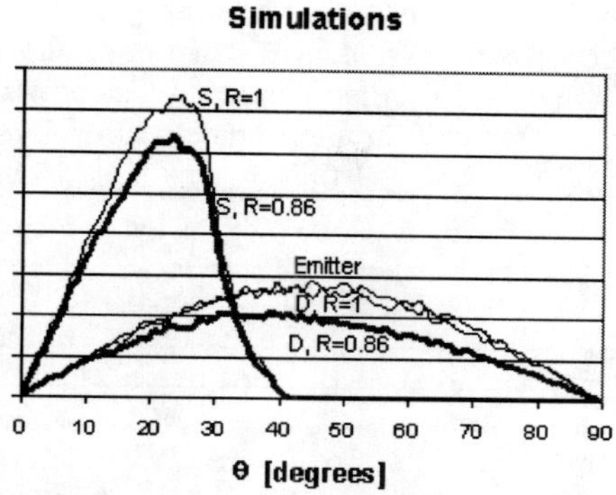

FIGURE 10. Examples of simulated angular distribution of power in Fabergé cones for $C = 4$, $\rho = 0.86$, and $\rho = 1$

Figure 10 shows angular distribution of power in a single and a double Fabergé cone when $\rho = 0.86$ and $\rho = 1$. For the latter case, the curve for the double cone is very close to the curve of the emitter.

Figure 11 shows curves of relative intensity as a function of radius length of the large opening of the single cones (where the edge filter would be placed) in order to display the intensity variation over the surface. It is clearly shown that for all geometries the intensity is lower in the center than at the rim. Similar simulations of the square cornets and circular cones show that these have even distribution of ray intensity.

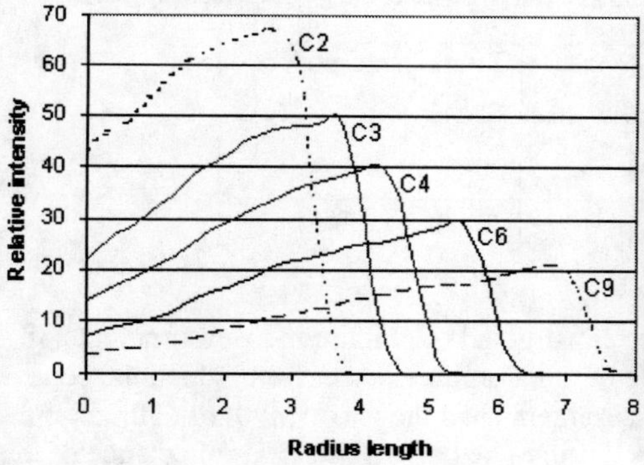

FIGURE 11. Ray intensities at the exit aperture of a single Fabergé cone displayed as a function of distance from centre.

FIGURE 12. The double Fabergé cone of the $C = 4$ geometry (and a cylinder used to connect the two halves during measurements).

5.2 Measurements of Fabergé cones

Prototypes of the Fabergé cone of $C = 4$ configuration have been constructed in aluminum, see Figure 12. The lathed surfaces were carefully polished before measuring. Figure 13 shows measured results of angular distribution of power for the aluminum Fabergé cones, as well as from the emulated emitter (wavelength of light 650 nm). The measured curves agree very well with the simulated curves of $\rho = 0.86$, see Figure 10.

However, a high optical efficiency requires higher reflectance in the inner wall of the Fabergé cone. An ideal surface would be gold, giving the reflectance of about 0.98 in infrared. The Fabergé cones have therefore been coated by gold using a Edwards E306 evaporation unit for thin film deposition, but the coating was not quite evenly distributed.

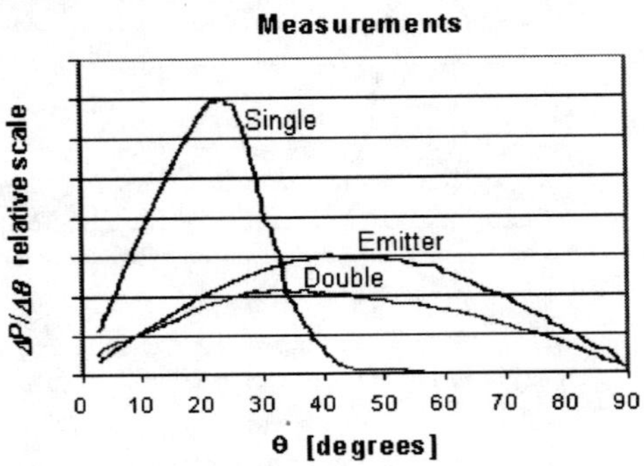

FIGURE 13. Measurent results of angular distribution of power from an emulated emitter and on aluminum Fabergé single and double cone.

GENERAL RESULTS AND CONCLUSIONS

Simulations and measurements have confirmed the hypotheses about the optical features of the Fabergé cones. Comparison with the circular cones shows improvement in optical efficiency, see Table 2.

TABLE 2. Comparison between circular cones and Fabergé cones regarding range of incident angles θ and transferred power P through double cones, measured results.

	Incident angles θ on second cone	Transferred power of double cones
Circular cones, C=4, H=13	22°±22°	0.45
Fabergé cones, C=4, H=8.4	22°±22°	Al: 0.73 Au: 0.85

As mentioned, these measured cones were constructed of aluminum. Measurements of reflectance have been performed on a sample of the aluminum sheet from which the cones were made. The light beam used for the measurements had the wavelength of 650 nm, the same as during the measurements of the rays through the cones. The total reflectance of the sample was measured as 85%, and the diffuse reflectance as 10%. The polished wall of the Fabergé cone could not be measured.

SUGGESTION ON A TPV GENERATOR PROTOTYPE WITH FABERGÉ OPTICS

Conclusions drawn from the present results lead to a concept of a TPV co-generation system which includes wood powder combustion at 1500K, ceramic emitter, Fabergé optics with interference filter, TPV cell array, recuperator, and heat exchanger, see Figure 14. After striking the emitter, remaining energy in the flue-gases is used to preheat combustion air before heating water in a heat exchanger. The Fabergé cones with the filter, as well as the TPV array, are cooled with water. The cone walls should be coated with gold in order to obtain high reflectance. A selectively reflecting filter designed to suit the emitter temperature 1500 K, and which is not too sensitive to uneven radiation, remains be procured.

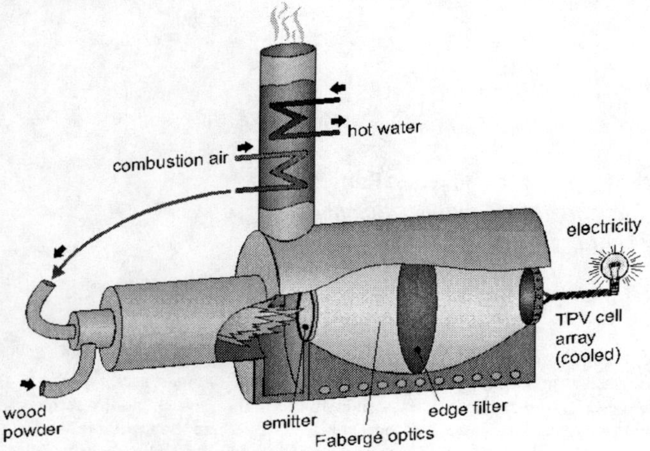

FIGURE 14. An impression of a TPV system.

TPV cells with bandgap $E_g = 0.6$ eV are available in research quantities today. TPV cells of GaSb with $Eg = 0.70$ eV are available in limited quantities, as well as edge filters fabricated with alternating layers of Si/SiO_2 tailored for incident angles of 20° and suitable bandgap [5]. The cells with the lower bandgap would fit the emitter temperature 1500 K.

ACKNOWLEDGEMENTS

The authors wish to thanks K. Jarefors, J. Marks, and J Back for help with using the TFCalc program, construction of the angle meter and monitoring program, and programming the user interface of the software for ray tracing analysis.

REFERENCES

1. Broman L, Jarefors K, Lindberg E, Marks J. Square Cones for TPV; Experiments and Computer Simulations. 4th NREL Conference on Thermophotovoltaic Generation of Electricity, AIP Conf. Proc. 460 (1999), pp 115-120
2. E. Lindberg. TPV Optics Studies – on the use of non-imaging optics for improvement of edge filter performance in thermophotovoltaic applications. Doctoral thesis, Swedish University of Agricultural Sciences, Uppsala 2002, ISBN 91-576-6328-9
3. Broman L, Jarefors K, Marks J, Lindberg E. Enhancing the optical efficiency of a thermophotovoltaic generator using square cones in combination with interference filters. Proc. Second World Conference on Photovoltaic Solar Energy Conversion, Vienna, Austria, pp 264-267 (1998)
4. Broman L, Rönnelid M, Binder B, Lindberg E. Use of Nonimaging Concentrators for Moderat Concentration of Sunlight onto Photovoltaic Cells. Proc. 9th E.C. PV Solar Energy Conf., Freiburg, (1989) pp 234-239
5. Fraas LM, Samaras JE, Huang HX, Minkin LM, Avery JE, Daniels WE, Hui S. TPV Generators Using the Radiant Tube Burner Configuration. Proc. the 17th European PV Solar Energy Conference and Exhibition, Munich, Germany, 22-26 October 2001

Micron-gap ThermoPhotoVoltaics (MTPV)

R.S. DiMatteo, P. Greiff, S.L. Finberg
The Charles Stark Draper Laboratory, Inc., Cambridge, MA 02139

K.A. Young-Waithe, H.K H. Choy, M.M. Masaki, C.G. Fonstad
Department of Electrical Engineering and Computer Science,
Massachusetts Institute of Technology, Cambridge, MA 02139

Abstract
To date, ThermoPhotoVoltaic (TPV) technology has been governed by Planck's Law. To increase the photonic flux from the emitter to the receiver in a TPV system the temperature must be increased. Higher temperatures lead to materials challenges and higher parasitic system level thermal losses. In addition they make it difficult for TPV to compete in many waste heat conversion applications where the available waste heat is at temperatures less than those required for TPV. [1] Planck indicated that his law applies only if the dimensions of the system are large compared to the wavelengths of the radiation. [2] The idea that the electromagnetic energy transfer in a ThermoPhotoVoltaic (TPV) system can be increased by 5 times or more beyond Planck's Law by placing the emitter in close proximity to the photovoltaic receiver has been proposed and studied theoretically, and has been referred to as Micron-gap TPV (MTPV). [3] In this paper we discuss the theory, the technology, and the experimental results that obtain when the distance between the emitter and receiver surfaces is reduced to sub-micron dimensions. The dramatic demonstration of this effect is the initial step in the development of a new class of energy conversion devices.

INTRODUCTION

Micron-gap ThermalPhotoVoltaics (MTPV) is a new technology that offers a potential breakthrough in the solid-state conversion of heat to electricity. Thermophotovoltaic devices were conceived in the early 1960's and consist of a heated black-body which radiates electromagnetic energy across a gap onto a photovoltaic device which converts the radiant power into electrical power. However, the amount of power out of a given TPV device area (I.e. power density) is constrained by the temperature of the hot side of the device (the emitter). Current implementations of TPV generally require a temperature in excess of 1.000C creating barriers to its practical use.

By contrast, Micron-gap TPV (MTPV) as proposed here provides a significant breakthrough in power source technology by transferring more power between the emitter and receiver without making the emitter hotter but only by reducing the size of the gap between them. By employing a submicron gap the achievable power density for MTPV can be increased by approximately an order of magnitude as compared to

CP653, *Thermophotovoltaic Generation of Electricity: 5th Conference*
edited by T. J. Coutts, G. Guazzoni, and J. Luther

conventional TPV. Or equivalently, for a given active area and power density, the temperature on the hot-side of an MTPV device can be reduced by about 45% allowing new applications for on chip power, waste heat power generation, and converter power

MTPV takes advantage of near-field (evanescent) radiation effects to increase power density. One mechanism of the phenomenon may be obtained from figure 1 which depicts a classical experiment demonstrating electromagnetic coupling across a gap. Calculations of the magnitude of the effect are shown in Figure 2. As shown a reduction of the gap to 0.1 micron is predicted to increase the power transfer by an order of magnitude. Sub-micron gaps have been obtained experimentally through use of MEMS technology. Work at Draper has analytically and experimentally demonstrated the increased energy density achievable for a MEMS-based MTPV cell.

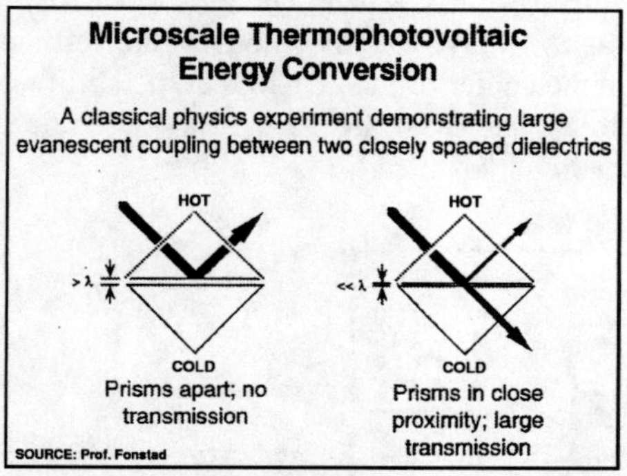

Figure 1 : A Classical Experiment

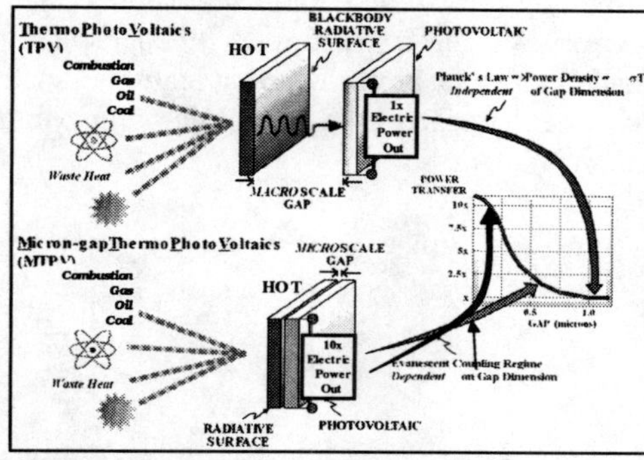

Figure 2: A Comparison between TPV and MTPV

FIRST EXPERIMENTS

Enhanced generation of carriers using submicron proximity to a heated surface was first demonstrated using custom-designed InAs photodiodes and special silicon-based heater chips produced using Microelectromechanical System (MEMS) techniques. The short circuit current of the photocells is shown to increase sharply (up to fivefold) when the spacing between the heater and photodiode surfaces is reduced below one micron. It is argued that the increase in short circuit current is due to increased evanescent coupling of black body radiation from the hot surface to the cold photocell, consistent with recent theoretical predictions.

The first experimental demonstration of the MTPV effect was done by placing a hot Silicon chip on a cooled InAs photovoltaic chip separated by SiO_2 spacers, which maintained the physical gap between the two surfaces as shown in Figure 3.

A 500-µm thick "heater chip," shown schematically in Fig. 3, was devised to provide the following functions: submicron gaps between it and the photovoltaic cell, a capacitor plate to allow measurement of the gap, and a controlled hot surface. The silicon heater chip was fabricated by MEMs techniques to leave a series of 6-µm diameter oxide mesas on the front, which acted as spacers to determine the separation of the heater and photovoltaic cell. Figure 4 shows two 1-µm high SiO_2 spacers similar to those used. The back of the chips was coated with a chemical vapor deposition (CVD) oxide insulator and patterned to give a contact window. The backs were also metallized with a thin film of titanium and platinum, [100 nm/200 nm]. The film was patterned to provide a contact to the bulk of the chip, which acted as the capacitor plate; a 180-ohm resistor, which was used for temperature measurement; and a 10-ohm resistor used as a heater. The perimeter of the chip was plasma etched down about 15 µm in the scribe streets so that any burrs that might occur during wafer sawing would not interfere with the gap spacing. The heater chip is 2.2 x 2.2 mm in size.

Custom TPV cells were designed and fabricated from InAs epitaxial layer structures grown by solid-source molecular beam epitaxy on p-type InAs substrates. [2]. The cells were etched-mesa structures with contacts to the n and p layers of the photovoltaic cell made in ring fashion to leave a featureless area in the center of the cell so that the ability to place the heater chip in submicron proximity to the surface of the TPV cell would not be obstructed.

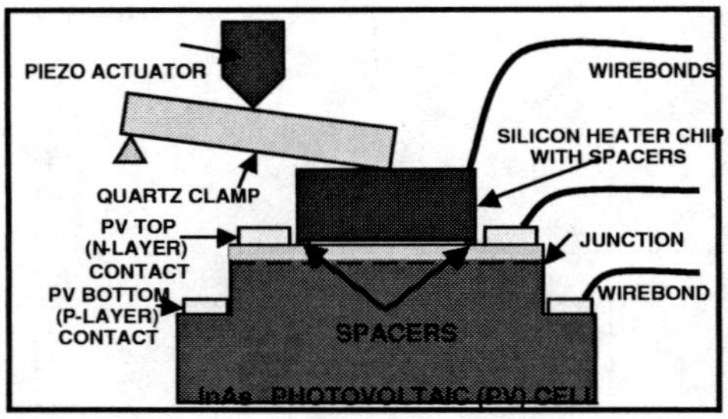

Figure 3 : A schematic of the experimental configuration

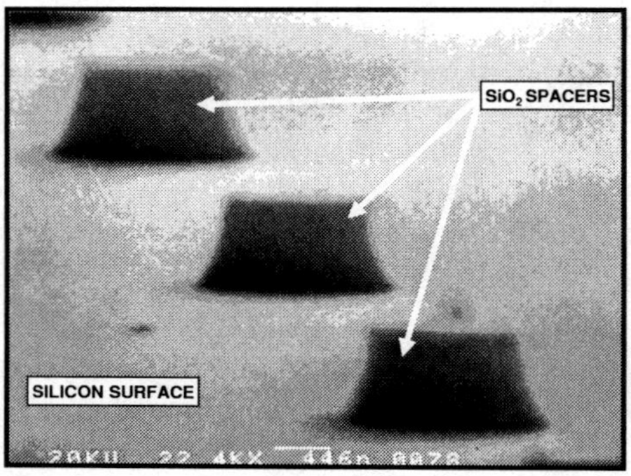

Figure 4: An SEM micrograph of the SiO2 spacers

The heater chip and photovoltaic cell were assembled as shown in Figs. 5-7. The cold side of the assembly consists of the photovoltaic cell mounted on a brass block, which is mounted on a thermoelectric cooler, which is mounted on a liquid cooling block. All are attached with thermal vacuum grease. The heater chip is held in place on top of the photovoltaic cell by a mechanical spring-loaded clamp. The heater chip and the photovoltaic cell are wire bonded to the modified flatpack chip carrier shown. Contacts are brought out from the photovoltaic cell and the following components on the heater chip: heater element, temperature sensor, and capacitive contact.

The heater/photovoltaic cell assembly was placed in a bell jar with the liquid chiller outside. A small turbo pump was used to provide the vacuum in the 40-mtorr range. A surface-mounted thermocouple was used to monitor the brass mounting block temperature, which was used as the approximate photovoltaic temperature.

Figure 5 : Heater chip clamped to PV cell

Figure 6: heater/PV clamped into package and wire bonded.

Figure 7 : Assembly mounted in bell jar

The results of the first set of experiments are shown in Fig. 8. The following procedure was used. A piezoelectric actuator was brought into contact with the mechanical clamp from above. Contact was sensed by an increase in capacitance. The capacitance, heater chip temperature, and PV short circuit current were recorded. The piezo actuator was then extended further, thereby flexing the heater chip by fractions of a micron, thereby decreasing the gap and increasing the capacitance. Upon decreasing the gap, two observations were apparent immediately--the short circuit current increased and the temperature of the heater chip decreased. Both are explained by the increased coupling of energy across the gap. In order to obtain constant temperature data, the circuitry was changed so that the power to the heater was increased to maintain the initial temperature. The capacitance and photovoltaic short circuit current were also recorded. Data were taken for the three temperatures indicated in Fig. 8, with multiple cycles among the temperatures with repeatable results. As shown there is a very strong relationship between gap and short circuit current over the range of measurements. As the gap is decreased the capacitance increases and so does the photocurrent. As predicted by MTPV theory a large multiple of the initial photocurrent is obtained.

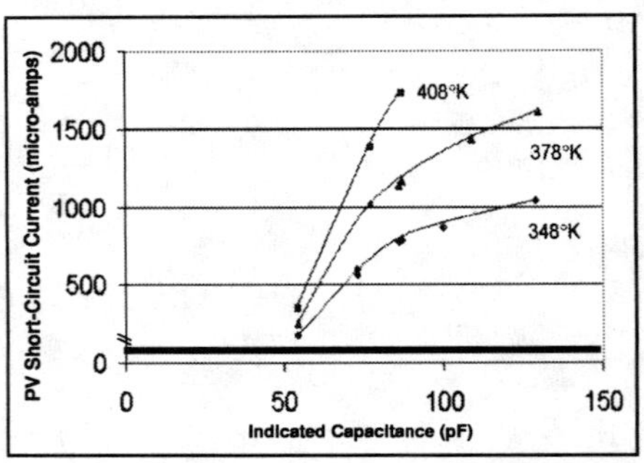

Figure 8 : First confirmation of MTPV effect

Because these initial results were so dramatic it became desirable to devise a test. to rule out the possible influence of stray thermal effects accounting for the current increase. The piezo actuator was oscillated up and down at a frequency of 200 Hz, which is beyond the frequency that thermal effects would be able to follow. The oscillating piezo was then brought into contact with the heater chip clamp, causing it, and therefore the gap, to oscillate. If the current were due to thermal effects, it would not be able to follow the gap oscillation frequency and the short circuit current would not be modulated by it. The in-phase nature of the response obtained as shown in figure 9 supports the arguments against a thermal effect.

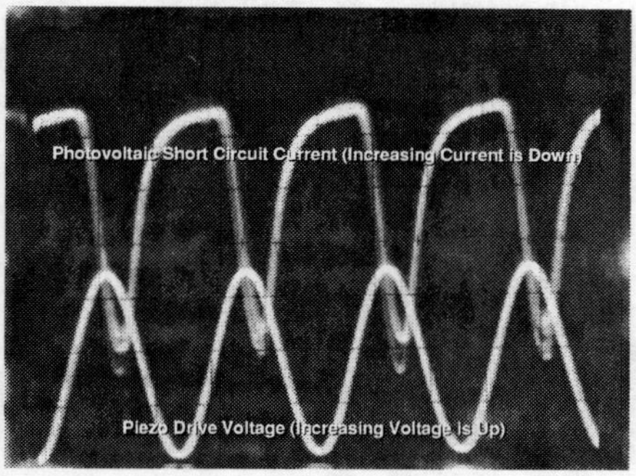

Figure 9 : bottom curve increasing piezo drive and decreasing gap. Top curve increasing photocurrent is down. 200Hz oscillation

Scaling up to a Large Size System

The 2.2mm x 2.2mm heater chip has turned out to be a convenient vehicle to evaluate the basic MTPV phenomenon. However, practical applications will require heater surfaces and photovoltaic cells in at least the 1 to 3 inch diameter size. A schematic of a possible system is shown in Fig 10. The heater portion of the system faces several major challenges in that it must maintain submicron contact with the PV cell and allow for a realistic amount of particulates and non-flatness of the surfaces. (It is assumed that the front surface electrodes of the PV cell will be buried beneath the surface). One approach to this problem is to build a tactile heater surface that will deform around defects. A concept of panels on springs is shown in Figures : 11 and 12.

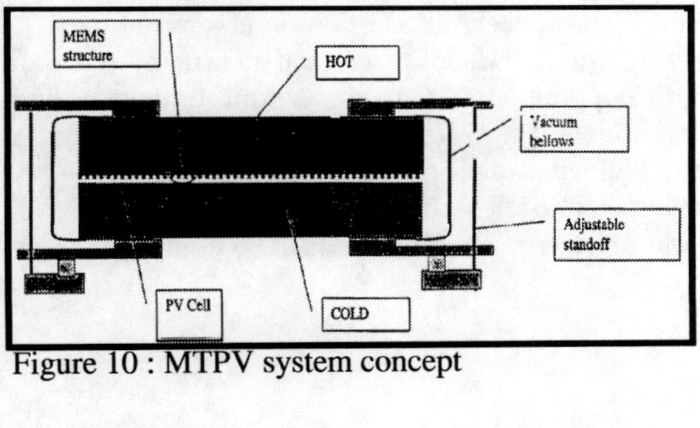

Figure 10 : MTPV system concept

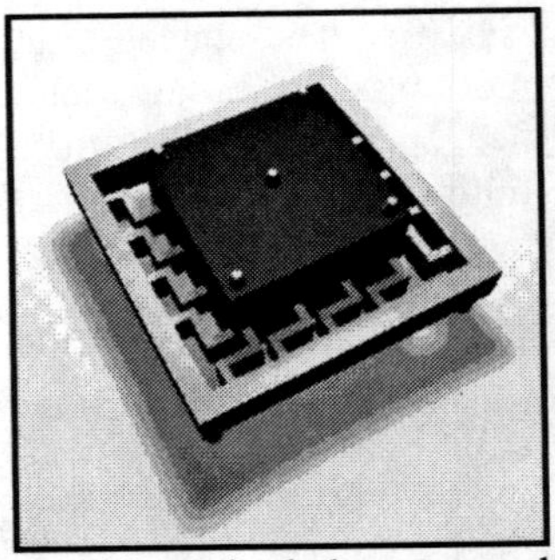

Figure 11 : a single heater panel supported on springs

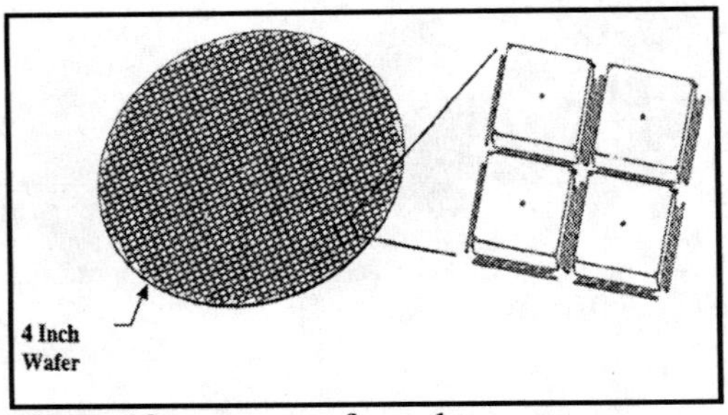

Figure 12 : an array of panels

A prototype array was fabricated from a four inch diameter silicon wafer using Mems technology; vertical plasma etching, selective doping, and anisotropic etching. Initial results were encouraging in that the design dimensions were reproduced. However, this approach has not been pursued as competing ideas are being evaluated. Figure 13 shows a photograph of a wafer scale fabrication. Figure 14 shows a close-up of several panels with the flexures viewed from the front side. The panels and surrounding frame are 300 microns thick and the flexures are 6 microns thick. Figure 15 is focused on the back side of the flexures. These have the typical speckled appearance of an etch stop which is based on a high concentration of boron. The front sides are smooth because the concentration of boron is at a maximum. The concentration of boron is lower at the interior and so the backs exhibit a rougher finish. Handling of such an array is delicate until a backing plate is applied and some popping out of panels was noticed in the as built array. The flexures were designed to accommodate five microns of deflection but it is easy to exceed this in handling and cleaning without proper procedures.

This phase of the work was undertaken to show the feasibility of scaling up a tactile array. Should this approach be selected much more needs to be done to achieve a working array.

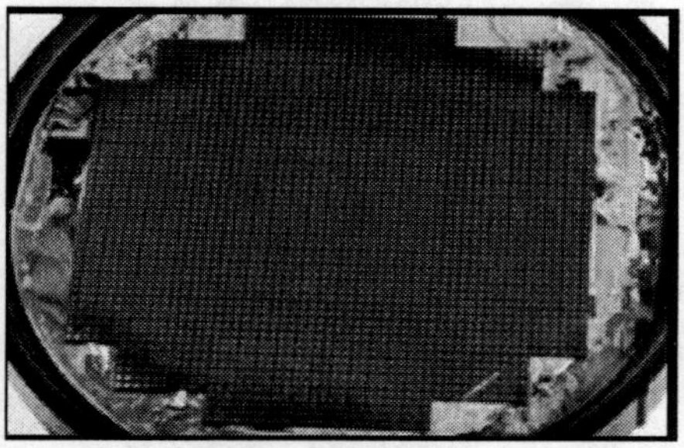

Figure 13 : wafer scale heater array

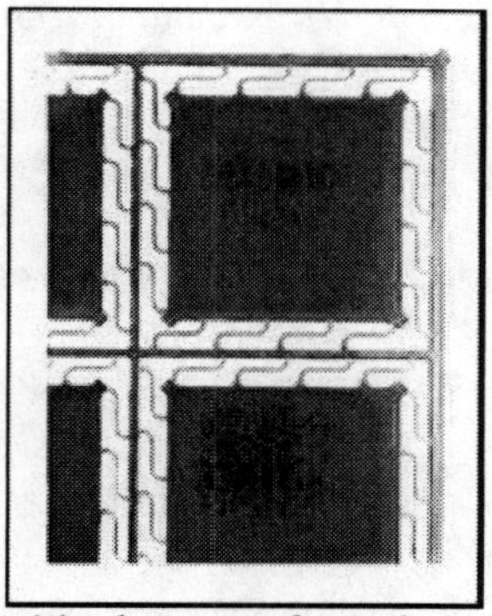

Figure 14 :close-up of panels on wafer scale array

Figure 15 : Flexures viewed from the back

CURRENT AND FUTURE WORK

More recent work has concentrated on automating the data taking procedure, extending the range of heater temperature and increasing the accuracy of measurement. We have also duplicated the effect using InGaAs photovoltaic cells. This work will be reported on in the near future. Recent work has also begun investigating the possibility of quantum coupling [5a] to achieve quasi-monochromatic electromagnetic transfer with potential for record efficiencies. Quasi-monochromatic transfer has also recently been reported in the literature through surface wave resonances in the near field.[5b,c,d] All of the above are examples of the unexplored physics in the MTPV configuration. Future work will be directed at expanding the data base, extending the theoretical application of the technology, and developing methods of practical application for power generation.

[1] Yamaguchi, Hiromi and Masafumi Yamaguchi, "Thermophotovoltaic Potential Applications for Civilian and Industrial Use in Japan", Thermophotovoltaic Generation of Electricity: Fourth NREL Conference, eds. T. J. Coutts, J. P. Benner, C. S. Allman, American Institute of Physics, 1999, p.17 – 29.

[2] Planck, M., The Theory of Heat Radiation, New York, Dover Publications, 1959, p. 2. (This book is "an unabridged and unaltered republication of the Masius translation" of Planck's second edition (1913) "as originally pub. in 1914 by P. Blakiston Son & Co.")

[3] a) DiMatteo, R.S., Enhanced Semiconductor Carrier Generation Via Microscale Radiative Transfer, Masters Thesis, Department of Electrical Engineering and Computer Science, and Technology and Policy Program, M.I.T., June 1996*; b) Whale, M. D., A Fluctuational Electrodynamic Analysis of Microscale Radiative Transfer and the Design of Microscale Thermophotovoltaic Devices,* Doctoral Thesis, Department of Mechanical Engineering, M.I.T., June 1997*; c) Raynolds, J.E., "Enhanced Electromagnetic Energy Transfer Between a Hot and Cold Body at Close Spacing due to Evanescent Fields,"* in Thermophotovoltaic Generation of Electricity: Fourth NREL Conference, ed. by T.J. Coutts et al., American Institute of Physics, Woodbury, N.Y., 1999, pp. 49-57*; d) Pan, J.L., Choy, H.K.H., and C.G. Fonstad, Jr., "Very Large Radiative Transfer over Small Distances from a Black Body for Thermophotovoltaic Applications,"* IEEE Transactions on Electron Devices, 47, 2000, 241-249.

[4] DiMatteo,Greiff,Finberg,Young-Waithe,Choy,Fonstad, Enhanced semiconductor carrier generation via coupling across a non-isothermal vacuum microcavity, Applied Physics Letters, 17, September 2001, Vol. 79, No12. ….

[5] a) P.L. Hagelstein, K.P. Sinha, A. Meulenberg, R.S. DiMatteo, C.G. Fonstad, "Power Enhancement from Quantum Coupling between Semiconductors for Photonic Based Energy Conversion", ps. 29-6 – 29-10, Research Laboratory of Electronics, MIT, Progress Report 2001, RLE No. 144, 2002; *b) Carminati, et.al.,* Ecole Centrale Paris, TNT2001, 9/3-7/01; *c) A. V. Schegrov, K. Joulain, R. Carminati, J-J Greffet, Near-Field Spectral Effects due to Electromagnetic Surface Excitations,* Physical Review Letters, p. 1548-1551, Vol. 85, Number 7, 14 August 2000; *d) J-J Greffet, Remi Carminati, K. Joulain, J-P Mulet, Y. Chen, Coherent Emission of Light by Thermal Sources,* Letter to Nature, ps. 61-64, Vol. 416, 7 March 2002.

Fundamental studies on non radiative processes in ceramic selective emitter materials for TPV applications.

G. Torsello[(1)], M. Lomascolo[(3)], A. Bianco[(2)], A. Licciulli[(2)], D. Diso[(3)], S. Tundo[(2)], M. Mazzer [(3)]

[(1)] *Dipartimento di Fisica, Università di Lecce - Campus Universitario, Via Arnesano - 73100 Lecce (I)*
[(2)] *Dip. di Ingegneria dell'Innovazione Università di Lecce - Campus Universitario, Via Arnesano - 73100 Lecce (I)*
[(3)] *CNR-CNR-IMM Campus Universitario, Via Arnesano, 73100 Lecce (I)*

Abstract. In order to engineer a rare-earth based selective emitter, having both good optical and mechanical properties, a detailed theoretical study of both the excitation and the decay processes is required at the microscopic level. In this work we present a theoretical model by which the high temperature quantum efficiencies can be evaluated on the basis of the correlation between the local symmetry of the potential around the rare earth atoms and the probability of their non-radiative decay. The model, based on a point charge approach, provides the electric field and the crystal field parameters (CFPs) and consequently, the non radiative decay rates for a few representative configurations. The main result is that different local symmetries around, for instance, Er^{3+} atoms may provide dramatically different non radiative decay probabilities. An experimental confirmation of the theoretical model was obtained by the analysis of a set of ceramic samples containing the same Erbium fraction and treated at different sintering temperature (in 780K –1400K range) in order to obtain different crystal phases. The radiative transition involving the orbital $^4I_{11/2}$ -> $^4I_{15/2}$ which is expected to be quite sensitive to the different ions configurations, has been studied by means of temperature dependent photoluminescence experiments (in the range of 20K-300K) performed on samples sintered at different temperatures. The results show that the decay rate of $^4I_{11/2}$ -> $^4I_{15/2}$ transition (not observable at room temperature) increases with increasing sintering temperature manly due to the development of a garnet configuration which strongly inhibits non radiative decay processes.

1. INTRODUCTION

The Rare Earth doped Selective Emitter (RESEs) have been investigated in details for their promising applications in thermophotovoltaic systems. In fact at high temperature (1000°K-1500°K) RESEs present characteristic sharp emission bands in the infrared spectral region around 1-2μm, and when they are used in conjunction with suitable photovoltaic cell, they are strategic materials to improve the performance of thermophotovoltaic generators. In order to design a selective emitter material having both good optical and mechanical properties, a detailed theoretical study of the rare earth ions excitation and the subsequent decay processes is required. In this work we present the results of a theoretical study which describes the non radiative decay processes occurring in RESEs, taking into account the local symmetry properties of the surrounding host atoms. These properties can directly be related to the sintering

CP653, *Thermophotovoltaic Generation of Electricity: 5th Conference*
edited by T. J. Coutts, G. Guazzoni, and J. Luther

condition of the RAREs systems. In order to check the predictions of this theory we have performed Cathodoluminescence (CL) and Photoluminescence (PL) experiments on Erbium doped Al_2O_3 samples sintered at different temperature.
The most important result of this work is that there is a clear asymmetry between thermal excitation and decay processes, that is two different RESEs can be equally good absorbers of radiation in a given wavelength range but their behaviour as selective emitters at high temperature may differ dramatically as a result of the local symmetry of the chemical bonds around the rare earth atoms.
This work is organised as follows: in paragraph 2 we discuss how the ionic electric field can induce a non radiative decay in the excited rare earth ions. In paragraph 3 we discuss cathodoluminescence (CL) and photoluminescence (PL) experimental results for Er doped Alumina samples sintered at different temperatures, T_S, to follow the evolution of non radiative decay processes. Er doped Alumina samples were prepared by slip casting [6]. Erbium is originally added to the slurry as a nitrate salt and then converted into Erbia and Erbium Aluminium garnet through subsequent thermal treatments. The strong improvement of the selective emission properties turns out to be correlated to the onset of the formation of the garnet in agreement with the prediction of the theory.

2. ELECTROSTATIC INTERACTION AND CRYSTAL FIELD PARAMETERS

Several theoretical approaches have been attempted to macroscopically describe the processes at the basis of selective emission.
In 2000-2001 D.L. Chubb and co-workers developed a complete theoretical model describing the temperature gradient across a selective emitter film. That model could predict the optimum film thickness to optimise selective emission from the optical point of view[1]. In 1996 V.V. Gogovlev and coworkers described the non radiative excitation and the decay processes through a rate equations scheme [2]. However these and other similar approaches do not describe the atomic processes which ultimately control selective emission, and therefore they do not explain how the microscopic structure around the rare earth ions influences the high temperature spectral emittance in a RESE. Our approach is based on the study of the local symmetry of the crystal field potential V_{CF} whose interaction with the 4f electron shell of rare earth ion can induce or inhibit non radiative decay channels. If $\vec{r}_e$ is the electron positional observable in a 4f shell, the crystal field potential V_{CF} of the material can be expressed using the following expansion [3]:

$$V_{CF}(\mathbf{r}_e, \{\mathbf{R}_i(t,T)\}) = \sum_{\substack{l=0\ldots\infty \\ m=-l\ldots l}} B_{lm}(t,T) C_{lm}(\vartheta_e, \varphi_e) r_e^l \quad (1)$$

were $B_{lm}(t,T)$ are the crystal field parameters which depend on time t and temperature T, $\{\vec{R}_i(t,T)\}$ is the set of ion coordinates in the host material, C_{lm} are the spherical harmonic function. The rare earth atomic state is defined by $|\gamma, J, M\rangle$ were J, M are quantum numbers of the angular momentum and γ represents the remaining quantum numbers.

In this frame the transition between two atomic states $|\gamma, L, M\rangle \rightarrow |\gamma', L', M'\rangle$ induced by the crystal field V_{CF} is completely characterised by the matrix element [4]:

$$\langle \gamma, J, M | V_{CF} | \gamma', J', M' \rangle = \sum_{\substack{l=0\ldots\infty \\ m=-l\ldots l}} B_{lm}(t,T)(-1)^{J-M} \begin{pmatrix} J' & l & J \\ -M' & m & M \end{pmatrix} \langle \gamma | r_e^l | \gamma' \rangle \quad (2)$$

were the Wigner symbol $\begin{pmatrix} J' & l & J \\ -M' & m & M \end{pmatrix}$ is non vanishing only for $m = M' - M = \Delta M$. As an important consequence we have that the multipole field component of $(l,m)^{\text{th}}$ order opens a non radiative decay channel for transitions having $\Delta M = m$ with a transition rate that, at the first order, can be expressed by $\hat{P}_{\Delta M}^{J,J'} = |\langle \gamma, J, M | V_{CF} | \gamma', J', M' \rangle|^2$, which in turn is linearly dependent on the $B_{lm}(t,T)$ coefficients. These coefficients depend on the local symmetry of the ions which, therefore, deeply affect the efficiency of the atomic transitions, the radiative decay channel being quenched in the presence of efficient non radiative decay processes. In rare earth based atomic systems we typically have $J = J'$ and $M \neq M'$ for any transition involved in selective emission. In particular in the case of Erbium the characteristic transitions have $\Delta M = 1$ ($^4I_{13/2}$ -> $^4I_{15/2}$) and $\Delta M = 2$ ($^4I_{11/2}$ -> $^4I_{15/2}$), corresponding to emission bands at 1.3-1.6 μm (very interesting for TPV applications) and 0.8-1.0 μm respectively.

3. EXPERIMENTAL RESULTS

In order to check the relationship (2) between non radiative decay and the ionic potential symmetries, we performed CL experiments at room temperature on several erbium doped Alumina samples synthesised at different temperatures, T_S. A Cambridge Stereoscan 360 SEM equipped with an Oxford Instrument mono-CL-system was used for all the CL experiments and the spectra are reported in Fig. 1 for increasing sintering temperatures. For T_S lower than 1200 °C only the $\Delta M = 1$ ($^4I_{13/2}$ -> $^4I_{15/2}$) transition is present. At higher sintering temperatures the $\Delta M = 2$ ($^4I_{11/2}$ -> $^4I_{15/2}$) transition appears and both bands evolve becoming more and more structured. This can be understood in the frame of our model. In fact by varying the sintering temperature, T_S, we obtain different atomic configurations around the Er ion; therefore the symmetry properties of the potential, V_{CF}, change thus producing a crucial modification of non radiative transition probabilities $\hat{P}_{\Delta M}^{J,J'}$.

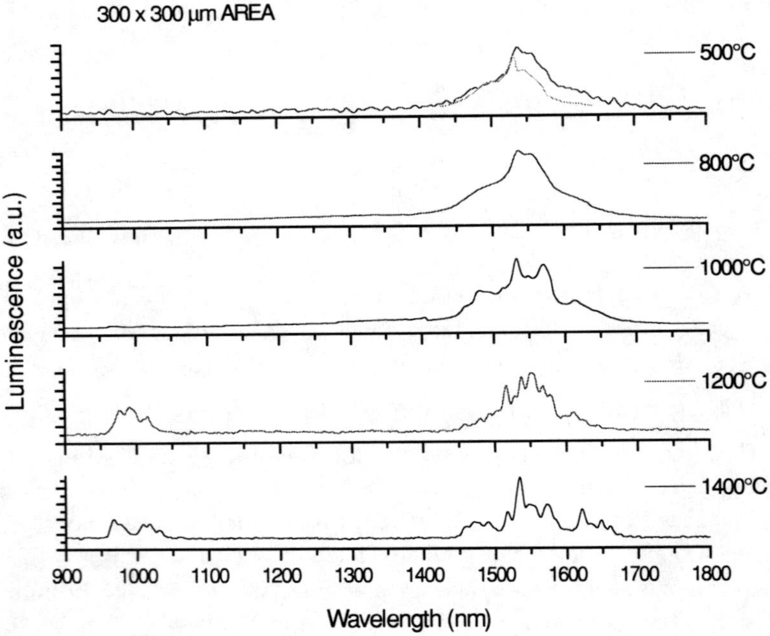

FIG.1 CL emission spectra of erbium doped Alumina sample with different sintering temperature

As pointed out previously, the non radiative decay probability $\hat{P}_{\Delta M}^{J,J'}$ depends on ΔM. For our samples the Er doped Alumina have two principal emission band. One ($^4I_{11/2}$ -> $^4I_{15/2}$) in the 0.8-1.0 μm range and corresponds to $\Delta M = 2$. For this transition the non radiative decay channel is activated only if the ionic crystal field contains a quadrupole component. The other transition ($^4I_{13/2}$ -> $^4I_{15/2}$), corresponds to $\Delta M = 1$ and the non radiative excitation channel is present if the electric field presents a dipole component.

Therefore the evolution of the efficiencies of this two transitions is different when the sintering temperature T_S, and consequently the bond symmetry change (Fig. 1). For $T_S < 1050°C$ the transition $^4I_{11/2}$ -> $^4I_{15/2}$ (0.8-1um) is completely absent whereas the transition $^4I_{13/2}$ -> $^4I_{15/2}$ reveals only an evolution of the structure. A theoretical calculation of the non radiative decay rate requires the knowledge of the coefficients $B_{lm}(t,T)$ which are functions of the coordinates of the ions. Using a simple point charge model we calculated the multipole components of the field produced by simple ionic configurations (the details of this model will be presented in a forthcoming paper). In particular we analysed a configuration with a erbium ion coordinated with six Oxygen ions (structure named A) which corresponds to samples sintered at temperatures below 1050°C and the same configuration inside a octahedric system of Aluminium ions (structure named B) corresponding to samples sintered at temperatures higher than 1050°C. Tab.1 shows the results of this numerical calculation in terms of the ratio between crystal field parameters corresponding to monopole field terms (in Er^{3+}).

Tab 1. CFPs numerical comparison.

	A	B
B_{l1}/B_{00}	10^{-8}	10^{-8}
B_{l2}/B_{00}	10^{-2}	10^{-8}

This numerical calculation demonstrates that when the ionic complex A is inserted in a distorted octahedric configuration of Aluminium ions, the non radiative decay in transitions with $\Delta M = 2$ is more efficient than in transitions with $\Delta M = 1$ (Tab.1 column 1). This explains the evolution of the CL spectra reported in Fig.1. With $T_S > 1050°C$, the "dopant" ions are arranged in a garnet structure with the Oxygen ions in a tetrahedric configuration and the Aluminium ions in a octahedric configuration. In the case of the garnet structure the non radiative coefficients for $^4I_{11/2}$ -> $^4I_{15/2}$ and $^4I_{13/2}$ -> $^4I_{15/2}$ are comparable (Tab.1 column 2) and we clearly observe two emission bands. At lower sintering temperature the distorted structure of the aluminium ions fail to provide a quadrupole term as in the case of the A structure. In this case, the transition $^4I_{13/2}$ -> $^4I_{15/2}$ with $\Delta M = 2$ exhibits a strong non radiative channel and it is not observed in the CL spectra.

To punctually confirm our theory we have carried out a PL experiment in cryogenic condition in order to suppress non radiative decay. An Argon laser beam (514.5 nm) was focused on the sample and the emitted PL signal was analysed by a monochromator TRIAX 320 and revealed by an Andor CCD (for the 0.8-1μm band) and by an Edinburgh Ge detector (for the 1.2-1.6 μm band).

Fig.2 and Fig.3 show the PL spectra of three erbium-doped Alumina samples (A,B,C) sintered at $T_S = 500°C$ (A), $T_S = 700°C$ (B), $T_S = 900°C$ (C) and cooled at

$T_C = 11°K$ (Fig.1) and $T_C = 200°K$ (Fig.2). It appears that the PL emission efficiency for $^4I_{13/2} \rightarrow {}^4I_{15/2}$ ($\Delta M = 1$) at 11K is comparable for all samples while at $T_C = 200°K$ samples A and B have an efficiency six times lower than sample C. These measurements clearly confirm the prediction of the theory on the effect of the ion symmetry on the emission efficiencies.

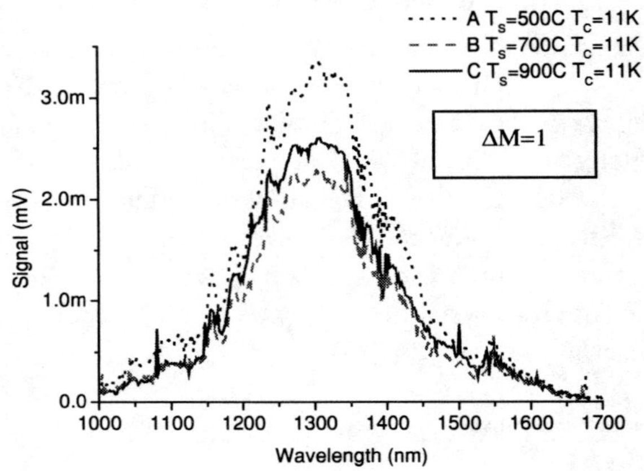

FIG.2 PL emission spectra samples A,B,C at $T_C = 11°K$.

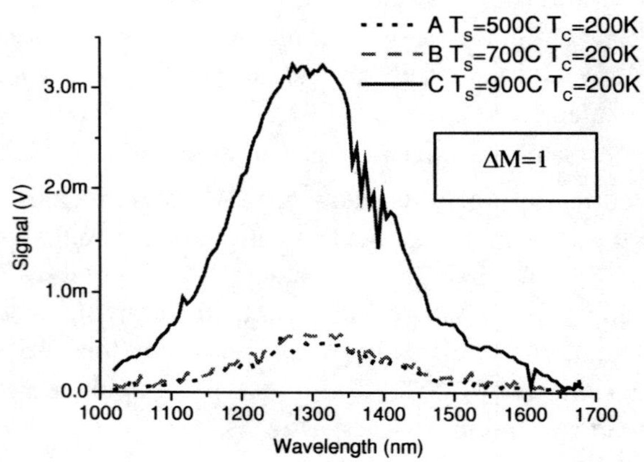

FIG.3 PL emission spectra samples A,B,C at $T_C = 200°K$

The X-ray analysis, not reported here, reveal that for $T_S > 1050°C$ Erbium Aluminium garnet is progressively more abundant. For this reason the dipole component in the electric field around the Er ion decreases with increasing sintering temperature. As a result the emission efficiency at 200K increases because the non radiative decay rate decreases. At $T_C = 11K$ the non radiative decay rate is negligible in all the samples and therefore, the emission efficiencies are comparable. A further confirmation of the model comes from the behaviour of sample C ($T_S = 900°C$) where the non radiative transition $^4I_{13/2}$ -> $^4I_{15/2}$ is quenched and the emission efficiencies turn out not to change between $T_C = 11K$ and $T_C = 200K$ while for the other transition, $^4I_{11/2}$ -> $^4I_{15/2}$ ($\Delta M = 2$) there is no difference with the other samples.

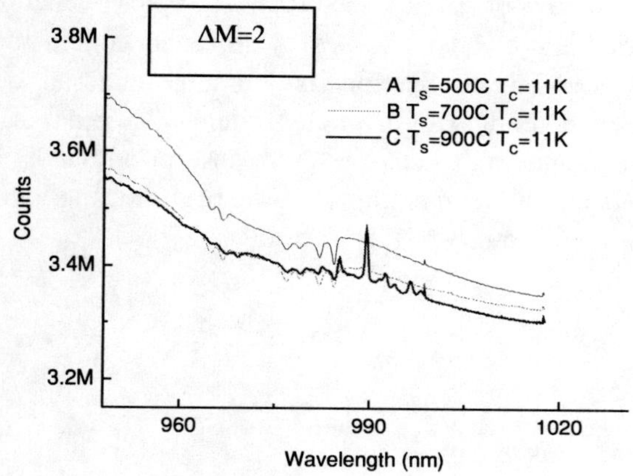

FIG.4 PL emission spectra of erbium doped Alumina samples A,B,C at 11°K.

Fig.4 shows the $^4I_{11/2}$ -> $^4I_{15/2}$ transition when the samples A,B,C are cooled at 11K. In this case ($\Delta M = 2$) the non radiative decay channel is efficient and it is revealed by the presence of a series of absorption lines floating on the background of the Alumina emission.

This non radiative dissipation confirms the existence of a strong coupling of RE ions with the electric field generated by the local environment. When $T_S > 1050°C$, the electric quadrupole moment of the electric field decreases because of the presence of an increasing fraction of garnet where the non radiative decay coefficients for $\Delta M = 2$ and $\Delta M = 1$ are comparable. This is the reason why two emission band are observable at room temperature in the CL spectra for $T_S > 1050°C$.

4. CONCLUSIONS

A theoretical model capable to explain how the efficiency of rare earth selective emitters can be optimised has been presented. The symmetry of the ionic potential around the rare earth atoms has been shown to be a crucial parameter to control the non-radiative decay processes which may prevent a good absorber to act as an efficient selective emitter at high temperature. The predictions of the model where confirmed by room temperature and cryogenic luminescence analysis designed to study the emission properties of the analysed RESE samples by switching on and off the non-radiative recombination channels. A very important conclusion from the point of view of thermophotovoltaic applications is that some materials with a particular symmetry of the electric potential (for example Er doped Zirconia) can exhibit a good optical absorption corresponding to the rare earth absorption bands, but they are not good selective emitters at high temperature because of the strong non radiative decay processes allowed by the symmetry of the potential.

REFERENCES

1. Donad L. Chubb, Anna Maria T.Pal, Martin O.Patton and Phillip P.Jenkins "*Rare Earth Doped High Temperature Ceramic Selective Emitters*" Journal of The European Ceramic Society **19** (1999) 2551-2562.
2. V.V. Gogovlev, C.H. Winston Chen and W.R. Garrett, *"Heat to light energy conversion by emitters doped with rare-earth metal ions"* Appl. Phys. Letters **69** (2), 8 July 1996 pp.280 –282.
3. Lidia SMENTEK, *"Theoretical description of the spectroscopic properties of rare earth ions in crystals"*, Physics Reports **297**, 1998 pp.155-237.
4. A.R. Edmonds, *"Angular Momentum in Quantum Mechanics"* edited by Princeton University Press 1960, Chapter 5, pp. 68-79.
5. J.A. Reissland, *"The Physics of Phonons"* edited by John Wiley & sons LTD. 1973.
6. A. Licciulli, A. Maffezzoli, G. Torsello, D. Diso, S. Tundo, M. Rella, M. Mazzer, *"Sol gel preparation of selective emiytters for thermophotovoltaic conversion"* Journal of Sol-Gel Science and Technology, **26**, (2002)

Rugate Technology For Thermophotovoltaic (TPV) Applications: A New Approach To Near Perfect Filter Performance

Ugur Ortabasi*, Bertrand Bovard**

*United Innovations, Inc., 1780 La Costa Meadows Dr. San Marcos, CA 92069, USA

**Rockwell Scientific, 1049 Camino Dos Rios, Thousand Oaks, CA 91360, USA

Abstract. Rugate filter technology is relatively unknown to the Solar PV- and the TPV community. The objective of this paper is to draw attention to Rugate filters as an enabling technology that may help to realize unprecedented conversion efficiencies in each respective field. Rugate filters are advanced optical interference filters with a continuously varied refractive index that changes periodically in the direction perpendicular to the film plane. The near noise free transmission and reflection quality of these filters, combined with their precision and the durability, promise significant improvements in the photon economy, performance and reliability of TPV devices. This paper reports the analytical modeling results for a set of Rugate filters designed for the subcells of a novel 4-bandgap cell system with a new architecture. The purpose of this new approach is to circumvent the limitations of multi-junction cells to achieve ultimate conversion efficiencies. The purpose of these near-to-perfect Rugate filters is to induce spectral splitting of concentrated solar radiation trapped in a Photovoltaic Cavity Converter by selective transmission and reflection of photons. Wideband Rugate filter designs developed for InGaP, GaAs, InGaAsP and InGaAs cells exhibit about 99% average transmission over the spectral response range of the respective cells, while reflecting more than 99 % of the solar photons outside their transmission band. Results of first filter deposition trials are briefly discussed in conjunction with TPV applications.

INTRODUCTION

One of the remaining challenges for TPV systems is the spectral matching of the emitter irradiance to the quantum efficiency of the photocell. The spectrum of broadband emitters operating at temperatures 1000–2000 K contain significant number of sub-bandgap photons with energies not sufficient enough to generate charge carriers in the semiconductor. In order to increase efficiency, these photons are reflected back to the radiation source so that their energy is recycled within the TPV system. Photons with energies above the bandgap can create only one electron-hole pair. Any excess energy is absorbed in the lattice as heat. This thermalization process represents another mechanism for energy loss. The efficiency of photon energy utilization reaches a maximum when the photon energy slightly exceeds the bandgap. A precise band-pass filter that minimizes the losses due to sub-bandgap photons and thermalization process increases the TPV performance significantly. Such a filter with perfect transmission and reflection characteristics transforms the energy contained in a broadband spectrum

CP653, *Thermophotovoltaic Generation of Electricity: 5th Conference*
edited by T. J. Coutts, G. Guazzoni, and J. Luther

to generate photons within a narrowband spectrum. Rugate technology reported in this paper offers the potential to manufacture such near-to-perfect filters.

In the past several filter types including dielectric stacks, plasma filters, resonant array filters or a combination thereof have been developed with a limited success. A comparative summary of these filter types is given in reference [1]. Recent experimental [2] and modeling studies [3] have shown that the filter performance dominates the system performance for any practical TPV systems and the application of a perfect band-edge filter may more than double the efficiency when compared to the same TPV system, without a filter. Thus there is a compelling need to improve the state-of-the-art filter technology in order to increase the overall efficiency of the present TPV devices.

The objective of this paper is to draw attention to Rugate filters as an enabling technology. The introduction of Rugate filters into TPV may help to realize unprecedented conversion efficiencies by improving the photon economy beyond the more familiar filter technologies mentioned above. The intrinsic accuracy of Rugate filter design principles could lead, in the near future, to the manufacture of efficient, near noise free, band-pass filters or band-edge filters for use with broadband and narrowband TPV systems.

In this paper we report our first modeling results involving Rugate filters designed to induce spectral splitting of concentrated solar radiation in a Photovoltaic Cavity Converter (PVCC) [4]. The splitting (more correctly stripping) of the solar spectrum is achieved by selective transmission and reflection of trapped photons during their random walk inside the cavity [4]. Present paper reports the progress made in designing four conjugate Rugate filters that are developed to match the internal quantum efficiencies of InGaP, GaAs, InGaAsP and InGaAs cells. In other words, each conjugate filter is designed to transmit all incident photons within the frequency band quantum efficiency of the respective cell and reflect all other photons outside this transmission band.

Above mentioned set of the III-V subcells form the basis of a novel 4-bandgap conversion system [4]. The analytical Rugate filter models developed for these subcells exhibit about 99% average transmission over the internal quantum efficiency range of the respective cells, while reflecting more than 99 % of the photons with energies outside their transmission band. Results of the first experimental deposition trials as well as some of the engineering challenges are discussed later in the paper.

Although the PVCC concept is based purely on the photovoltaic conversion process, which takes place at low temperatures (close to ambient), the basic principles of the photon recycling mechanism involved in PVCC and in TPV devices are analogous. Both systems are optimized to reutilize photons which energy wise do not match the quantum efficiency of the semiconductor(s) in use. In the case of TPV the non-matching photons are reflected back to the heat source by reflection from a selective mirror and their energies are conserved to generate new matching photons. In the case of PVCC the non-matching photons are reflected (by selective Rugate filters) back into the cavity where they have a high probability to strike and enter a matching cell.

RUGATE FILTER TECHNOLOGY

Rugate filters are interference thin film filters. Let's consider a plane wave traveling through air and encountering a thin layer of transparent dielectric material of given thickness and refractive index (RI). At the interface where the RI changes occur some of the wave is reflected while the rest of it is transmitted through the medium. If the thickness of the slab is on the order of wavelength of the incident light, the reflected wave will interfere with the incoming wave. Maximum destructive interference is achieved at quarter of the optical thickness, when the reflected wave is 180 degrees out of phase with the incident wave. Optical thickness is defined as the product of the physical film thickness and the refractive index of the thin film. Thus by stacking materials of different refractive indices one may greatly influence the amount of light at a specific wavelength passing through the materials. The simplest approach for this is to use two different materials, alternatively stacked upon each other (multi-layer interference filters). Such constructions are used extensively for a wide range of applications. However there are limitations to this method. Such filters tend to exhibit secondary lobes just outside the stop-band and harmonics at shorter wavelengths of the rejected wavelength. As a result their out-of-band transmittance integrated over a wide wavelength range is reduced. One may reduce these side lobes and secondary harmonics by smoothing out the transition of the RI between the interfaces. This is accomplished by including additional materials with intermediate values of the RI, i.e., mixtures or alloys of the two end materials. By adding more types of films with different RI values, one can eventually approach a continuous change in the RI profile. Thus, a continuous profile wave function (say, a sine wave) replaces the binary low/high RI profile of the simple stack case. The continuously varied refractive index profile design is known as the Rugate filter.

The word Rugate is actually an adjective, having Latin origins "wrinkled or creased". Thus, generally stated, Rugate filters are inhomogeneous optical interference filters with a continuously varied refractive index gradient that changes periodically in the direction perpendicular to the film plane.

Reference [5] by Bovard gives an extensive overview of the Rugate filter theory. A variety of such Rugate filters have received a lot of attention during the past 20 years, and new deposition processes aimed at precisely controlling RI profiles have been developed to produce single-line and multi-line, high optical density Rugate filters [6–9]. However, less progress has been made to integrate inhomogeneous layers in structures such as broadband pass filters [10-11], beam splitters and non-polarizing filters.

To complete the design, high performance anti-reflection coatings are applied to prevent interferences between the Rugate filter and the outer media, typically air and a substrate such as glass.

To the knowledge of the authors, the present work is the first attempt to integrate wide band Rugate filters with integral AR coatings into High Concentration Photovoltaics and Thermophotovoltaic designs.

PHOTOVOLTAIC CAVITY COVERTER (PVCC) SYSTEM USING PHOTON RECYCLING

The PVCC approach, to maximize the utilization of the solar spectrum by means of a multi-bandgap subcell system, is the same underlying principle as with the well-established multi-junction (MJ) solar cells. By design, incident photons from an emitter (sun, IR source, etc.,) are selectively absorbed by the subcells, commensurate to their energies. The system is optimized by minimizing the absorption of the below bandgap photons by those subcells which have higher bandgap energy.
The difference between the two multi-bandgap cell systems i.e. MJ versus PVCC is the "architecture" that determines how the "subcells" are arranged to construct the multi-bandgap device. Both systems are shown in Fig. 1. (The size of the PVCC system is miniaturized to the scale of the MJ cell for the purposes of illustration).

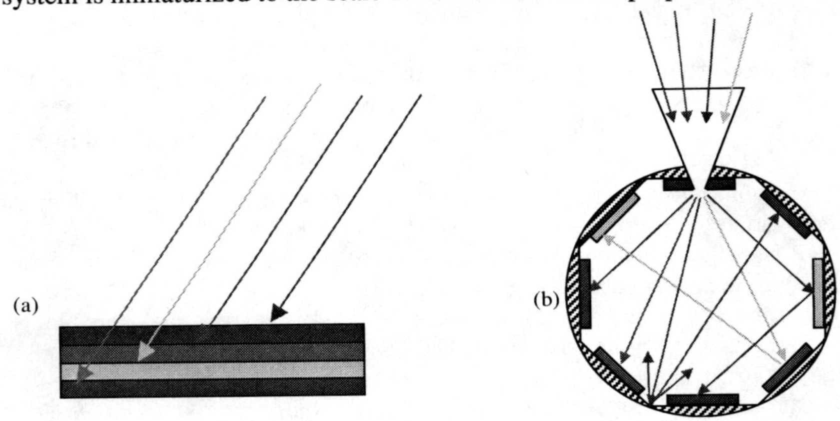

FIGURE 1. Principles of spectral stripping by multi-bandgap systems; (a) Vertically grown or stacked MJ, (b) Independent single-junction cell systems in a cavity. Cells are covered with Rugate filters

MJ's "vertical" architecture is formed by growing monolithically or by stacking different bandgap materials on top of each other. The PVCC in contrast has a "lateral" architecture" that is formed by arranging single-junction cell groups in a 2-D array on the interior wall of the cavity. This change in the architectural configuration allows the subcells in the PVCC directly interact with the filtered light by the respective Rugate filters. This way many of the well known concerns or limitations associated with MJ cells (monolithic or stacked) are circumvented. Some of the generic problems MJ systems face are: lattice mismatch-, current limitation-, tunnel junctions, optical interfaces, undesirable photon absorption, material choice limitations, overly complicated manufacturing issues and thermal management difficulties. Some of these problems become more significant when systems are designed to operate at concentrations exceeding 400 suns i.e., at about 30 watts/cm^2.

As a trade-off the PVCC approach requires a highly efficient "external" spectral splitting (stripping) capability that is strongly dependent on the quality of the Rugate filters and the reflectance of the exposed lambertian surface between the cells.

Spectral Splitting Process Inside The PVCC Cavity

Fig. 2 shows the principles of the "Spectral Splitting by Selective Reflection and Transmission" method developed under this project.

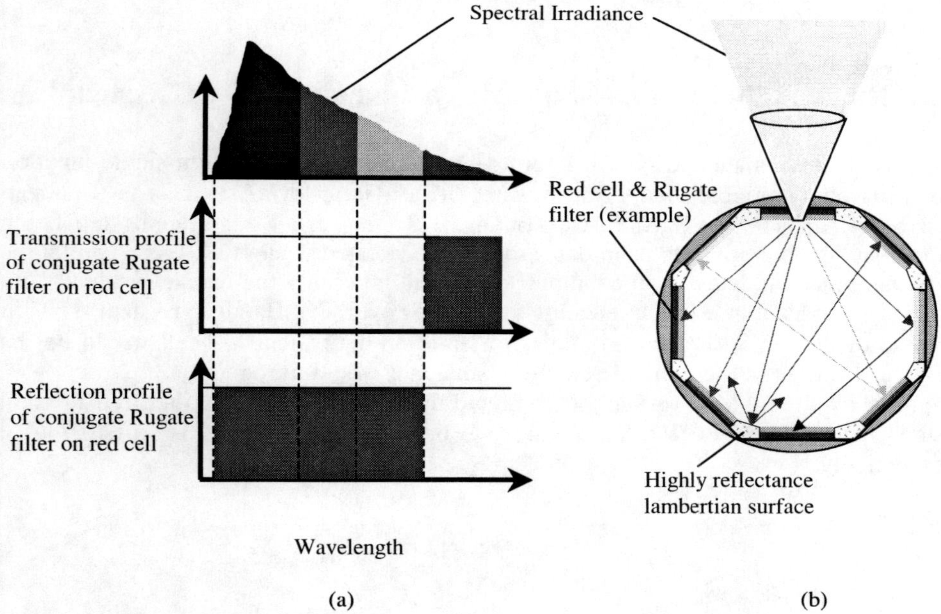

FIGURE 2. Principles of spectral stripping by selective transmission and reflection inside a PVCC cavity

The PVCC shown in Fig. 2b illustrates a 4-bandgap system consisting of four different subcell groups (red cells, blue cells, etc.,) which have different but adjacent internal quantum efficiencies. As shown in Fig. 2a the four groups of subcells, together span congruently the total solar spectrum. Thus full utilization of the solar spectrum energy content becomes possible.

Each subcell is covered with a matching (conjugate) Rugate filter to control the transmission or reflection of striking photons depending on their wavelength. The rest of the unoccupied interior is covered with a highly reflective lambertian surface (Spectralon in our case). Now consider a red photon that enters the cavity. This sample photon may undergo the following interactions: a) it may strike a red cell directly and enters the cell; b) it may strike the cavity wall first and is reflected diffusely; c) it may strike one of the other subcell types and is reflected specularly. In the case of c) and b) the reflected red photon continues its random walk until it strikes and enters a red cell where it ultimately contributes to the photocurrent with a high probability. As a result of the described **photon recycling** process, most of the photons enter ultimately their respective conjugate subcells, less they are absorbed beforehand. The probability with which a once reflected photon enters a matching cell is called the **Photon Utilization Factor (PUF).** Extensive ray tracing studies have shown that a PUF of 0.9 can be realized if the average reflectance and transmittance of the Rugate filters as well as the reflectance of the lambertian surface are about of 99%. As it will be discussed later in the paper a PUF of 0.9 is necessary to realize a collective cell conversion efficiency of ~ 50%, Thus for the PVCC to reach high conversion efficiencies the near-to-perfect characteristics of the Rugate filters is a prerequisite.

Rugate Filter Design Studies for a 4-bandgap PVCC System

The UI^2 team has studied two different 4-bandgap systems based on single-junction cells that are available today. These were: InGaP-GaAs-InGaAsP-InGaAs (Bandgap System I) and InGaP-GaAs-Si-Ge (Bandgap System II). Both systems are being considered as the possible candidates to be incorporated in the first PVCC prototype now being assembled. As an example we present here only the results obtained with the Bandgap System I. It is interesting to note however, that Bandgap System II has Si as one of the subcells. Incorporation of Si into an equivalent MJ cell would not be feasible because of the dramatic lattice mismatch it would introduce.
Fig. 3 shows the AM 1.5 solar spectrum and the conjugate Rugate filter transmission bands for InGaP (1.86eV), GaAs(1.42eV), InGaAsP(1.10eV) and InGaAs(0.74eV), respectively.

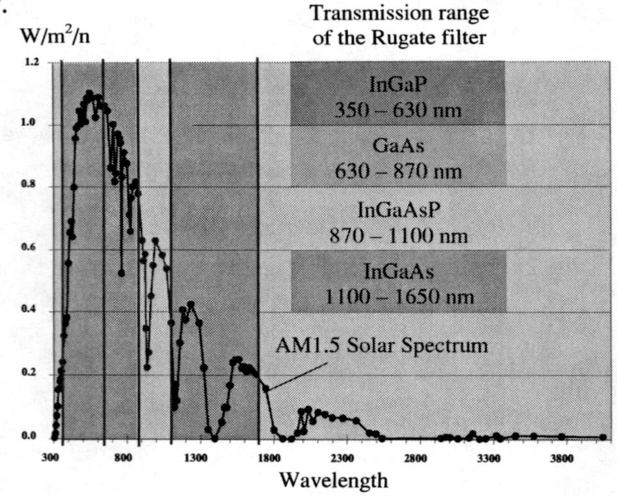

FIGURE 3. AM 1.5 solar spectrum and the position of the selected Rugate transmission bands for the conjugate cells

Sensitivity studies have shown that maximum collective cell efficiency is achieved if the Rugate transmission band edges are lined up with the energy bandgap frequencies of the respective cells. Extensive design studies have resulted in excellent filter characteristics for all III-V subcells selected for the Bandgap System I. For the sake of brevity only the reflectance/transmittance profile of the conjugate Rugate filter for GaAs is shown in Fig. 4. The Rugate filters developed for this phase of the project were designed to be deposited on fused silica substrates that is suspended over the solar cell with an air gap in between (Fig. 5). Anti-reflective coatings are applied at all optical interfaces as a part of the deposition process to avoid Fresnel losses. The average reflection as well as the average transmission of the Rugate filter shown in Fig. 4 is about 99%. The results for the other cells are also very similar. In all cases the average transmission and reflection profiles are symmetrical, and the up-ramp and down-ramp slopes are very steep. No lower harmonics or side lobes have been observed.

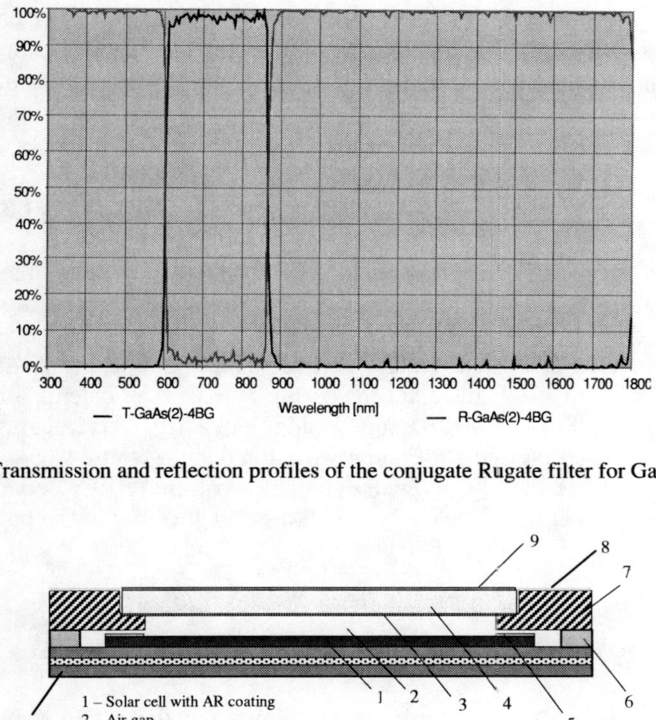

FIGURE 4. Transmission and reflection profiles of the conjugate Rugate filter for GaAs

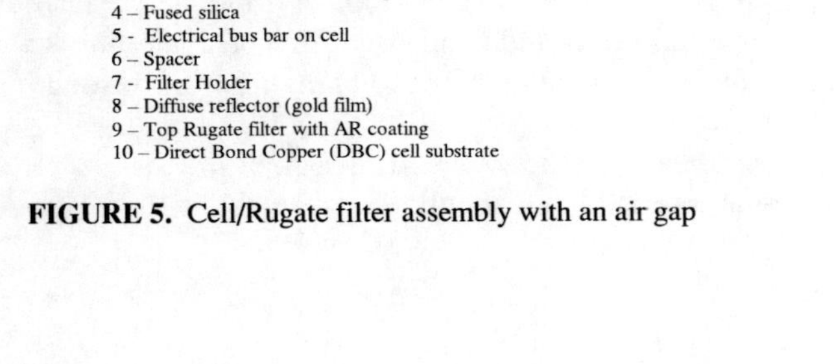

FIGURE 5. Cell/Rugate filter assembly with an air gap

The manufacturing of filters that exhibit optical performances of the type presented in figure 4 is not without challenges. Depending on the thin material film used, the steepness of the transitions between highly reflective and highly transmissive, the range of angles over which the filter must perform and the aggregate width of the reflection region, the total physical thickness of the Rugate filter can exceed 50 microns. In our design studies, the total physical thickness was minimized and kept between 10 and 35 microns for all filters. Such film thicknesses are not completely unusual. For example, state-of-the-art filters produced for dense wavelength division multiplexing in fiber optics communications are typically 20 to 30 micron thick. Manufacturing of thicker filters demands tight process control over extended deposition time (12 to 36 hours, depending on the process used and the exact thickness).

Our preliminary experimental results show that filters with broad out-of-band reflection regions with excellent reflectance can be manufactured reliably. The principal challenge is to maximize the in-band transmittance by minimizing reflectance and absorption. Reflectance losses come from manufacturing errors in the as deposited refractive index profile of the Rugate filters. Absorption comes from residual impurities (typically metallic) present in the film. Both of these limitations can be addressed by improving process control and further optimization of the deposition conditions.

Analytical Performance Results for PVCC

Extensive ray tracing studies were performed for the cavity interior based on the theoretical filter designs developed under this project. The results predicted an excellent Photon Utilization Factor (PUF) of 0.9. This PUF value was then utilized to develop a cell model to calculate performance of the selected 4-bandgap System I (InGaP-GaAs-InGaAsP-InGaAs) inside the PVCC. The calculations assumed a flux concentration of 100 AM1.5D suns inside the cavity. The calculated lower limit for the **collective conversion efficiency** for Bandgap System I was **49.1%** [4]. These analytical results clearly show the potential of the PVCC concept however, the complete verification of the concept will await the testing of the prototype at NREL's High Flux Solar Furnace (HFSF) facility by the end of 2002.

RELEVANCE OF RUGATE FILTERS TO TPV

PVCC results reported in this paper are limited to the useful range of AM 1.5 solar spectrum ranging from 350 nm to 1800 nm. In principle we have analytically shown that very efficient, band-pass and band-edge filters Rugate filters can be designed that meet the broadband requirements of PVCC ranging from several hundred nanometers to over 1000 nm.

For the TPV applications the range of spectral interest is red shifted. Efficient selective filters are needed from about 500 nm to about 10000 nm. Broadband Rugate filter systems developed under this project are definitively usable for narrowband emitters up to 2000 nm.

Although in principle Rugate filters can be designed for emitter spectra extending beyond 2 microns, there is little published data on the broadband IR- and far IR Rugate filters. Materials used in the present application can be used beyond 2 microns (up to 4-5 microns). In our design efforts we limited the reflectance band to 1800 nm in order to keep the total thickness of the film to a minimum and also because of the diminishing energy content of the AM 1.5 solar spectrum that did not justify the added complexity. Rockwell Scientific makes infrared Rugate filters for applications in the MWIR and LWIR ranges spanning from 2 to 15 microns. Rugate filter manufacturing results for this project will be reported in the near term.

CONCLUSIONS

Rugate filter technology, relatively unknown to the PV- and TPV community, has the potential to significantly increase the conversion efficiencies in both fields. This paper reports, as an illustration, the recent advances made in the field of High Concentration Photovoltaics (HCPV) where Rugate filters are used for inducing spectral splitting in a Photovoltaic Cavity Converter (PVCC). Analytical studies based on Rugate filter designs with AR coatings, developed for a 4-bandgap system, predicted a collective conversion efficiency of 49.1% under a concentration of 100 AM 1.5. If this efficiency level can be verified by the experiments, planned for the near term, the respective improvement over the world record (34 %) would be 15% absolute or 44% relative.

The similarity of photon recovery requirements between PVCC and the TPV suggests that Rugate filter technology can benefit TPV also dramatically. Present work on Rugate filters for the PVCC is limited to the solar spectrum from 350 nm to 1800 nm. In principle however, broadband Rugate filter technology can be expanded beyond 2 microns successfully.

ACKNOWLEDGMENTS

This work was supported in part by U.S. Department of Energy through the National Renewable Energy Laboratory as a part of "Concentrating Photovoltaics: Module Design for Near Term Demonstration and Test" program, under contract No. ADC –0-30443-1 and "Photovoltaic Technologies Beyond the Horizon" program, under contract No. NCQ-130619-15.

REFERENCES

1. Archer, M. D., and Hill, R., *Clean Electricity from Photovoltaics,* Imperial College Press, London, 2001 Vol. 1, pp. 491-496.

2. Beausang, J. and Reynolds J.E., "Thermodynamic Analysis of Thermophotovoltaic Efficiency and Density Tradeoffs", *The Fifth Conference on Thermophotovoltaic Generation of Electricity (TPV5),* Conference Paper Abstracts, 2002, pp. 6.
3. Bitnar, B., Durisch, W., Mayor, J-C., Sigg, H., Tschudi, H. R., Palfinger, G., Gobrecht, J., "Record gas-to-electricity power efficiency of a silicon solar cell based TPV system", *The Fifth Conference on Thermophotovoltaic Generation of Electricity (TPV5),* Conference Paper Abstracts, 2002, pp. 8.
4. Ortabasi, U., Lewandowski, A., Aiken, D. J., Sharps, P. L., Bovard, B.G., "Dish/Photovoltaic Cavity Converter (PVCC) System For Ultimate Solar-to-Electricity Conversion Efficiency: General Concept and First Performance Predictions", *29th IEEE Photovoltaic Specialist Conference,* 2002, IEEE Conference Proceedings, to be published.
5. Boward, B. G., *Appl. Ops.,* **32, no. 28**, 5427-5442, (1993).
6. Sankur, H., Gunning, W. G., *Appl. Opt.,* **27,** 1564-1567, (1988).
7. Donovan, E. P., and DeNatale, J. F., *Appl. Opt.,***28,** 2940-2944, (1989).
8. Gunning, W. J., Hall, R. L., Woodberry, F.J., Southwell, W. H., and Gluck, N. S., *Appl. Opt.,* **28,** 2945-29-48, (1989).
9. Hall, R. L., " Gradient index bandpass filters", *Optical Interference Coatings,* Vol. **15** of 1992 OSA Technical Digest Series (Optical Society of America, Washington D.C., 1992), pp. 116-118.
10. Willey, R. R., "Rugate broadband antireflection coating design", *Current Developments in Optical Engineering and Commercial Optics,* R. E. Fisher, H. M. Pollicove, and W. J. Smith, Proc. Soc. Photo-Opt. Instrum. Eng., **1168**, 224-228 (1989).
11. Willey, R. R., Verly, P. G., and Dobrowolski, "Design of wideband antireflection coating with the Fourier transform method, *Thin Films and Applications,* R. Herrmann, Proc. Soc. Photo-Opt. Instrum. Eng., **1270,** 36-44 (1990).

A parametric design study of InGaAs micro-thermophotovoltaic cells coupled with various emitters at near and far spacing

Jeffrey M. Bierley & Ashok Gopinath *
Department of Mechanical Engineering, Code ME
Naval Postgraduate School, Monterey, California 93943, USA

Abstract

A design study was conducted to determine the effect of various parameters on the performance of InGaAs thermophotovoltaic devices. The electrical power output density of the device, and the device efficiency were analyzed by considering different emitter temperatures and materials, various receiver band gap levels, and a range of gap spacings between emitter and receiver. The far spacing model, or spacing of magnitude much greater than the wavelength of the emitted radiation, was based on classical radiative heat transfer equations and standard photovoltaic cell equations. The close spacing model, or spacing with a magnitude on the order of magnitude of the wavelength of the emitted radiation, was based on the fluctuation-dissipation theorem whereby the energy is transmitted by evanescent waves between the emitter and receiver. The analytical models used realistic property values for both the receiver and the emitter, and the results show the superiority of tungsten and rhenium as emitters for the far spacing case. At close spacing, silicon carbide exceeds the performance of the refractory metals down to the spacing limited by current technology. However, if spacing control technology advances and smaller gaps become feasible, rhenium emitter devices could provide a significant increase in power density, and with similar efficiencies as the far spacing case.

Introduction

Recently, great interest has developed in the research and development of thermophotovoltaic (TPV) technology. This technology provides significant improvements over existing energy conversion systems through the use of fewer components, fewer moving parts, and a simpler overall design.

A typical TPV device consists of a heat source, an emitter which transfers the energy of the heat source by radiation at some desired wavelength or range of wavelengths, a semiconductor receiver in close proximity to the emitter which is similar to a traditional solar cell except that it contains different materials which help tune the receiver to the thermal vice solar wavelength radiation, and some sort of electrical distribution system to transmit the electrical current developed by the receiver. Figure 1 shows a typical TPV device.

This paper is primarily concerned with the emitter and receiver portions of the cell. The geometry of this design study is limited to two parallel flat plates, one being the high temperature emitter and the other being the low temperature receiver. Current work in the field has focused on constant temperature emitter and receiver, and this paper continues with that assumption. This paper also uses InGaAs receivers exclusively because current work in the field to produce high efficiency high power density cells has

CP653, *Thermophotovoltaic Generation of Electricity: 5th Conference*
edited by T. J. Coutts, G. Guazzoni, and J. Luther
2003 American Institute of Physics 0-7354-0113-6

shown this to be the superior type of cell and as a result there is much available technical information available on this type of receiver.
Most recently, Shi et al [1] performed a low-temperature far-spacing analysis of an InGaAs device at a band gap of 0.48 eV with a broadband grey emitter with emissivities of 0.1 and 1.0. They also used a 40% Er-15% Ho YAG emitter and an ideal emitter. Whale [2] analyzed InGaAs devices at several different band gaps, temperatures, and spacings. For the emitter, however, he uses a Drude model. Two other important works in this area include Raynolds [3] and Loomis and Maris [4]. Loomis and Maris develop an equation for the heat transfer between two bodies by evanescent electromagnetic waves, which is the dominant phenomenon at close spacing. Raynolds uses their method to illustrate the effect of spacing on the heat transfer between two bodies. The optimum property values he discovered for the metal in his paper are used to bracket the close spacing results in this paper as an "ideal" metal emitter. DiMatteo et al [5] have demonstrated a factor of 5 increase in power density at submicron spacing.
This paper adds to the current work in the field by exploring many different types of emitters, including rare earth aluminum garnets, refractory metals, and silicon carbide. By using actual measured material properties for the emitter materials, this paper gives a more realistic assessment of TPV performance than previous works which used more theoretical models for the emitter material.

Modeling Considerations

Far spacing modeling consists of application of the classical radiative heat transfer equations. The geometry considered in this paper is two semi-infinite parallel plates representing the emitter and receiver. The separation between the plates is assumed to be of much greater magnitude than the wavelength of the radiation but much smaller magnitude than the size of the plate. Figure 2 shows the geometry of the problem.

The spectral, blackbody emissive power is given by [1]

$$E_{\lambda,b} = \frac{2\pi h c_0^2}{\lambda^5\left[\exp(hc_0/\lambda k_b T)-1\right]} \qquad (1)$$

The spectral net heat transfer between two semi-infinite plate is given by [1]

$$q_\lambda = \frac{E_{\lambda,b1}-E_{\lambda,b2}}{1/\in_{\lambda 1}+1/\in_{\lambda 2}-1} \qquad (2)$$

To calculate the power output of the receiver, the heat flux incident on the TPV cell must be calculated. For the geometry of Figure 2, this incident flux is the radiosity from the emitter[1]

$$q_{\lambda,o1} = E_{\lambda,b1} - \frac{1-\in_{\lambda 1}}{\in_{\lambda 1}} q_\lambda \qquad (3)$$

The short circuit current of the cell is given by the integral of the product of the incident flux and the quantum efficiency divided by the energy of a photon and is given by the following equation[1],

$$I_{sc} = \frac{e}{hc_o} \int_{\lambda_1}^{\lambda_2} \lambda Q_E(\lambda) q_{\lambda,o1}(\lambda) d\lambda \qquad (4)$$

The remaining equation development is from standard photovoltaic cell relations and is the same for both the far spacing and the closing spacing cases. Shi et al[1] use an approximate equation for dark current given by Green[6], but a more accurate experimentally determined variation of dark current can be found in Whale[2]. The relationship between dark current and band gap is curve fit through experimental data and this data was used for calculations in this paper.

Once the dark current is known, the open circuit voltage can be found from the following equation[1],

$$V_{oc} = \frac{k_b T_2}{e} \ln\left(\frac{I_{sc}}{I_d} + 1\right) \qquad (5)$$

The maximum electrical power output of the cell can be obtained from the short-circuit current, the open-circuit voltage, and a parameter known as the fill factor which is defined as[1]

$$FF = \frac{v - \ln(v + 0.72)}{v + 1} \qquad (6)$$

where

$$v \equiv \frac{V_{oc} e}{k_b T_2}$$

Now, electrical power output can be calculated from the following equation [1],

$$P_{out} = I_{sc} V_{oc} FF \qquad (7)$$

The efficiency of the device is the electrical power output over the total incident flux given by the following equation[1]

$$\eta = \frac{P_{out}}{q} \qquad (8)$$

where the total incident heat flux is given by[1]

$$q = \int_0^\infty q_\lambda d\lambda \qquad (9)$$

The close spacing model allows the spacing between the emitter and receiver to be of the same order of magnitude or less than the incident radiation. The primary contributor to the energy transfer at close spacing is the evanescent field. The fluctuating electromagnetic field extending beyond the surface of a body is composed of normally propagating blackbody radiation and evanescent waves. The evanescent effect also known as radiation tunneling makes it possible for additional energy to be transferred from a material if another material is in close enough proximity. In a far spaced device, this evanescent field exists but it decays rapidly with distance from the body so the other body cannot be affected by the field. At submicron gap spacings, the evanescent field interacts with the adjacent material allowing increased energy transfer. The method used to derive the heat transfer at close spacings is known as the fluctuation-dissipation theorem. Loomis and Maris[4] derive the following equation for heat flux between two bodies at close spacing

$$P = -\frac{\hbar}{c_0^2\pi^2}\int_0^\infty \omega^3 d\omega\left(\frac{1}{\exp(\hbar\omega/k_bT_1)-1}-\frac{1}{\exp(\hbar\omega/k_bT_2)-1}\right)$$
$$\times\int p^3 dp\left(\begin{array}{c}\dfrac{(s_1+s_1^*)(s_2+s_2^*)}{\left|(p+s_1)(p+s_2)-(p-s_1)(p-s_2)e^{2ip\omega d/c_0}\right|^2}+\\ \dfrac{(\varepsilon_1^*s_1+\varepsilon_1 s_1^*)(\varepsilon_2^*s_2+\varepsilon_2 s_2^*)}{\left|(\varepsilon_1 p+s_1)(\varepsilon_2 p+s_2)-(\varepsilon_1 p-s_1)(\varepsilon_2 p-s_2)e^{2ip\omega d/c_0}\right|^2}\end{array}\right)\left|e^{2ip\omega d/c_0}\right| \qquad (10)$$

where
$$s_1 = \sqrt{\varepsilon_1 - 1 + p^2}$$
$$s_2 = \sqrt{\varepsilon_2 - 1 + p^2}$$

The p integration goes from 1 to 0 along the real axis and then 0 to ∞ along the imaginary axis. Whale[2] mentions the validity of equation (10) but points out that further simplification of equation (10) performed by Loomis and Maris[4] by assuming the wave vector in the material is equal to the wave vector in vacuum limits the validity of the approximation to extremely close spacings. Note that equation (10) which was used in this study makes no such approximation and the far spacing results show good correlation with Whale.

Whale proposes the following equation for short-circuit equation at close spacing to account for the eventual limit in performance as near-field flux greatly increases[2],

$$I_{sc} = \frac{e\Phi_p}{\left[1+\left(\dfrac{\Phi_p\tau_t}{L_o N_D}\right)^{2/3}\right]^{1/3}} \qquad (11)$$

where,

$$\Phi_p = \int_{E_g}^{\infty}\Phi_{p\lambda}Q_E dE \qquad (12)$$

and $\Phi_{p\lambda}$ is equation (10) divided by the energy of a photon. In the notation of equation (10) this is $\hbar\omega$. Equation (11) shows that now more information about the receiver is required to accurately model it. These values are very dependent on cell fabrication and will be discussed along with material property modeling at the end of this section.

To calculate the performance of the cell equations (5) through (8) are used as before in the far spacing case. Equation (10) is the close spacing equivalent to equation (9) for the spacing case.

Both models were incorporated into MATLAB® programs and solved using the QUAD function. The double integral in equation (10) was solved on the inner integral by QUAD and the outer integral by Simpson's Rule due to the complexity of the equation and the fact that the effective upper limit of the p integral changes at each value of frequency.

Accurate modeling of material properties is integral in the realistic modeling of a TPV device. The important parameters for materials include spectral variation of emissivity for the far spacing case, spectral variation of permittivity for the close spacing case, and spectral variation of quantum efficiency for both cases. Additional cell device parameters such as recombination lifetime, cell diffusion length, and dopant concentration are needed to accurately predict cell output power for the close spacing case.

For the far spacing model, spectral variation of emissivity is needed for both the emitter and receiver. The emissivity is related to the optical constants n and k of the material by the following equation where n is the index of refraction and k is the extinction coefficient[7],

$$\epsilon_{\lambda}=\frac{1}{2}\left(\begin{array}{c}\frac{8n}{n^2+k^2}\left(1-\frac{n}{n^2+k^2}\ln\left[(n+1)^2+k^2\right]+\frac{n^2-k^2}{k(n^2+k^2)}\tan^{-1}\frac{k}{n+1}\right)+\\ 8n\left(1-n\ln\frac{(n+1)^2+k^2}{n^2+k^2}+\frac{n^2-k^2}{k}\tan^{-1}\frac{k}{n(n+1)+k^2}\right)\end{array}\right) \tag{13}$$

For the silicon carbide and refractory metal emitters, the optical constants n and k were taken from the Electronic Handbook of Optical Constants©[8]. The spectral variation of emissivity was then calculated from equation (13). The rare earth Er-YAG emitter emissivity was taken from measured values from Lowe et al[9]. For the ideal emitters, the blackbody was modeled with a constant emissivity of one at all wavelengths. The ideal emitter was modeled with an emissivity of one over the range where the quantum efficiency of the receiver is non-zero and zero at all other wavelengths. Figure 3 shows an ideal emitter emissivity with a typical receiver quantum efficiency curve.

These two ideal cases were used to bracket the far spacing results. The receiver material properties were taken from Whale[2].

The close spacing model requires the permittivity of the material vice the emissivity which is related to the optical constants through the following equations[7],

$$\begin{aligned}\varepsilon&=\varepsilon'-i\varepsilon''\\ \varepsilon'&=n^2-k^2\\ \varepsilon''&=2nk\end{aligned} \tag{14}$$

The same optical constants used in the far spacing case were used in the near spacing case, except they were converted to permittivity via equation (14) vice emissivity. The rare earth emitter was not analyzed in the close spacing case because there was no data available to convert to permittivity, only measured values of emissivity were given for this material. Additionally to bracket the close spacing results, some ideal emitter was needed. The blackbody and ideal emitters of the far spacing model could not be used since only emissivity information was available. Raynolds suggests optimum parameters for a metal in his paper. He uses a Drude model with the following relation for permittivity[3],

$$\begin{aligned}\varepsilon(\omega)&=1+\frac{4\pi\sigma}{\omega}i\\ \sigma&=6.67\times10^{13}\ \text{sec}^{-1}\end{aligned} \tag{15}$$

Equation (15) provides an ideal metal emitter permittivity and is used as a bracket on the close spacing results.

Most of the optical constant property data used was measured at room temperature. Since the emitter temperatures at which meaningful output can be achieved are between 1000-2000 K, some scaling of the properties was necessary. The Hagen-Rubens model for the optical constants of a metal states[7]

$$n = k = \sqrt{30\lambda\sigma_{DC}} \tag{16}$$

where

$$\sigma_{DC} = \sigma_{ref}\frac{T_{ref}}{T} \tag{17}$$

Thus, for a Hagen-Rubens material, the optical constants scale with the inverse square root of temperature. The refractory metal properties used in this paper were experimentally measured but over much of the frequency range of interest, n and k are of the same order of magnitude. To account for temperature variation of optical constants of the refractory metals, the data was scaled exactly as in equation (16) and (17). The effect of temperature on optical properties of non-metals is much harder to quantify. The Electronic Handbook of Optical Constants©[8] lists experimentally determined variation of index of refraction with temperature for various semiconductors. While SiC is not present in the list, all of the variations of $\frac{dn}{dT}$ are of the order of magnitude of $1\times10^{-4}/K$ so this was used as a representative scaling factor to scale the optical properties of silicon carbide. This scaling has very little effect on the performance with a SiC emitter. The Er-YAG emitter data was measured at 1500 K and no scaling was done to this data since it was already given in emissivity. Demonstration of the scaling effect for both SiC and the refractor metals is contained in the Results section.

Equally as important as the material properties are the device parameters, especially the quantum efficiency. Quantum efficiency data was taken from Whale[2]. The quantum efficiency is non-zero from the wavelength corresponding to the band gap energy down to some lower wavelength (higher energy). The data in Whale[2] for quantum efficiency was incomplete in that the data was not given for quantum efficiency up to the band gap wavelength. Only the 0.73 eV data was complete. The quantum efficiency must go to zero at the band gap since below the band gap energy (or above the band gap wavelength) the incident photons cannot put electrons into the conduction band. To account for the fact that quantum efficiency must go to zero at the band gap energy, the data was extrapolated out by a straight line from the last data point to zero at the wavelength corresponding to the band gap energy.

The close spacing analysis requires several additional device parameters, specifically the trap recombination lifetime, the diffusion length of the cell at far spacing, and the dopant concentration. To keep the focus of this paper on cell performance as function of primarily emitter parameters, the same values of these parameters were used in this paper as in Whale[2].

Results and Discussion

Electrical power output density and efficiency were calculated for emitter temperatures of 1000 K, 1500 K and 2000 K at various band gaps for the following emitters: SiC, Er-YAG, tungsten, rhenium, a blackbody emitter, and an ideal emitter which has an emissivity of one over the range where quantum efficiency is non-zero. For all of the results, both far and close spacing, the receiver temperature is 300 K. Figures 4 and 5 shows the variation of power output and efficiency with temperature and band gap for SiC and InGaAs at 300 K.

A comparison of all emitters at each temperature was also conducted. The comparison of power output and efficiency for 2000 K emitters is shown in Figures 6 and 7. The numbers in Figures 6 and 7 are the same order of magnitude of experimentally measured values in the field. Reported values of power output have ranged from 1.5 – 3 W/cm^2 for a 1700 K emitter[10]. Coutts also mentions that typical TPV output values may be as large as 5 W/cm^2[11]. Whale reports calculated power output values for his low conductivity Drude model metal on the same order of magnitude (between 1 –10 W/cm^2) at 2000 K with efficiencies between 10 and 20% at 2000 K[2]. Figure 8 shows a comparison of Whale's[2] results for an unspecified Drude model emitter of low conductivity at 2000 K with a rhenium emitter also at 2000 K. Both emitters are coupled with a 0.47 eV InGaAs receiver at 300 K.

The first goal of the close spacing analysis was to identify the most promising emitter materials from a large group of choices and then perform a more detailed analysis on those few materials. Figure 9 shows the power output for 2000 K emitters with a 300 K 0.47 eV InGaAs receiver. The emitters were modeled using 300 K optical property data since this graph is meant to only identify trends and pick out the most promising materials. Scaling will shift all values upward and subsequent more detailed calculations are scaled for temperature. Figures 10 and 11 show the scaling effect for tungsten and silicon carbide.

The most promising metals based on the results in Figure 9 are tungsten and rhenium. These two metals and SiC were further analyzed with scaled optical property values and with many spacing data points to get smooth curves. Figures 12 and 13 show the results for tungsten, rhenium, and SiC at 2000 K with optical properties scaled to 2000 K coupled with 0.47 eV InGaAs at 300K.

The receiver band gap was then varied with a rhenium emitter. Figures 14 and 15 show the variation of power output and efficiency with band gap for rhenium at 2000 K and InGaAs at 300 K. Figure 14 shows that power output does not significantly change with band gap. Efficiency, however is very sensitive to band gap. At the optimum band gap of 0.64 eV, it is interesting to compare rhenium with SiC. Figures 16 and 17 show the power output and efficiency for emitters at 2000 K with 0.64 eV InGaAs at 300 K.

Conclusions

The far spacing results show the superiority in efficiency of tungsten as an emitter for a TPV device. Both rhenium and tungsten show outputs of around 4 W/cm^2 at between 15 –20% efficiency. SiC had the highest power output density.
Close spacing results have shown the tremendous power density increase that MTPVs are capable of achieving due to evanescent field energy transfer. Almost a 1000X increase in power density could be achieved if gap spacing control technology advances. With current technology, the gap spacing becomes limiting at around 0.05 microns[12], and at this spacing, SiC outperforms rhenium. At smaller gaps however, rhenium greatly exceeds SiC as an emitter candidate. It is important to realize that while MTPVs offer incredible increases in power output, it comes at the cost of significant increases in required incident heat flux, and as much as a 1000X increase in incident heat flux may be required. Several important facts arise from this point. The heat source for a close spacing MTPV will need to be capable of creating very large heat fluxes and higher device efficiencies are needed since the power output is so much larger than the far spacing case to reduce the load on the heat source. Additionally, the additional heat flux will create a tremendous cooling load to maintain the receiver at a low temperature.

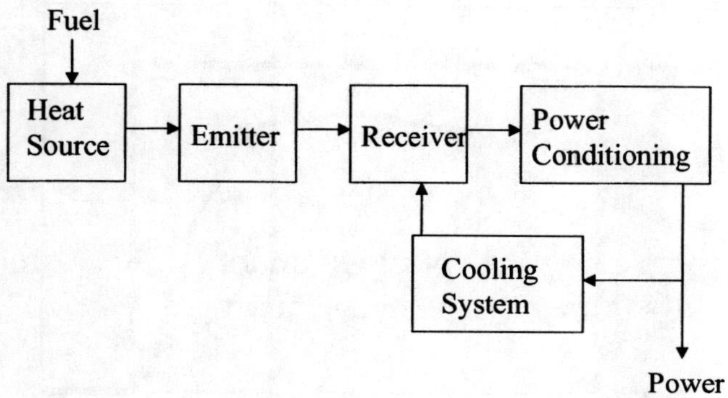

Figure 1. Typical TPV Device

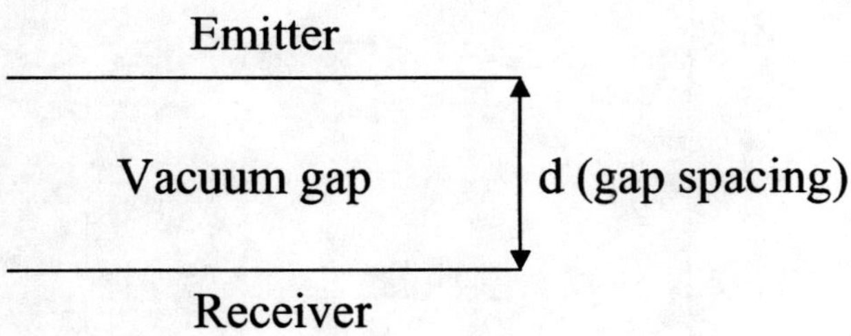

Figure 2. TPV device geometry

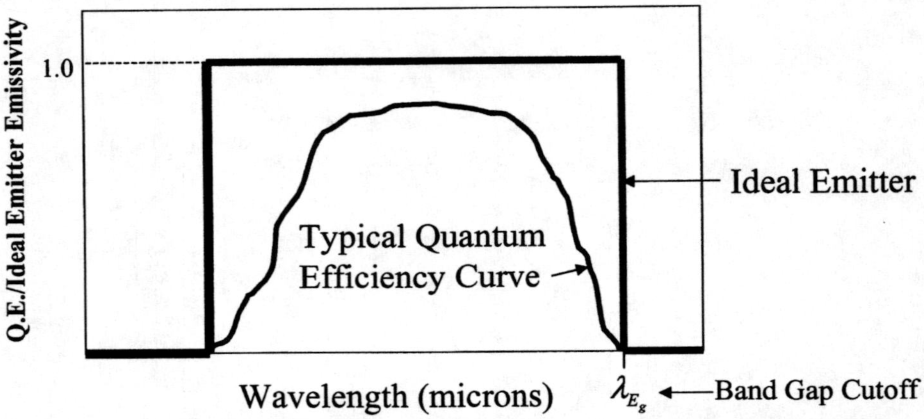

Figure 3. Ideal Emitter Emissivity and a Typical Receiver Quantum Efficiency Curve

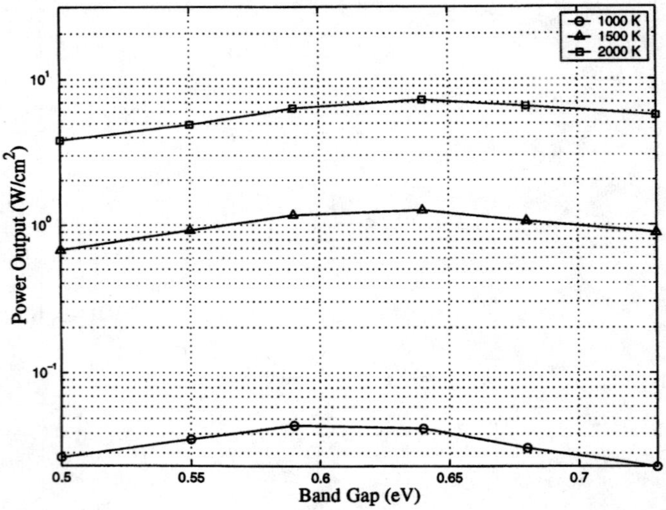

Figure 4. SiC Emitter Temperature Effect on TPV Device Power Output

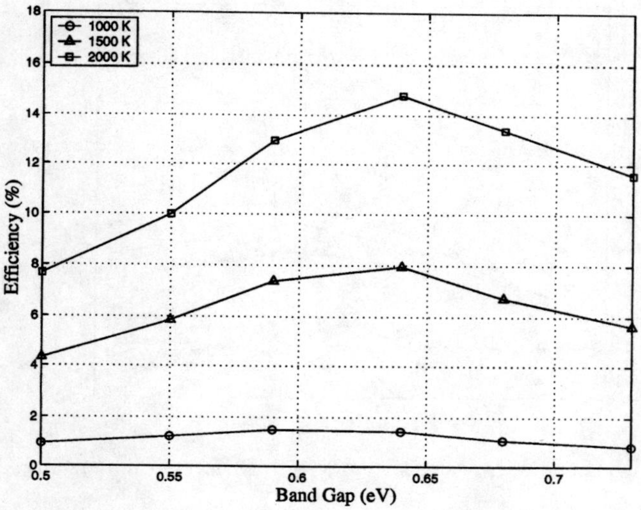

Figure 5. SiC Emitter Temperature Effect on TPV Device Efficiency

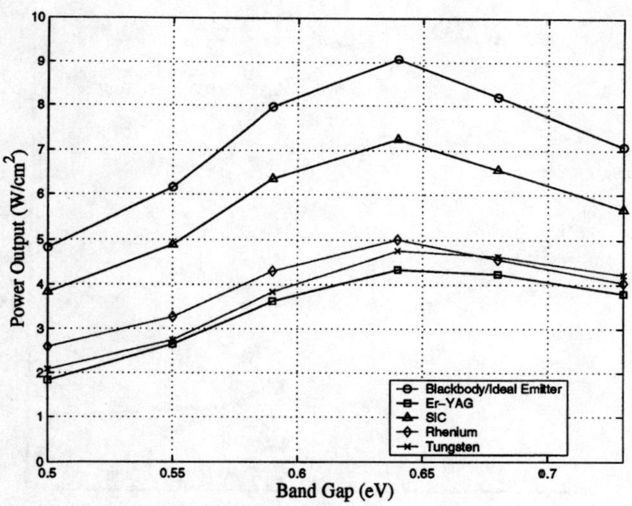

Figure 6. Comparison of TPV Device Power Output with Different Emitters at 2000K

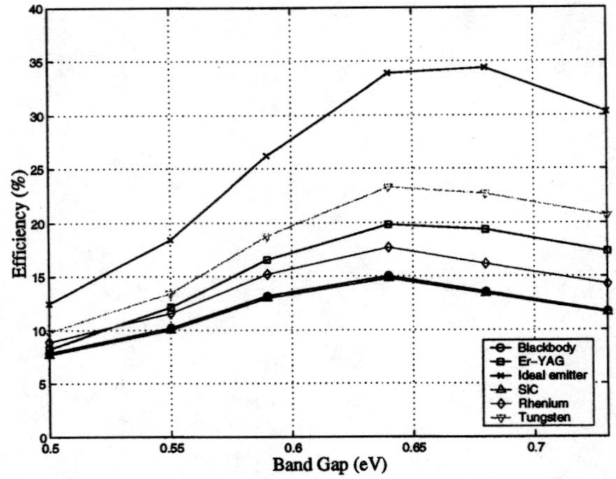

Figure 7. Comparison of TPV Device Efficiency with Different Emitters at 2000 K

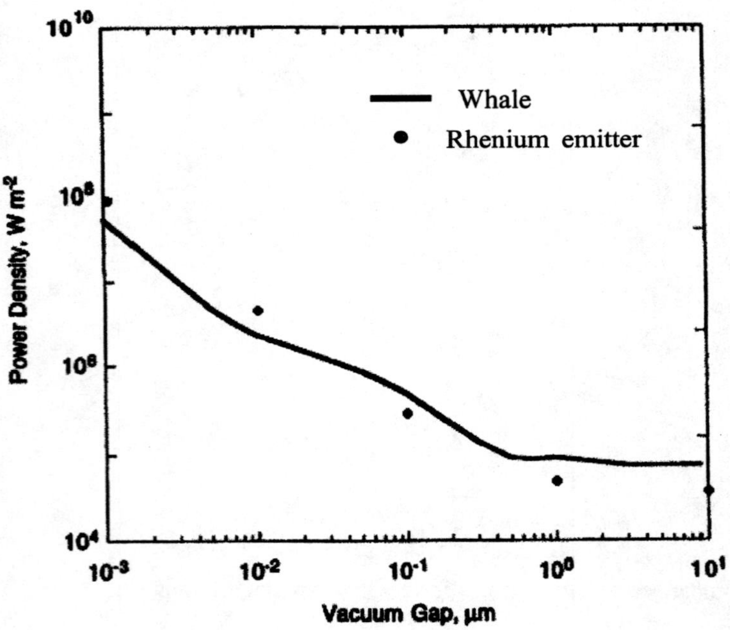

Figure 8. Comparison of Whale's Results for Drude model emitter with Rhenium emitter (both at 2000 K) coupled with 0.47 eV InGaAs at 300 K (adapted from Whale[2], Figure 40, p. 134)

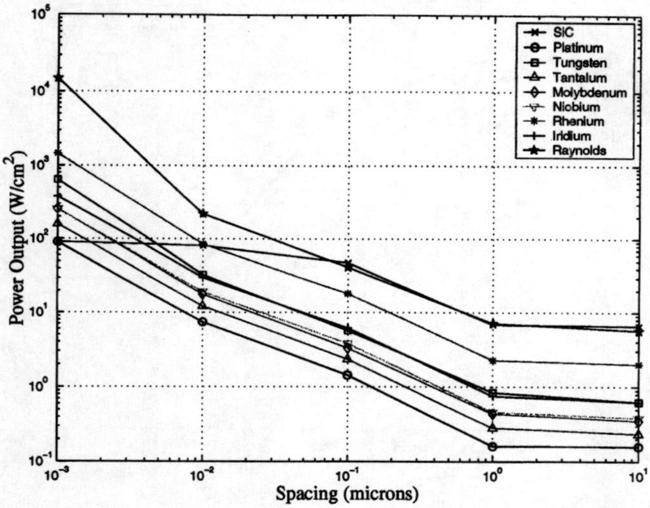

Figure 9. Comparison of MTPV Device Power Output with Different Emitters at 2000 K with 0.47 eV InGaAs Receiver at 300K

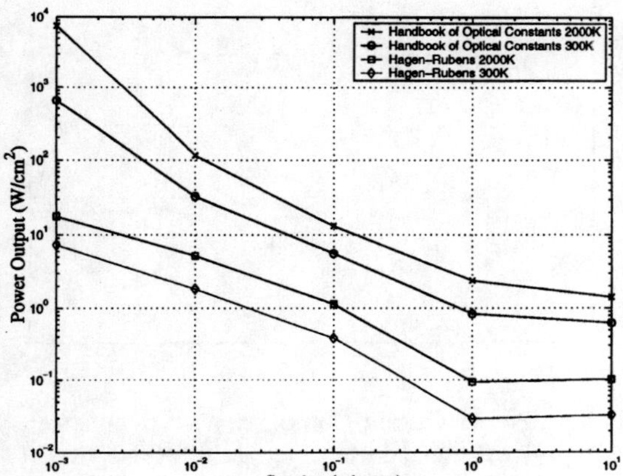

Figure 10. Demonstration of Scaling Effect for MTPV Device with Tungsten Emitter at 2000 K with 0.47 eV InGaAs Receiver at 300 K

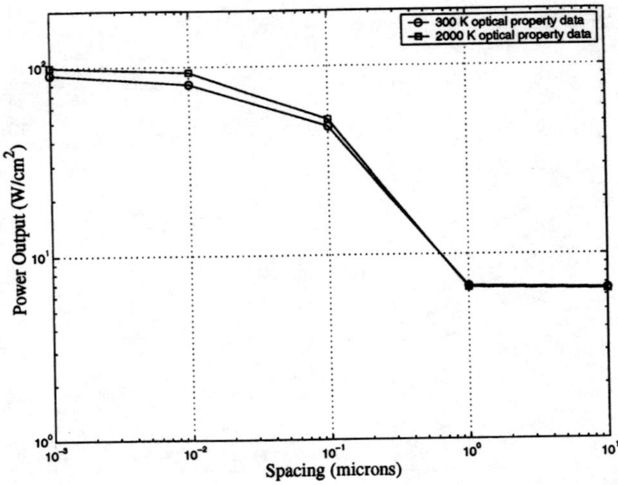

Figure 11. Demonstration of Scaling Effect for MTPV Device with SiC Emitter at 2000 K with 0.47 eV InGaAs Receiver at 300 K

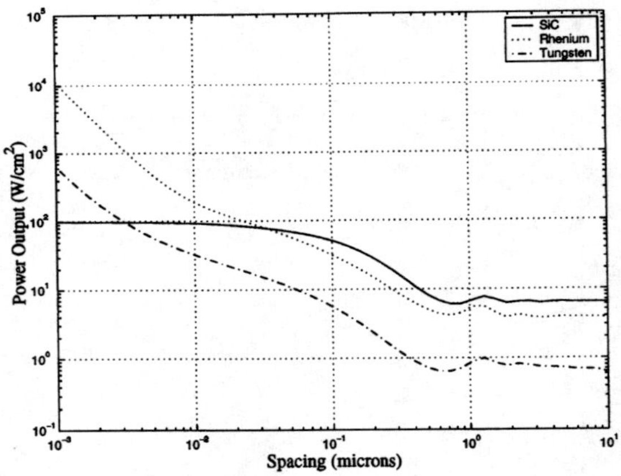

Figure 12. MTPV Device Power Output Comparison with Tungsten, Rhenium, and SiC Emitters at 2000 K with 0.47 eV InGaAs Receiver at 300 K

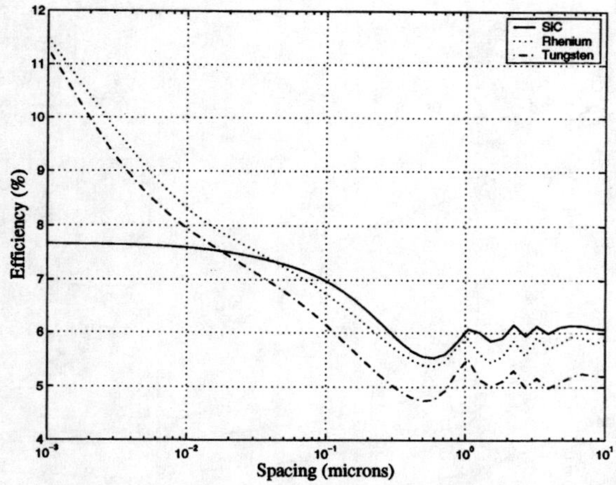

Figure 13. MTPV Device Efficiency Comparison with Tungsten, Rhenium, and SiC Emitters at 2000 K with 0.47 eV InGaAs Receiver at 300 K

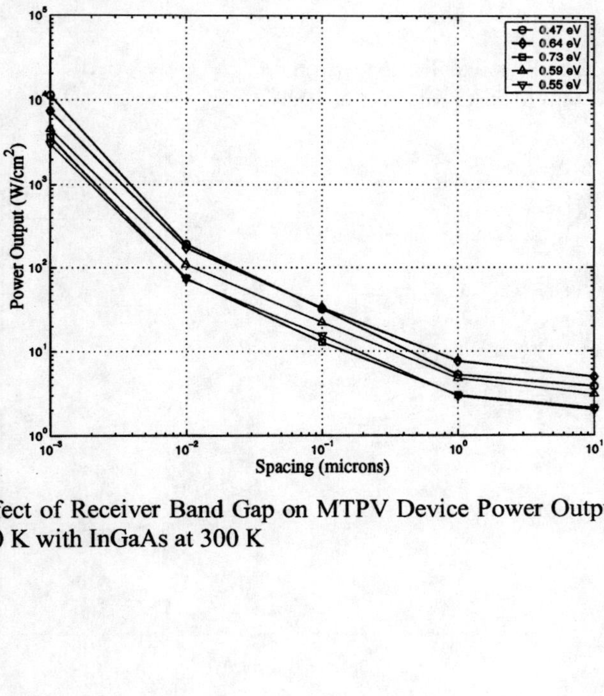

Figure 14. Effect of Receiver Band Gap on MTPV Device Power Output with Rhenium Emitter at 2000 K with InGaAs at 300 K

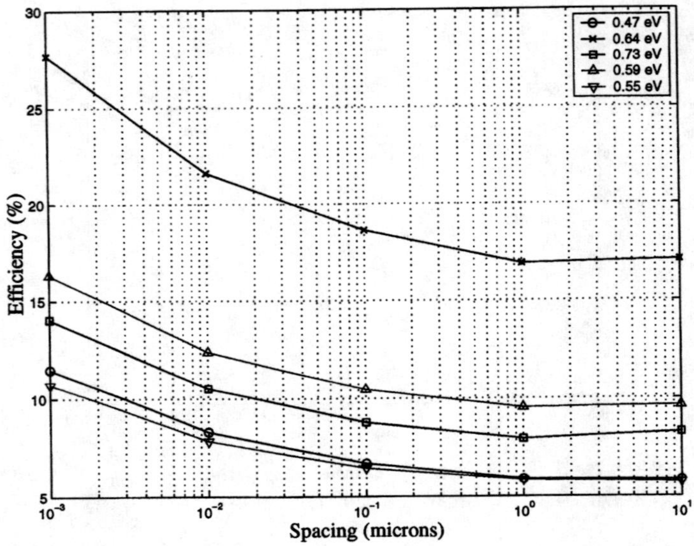

Figure 15. Effect of Receiver Band Gap on MTPV Device Efficiency with Rhenium Emitter at 2000 K with InGaAs Receiver at 300K

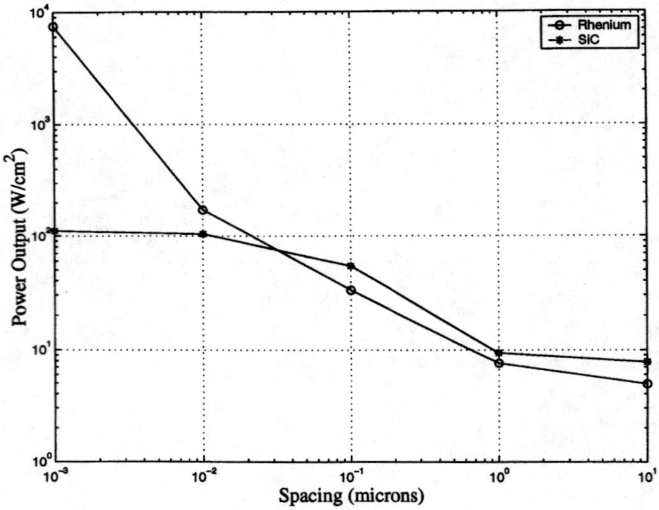

Figure 16. Comparison of MTPV Device Power Output with Rhenium and SiC emitters at 2000 K with 0.64 eV InGaAs Receiver at 300 K

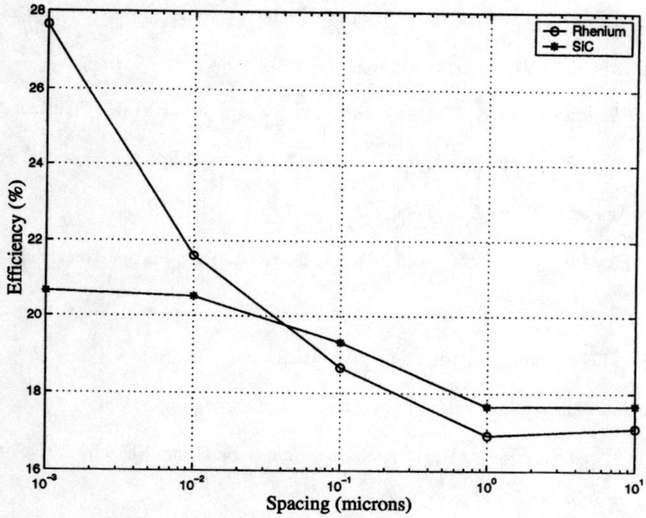

Figure 17. Comparison of MTPV Device Efficiency with Rhenium and SiC emitters at 2000 K with 0.64 eV InGaAs Receiver at 300 K

REFERENCES

[1] Shi, L.Z., Spencer, Y. Xu, and Boehm, R.F., 2001, "Analysis of a Single-Layer Thermophotovoltaic System at Moderate Temperatures," *Journal of Thermophysics and Heat Transfer,* Vol. 15(4), 453

[2]Whale, M.D., 1997, A Fluctuational Electrodynamic Analysis of Microscale Radiative Transfer and the Design of Microscale Thermophotovoltaic Devices, Ph.D. Dissertation, Department of Mechanical Engineering, Massachusetts Institute of Technology.

[3] Raynolds, J.E., 1999, "Enhanced Electro-Magnetic Energy Transfer Between a Hot and Cold Body at Close Spacing Due to Evanescent Fields," *AIP Conference Proceedings 460 – Thermophotovoltaic Generation of Electricity, 4th NREL Conference,* pp. 49-57.

[4] Loomis, J.J. and Maris, H.J., 1994, "Theory of Heat Transfer by Evanescent Electromagnetic Waves," *Physical Review B,* Vol. 50(24), pp. 18517-18524.

[5] DiMatteo, R.S., Greiff, P., Finberg, S.L., Young-Waithe, K.A., Choy, H.K.H., Masaki, M.M., and Fonstad, C.G., 2001, "Enhanced Photogeneration of Carriers In a Semiconductor via Coupling Across a Nonisothermal Nanoscale Gap," *Applied Physics Letters,* Vol. 79(12), pp. 1894-1896.

[6] Green, M.A., 1982, Solar Cells: Operating Principles, Technology and System Applications, Prentice-Hall, Inc., Englewood Cliffs, NJ, Chapters 4 and 5.

[7] Modest, M.F., 1993, *Radiative Heat Transfer,* McGraw-Hill, Inc., New York, Chapters 2 and 3.

[8] Palik, E.D., ed., 1999,"Electronic Handbook of Optical Constants of Solids," Version 1.0, Copyright ©1999 SciVision-Academic Press, San Diego, CA.

[9] Lowe, R.A., Chubb, D.L., Farmer, S.C., and Good, B.S., 1994, "Rare-Earth Garnet Selective Emitter," *Applied Physics Letters,* Vol. 64(26), pp. 3551-3553.

[10] Krist, K., 1995, "GRI Research on Thermophotovoltaics," *AIP Conference Proceedings 321 –The First NREL Conference on Thermophotovoltaic Generation of Electricity,* pp. 54-66.

[11] Coutts, T.J., 2001, "An Overview of Thermophotovoltaic Generation of Electricity," *Solar Energy Materials and Solar Cells,* Vol. 66, pp. 443-452.

[12] Whale, M.D., 2001, "The Influence of Interference and Heterojunctions on the Performance of Microscale Thermophotovoltaic Devices," *Microscale* Thermophysical Engineering, Vol 5, pp. 89-106.

Optical Cavity Effects on TPV Efficiency and Power Density

B. Wernsman, R. G. Mahorter, and R. M. Thomas

Bechtel Bettis, Inc., West Mifflin, PA 15122

Abstract. Analytical expressions have been derived for the effective radiator emissivity and view factor of a given optical cavity geometry as a function of radiator emissivity and reflectance of the shield and TPV device. Using these expressions, accurate predictions of the device optical response (J_{SC} and Q_{abs}) have been made for two different cavity configurations (Au shield and TPV device shield). Also, the measured and calculated results show that placing a TPV device into an array of TPV devices will lower its output performance by at least 1.7% (relative) compared to that measured using a Au shield.

INTRODUCTION

Two important performance parameters of photovoltaic (PV) and thermophotovoltaic (TPV) devices are efficiency and power density. Since the optical cavity for PV devices is well understood, comparisons between the efficiency and power density of different devices can be easily made. However for TPV devices, the optical cavity that consists of the radiator, TPV devices and other exposed surfaces (i.e. shields) in a given geometry can vary between systems [1-4]. Therefore, the interpretation of experimentally measured results must account for the optical cavity, especially when comparing results from one measurement system to another. Also, the behavior of the TPV device when placed into an array consisting of many TPV devices has to be understood.

Modeling of cavity effects can greatly increase the understanding of TPV device performance parameters. Ray tracing codes have been used to accurately model the optical cavity effects on device performance [2]. Because these codes require extensive set-up and run times, running parametrics to determine the effects of slight variances in device or system parameters (e.g. device reflectance, shield reflectance and radiator temperature) is time intensive. The effect of changes in system parameters on device operation can be modeled more swiftly by performing calculations using analytical expressions. However, these calculations need to accurately predict the optical cavity behavior. The methodology used to formulate the analytical expressions is given below.

CP653, *Thermophotovoltaic Generation of Electricity: 5th Conference*
edited by T. J. Coutts, G. Guazzoni, and J. Luther

METHODOLOGY

Modifications of the radiator gray-body emission properties can be used to effectively predict the behavior of TPV devices in an optical cavity. The system efficiency (η) of TPV devices is given by:

$$\eta = \frac{Q_e^{max}}{Q_{abs}}, \tag{1}$$

where Q_e^{max} is the maximum electrical output power density, Q_{abs} is the total heat absorbed by the device and is given by:

$$Q_{abs} = \int_0^\infty (1 - R)\rho_{hv} d\lambda, \tag{2}$$

J_{SC} is the device short circuit current density that is given by:

$$J_{SC} = q\int_0^{\lambda_g} QE_{ext}\Gamma_{hv} d\lambda, \tag{3}$$

R is the device reflectance, $\rho_{hv}d\lambda$ is the photon energy flux impinging on the device (Planckian distribution), q is the electronic charge, QE_{ext} is the device external quantum efficiency, $\Gamma_{hv}d\lambda$ is the photon flux impinging on the device, λ_g is the wavelength of light that corresponds to the bandgap energy and λ is the wavelength of light. ρ_{hv} is given by:

$$\rho_{hv} = F_{eff}\varepsilon_{eff}\frac{2\pi hc^2}{\lambda^5\left(e^{\frac{hc}{\lambda kT_{rad}}} - 1\right)}, \tag{4}$$

where F_{eff} and ε_{eff} are the radiator effective view factor and effective emissivity, respectively, h is Planck's constant, c is the speed of light, k is Boltzmann's constant and T_{rad} is the radiator temperature. Γ_{hv} is found by multiplying Eq. (4) by $\lambda/(hc)$ and is given by:

$$\Gamma_{h\nu} = F_{eff}\varepsilon_{eff}\frac{2\pi c}{\lambda^4\left(e^{\frac{hc}{\lambda kT_{rad}}}-1\right)}. \tag{5}$$

As Eqs. (4) and (5) show, $\rho_{h\nu}$ and $\Gamma_{h\nu}$ are both functions of the radiator effective view factor and effective emissivity. These effective parameters are dependent on the optical cavity of the system where the device is used.

The radiator effective emissivity is defined as:

$$\varepsilon_{eff} = \frac{E_{rad}^{Device}}{E_{BB}^{Device}}, \tag{6}$$

where E_{rad}^{Device} and E_{BB}^{Device} are the total amount of light energy absorbed by the device when illuminated by the radiator and a blackbody, respectively. These two energy parameters need to be derived to formulate ε_{eff}.

Figure 1 shows a simplified radiator/device cavity system used to derive the energy parameters needed to calculate ε_{eff}. This calculation assumes that the spacing between the radiator and TPV device is approximately an order of magnitude smaller than the size of the radiator or TPV device. As shown in Fig. 1, the total amount of energy absorbed in the device is the sum of all the energy that travels across surface 1 toward the device. To calculate this sum, it is necessary to trace the light rays as they propagate through the cavity. A light ray is emitted into the cavity via the radiator at an initial energy of E_0. This ray travels through surface 1 toward the device. At the device, a portion of this ray is reflected while the rest of it is absorbed. The energy of the reflected portion is RE_0. This reflected ray travels through surface 1 toward the radiator. At the radiator, this ray can be either absorbed or reflected. The energy of the reflected

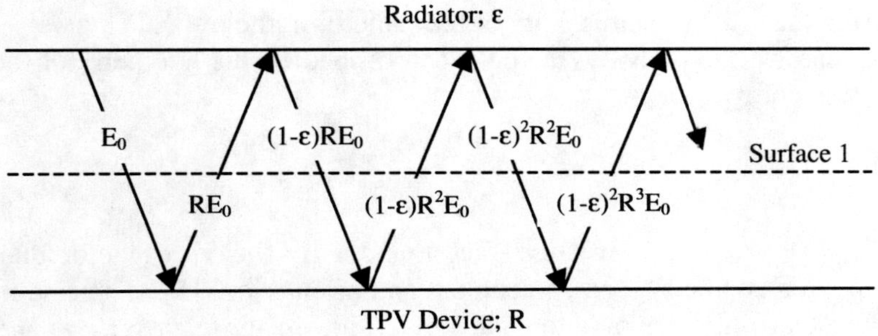

FIGURE 1. Simplified Schematic of Optical Cavity Showing Rays Bouncing Between TPV Device and Radiator.

portion is (1-ε)RE_0 where ε is the radiator emissivity. This reflected ray then behaves in the same manner as the initial ray and the process of rays bouncing between the radiator and TPV device continue until all the energy of the ray is absorbed. The formulation for the net energy absorbed by the MIM using the ray tracing method is given by:

$$E_{rad}^{Device} = E_0 - RE_0 + (1-\varepsilon)RE_0 - (1-\varepsilon)R^2E_0 + (1-\varepsilon)^2R^2E_0 - (1-\varepsilon)^2R^3E_0 \ldots \quad (7)$$

This equation simplifies to:

$$E_{rad}^{Device} = \frac{E_0(1-R)}{1-R(1-\varepsilon)}. \quad (8)$$

The initial ray energy (E_0) is given by:

$$E_0 = \varepsilon\sigma T_{rad}^4, \quad (9)$$

where σ is Steffan-Boltzmann's constant. Substituting Eq. (9) into Eq. (8) gives:

$$E_{rad}^{Device} = \frac{\varepsilon\sigma T_{rad}^4(1-R)}{1-R(1-\varepsilon)}. \quad (10)$$

This equation can be used to calculate E_{BB}^{Device} by making ε = 1. Thus, ε_{eff} (Eq. (6)) becomes:

$$\varepsilon_{eff} = \frac{\varepsilon}{1-R(1-\varepsilon)}, \quad (11)$$

formulating the effective emissivity of the radiator in the optical cavity.

The radiator effective view factor is determined by the geometry of the cavity system. It is defined as:

$$F_{eff} = C_{cav}F, \quad (12)$$

where C_{cav} is a cavity correction factor and F is the view factor due to the radiator/device geometry (size, placement and orientation) [5]. The calculation of F assumes that only the radiator and device are in the cavity [5]. The cavity correction factor is used to compensate for other non-absorbing surfaces in the cavity that allow more energy to strike the TPV device, making it dependent on the cavity geometry. Thus, this correction factor is given by:

$$C_{cav} = \frac{Q_{abs}^{wc}}{Q_{abs}^{woc}}, \tag{13}$$

where $Q_{abs}{}^{wc}$ and $Q_{abs}{}^{woc}$ is the amount of heat absorbed in the TPV device with and without the cavity effects, respectively.

A Monte Carlo, ray tracing analysis, performed by TracePro$^{(TM)}$, is used to calculate C_{cav}. This analysis consists of computing the amount of heat absorbed by a perfectly absorbing TPV device with and without the extra reflecting cavity surfaces. By making the device a perfect absorber, extra reflections between the radiator and device were eliminated, making ε_{eff} equal ε. Thus, by using this methodology, analytical expressions for ρ_{hv} and Γ_{hv} that are dependant upon the optical cavity components (shield reflectance, radiator emissivity and temperature, and device reflectance) for a given cavity geometry can be found. By using these analytical expressions, the effects of slight variations of these cavity components on device performance can be determined more easily. To verify this methodology, the performance of a TPV device under different cavity conditions was measured.

EXPERIMENTAL SETUP

Figure 2 shows a simplified schematic of the optical cavity of the experimental system. The square SiC radiator is 54 mm on a side. The water cooled shield is a circular disk with a diameter of 63.5 mm and a 19.6x19.6 mm^2 square hole in its center. The TPV device is a 0.6 eV InGaAs/InPAs double heterostructure n/p/n 30 junction 2x2 cm^2 monolithic interconnected module (MIM) using transmissive integrated spectral control [6,7]. The spacing between the radiator and shield is 2 mm while the spacing between the radiator and MIM is 2.5 mm.

The MIM is glued to an aluminum nitride saddle with metal contact pins manufactured in it. Sputtered gold between the MIM busbars and the saddle pins facilitates electrical connections to the MIM. Also, a thermocouple hole has been placed in the saddle to monitor the temperature of the MIM. It is important to note that, for the heat flux through the saddle, the temperature difference between the saddle and MIM is calculated to be $< 1^{\circ}C$. In a similar manner to Ref. [2], the MIM/saddle device is placed on top of a copper pedestal to measure the heat rejected by the device. A proportional-integral-derivative (PID) controlled thermoelectric heater/cooler is placed between the copper pedestal and the cooling water to maintain the saddle temperature at a selected value.

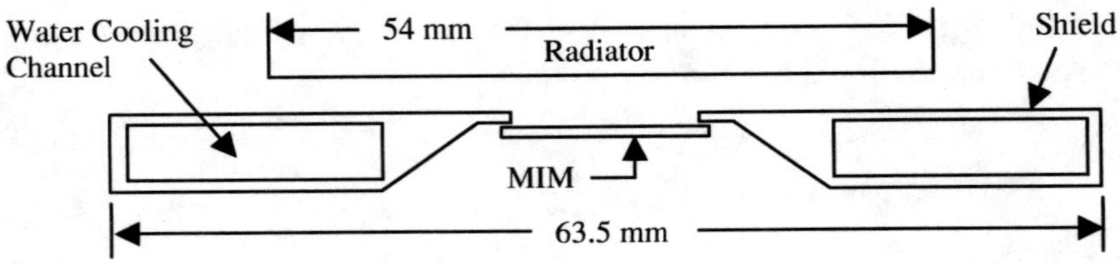

(Not to Scale)

FIGURE 2. Simplified Schematic of Optical Cavity from Experimental System.

RESULTS AND DISCUSSION

Using the above methodology and experimental setup, calculations of C_{Cav} were made for various radiator emissivities and shield reflectances (R_{Sh}). Results from these calculations are given in Fig. 3. As shown, C_{Cav} increases as either R_{Sh} is increased or ε is decreased, as expected. Also, for this system, F is calculated to be 0.9896. Using the results from Fig. 3, an analytical expression for C_{Cav} can be found that is given by:

$$C_{cav} = \begin{cases} (-1.438\varepsilon + 0.8136)R_{Sh}^2 + (-0.2249\varepsilon - 0.01764)R_{Sh} - 0.01378\varepsilon + 1.007; & \varepsilon \le 0.5 \\ (-0.1888\varepsilon + 0.1888)R_{Sh}^2 + (-0.1896\varepsilon - 0.1896)R_{Sh} - 0.001\varepsilon + 1.001; & \varepsilon > 0.5 \end{cases}. \quad (14)$$

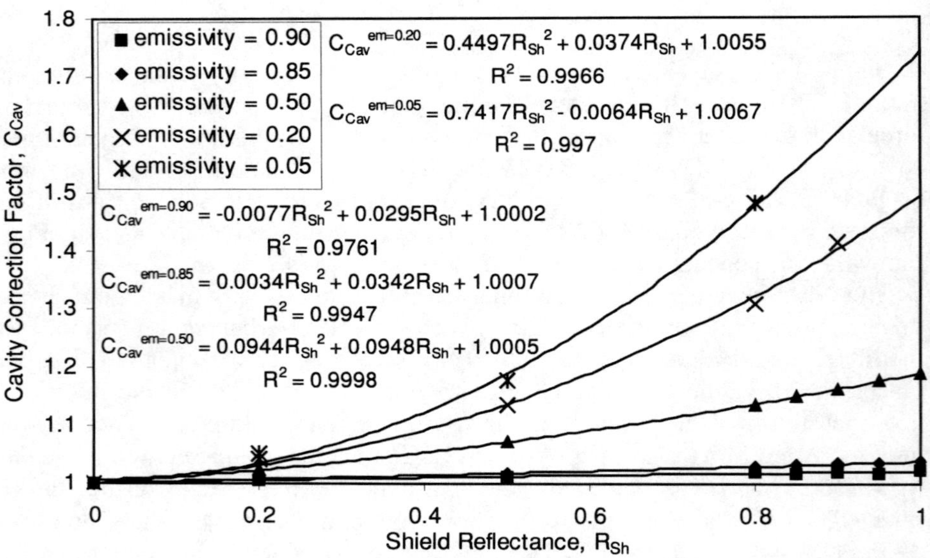

FIGURE 3. Calculated C_{Cav} as a Function of R_{Sh} and ε for the Experimental Setup.

The performance of a MIM was measured under different cavity conditions (different R_{Sh} values) to verify this C_{Cav} expression. The electrical performance of the MIM can be summarized in its maximum output power density (Q_e^{MAX}) as a function of short circuit current density (J_{SC}) and its external quantum efficiency (QE_{ext}), that are given in Figs. 4 and 5, respectively. A SiC radiator with an emissivity shown in Fig. 6 is used along with two shields. The first shield was coated with Au while the second shield had MIMs (8, 2x2 cm^2, 30 junction MIMs) placed around the MIM under test to understand the effect of placing a device in a large array. The reflectance of the shield and MIMs (both for the device under test and the shield) are also given in Fig. 6. By using Eq. (14) and the curves given in Fig. 6, C_{Cav} can be calculated as a function of wavelength for each type of cavity (Au shield and MIM shield) as shown in Fig. 7. Using this C_{Cav} curve, the calculated F, Q_e^{MAX} from Fig. 4, QE_{ext} from Fig. 5, the MIM reflectance from Fig. 6 and Eqs. (1)-(5), (11) and (12), the performance of the MIM for the given optical cavity can be calculated.

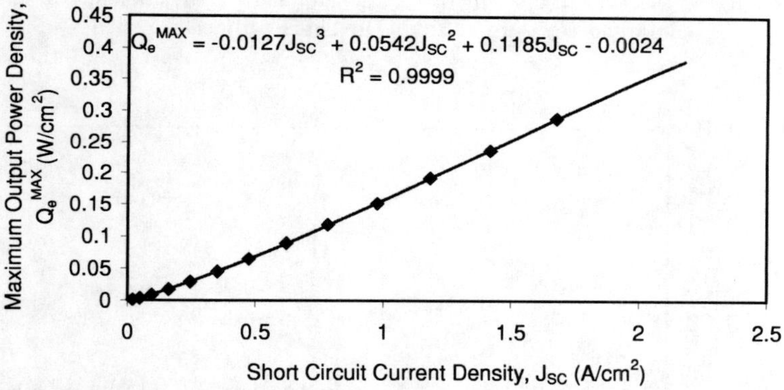

FIGURE 4. Q_e^{MAX}-J_{SC} Curves Showing Electrical Performance of the MIM.

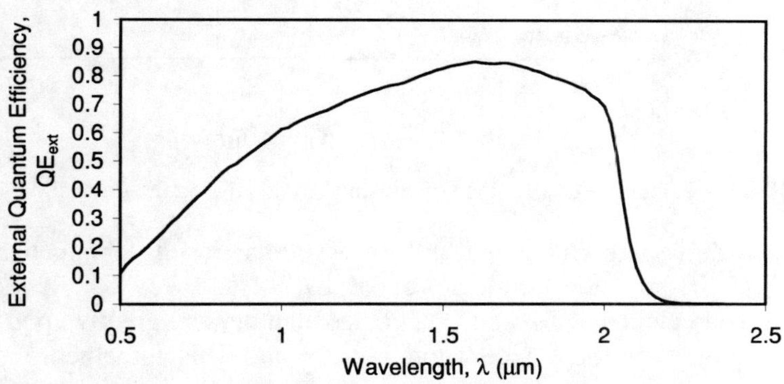

FIGURE 5. QE_{ext} Response Showing Electro-Optic Performance of the MIM.

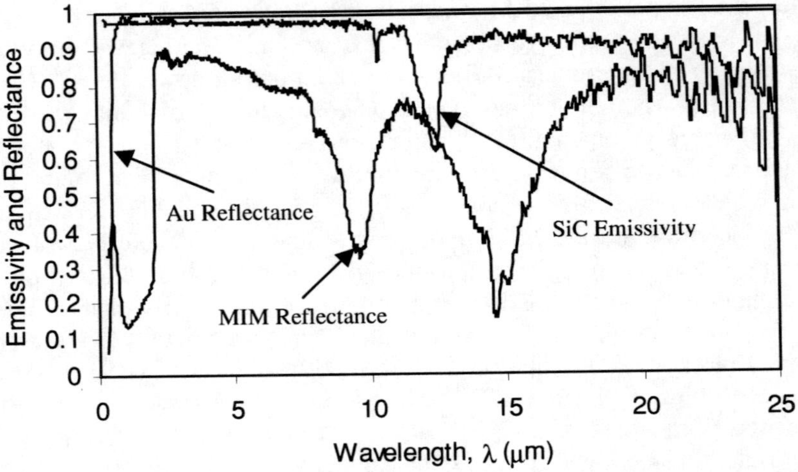

FIGURE 6. Measured Emissivity of SiC Radiator and 10° Directional Hemispherical Reflectance of Au and MIMs Used in Experimental System.

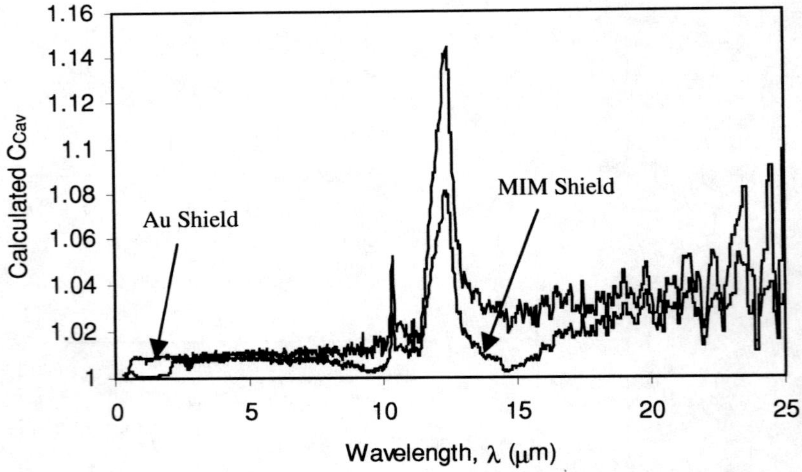

FIGURE 7. Calculated C_{Cav} for the Experimental System Using the Au and MIM Shields.

Table 1 shows the calculated and experimental results obtained for the two cavities at SiC radiator temperatures of 872°C and 929°C. As shown, the measured and calculated J_{SC} and Q_{abs} values, that depend greatly upon the photon flux impinging upon the device, agree rather well. This agreement indicates that the analytical expressions derived using the above methodology accurately predict the effects of the cavity on the optical response of the device. However, the measured electrical response given by Q_e^{MAX} for the MIM shield case is much lower than the calculated values, making the measured and calculated system efficiency for the MIM shield case different. This difference between the measured and calculated Q_e^{MAX} values may be due to the placement of the MIM under the MIM shield, requiring further investigation. However, even with this discrepancy, the measured and calculated values show that by replacing the Au shield with the MIM shield the MIM output performance decreases (Table 2). Thus, as shown in Table 2, the system efficiency and power density of a MIM measured using the Au shield will decrease by at least 1.7% (relative) when placed in an array of MIMs. Therefore, the MIM output performance is affected by variation in the optical cavity.

TABLE 1. Measured and Calculated MIM Output Performance Parameters.

Operating Conditions			MIM Output Performance							
T_{Rad} (°C)	T_{Saddle} (°C)	Shield	J_{SC} (A/cm²)		Q_e^{MAX} (W/cm²)		Q_{abs} (W/cm²)		η (%)	
			M*	C*	M*	C*	M*	C*	M*	C*
872	25	Au	1.41	1.40	0.234	0.235	2.46	2.45	9.51	9.59
872	25	MIM	1.39	1.38	0.221	0.230	2.44	2.44	9.06	9.43
929	26	Au	1.99	1.99	0.351	0.348	3.11	3.12	11.3	11.2
929	26	MIM	1.95	1.96	0.330	0.342	3.10	3.10	10.6	11.0

* M = Measured; C = Calculated.

TABLE 2. Relative Change in Measured and Calculated MIM Output Performance Parameters when the Au Shield Replaces the MIM shield.

Operating Conditions		Relative Change in MIM Output Performance (%)							
T_{Rad} (°C)	T_{Saddle} (°C)	J_{SC}		Q_e^{MAX}		Q_{abs}		η	
		M*	C*	M*	C*	M*	C*	M*	C*
872	25	-1.42	-1.43	-5.56	-1.70	-0.81	-0.41	-5.15	-1.67
929	26	-2.01	-1.51	-5.98	-1.72	-0.32	-0.64	-5.31	-1.79

* M = Measured; C = Calculated.

CONCLUSION AND FUTURE WORK

Analytical expressions for the effective radiator emissivity and view factor as a function of radiator emissivity and reflectance of the shield and MIM, for a given optical cavity geometry, have been derived. Using these expressions, accurate predictions of the device optical response (J_{SC} and Q_{abs}) have been made for two different cavity configurations (Au shield and MIM shield). However, since the calculated electrical response (Q_e^{MAX}) did not correspond well with that measured, the measured and calculated η did not match well. This variation between the measured and calculated Q_e^{MAX} may be due to the placement of the MIM under the MIM shield. The measured and calculated results show that by placing a MIM into an array of MIMs will lower its output performance by at least 1.7% (relative) compared to that measured using a Au shield. Therefore, care must be used when interpreting TPV device output performance information since the optical cavity can affect its output performance.

Efforts are underway to understand the discrepancy between the measured and calculated Q_e^{MAX} values. Also, to further verify the derived analytical expressions, TPV output performance measurements will be made using radiators with lower and wavelength dependent emissivities.

REFERENCES

1. Fraas, L., Groeneveld, M., Magendenz, G., and Custard, P., "A Single TPV Cell Power Density and Efficiency Measurement Technique," in *Thermophotovoltaic Generation of Electricity: Fourth NREL Conf*erence, edited by T. J. Coutts et al., AIP Conference Proceedings 460, New York, 1999, pp. 312-316.
2. Gethers, C. K., Ballinger, C. T., and DePoy, D. M., "Lessons Learned on Closed Cavity TPV System Efficiency Measurements," in *Thermophotovoltaic Generation of Electricity: Fourth NREL Conf*erence, edited by T. J. Coutts et al., AIP Conference Proceedings 460, New York, 1999, pp. 335-348.
3. Good, B. S. and Chubb, D. L., "Effects of Geometry on the Efficiency of TPV Energy Conversion," in *Thermophotovoltaic Generation of Electricity: Third NREL Conf*erence, edited by T. J. Coutts et al., AIP Conference Proceedings 401, New York, 1997, pp. 487-503.
4. Fraas, L., Ballantyne, R., Samaras, J., and Seal, M., "Electric Power Production Using New GaSb Photovoltaic Cells with Extended Infrared Response," in *Thermophotovoltaic Generation of Electricity: First NREL Conf*erence, edited by T. J. Coutts et al., AIP Conference Proceedings 321, New York, 1995, pp. 44-53.
5. Holman, J. P., *Heat Transfer*, McGraw-Hill, Inc., New York, 1981.
6. Wilt, D. M., Murray, C. S., Fatemi, N. S., and Weizer, V., "n/p/n Tunnel Junction InGaAs Monolithic Interconnected Module (MIM)," in *Thermophotovoltaic Generation of Electricity: Fourth NREL Conf*erence, edited by T. J. Coutts et al., AIP Conference Proceedings 460, New York, 1999, pp. 152-160.
7. Siergiej, R. R., Wernsman, B., Derry, S. A., Mahorter, R. G., Wehrer, R. J., Link, S. D., Palmisiano, M. N., Messham, R. L., Murray, S., Murray, C. S., Newman, F., Hills, J., and Taylor, D., "20% Efficient InGaAs/InPAs Thermophotovoltaic Cells," in *Thermophotovoltaic Generation of Electricity: Fifith NREL Conf*erence, to be published.

SESSION 5

Cell Technology

Chair: Ben Wernsman

An Overview of TPV Cell Technologies

V.M.Andreev

Ioffe Physico-Technical Institute, 26 Polytechnicheskaya, 194021, St.Petersburg, Russia

Abstract. Results of growth, material characterization and device performance of TPV cells based on Ge, Si, InGaAs/InP, GaSb, InGaAsSb, AlGaAsSb, InAsSbP and InAs fabricated by LPE, MOCVD, MBE and diffusion methods are presented. The highest efficiencies have been obtained in TPV cells based on GaSb and InGaAs (lattice matched to InP substrate): external quantum yield as high as 90% in IR-part of photosensitivity spectrum; V_{oc}=0.45-0.52 V; FF=0.7-0.8 at photocurrent densities of 1-5 A/cm^2 and efficiency of more than 10% under available matched radiators. A low-cost Zn-diffusion technology for reproducible fabrication of *p-n*-GaSb structures has been developed for producing the high efficiency TPV cells being used for TPV generators. Growth of epitaxial lattice-matched AlGaAsSb wide-bandgap windows and InGaAsSb low-bandgap (0.5-0.6 eV) photoactive layers on GaSb substrates, as well as a fabrication of tandem TPV devices based on GaSb top cells and InGaAsSb bottom cells should allow to improve the GaSb-based cell performance. Another approach for fabrication of the perspective TPV devices with bandgaps of 0.55-0.74 eV has been realized by growth of the lattice-matched and mismatched InGaAs layers on InP substrates. Monolithic interconnected modules (MIMs) fabricated on semi-insulated InP-substrates by the MOCVD method ensure a decrease of the Joule losses and an increase of the sub-bandgap photon reflection in the structures with a back-surface reflector that should contribute to TPV system efficiency owing to photon recirculation. TPV cells based on InAsSbP/InAs are capable to convert radiation with wavelengths up to 2.5-3.5 μm and ensure operation with IR emitters heated to lower temperatures.

INTRODUCTION

The investigations in the field of thermophotovoltaics were started in the early 1960s [1], but the real advantage of TPV approach was demonstrated only in the beginning of 1990s [2]. Theoretical and semi-empirical modeling [3-11] has shown that the optimal bandgap energy is in the range of 0.4 – 0.6 eV in TPV cells designated for operation with blackbody (graybody) emitters at the temperatures of 1200-1500 °C. Germanium and silicon were the materials at first suggested and applied to thermophotovoltaic conversion of radiation from fuel-fired emitters. However, advantages such as low cost and commercial availability of the first TPV systems based on these materials have not been realized. Among the III-V compounds, gallium antimonide (0.72 eV) was the first semiconductor widely used in TPV devices. InGaAsSb (0.5 – 0.6 eV) solid solutions lattice matched to GaSb as well as InGaAs

CP653, *Thermophotovoltaic Generation of Electricity: 5th Conference*
edited by T. J. Coutts, G. Guazzoni, and J. Luther

lattice matched (E_g = 0.74 eV) and mismatched (0.5-0.6 eV) to InP substrates have been developed for TPV applications as well. Semiconductors with bandgaps wider than 0.75 eV, such as Si, can be also applied to TPV systems with selective emitters, in which the radiation maximum is shifted to the short-wavelength part of spectrum (to the photosensitivity region of these semiconductors).

The reviews of the works in the field of TPV cells including the papers presented at the Third NREL TPV Conference were published in [7-10]. Our paper provides an overview of the recent developments in TPV cell technology.

GERMANIUM TPV CELLS

The intensive investigations of Ge-based PV cells have been carried out recently [13-23]. This was motivated by that Ge had become the basic material (substrate and bottom cell) for triple-junction InGaP/GaAs/Ge solar cells [20-22] widely used already for fabrication of high-efficiency radiation-stable space arrays. These cells were applied recently for terrestrial concentrator photovoltaics and for thermophotovoltaics as well.

Monolithic GaAs/Ge [15] and $Al_{0.08}Ga_{0.92}As$/Ge [16] tandems with bottom Ge cells were developed for solar arrays. The increase of V_{oc} up to 0.32 V was achieved in the bottom Ge cell under 100 suns (AMO). Efficiencies of 3.5-4.6 % were obtained [17,18] under 10-170 suns (AMO) in the single junction cells based on *p*-GaAs/*p*-Ge/*n*-Ge. The *p*-GaAs top layer was epitaxially grown by MOCVD on *n*-Ge. This GaAs growth creates a *p-n* junction in Ge wafers through Ga and As in-diffusion. SIMS profiles show that the surface gallium-atom concentration in Ge is more than 10^{20}cm^{-3} and much higher than the arsenic-atom concentration (about 10^{19}cm^{-3}).

Chemical growth process of oxide layers on Ge-substrates was developed at NASA Glenn RC [19] for surface passivation on n^+pp^+Ge (As, In) duffused cells. The V_{oc} = 0.24 V and the short circuit current density of 55.5 mA/cm^2 were achieved in these cells under AMO, 25 ^{0}C conditions.

Ge PV cells were fabricated at NREL [21] by growing *n*-type GaInP epilayers on *p*-type Ge substrates by MOCVD. The resulting junctions were found to be diffused *n*-GaInP/*n*-Ge/*p*-Ge homojunctions. The open circuit voltage in these cells increases from 0.25 V at I_{sc} = 0.1 A /cm^2 up to 0.4 V at 30 A/cm^2.

Back-surface reflection of non-absorbed sub-bandgap photons from a TPV cell allows maximizing the efficiency of a TPV system owing to possible reabsorbtion of these photons in the radiator (Figure 1). Back contacts with increased reflectivity for photons with $h\nu < E_g$ were developed [23] for the cells based on *n*-Ge ($n = 10^{17}$ cm^{-3}, 250 μm thick). The back-surface mirror consists of MgF_2 and Au or Ag layers. Back ohmic contacts Au/Ge/Ni/Au are formed in the holes in the MgF_2 layer (Figure 1) with total holes "shadowing" of 3% and annealed at 400 °C. The cell reflectance at wavelengths longer than 1.9 μm is about 82-87% for Ge cells (Figure 2, curve 2). The external quantum yield on the free from contact area as high as 0.9-0.95 (Figure 2, curve 1) was measured in these Ge cells.

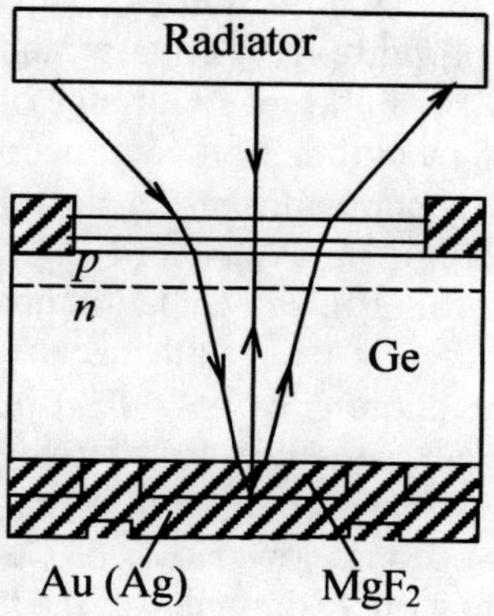

FIGURE 1. Schematic drawing of the sub-bandgap radiation reabsorption in a Ge TPV cell with a back-surface "mirror" contact.

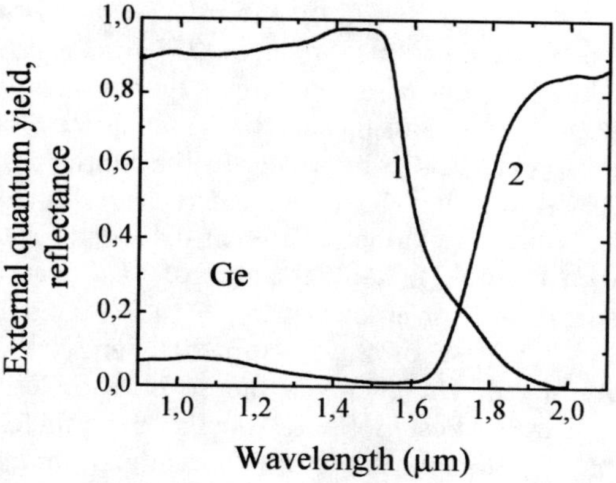

FIGURE 2. Spectral responses calculated on the photoactive area (curve 1) and reflectance (curve 2) for a Ge TPV cell with a back contact fabricated through the holes in MgF_2 layers.

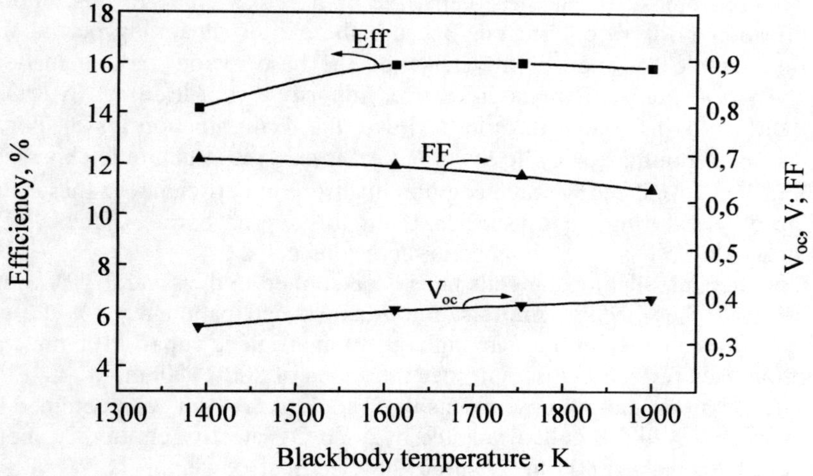

FIGURE 3. Calculated GaAs/Ge TPV cell efficiencies as a function of a blackbody IR-emitter temperature. BB spectrum is cut off at λ>1820 nm. 100% reflection of sub-bangap photons from the cell to the emitter is assumed.

TPV cells based on the *p*-GaAs/*p*-Ge/*n*-Ge heterostructure with a thin (0.1 μm) GaAs layer were fabricated [23] by the low-temperature liquid-phase epitaxy and additional Zn-diffusion. High external quantum yield (85-90% in the spectral range of 900-1550 nm) in these cells demonstrates low densities of recombination centers on the *p*-GaAs/*p*-Ge interface. In the fabricated GaAs/Ge cells, photocurrent density of 3.8 A/cm^2 has been estimated under blackbody irradiation with T_{BB} = 1473 K. Measured V_{oc} values as high as 0.4V and *FF* =0.7 were used for efficiency calculations (see Figure 3). PV efficiency of 16% was estimated for the Ge cells with a back-surface mirror assuming T_{BB} = 1473 K and 100% returning photons with $\lambda > 1.8$ μm. For the case of 85% return efficiency, PV efficiency was estimated to be 9.8% Achieved efficiencies of the developed Ge-based TPV-cells is a little bit worse than that in the best GaSb TPV cells. Nevertheless, taking into account lower cost of Ge-based cells and the high sub-bandgap photon reflection the conclusion can be made that these Ge cells are promising for TPV applications.

SILICON TPV CELLS

25% efficiency was achieved [24] under 100 suns (AM1.5) in the concentrator solar cells based on boron-doped silicon with low resistivity (0.1-0.2 Ohm·cm) and long minority carrier lifetimes. Multijunction silicon cells with point contacts of either polarity formed on the back surface have been developed [25] for operation at high illumination intensities. In these cells, the optical losses due to obscuration by the metallization grid are eliminated. Because the carrier separation occurs within *p-n* junctions at the backside of the structure, the base region may be made of high-resistivity float zone silicon having longer minority carrier lifetimes (over 3 ms). The front surface is passivated in order to lower the recombination losses. For the same purpose and, simultaneously, to lower optical losses, the structure thickness is reduced to about 100 μm. The long-wavelength light absorption efficiency in these structures is enhanced by texturing the top surface. Using this approach, the efficiency of 27.5% at 100 suns, AM1.5 has been obtained in these solar cells.

Commercial silicon solar cells have been applied in the several TPV systems [26-28] with Yb_2O_3 selective emitters. Important Si cell parameters for the use in the system with the Yb_2O_3 emitter are high quantum efficiency up to 1100 nm wavelength and minimised reflection losses in spectral region of 800-1100nm.

The maximum output power density up to 81.2 mW/cm^2 was obtained [27] from commercially available cells irradiated by a Yb_2O_3 selective emitter. A photoreceiver with a total area of 380 cm^2 ensured the electrical power of 21 W with 1.35 kW thermal input power. A larger PV receiver with an area of 2000 cm^2 has generated the 120 W electricity at 12 kW thermal input power. Recently [28] a new, somewhat larger photocell generator consisting of high efficiency silicon cells provided a system efficiency of 2.8% with preheated (370 °C) combustion air.

GaSb TPV CELLS

High efficiency GaSb cells have been developed at first as a bottom cells for concentrator mechanically stacked GaAs/GaSb tandems [29-34]. Recently the investigations of GaSb TPV devices have been considerably extended [35-53]. It has been shown [29-30] that surface recombination rate for *p*-type GaSb was lower than that for *p*-type GaAs. For this reason GaSb TPV cells with a high quantum yield were fabricated using a simple Zn-diffusion method without AlGaSb or AlGaAsSb window layers. The pseudo-closed-box diffusion technique was used. The mixtures of zinc and antimony, zinc and gallium or zinc-gallium-antimony were used as the diffusion sources.

It was found that the surface Zn concentration of 10^{20} cm^{-3} is independent of the diffusion temperature at the selected zinc-antimony source and that a sharp Zn concentration gradient takes place in these structures. The free-carrier (holes) distributions were recorded by the Raman scattering spectroscopy method accompanied by the precise anodic oxidation of the structure [31, 33]. Comparison of SIMS and Raman profiles shows a good concentration correspondence of free carriers and Zn-atoms. This fact indicates that almost all diffused Zn atoms are electrically active.

JX Crystals Inc. [2,36,37,45] has organized the production of high power TPV cells based on the diffusion technology. The following performance has been achieved in the flash tested cells with a total area of 1.5 sq. cm [36]: $FF = 0.84$, $V_{oc} = 0.52$ V, $I_{sc} = 6$ A, $P_{max} = 2.6$ W. The water cooled cells in continuous operation in front of the heated SiC emitter operating at 1380^{0}C ensured the following output parameters: $FF = 0.67$, $V_{oc} = 0.44$ V, $P_{max} = 1.76$ W. Several TPV photoreceivers based on developed cells have been fabricated [37, 45] for TPV generators. Water cooled cylinder [37] contains 380 GaSb cells arrayed on the inside surface of the cylinder of 5.3" in diameter. The cells were mounted on twenty circuits with 19 cells per circuit. Each circuit had length of 8.5". Tested with a 1600 °C glowbar the cylinder generated 900 Watts with an electric power density of 2.5 W/cm^2. The advanced GaSb TPV shingle circuit was designed for the Midnight Sun stove panel [45]. This circuit utilizes a terraced AlSiC substrate having a thermal expansion coefficient close to that of GaSb. TPV cells are connected in series in a shingle pattern. The fabricated circuit had 22 cells in series and a total active area of 30 cm^2. This circuit produced the electric power up to 44 Watts.

To reduce the TPV system cost, GaSb cells based on polycrystalline material have been developed [45, 46]. The achieved polycrystalline cell average performance is down from the single crystal case by about 10%. However the potential cost savings are expected to dramatically out weigh this small performance drop.

The concept developed at the Fraunhofer ISE [43-44, 47-48] for GaSb TPV cells is the use of strong built-in electric fields (up to 8 kV/cm) in the *p*-type emitter formed due to a specific steep Zn diffusion profile regulated through the Zn diffusion parameters and precise etching. It leads to an improvement of the device performance. Formation of strong built-in electrical fields ensures a decrease of the surface recombination influence, which results in an increase of the generated current and voltage in the TPV cells. PV cells with the internal quantum yield exceeding 90% and the open-circuit voltage larger than 0.5 V at current densities > 3 A/cm^2 were fabricated [47,48] with using this approach. Calculation [48] based on these cell measurements shows that efficiencies up to 30% could be achieved assuming a blackbody temperature of 1300-1500 K and a perfect band edge filter.

MOCVD-grown AlGaAsSb/GaSb heterostructures with a window AlGaAsSb (0.72-0.9 eV) layer were fabricated in FhG-ISE [61]. It was found that *p*-GaSb/*n*-AlGaAsSb heterostructure ensures excellent quantum efficiency as a corresponding homojunction devices while the open-circuit voltage increased by approximately 20 mW due to a reduced dark current in the device.

LPE growth of Te-doped GaSb layers was carried out in the IOFFE Institute [49-53] in attempts to prepare a material of a higher crystal quality. Ga-, Sb- and Pb-rich melts were used in the LPE processes. A maximal external quantum yield was obtained in the cells, which were prepared using the two-stage diffusion of Zn to epilayers grown from Ga-rich melt (Figure 4). Two-stage diffusion process was developed to improve GaAs cell performance [49-53]. At the first stage, GaSb wafers were exposed to form a shallow *p-n* junction in the photoactive area of a cell. At the second stage, a deep *p-n* junction (1-1.5 μm) was formed by additional spatially selective diffusion process to reduce the current leakages under contact grid fingers (Figure 5). Then, anodic oxidation and selective etching were employed for precise thinning (to 0.2-0.3 μm) the photoactive diffused *p*-GaSb layer. A maximum photocurrent of up to 30 mA/cm^2 under the sunlight with AMO spectrum cut-off at λ < 900 nm was obtained from the cells with the *p*-region thickness in the range of 0.15-0.3 μm. That explains the necessity for the second deep selective diffusion process.

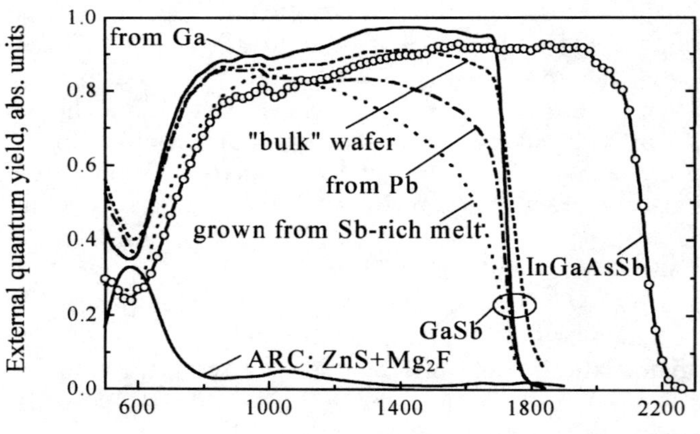

FIGURE 4. Spectral responses (on the photoactive area) of the TPV cells based on GaSb and lattice-matched InGaAsSb epitaxial layer. Reflectance from a GaSb cell with ZnS+MgF_2 antireflection coating is shown as well.

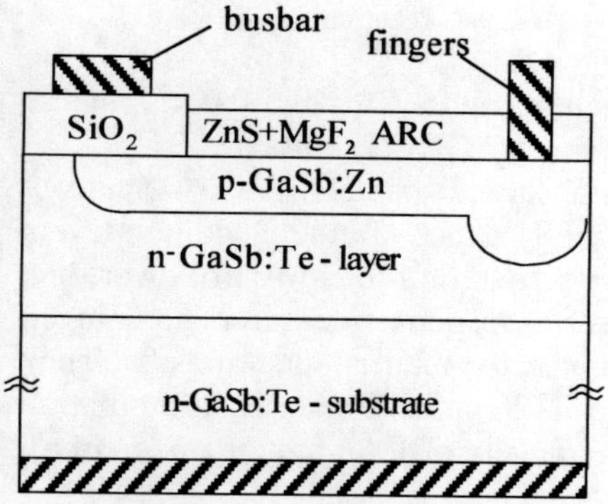

FIGURE 5. Cross section of epitaxial GaSb cell fabricated by two stage Zn-diffusion method.

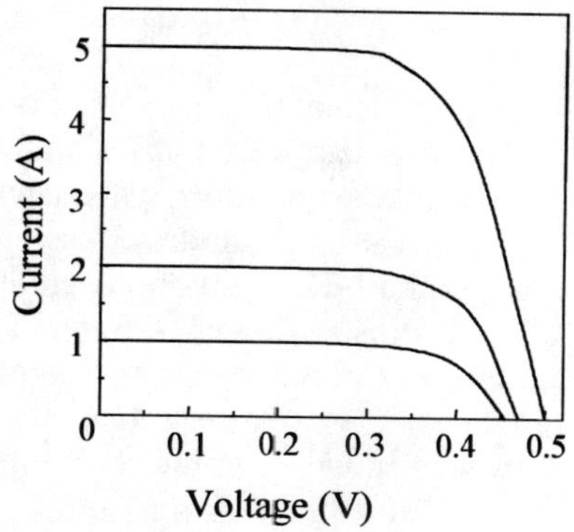

FIGURE 6. Illuminated current-voltage characteristics of a 1cm^2 GaSb TPV cell fabricated by two-stage diffusion process.

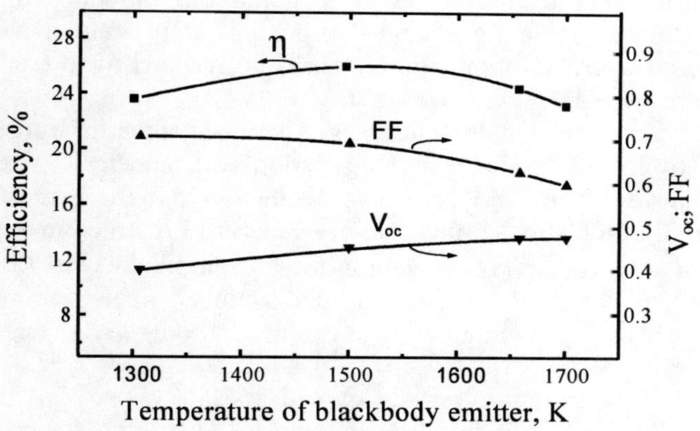

FIGURE 7. GaSb TPV cell efficiency (η),fill factor (FF) and open circuit voltage (V_{oc}) as a function of blackbody emitter temperature. Efficiency calculated for radiation spectra cut-off at λ > 1820 nm.

Using the developed technologies, the cells of 1 cm^2 and 2 cm^2 in area were fabricated for TPV generators. The deviation of V_{oc} and FF in the 1 cm^2 cells at I_{sc} = 1 A is in the following ranges: V_{oc} = 0.42-0.45 V and FF = 0.70-0.72 (Figure 6). The cells of 2 cm^2 in area generate a photocurrent up to 9 A at V_{oc} = 0.52 V. Calculated efficiency exceeding 25 % were obtained in the developed GaSb TPV cells assuming a blackbody emitter temperature of 1400-1600 K and radiation spectrum cut-off at λ > 1820 nm (Figure 7).

InGaAsSb/GaSb TPV CELLS

Lattice matched InGaAsSb/GaSb heterostructures with E_g = 0.5-0.6 eV in the photoactive cell area were fabricated by LPE, MBE and MOCVD epitaxies.

Double heterostructures with a wide-bandgap, lattice matched AlGaAsSb front surface passivating "window" layers and back surface field cladding layers surrounding the InGaAsSb photoactive layer were grown by LPE in AstroPower, Inc. [54, 55]. The absorption edge implied by the spectral response measurements ranged from 2250 to 2300 nm. At the wavelength of 2000 nm, the internal quantum efficiency as high as 95% was measured in the fabricated TPV cells. Under an illumination intensity corresponding to the short-circuit current density of 1 A/cm^2, the open-circuit voltage of 0.26 V [54] was measured.

As an alternative TPV structure, epitaxial layers of n-$In_{0.15}Ga_{0.85}As_{0.17}Sb_{0.83}$ (0.55 eV)/n-GaSb were diffused with zinc to form a p-n homojunction [55-57]. It was determined that, similar to GaSb cells, strong built-in electric fields near the surface lead to a reduction of the saturation value of the injection component of the dark current of $4 \cdot 10^{-6}$ A/cm^2 and, hence, to the increase of the open-circuit voltage. Moreover, built-in electric fields due to reducing the influence of the surface recombination allowed obtaining good PV cell parameters without passivating the AlGaAsSb window layer. At illumination intensity corresponding to I_{sc} = 3 A/cm^2, the open circuit voltage was increased up to 0.36 V.

A similar Zn-diffusion technique has been developed at the Ioffe Institute [39-41] for fabrication of InGaAsSb/GaSb single-junction and tandem TPV cells. A high (90%) external quantum yield (see Figure 4) was measured in the TPV cells fabricated by Zn-diffusion in to n-GaSb – n-InGaAsSb – n-GaSb LPE structure with a thin (0.3 μm) GaSb front passivating layer. Monolithic two-terminal GaSb (top)/InGaAsSb (bottom) TPV tandems with an internal connecting tunnel junction were also fabricated [40, 41]. The open circuit voltage was increased up to 0.6 V at I_{sc} = 0.5 A in these tandems.

The molecular beam epitaxy was applied [58, 59] also for fabrication of InGaAsSb/GaSb TPV cells. It has been shown [59] that the transport of photo-generated electrons in p-InGaAsSb is very efficient. Diodes based on the structure, which consists of n-InGaAsSb base, p-InGaAsSb thick emitter and p^+ GaSb thin window layer, exhibit the electron diffusion length of 29 μm in an 8 μm wide emitter and have internal quantum efficiencies of 95% at wavelength of 2000 nm that are well-suited to efficient TPV diode operation.

Antimonide based TPV cells have been fabricated using the quaternary layers of GaInAsSb [60] grown also by MOCVD on GaSb substrates. Improved devices are obtained [60] when optical absorption occurs in the p-layer due to the longer minority carrier diffusion length. It was shown that thin n-emitter/thick p-base devices are more promising since surface passivation is less critical than for p-emitter devices.

InGaAs/InP TPV CELLS

Lattice matched $In_{0.53}Ga_{0.47}As/InP$ heterostructures were developed initially for fabrication of infra-red photodetectors. Then, these diodes were used as a laser power converters [64] and as bottom cells for monolithic InP(top)/InGaAs(bottom) [62] or mechanically stacked GaAs (top)/InGaAs(bottom) [31, 32] tandems. InGaAs layers were grown on InP substrates by LPE [31, 32, 63] and by MOCVD [62] methods. 6% efficiency (AM0, 100 suns) was measured [31] in the LPE cells illuminated through the InP substrate and the GaAs IR transparent filter. TPV cells fabricated by simultaneous Zn- and P-diffusion into the LPE grown InGaAs layer have demonstrated V_{oc} = 0.42-0.5 V and FF = 0.75-0.78 at I_{sc} = 1-15 A/cm^2 [63].

Lattice-matched (or mismatched) InGaAs/InP monolithic interconnected modules (MIMs) were developed [64-67] for TPV applications. In this approach, InGaAs structure was grown by MOCVD on a semi-insulating InP substrate. Then the structure was divided into several subcells, which were interconnected in series. This has the effect of reducing the current output of a single cell by a factor equal to the number of subcells, which reduces (in principle) the Joule losses and increases the power output at high illumination intensity. A MIM has other advantages over conventional one-junction cells, such as simplified array interconnections and heat-sinking, and radiation recycling capability using a back surface reflector.

Semi-insulating substrates are transparent to infrared radiation that allows obtaining extremely high sub-bandgap photon reflection in the structures with a high-reflective back-surface reflector. Reflectivity of about 95% for photons with $\lambda > 1.7$ μm was obtained [65] from the back reflector made of Ag or Au deposited right to back surface of semi-insulating InP. Reflectivity was enhanced up to 97-98% by interposing a MgF_2 layer between InP and Au or Ag. The measured reflectivity from the MIM TPV cells was 80-85% in the spectral range of 1.8-2.7 μm and increased at $\lambda > 2.7$ μm [66]. The external quantum efficiency of about 90% was achieved in spectral range of 0.6-1.5 μm that ensures I_{sc} = 56 mA/cm^2 under AM0 spectrum. V_{oc} = 0.4 V, FF = 0.71 and efficiency of 11.7% (AM0) were achieved [66] in the developed MIM devices with the grid shadowing of 16%.

High-performance, lattice-mismatched InGaAs/InP MIMs with InGaAs bandgap energy of 0.6 eV [68, 69] and 0.55 eV [70] were developed to make spectral response of the TPV cells with lower bandgaps better matched to the graybody spectral power density profile for low to moderate emitter temperatures (i.e. 1000-1200 ^{o}C). The lattice-mismatched structures were grown by MOCVD using a compositionally step-graded InPAs buffer to accommodate a lattice mismatch of 1.1% between the active InGaAs cell structure and InP substrate [68]. Internal quantum efficiency in 0.6 eV MIMs with a back surface reflector was estimated to be near unity close to the bandedge.

Good performance characteristics were demonstrated also for TPV converters made of a lattice-matched $InAs_{0.32}P_{0.68}$(0.96eV)/$Ga_{0.32}In_{0.68}As$(0.6eV)/$InAs_{0.32}P_{0.68}$ double-heterostructure, which was grown lattice-mismatched on an InP substrate, using the step graded region of InPAs [60]. The average V_{oc} value for subcells is 0.39

V, which yields a voltage factor of 0.64 V/eV. High reflectances (80-90%) were achieved for $\lambda = 2$-12 μm, which indicates that sub-bandgap photon recirculation will be effective in the TPV system based on developed MIMs.

MIM devices based on InGaAs with a bandgap reduced to 0.55 eV were developed [70] as well. Novel buffering schemes improved cell architectures, and a variety of interconnected schemes were investigated in order to maximize the device power density and efficiency. The external quantum efficiency of the device with no antireflective coating was found to be 0.7 and the dark current density was found to be 40 μA/cm^2. The 30-junction MIM had $V_{oc} = 8.3$ V and $FF = 0.61$ at the short circuit current density of 1.15 A/cm^2. Such a type of cells is very promising for the TPV systems operating with low temperature (1000-1100 °C) emitters.

LOW-BANDGAP InAsSbP/InAs TPV CELLS

Epitaxial InAsSbP TPV cells were fabricated [55, 71, 72] from *p*-InAsSbP/*n*-InAsSbP/*n*-InAs heterostructures grown on *n*-InAs substrates. Epitaxial growth of InAsSbP quaternary layers lattice matched to InAs was carried out by the liquid phase epitaxy. A p-n junction was formed during the epitaxy [55, 71] or by Zn-diffusion [72] in n-InAsSbP layers.

The low-temperature (340-350 °C) pseudo-closed box Zn diffusion process was applied [72] to form a *p-n* junction in the quaternary InAsSbP (0.45-0.48 eV) alloy. A patterned silicone nitride or SiO_2 dielectric layers on the top surface were used as a diffusion mask, so that the junction was formed only within the designated illumination area, which allows to reduce current leakages. A busbar was deposited on this dielectric layer and only contact grid fingers contacted to the junction region. Figure 8, curve 1 shows the spectral response in *p*-InAsSbP/*n*-InAsSbP/*n*-InAs TPV cell. Reduction of the photosensitivity with the wavelength increase may be explained by a not sufficient diffusion length for photogenerated holes in the obtained epitaxial n-InAsSbP base layer.

P-n InAs TPV cells were fabricated [72] by using wafers prepared from ingots grown by the Czochralski technique, Te-doped. The InAs wafers were subjected to the zinc diffusion procedure by means of the "pseudo-closed box" technique to form a *p-n* junction in photoactive area of the cells from a pure zinc source. Figure 8, curve 2 represents the spectral response of a *p*-InAs/*n*-InAs TPV cell, which demonstrates rather high photosensitivity up to 3.5 μm at room temperature. Photosensitivity in the short-wavelength ($\lambda < 2.5$μm) part of the spectrum can be increased by growth of a wide-bandgap lattice-matched window InAsSbP layer on InAs.

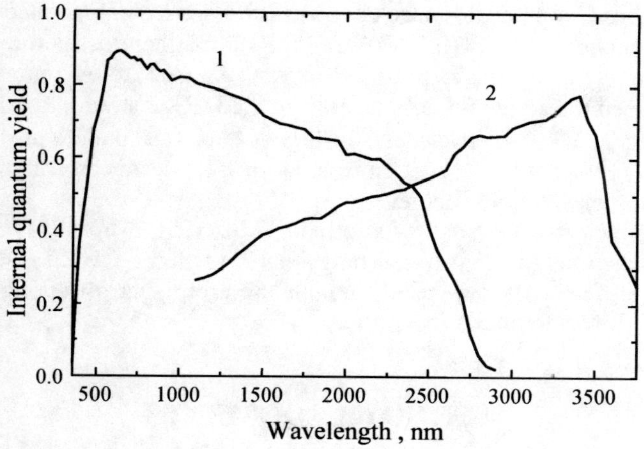

FIGURE 8. Spectral responses of the TPV cells based on: *p*-InAsSbP/*n*-InAsSbP/*n*-InAs heterostructure (curve 1), and *p-n* InAs fabricated by Zn diffusion process (curve 2).

SUMMARY

GaSb and low-bandgap ($E_g < 0.6$ eV) antimonide-based devices continue to make a sufficient progress. High productive with a good reproducibility diffusion technology of high efficiency GaSb-based TPV cells were developed by several teams. Formation of the sharp Zn concentration gradient in the thin front diffused layers and a strong buit-in electrical field, the use of the epitaxial and two-stage diffusion technologies ensured the cell performance improvement. Efficiencies exceeding 25% for in-band ($\lambda < 1800$ nm) blackbody (1200-1700 K) radiation were achieved in advanced GaSb TPV cells. To reduce the TPV system cost, the polycrystalline GaSb material was used for fabrication of TPV cells. The potential cost saving compensates a small performance drop in polycrystalline cells. Using the Zn-diffusion into epitaxially-grown InGaAsSb (0.55 eV) lattice-matched to GaSb, the saturation value of the injection component of the dark current was reduced to $4 \cdot 10^{-6}$ A/cm^2 and the open-circuit voltage was increased up to 0.36 V at illumination intensity corresponding to $I_{sc} = 3$ A/cm^2.

High efficiency InGaAs TPV cells based on $In_{0.53}Ga_{0.47}As/InP$ lattice matched heterostructures, fabricated by both LPE and MOCVD methods have been developed. MOCVD growth of monolithic interconnected modules based on both lattice-matched and mismatched InGaAs/InP structures were made more sophisticated during the last years. High sub-bandgap photon reflection has been realized in these structures with a back-surface reflector owing to high transparency of semi-insulating InP substrates to infrared radiation. Good performance characteristics were demonstrated for TPV cells based on mismatched InGaAs (0.55-0.6 eV)/InP double-heterostructures, which were grown by MOCVD using the step graded region of InAsP to accommodate a lattice mismatch between the active InGaAs layers and the InP substrate.

In Ge-based TPV cells, efficiencies of higher than 16% under cut-off at $\lambda > 1800$ nm blackbody (1400-1900 K) spectrum at I_{sc} = 3-20 A/cm^2 and high sub-bandgap photon reflection of 85% were achieved.

Photosensitivity of TPV cells based on *p*-InAsSbP/*n*-InAsSbP/*n*-InAs and *p*-InAs/*n*-InAs structures in the infrared range was widened up to 2.5-3.4 micrometers, correspondingly. These PV low-bandgap cells are promising for applications in TPV generators with lower-temperature emitters.

ACKNOWLEDGMENTS

The author would like to thank the colleagues at Photovoltaics Laboratory of Ioffe Institute V.D.Rumyantsev, V.P.Khvostikov, V.A.Grilikhes, V.R.Larionov, M.Z.Shvarts, L.B.Karlina, G.V.Il'menkov and N.Kh.Timoshina for help and for useful discussions.

REFERENCES

1. Broman, L., Thermophotovoltaic bibliography, *Progress in Photovoltaics* **3**, 65-74 (1995).

2. Fraas, L.M., Ballantyne, R., Samaras, J., Seal, M., "A thermophotovoltaic electric generator using GaSb cells with a hydrocarbon burner," in *First World Conference on Photovoltaic Energy Conversion*, Hawaii, 1994, pp. 1713-1716.

3. Woolf, L.D., "Optimum efficiency of single and multiple band gap cells in TPV energy conversion," in *Conference Record of the 18th IEEE PVSC*, 1985, pp. 1731-1732.

4. Wanlass, M.W., Ward, J.S., Emery, K.A., and Coutts, T.J., "$Ga_xIn_{1-x}As$ thermophotovoltaic converters," in *First Wold Conference on Photovoltaic Energy Conversion*, Hawaii, 1994, pp. 1685-1691.

5. Gray, J.L., and El-Husseini, A., "A simple parametric study of TPV system efficiency and output power density including a comparison of several TPV materials," in *Second NREL Conference on Thermophotovoltaic Generation of Electricity –1995*, AIP Conference Proceedings 358, New-York, 1996, pp. 3-15.

6. Iles, P.A., Chu, C., and Linder, E., "The influence of bandgap on TPV converter efficiency," in *Second NREL Conference on Thermophotovoltaic Generation of Electricity – 1995,* edited by T.J.Coutts et al., AIP Conference Proceedings 358, New-York, 1996, pp. 446-457.

7. Coutts, T.J., Wanlass, M.W., Ward, J.S., and Johnson, S., " A review of recent advances in thermophotovoltaics" in *Conference Record of 25th IEEE PVSC*, Washington, DC, 1996, pp. 25-30.

8. Coutts, T.J., "A review of progress in thermophotovoltaic generation of electricity," *Renewable and Sustainable Energy Reviews* **3**, 77-184 (1999).

9. Cody, G.D., "Theoretical maximum efficiencies for thermophotovoltaic devices," in *Fourth NREL Conference on Thermophotovoltaic Generation of Electricity – 1998*, AIP Conference Proceedings 460, New York, 1999, pp. 58-67.

10. Coutts, T.J., "An overview of thermophotovoltaic generation of electricity," in *Technical Digest of the International PVSEC –11*, Sapporo, 1999, pp. 137-140.

11. Andreev, V.M., Khvostikov, V.P., Larionov, V.R., Rumyantsev, V.D., Sorokina,S.V., Shvarts,M.Z., Vasil'ev, V.I., Vlasov, A.S., "Tandem GaSb/InGaAsSb thermophotovoltaic cells," in *Conference Record of 26th IEEE PVSC,* Anaheim, 1997, pp. 935-938.

12. Andreev, V.M., Grilikhes, V.A., Rumyantsev, V.D., *Photovoltaic Conversion of Concentrated Sunlight,* J. Wiley&Sons, Chichester, 1997.

13. Venkatasubramanian, R., Timmons, M.L., Pickett, R.T., Colpitts, T.S., Hills, J.S. and Hutchby, J.A., " Development of low-bandgap Ge and $Si_{0.07}Ge_{0.93}$ solar cells for monolithic and mechanically-stacked cascade applications," in *Conference Record of the 21th IEEE PVSC,* Piscataway, 1990, pp. 73-78.

14. Bullock, J.N., "Interface contribution to GaAs/Ge heterojunction solar cell efiiciency," Transactions on Electron Devices **36**, 1238-1243 (1989).

15. Tobin, S.P., Vernon, S.M., Bajgar, C., Haven, V.E., Geoffroy, L.M., Sanfacon, M.M., Lillington, D.R., Hart, R.E., Emery, K.A., Matson, R.J., " High efficiency GaAs/Ge monolithic tandem solar cells," in *Conference Record of the 20th IEEE PVSC,* Las Vegas, 1988, pp. 405-410.

16. Timmons, M.L., Hutchby, J.A., Wagner, D.K., and Tracy, J.M., "Monolithic AlGaAs/Ge cascade solar cells," in *Conference Record of the 20th IEEE PVSC,* Las Vegas, 1988, pp. 602-606.

17. Wojtczuk, S.J., Tobin, S.P., Keavney, C.J., Bajgar, C., Sanfacon, M.M., Geoffroy, L.M., Dixon, T.M., Vernon, S.M., Scofield, J.D., and Ruby, D.S., "GaAs/Ge tandem-cell space concentrator development," *IEEE Transactions on Electron Devices* **37**, 455-462 (1990).

18. Wojtczuk, S., Tobin, S., Sanfacon, M., Haven, V., Geoffroy, L., and Vernon, S., "Monolitic two-terminal GaAs/Ge tandem space concentrator cells," in *Conference Record of the 22nd IEEE PVSC,* Las Vegas, 1991, pp. 73-79.

19. Faur Mircea, Faur Maria, Bailey, S.G., Flood, D.J., Brinker, D.J., Wheeler, D.R., Alterovitz, S.A., Scheiman, D., Mateescu, G., Faulk, J., Goradia, C., Goradia, M., "Front surface engineering of high efficiency Si solar cells and Ge TPV cells," in *Conference Record of the 26th IEEE PVSC,* Anaheim, 1997, pp. 847-851.

20. Karam, N.H., King, R.R., Cavicchi, B.T., Krut, D.D., Ermer, J.H., Haddad, M., Cai, L., Joslin, D.E., Takahashi, M., Eldredge, J.W., Nishikawa, W.T., Lillington, D.R., Keyes, B.M., Ahrenkiel, R.K., "Development and characterization of high-efficiency $Ga_{0.5}In_{0.5}P$/GaAs/Ge dual- and triple-junction solar cells," *IEEE Transactionsons Electron Devices* **46**, 2116-2125 (1999).

21. Friedman, D.J., Olson, J.M., Ward, S., Moriarty, T., Emery, K., Kurtz, Sarah, Duda, A., King, R.R., Cotal, H.L., Lillington, D.R., Ermer, J.H., Karam, N.H., "Ge concentrator cells for III-V multijunction devices," in *Conference Record of the 28th IEEE PVSC,* Anchorage, 2000, pp. 965-967.

22. Cotal, H.L., Lillington, D.R., Ermer, J.H., King, R.R., Karam, N.H., Kurtz, S.R., Friedman, D.J., Olson, J.M., Ward, J.S., Duda, A., Emery, K.A., Moriarty, T., "Triple-junction solar cell efficiencies above 32%: the promise and challenges of their application in high-concentration – ratio PV systems," in *Conference Record of the 28th IEEE PVSC,* Anchorage, 2000, pp. 955-960.

23. Andreev, V.M., Khvostikov, V.P., Khvostikova, O.V., Oliva, E.V., Rumyantsev, V.D., Shvartz, M.Z., "Low-bandgap and InAsSbP/InAs-based TPV cells", in *Fifth Conference on Thermophotovoltaic Generation of Electricity,* Rome, *2002,* this volume.

24. M.A.Green, Z.Jianhua, A.W.Blakers, M.Taouk and S.Narayanan, "25-percent efficient low-resistivity silicon concentrator solar cells", *IEEE Electron. Dev. Lett.,* **EDL-7**, 583-585 (1986).

25. Sinton, R.A., Kwark, Y., Gan, J.Y., and Swanson, R.M., "27.5-percent silicon concentrator solar cells," *IEEE Electron. Dev. Lett.,* **EDL-7**, 567-569 (1986).

26. Kushch, A.S., Skinner, S.M., Brennan, R., Sarmiento, P.A., "Development of a cogenerating thermophotovoltaic power combination hot water heater/hydronic boiler," in *Third NREL Conference on Thermophotovoltaic Generation of Electricity –1997,* AIP Conference Proceedings 401, 1997, pp.373-386.

27. Bitnar, B., Durisch, W., Grutzmacher, D., Mayor, J.-C., Muller, C., von'Roth, F., Selvan, J.A.A., Sigg, H., Tschudi, H.R., Gobrecht, J., "A TPV system with silicon photocells and a selective emitter", in *Conference Record of the 28th IEEE PVSC,* Anchorage, 2000, pp. 1218-1221.

28. Durish, W., Bitnar, B., Mayor, J.-C., Fritz von Roth, Sigg, H., Tschudi, H. R., Palfinger, G., "Progress in the development of small thermophotovoltaic prototype systems", in *Proceedings of 17th Photovoltaic Solar Energy Conference and Exhibition*, Munich, 2001 (to be published).

29. Fraas, L. M., Girard, G.R., Avery, J.E., Arau, B.A., Sundaram, V.S., Thompson, A.G., "GaSb booster cells for over 30% efficient solar-cell stacks", J. Appl. Phys. **66**, 3866-3870 (1989).

30. Fraas, L.M, Avery, J.E., Martin, J., Sundaram, V.S., Girard, G., Dinh, V.T., Davenport, T.M., Yerkes, J.W., and O'Neill, M.J., "Over 35-percent efficient GaAs/GaSb tandem solar cells," *IEEE Transactionsons Electron Devices* **37**, 443-449 (1990).

31. Andreev, V.M., Karlina, L.B., Kazantsev, A.B., Khvostikov, V.P., Rumyantsev, V.D., Sorokina, S.V., Shvarts, M.Z., "Concentrator tandem solar cells based on AlGaAs/GaAs-InP/InGaAs (or GaSb) structures," in *First World Conference on Photovoltaic Energy Conversion*, Hawaii, 1994, pp. 1721-1724.

32. Andreev, V.M., Karlina, L.B., Khvostikov, V.P., Rumyantsev, V.D., Sorokina, S.V., Shvarts, M.Z., "Concentrator tandem solar cells based on A^3B^5 compounds," in *13th European Photovoltaic Solar Enrgy Conference*, Nice, 1995, pp.329-332.

33. Sulima, O.V., Faleev, N.N., Kazantsev, A.B., Mintairov, A.M., Namazov, A., "Low-temperature Zn diffusion for GaAs solar cells structures fabrication," in *Fourth European Space Power Conference*, Poitiers, 1995, pp. 641-643.

34. Andreev, V.M., Khvostikov, V.P., Paleeva, E.V., Rumyantsev, V.D., Sorokina,S.V., Shvarts, M.Z., Vasil'ev, V.I., "Tandem solar cells based on AlGaAs/GaAs and GaSb structures," in *23rd Int. Symp. Compound Semiconductors-1996,* Inst. Phys. Conf. Ser. 155, 1996, pp. 425-428.

35. Dutta, P.S., and Bhat, H.L., "The physics and technology of gallium antimonide: An emerging optoelectronic material," *J. Appl. Phys.* **81**, 5821-5870 (1997).

36. Fraas, L.M., Huang, H.X., Ye, S.-Z., Hui, S., Avery, J., and Ballantyne, R., "Low cost high power GaSb photovoltaic cells," in *Third NREL Conference on Thermophotovoltaic Generation of Electricity – 1997,* AIP Conference Proceedings 401, 1997, pp. 33-40.

37. Fraas, L., Samaras, J., Huang, H.-X., Seal, M., West, E., "Development status on a TPV cylinder for combined heat and electric power for the home," in *Fourth NREL Conference on Thermophotovoltaic Generation of Electricity-1998,* AIP Conference Proceedings 460, New York, 1999, pp. 371-383.

38. Bett, A.W., Faleev, N.N., Mintairov, A.M., Namazov, A., Sulima, O.V., Stollwerck, G., "Optimization of low-temperature Zn diffusion for GaSb solar cell structures fabrication," in *13th European Photovoltaic Solar Energy Conference,* Nice, 1995, pp. 88-91.

39. Andreev, V.M., Khvostikov, V.P., Paleeva, E.V., Sorokina, S.V., and Shvarts, M.Z., "GaAs and GaSb based solar cells for concentrator and thermophotovoltaic applications," in *Proc. 25th IEEE PVSC,* Washington, DC, 1996, pp. 143-146.

40. Andreev, V.M., Khvostikov, V.P., Sorokina, S.V., Shvarts, M.Z., and Vasil'ev, V.I., "GaSb based PV cells for solar and thermophotovoltaic applications," in *Proceedings of 14th European Photovoltaic Solar Energy Conference*, Barcelona, 1997, pp.1763-1766.

41. Andreev, V.M., Khvostikov, V.P., Larionov, V.R., Rumyantsev, V.D., Sorokina, S.V., Shvarts, M.Z., Vasil'ev, V.I., Vlasov, A.S., "Tandem GaSb/InGaAsSb thermophotovoltaic cells," in *Conference Record of the 26th IEEE PVSC,* Anaheim, 1997, pp. 935-938.

42. Bett, A.W., Keser, S., Stollwerck, G., and Sulima, O.V., "Large-area GaSb photovoltaic cells," in *Third NREL Conference on Thermophotovoltaic Generation of Electricity – 1997,* AIP Conference Proceedings 401, New York, 1997, pp. 41-53.

43. Bett, A.W., Keser, S., Stollwerck, G., and Sulima, O.V., "Recent progress in developing of GaSb photovoltaic cells," in *14th European Photovoltaic Solar Energy Conference,* Barcelona, 1997, pp. 993-998.

44. Bett, A.W., Keser, S., and Sulima, O.V., "Study of Zn diffusion into GaSb from the vapour and liquid phase," *Crystal Growth* **181**, 9-16 (1997).

45. Fraas, L.M., Ballantyne, R., Ye, S.-Z., Hui, S., Gregory, S., Keyes, J., Avery, J., Lamson, D., and Daniels, B., "Commercial GaSb cell and circuit development for the midnight sun TPV stove," in

Fourth NREL Conference on Thermophotovoltaic Generation of Electricity-1998, AIP Conference Proceedings 460, New York, 1999, pp. 480-487.

46. Sulima, O.V., Bett, A.W., Dutta, P.S., Ehsani, H., and Gutmann, R.J., "Thermophotovoltaic cells on zinc diffused polycrystalline GaSb," in *16th European Photovoltaic Solar Energy Conference*, Glasgow, 2000.

47. Stollwerck, G., Sulima, O.V., Bett, A.W., "Characterization and simulation of GaSb device-related properties," *IEEE Transaction on Electron Devices* **47**, 448-457 (2000).

48. Sulima, O.V., Belt, A.W., "Fabrication and simulation of GaSb thermophotovoltaic cells," in *Technical Digest of 11th International Photovoltaic Science and Engineering Conference*, Sapporo, 2000, pp. 441-442.

49. Rumyantsev, V.D., Khvostikov, V.P., Sorokina, S.V., Vasil'ev, V.I., and Andreev, V.M., "Portable TPV generator based on metallic emitter and 1.5-Amp GaSb cells," in *Fourth NREL Conference on Thermophotovoltaic Generation of Electricity – 1998*, AIP Conference Proceedings 460, New York, 1999, pp.384-391.

50. Andreev, V.M., Khvostikov, V.P., Larionov, V.R., Rumyantsev, V.D., Sorokina, S.V., Shvarts, M.Z., Vasil'ev, V.I., Vlasov, A.S., "Single and dual junction GaSb/InGaAsSb TPV cells," in *2nd World Conference and Exhibition on Photovoltaic Solar Energy Conversion*, Vienna, 1998, pp. 330-333.

51. Khvostikov, V.P, and Rumyantsev, V.D., "Revealing the distributed resistance components in GaSb TPV cells by experimental technique including electrolumescence," in *2nd World Conference and Exhibition on Photovoltaic Solar Energy Conversion*, Vienna, 1998, pp. 120-123.

52. Andreev, V.M., Vasil'ev, V.I., Khvostikov, V.P., Rumyantsev, V.D., Sorokina, S.V., Shvarts, M.Z., "GaSb/InGaAsSb cells for use in TPV generators," in *Technical Digest of 11th International Photovoltaic Science and Engineering Conference*, Sapporo, 2000, pp. 443-444.

53. Andreev, V.M., Khvostikov, V.P., Rumyantsev, V.D., Sorokina, S.V., Shvarts, M.Z., "Single-junction GaSb and tandem GaSb/InGaAsSb & AlGaAsSb/GaSb thermophotovoltaic cells," in *Conference Record of 28th IEEE PVSC*, Anchorage, 2000, pp. 1265-1268.

54. Shellenbarger, Z.A., Mauk, M.G., Cox, J.A., Gottfried, M.I., Sims, P.E., Lesko, J.D., McNeely, J.B., and DiNetta, L.C., "Improvements in GaSb-based thermophotovoltaic cells," in *Third NREL Conference on Thermophotovoltaic Generation of Electricity-1997*, AIP Conference Proceedings 401, New-York, 1997, pp. 117-128.

55. Mauk, M.G., Shellenbarger, Z.A., Cox, J.A., Sulima, O.V., Bett, A.W., Mueller, R.L., Sims, P.E., McNeely, J.B., DiNetta, L.C., "Liquid-phase epitaxy of low-bandgap III-V antimonides for thermophotovoltaic devices," *Journal of Crystal Growth* **211**, 189-193 (2000).

56. Bett, A.W., Ber, B.Y., Mauk, M.G., South, J.T., Sulima, O.V., "Pseudo-closed box diffusion of Zn into InGaSb and AlGaSb for TPV devices," in *Fourth NREL Conference on Thermophotovoltaic Generation of Electricity-1998*, AIP Conference Proceedings 460, New York, 1998, pp. 237-246.

57. Sulima, O.V., Beckert, R., Bett, A.W., Cox, J.A., and Mauk, M.G., "InGaAsSb photovoltaic cells with enhanced open-circuit voltage," *IEE Proc.-Optoelectr.*, **147**, 199-204 (2000).

58. Wang, C.A., Choi, H.K, Turner, G.W., Spears, D.L., and Manfra, M.J., "Lattice-matched epitaxial GaInAsSb/GaSb thermophotovoltaic devices," in *Third NREL Conference on Thermophotovoltaic Generation of Electricity-1997*, AIP Conference Proceedings 401, New York, 1997, pp.75-87.

59. Martinelli, R.U., Garbuzov, D.Z., Lee, H., Morris, N., Odubanjo, T., Taylor, G.C., and Connoly, J.C., "Minority-carrier transport in InGaAsSb thermophotovoltaic diodes," in *Third NREL Conference on Thermophotovoltaic Generation of Electricity-1997*, AIP Conference Proceedings 401, Now York, 1997, pp. 389-395.

60. Hitchcock, C.W., Gutmann, R.J., Borrego, J.M., Bhat, I.B., and Charache, G.W., "Antimonide-based devices for thermophotovoltaic applications," *IEEE Transactions on Electron Devices* **46**, 2154-2161 (1999).

61. Agert, C., Beckert, R., Hinkov, V., Sulima, O.V., Bett, A., "Novel device structures for infared solar cells based on GaSb: pn-homojunctions, pn-heterojunctions, and tandem cells", in *Proceedings of 17th European Photovoltaic Solar Energy Conference and Exhibition*, Munich, 2001 (to be published).

62. Wanlass, M.W., Ward, J.S., Emery, K.A., Gessert, T.A., Osterwald, C.R. and Coutts, T.J. "High performance concentrator tandem solar cells based on IR-sensitive bottom cells", *Solar cells,* **30**, 363-371 (1991).

63. Karlina, L.B., Ber, B.A., Kulagina, M.M., Kovarsky, A.P., Vargas-Aburto, C., Uribe, R.M., Brinker, D., Scheiman, D., "Simultaneous diffusion of Zn and P in InGaAs and GaSb cells for solar and thermophotovoltaic space applications," in *Conference Record of 28th IEEE PVSC*, Anchorage, 2000, pp. 1230-1233.

64. Wojtczuk, S., Parados, T., Walker, G. "P/N In(Al)GaAs multijunction laser power converters," in *Proceedings of the XIII Space Photovoltaic Research and Technology Conference*, Cleveland, 1994, pp. 363-371.

65. Fatemi, N.S., Wilt, D.M., Jenkins, P.P., Hoffman, R.W., Weizer, V.G., Murray, C.S., Riley, D. "Materials and process development for the monolithic interconnected module (MIM) InGaAs/InP TPV devices", in *Third NREL Conference on the Thermophotovoltaic Generation of Electricity 1997*, AIP Conference Proceedings 401, New York, 1997, pp. 249-262.

66. Wilt, D.M., Fatemi, N.S., Jenkins, P.P., Weizer, V.G., Hoffman, R.W., Jain, R.K., Murray, C.S., Riley, D.R. "Electrical and optical performance characteristics of 0.74 eV *p*/n InGaAs monolithic interconnected modules," in *Third NREL Conference on the Thermophotovoltaic Generation of Electricity-1997*, AIP Conference Proceedings 401, New York, 1997, pp. 237-247.

67. Ward, J.S., Duda, A., Wanlass, M.W., Carapella, J.J., Wu, X., Matson, R.J., Coutts, T.J., Moriarty, T., Murray, C.S., Rilay, D.R. "A novel design for interconnects in a TPV converter," in *Third NREL Conference on the Thermophotovoltaic Generation of Electricity – 1997*, AIP Conference Proceedings 401, New York, 1997, pp. 227-236

68. Fatemi, N.S., Wilt, D.M., Hoffman, R.W., Stan, M.A., Weizer, V.G., Jenkins, P.P., Khan, O.S., Murray, C.S., Scheiman, D., Brinker, D., "High-performance, lattice-mismatched InGaAs/InP monolithic interconnected modules (MIMs)", in *Fourth NREL Conference of Thermophotovoltaic Generation of Electricity – 1998,* AIP Conference Proceedings 460, New York, 1999, pp. 121-131.

69. Wanlass, M.W., Carapella, J.J., Duda, A., Emery, K., Gedvilas, L., Moriarty, T., Ward, S., Webb, J.D., Wu, X., "High-performance, 0.6 eV $Ga_{0.32}In_{0.68}As/InAs_{0.32}P_{0.68}$ thermophotovoltaic converters and monolithically interconnected modules", in *Fourth NREL Conference of Thermophotovoltaic Generation of Electricity – 1998,* AIP Conference Proceedings 460, New York, 1999, pp. 132-141.

70. Murray, C.S., Fatemi, N., Stan, M., Wernsman, B., Wehrer, R.J. "Growth processing and characterization of 0.55 eV n/p/n monolithic interconnected modules", in *Conf. Record of 28th IEEE PVSC*, Anchorage, 2000, pp. 1238-1241.

71. Mauk, M.G., Shellenbarger, Z.A., Cox, J.A., Tata, A.N., Warden,T.G., DiNetta, L.C., "Advances in low-bandgap InAsSbP/InAs and GaInAsSb/GaSb thermophotovoltaics", in *Proceedings of 28th IEEE PVSC, Anhorage*, 2000, pp. 1028-1031.

72. Khvostikov, V.P., Khvostikova, O.A., Oliva, E.V., Rumyantsev, V.D., Shvarts, M.Z., Tabarov, T.S., Andreev, V.M., "Zn-diffused InAsSbP/InAs and Ge TPV cells", in *Proceedings of 29th IEEE PVSC*, New Orleans, 2002 (to be published).

Development of InGaAsSb Based Monolithic Interconnected Modules (MIMs)

M.N. Palmisiano*, R.M. Biefeld⁺, J.G. Cederberg⁺, M.J. Hafich⁺, and G.M. Peake⁺

**Bechtel Bettis, Inc., West Mifflin, PA 15122*
⁺Sandia National Laboratory, Albuquerque, NM 87185

Abstract. The quaternary Sb-based materials are attractive for use in TPV devices due to the versatility associated with being able to grow InGaAsSb and AlGaAsSb material with a wide range of band gaps lattice matched to a GaSb substrate. Photodiodes (0.60 eV) consisting of an InGaAsSb p/n junction grown on GaSb were fabricated and characterized to establish baseline electrical performance of the material. In order to fabricate monolithic interconnected modules (MIMs) from these photodiodes using a conductive GaSb substrate, an electrical isolation layer must be added between the active layers and the substrate. For this purpose, AlGaAsSb cell isolation diodes (CIDs) were developed and characterized with respect to their ability to block current in reverse bias. These structures were combined in the first MIM structures grown and fabricated from Sb-based materials on GaSb substrates.

INTRODUCTION

InGaAsSb based TPV cells grown on GaSb substrates have been under development for the past few years [1]. This quaternary material system allows for the growth of a continuous range of band gaps (E_g) from 0.25-0.72 eV, lattice matched to GaSb. As well, higher band gap AlGaAsSb (0.72-1.65 eV) can also be grown lattice matched to GaSb. In this way, complicated TPV structures incorporating confinement layers, multiple junctions, quantum wells, etc. can be grown without the strain relaxation layers that are necessary in the lattice mismatched InGaAs on InP system (E_g<0.74 eV). In addition to simplifying design and growth, the absence of these buffer layers and associated dislocations may improve material electrical quality and uniformity. Single celled InGaAsSb based TPV devices have exhibited high open circuit voltages (0.290 V at 1 A/cm^2, for E_g=0.54 eV), and internal quantum efficiency greater than 90% [2].

However, in order to produce large area arrays more easily, a monolithic interconnected module (MIM) approach is attractive [3]. In a MIM, the TPV device consists of series connected cells on a common insulating substrate. Because these small cells are connected in series, voltage builds across the device while current

CP653, *Thermophotovoltaic Generation of Electricity: 5th Conference*
edited by T. J. Coutts, G. Guazzoni, and J. Luther

remains constant, leading to smaller I^2R losses for a given area device. For this reason, the MIM itself can become large without any outside interconnections – the size of the MIM is only limited by the size of the substrate that can be grown upon. Additionally, fabrication of a MIM is accomplished by standard IC industry photolithographic techniques which allow for changes in circuit design by simply changing the photomasks and device architecture, not external interconnection and packaging schemes. In this way it becomes much easier to optimize a MIM or an array of MIMs for a radiator of a given geometry and temperature profile. The MIM concept, as applied to the ternary InGaAs/InP materials system, has already demonstrated large area (> 2 cm^2) devices with greater than 20% efficiency [4].

The major difficulty in making MIMs from the InGaAsSb/GaSb system lies in the fact that the GaSb substrate is conductive. Therefore, some kind of isolation layer or layers must be grown between the substrate and the TPV device layers. To accomplish this, a wider band gap, AlGaAsSb n/p diode having reverse polarity with respect to the active layers has been designed and grown on the GaSb substrate. This cell isolation diode (CID) effectively acts to block current from leaking into the substrate and must be etched through in order to isolate adjacent cells in the MIM [5].

MIM DESIGN

Figure 1 is a schematic of the initial quaternary Sb-based MIM structure. It consists of an InGaAsSb p/n junction flanked by wider band gap AlGaAsSb or GaSb window and back surface field (BSF) layers. The window and BSF layers are meant to provide

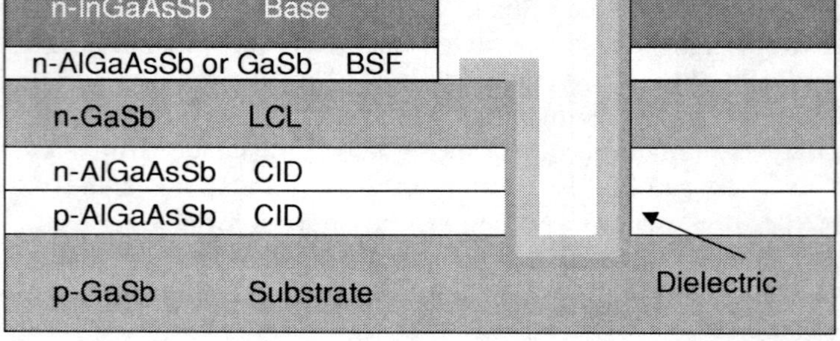

Figure 1. InGaAsSb based MIM structure containing AlGaAsSb CID grown on top of conducting GaSb substrate.

electrical confinement of electron hole pairs to the InGaAsSb active region. A highly doped, n-GaSb lateral conduction layer (LCL) is placed below the active region to facilitate current conduction between the cell itself and the metal interconnects [3]. This entire structure is grown on top of an AlGaAsSb CID, which is meant to block current from leaking into the conductive p-GaSb substrate. Because the performance of both the active region and the CID is critically important to the output of the device, evaluation of a full MIM structure is complicated. Information on internal loss in the photodiode is convoluted with current leakage across the CID. For this reason, considerable effort has been taken to develop and characterize both the photodiode and the CID separately before combining them into a MIM.

DEVICE FABRICATION

All photodiodes, CIDs, and MIMs were grown by metal organic chemical vapor deposition (MOCVD) in EMCORE D75 and D125 reactors. The precursors used were triethylgallium, ethyldimehtylamine-alane, triethylantimony, and arsine for the AlGaAsSb layers, triethylgallium, trimethylantimony, tertiarybutylarsine, trimethyl-indium for the InGaAsSb layers, and diethyltelerium, and diethylzinc or dimethylzinc for the n and p type dopants, respectively.

To fabricate MIMs (as shown in Figure 1), the top of the lateral conduction layer is etched to in order to make contact to the n-side of the device. A second etch is made through the CID in order to isolate individual cells in the MIM as well as the MIM from the remainder of the wafer. A citric acid/hydrogen peroxide/hydrofluoric acid etch (30/1/0.1 vol.%) with an etch rate of 1.8 μm/min was used for these etch steps. A pre etch surface treatment in HCl:H_2O was necessary prior to each etch in order to remove surface oxide layers that lead to surface roughening during etching. Silicon dioxide dielectric was sputter deposited in order to protect the exposed diode side walls of the MIMs (dielectric was not necessary on single junction photodiodes and CIDs). The metallization scheme for the cell interconnects consists of a multilayer stack of Pd/Ge/Au/Ti/Au. This metallization scheme has been shown to produce ohmic contacts to both n and p type GaSb with unannealed contact resistivity in the range of $8x10^{-5}$ - $1x10^{-4}$ Ω-cm^2 [6].

CID AND PHOTODIODE DEVELOPMENT

The structure used for CID development, shown in Figure 2, consists of an n-GaSb cap layer and an AlGaAsSb or AlAsSb n/p junction grown on p-GaSb. The most

n-GaSb:Te	1 μm,	$2x10^{18}$ cm^{-3},	top contact
n-AlGaAsSb	0.2 μm,	$5x10^{17}$ cm^{-3},	emitter
p-AlGaAsSb	0.9 μm,	$1x10^{17}$ cm^{-3},	base
undoped GaSb substrate			

Figure 2. CID developmental structure with layer thickness and dopant density to right of layer.

important parameters in defining CID performance in a MIM are the reverse breakdown voltage (V_{rb}) and the leakage current density (J_L). For this work, the reverse breakdown voltage has been defined as the voltage in reverse bias at which the current density equals -0.1 A/cm^2. This corresponds to the voltage at which 10% of the photogenerated current will leak into the GaSb substrate, effectively bypassing the p/n junctions in the MIM. Because a MIM voltage varies based on the number of cells connected in series, there is no definite voltage that the CID must hold off. For this reason, there is a trade off in MIM design in this materials system between increasing MIM size, and hence number of cells, and having to increase the V_{rb} requirement of the CID. To further characterize the CID, a leakage current density has been defined at typical cell operating voltage of 0.3 V (once again, reverse bias on the CID). According to in house modeling of the MIM circuit, J_L must be less than 0.1% of the photogenerated current (0.001 A/cm^2) for there to be no measurable degradation in device performance (V_{oc}, FF). By looking at the CID performance in these two different regimes of the current density-voltage curve, a more complete understanding of the material and electrical properties of various CIDs can be obtained.

Current density-voltage (J-V) curves for AlGaAsSb and AlAsSb CIDs are shown in Figure 3, with the above parameters noted. As expected, the wider band gap AlAsSb CID (1.65 eV) has a larger reverse breakdown voltage than the AlGaAsSb

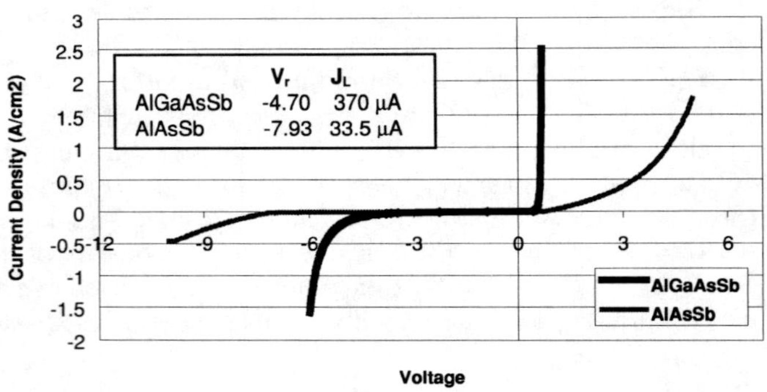

Figure 3. Current density-voltage characteristics for AlGaAsSb and AlAsSb CIDs, showing superior breakdown performance for the AlAsSb device.

(1.1 eV), due primarily to the larger built in voltage for the AlAsSb diode. The leakage current is quite low in both devices. As more CIDs are grown and compared, this leakage current measurement may provide a great deal of information concerning the effects of various material and processing induced defects on the overall leakage current of the device. While there is room for improvement, these CIDs are adequate for use in an initial quaternary MIM structure.

In addition to evaluating the performance of the CID, the InGaAsSb photodiode was evaluated prior to incorporation in a MIM. The photodiode test structure, shown in Figure 4, consists of a p-GaSb cap layer, an InGaAsSb p/n junction and an n-GaSb contact layer grown on n-GaSb. In this way, a baseline for material growth and

p-GaSb:Zn,	0.1 μm,	1×10^{18}	cm^{-3},	top contact
p-InGaAsSb	3 μm,	1×10^{17}	cm^{-3},	emitter
n-InGaAsSb	1 μm,	1×10^{18}	cm^{-3},	base
n-GaSb:Te	1 μm,	1×10^{18}	cm^{-3},	back contact
n-GaSb:Te substrate				

Figure 4. Photodiode test structure with layer thickness and dopant density to right of layer.

performance could be established, so that any degradation in electrical properties due to growth on top of the CID could be readily recognized. For all photodiode samples grown and studied the band gap was 0.60 eV. The best photodiode devices exhibited open circuit voltage (V_{oc}) in the range of 0.240 – 0.260 V with fill factors ranging from 57 – 63% at 1 A/cm^2. Using correlations developed by Hovel, the theoretical maximum V_{oc} for 0.60 eV material at this current density is 0.342 V [7]. Figure 5 shows a plot of J_{sc} vs. V_{oc} for two of the better performing photodiodes. The data

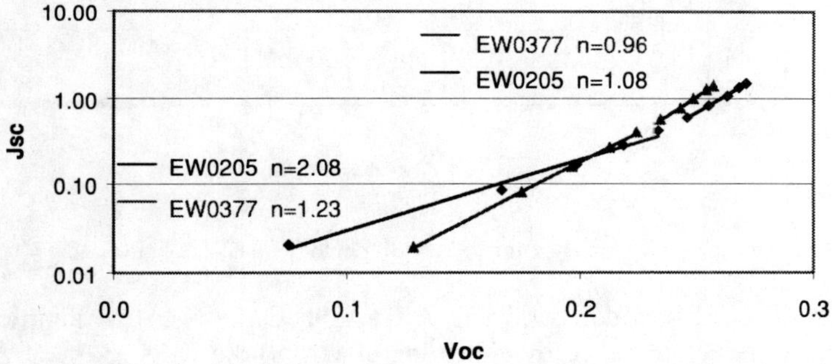

Figure 5. Flash illuminated J_{sc} vs. V_{oc} data for EW0205 and EW0377 photodiode samples showing distinct n=1 and n=2 regimes for EW0205 but only n=1 behavior for EW0377. Values of n are calculated from fitting the data to the ideal diode equation.

points were extracted from flash illuminated J-V data taken at various lamp intensities. While both devices have nearly the same V_{oc} at 1 A/cm^2, sample EW0205 exhibits distinct regions of n=1 and n=2 type behavior, while sample EW0377 shows only n=1 behavior. This suggests that there are significant generation and recombination events occurring in the space charge region for EW0205 but not for EW0377 [8]. While both samples had the same target doping levels on both sides of the junction, it appears that EW0205 may be more lightly doped, leading to a wider space charge region. Considering that the p-type emitter is unintentionally doped, unintended changes in growth conditions could have changed the background doping levels. The n=2 type behavior may significantly alter the performance of these devices and needs to be investigated further. The photodiodes as a group exhibit high external quantum efficiency (>60%) as shown in Figure 6 for EW0205. The band gap of the material is 0.600 eV, which matches the targeted band gap exactly, and the band cut off is very sharp, indicating good crystalline uniformity over the 0.00316 cm^2 area probed during the QE analysis. By taking into account the reflectance of the device, the internal QE of these devices is over 95% which compares favorably to reported values [2]. Based on this electrical and optical performance, these photodiodes should be adequate for incorporation into and evaluation of MIMs.

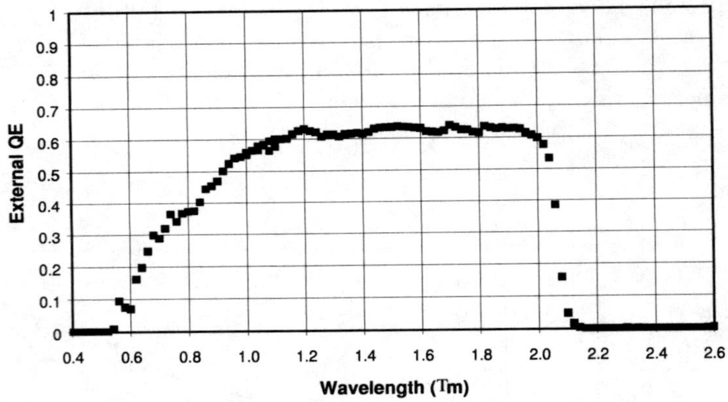

Figure 6. External quantum efficiency for InGaAsSb photodiode EW 205, band gap = 0.600 eV.

The addition of a wider band gap AlGaAsSb window to the simple photodiode may increase its performance by confining electron hole pairs to the active region of the device. However the initial attempts to include these materials in the structure resulted in very poor device quality, with V_{oc} of 0.060-0.080 V at 0.5 A/cm^2, compared to 0.220-0.230 V without the window. These samples had a rough surface morphology and SIMS analysis indicated high levels of oxygen impurities in the structure. This could lead to high surface recombination velocity that could explain the degraded device performance observed. Further development of these materials is ongoing.

MIM DEVELOPMENT

With the key components of the MIM, as well as reliable fabrication techniques in place, full MIM structures could be grown and processed. The first quaternary MIM structure consists of the InGaAsSb photodiode described above (Figure 4) grown on an AlGaAsSb CID (Figure 2) on a GaSb substrate. Due to reactor contamination issues, the CID was grown in one reactor and capped with GaSb, then transferred to another reactor in which the photodiode was grown after in-situ surface cleaning. The MIMs were processed as described earlier using a photomask set containing 1, 2, 4, 8 , 15, and 30 junction MIMs so that the affect of building voltage on MIM and CID performance could be ascertained. Figure 7 is a photograph of an InGaAsSb MIM processed with this photomask set.

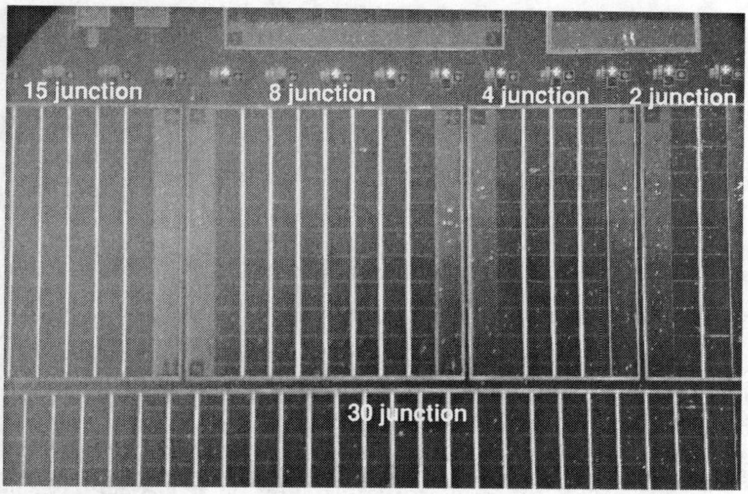

Figure 7. InGaAsSb based MIM with AlGaAsSb CID showing MIMs with various numbers of junctions.

As grown, these materials exhibited smooth mirror like surface morphology, close lattice matching to the GaSb substrate (as measured by XRD), and strong photoluminescence signal. Single junction devices with the MIM structure showed V_{oc} = 0.199-0.238 V at 0.56 A/cm^2, as good as any of the photodidodes grown without the CID. Hence, the AlGaAsSb based CID does not have any deleterious effects on subsequent growth of the photodiode. This is especially encouraging considering the degradation in device properties observed when AlGaAsSb was added to the photodiode as the window layer.

Figure 8 shows the open circuit voltage of the device as a function of the number of junctions for an InGaAsSb MIM and an equivalent MIM structure with no CID processed with the same mask set. As can be seen, the MIM without the CID builds no voltage – after the first junction, all of the current leaks into the substrate and is removed on the other bus bar of the device. The MIM with a CID, on the other hand, does show an increase in voltage with the number of cells, demonstrating that the CID concept works – voltage can build for TPV devices grown on a conductive substrate. However, the overall MIM with CID V_{oc} does not scale as a multiple of the single

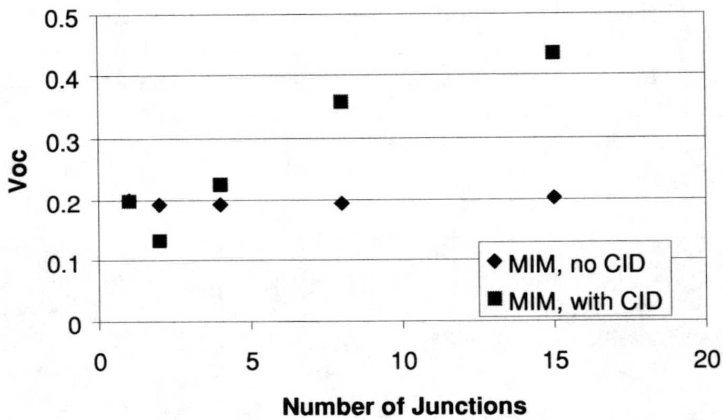

Figure 8. Open circuit voltage (V_{oc}) vs. number of junctions for InGaAsSb MIM with AlGaAsSb CID and InGaAsSb photodiode without CID showing voltage building for the MIM but not the photodiode.

junction V_{oc}. This is due to rectifying contacts between the bottom of the active diode and the cell interconnect leading to a large loss of voltage. The poor performance of these contacts can be attributed to the fact that the contacts were made to the n-InGaAsSb base, not GaSb. Unfortunately it was necessary to contact to this layer due to an error in the growth sequence that made the lateral conduction layer 0.1 μm instead of the designed 1 μm. The next generation of InGaAsSb based MIMs will have the correct thickness GaSb LCL, so the layer can be reliably etched to and ohmic contacts made at the LCL/metal contact interface.

SUMMARY

Quaternary Sb-based MIMs are an attractive TPV device due to the versatility associated with being able to grow InGaAsSb and AlGaAsSb materials with a wide range of band gaps lattice matched to commercially available GaSb substrates. To enable fabrication of these devices, a cell isolation diode was developed to block current produced in the active region from leaking into the conductive GaSb substrate. InGaAsSb photodiodes were grown, characterized, and used as a baseline for comparison to full MIM devices. It was shown that the growth of the photodiode on and AlGaAsSb CID does not lead to any degradation in the open circuit voltage of a single junction device. Voltage has been shown to build with the number of junctions in MIM devices consisting of InGaAsSb photodiodes grown on AlGaAsSb CIDs.

ACKNOWLEDGMENTS

We would like to thank Ben Wernsman, Greg Schmuck, and Becky Wehrer, all of Bechtel Bettis, for their technical support.

REFERENCES

1. Wang, C.A. et al., "Extending the Cutoff Wavelength of Lattice Matched GaInAsSb/GaSb Thermophotovoltaic Devices" in *Thermophotovoltaic Generation of Electricity, 4th NREL Conference,* edited by T.J. Coutts et al., AIP Conference Proceedings 460, New York, 1998, pp 256-265.

2. Wang, C.A. et al., "Recent progress in GaInAsSb thermophotovoltaics grown by organometallic vapor-phase epitaxy," *J. Crystal Growth* **195**, 346-355 (1998).

3. Wilt, D.M. et al., "Monolithically Interconnected InGaAs TPV Module Development," in *25th IEEE Photovoltaic Specialists Conference Proceedings,* 1996.

4. Siergiej, R.R. et al., "20% Efficiency InGaAs/InPAs Thermophotovoltaic Cells" to be published in *Thermophotovoltaic Generation of Electricity, 5th NREL Conference. (2002).*

5. Wanlass, M.W., "Electrical Isolation of Component Cells in Monolithically Interconnected Modules," US Patent #6,239,354, 2001.

6. Peake, G.M. et al., "Common n- and p- Contact for Serial Wiring of Quaternary Antimonide Monolithic Interconnected Module Thermophotovoltaic Devices," *J. Electrochem. Soc.* **149**, G416-G419 (2002).

7. Hovel, H.J., *Semiconductors and Semimetals – Volume 11, Solar Cells,* Academic, New York, 1975, pp. 49-53.

8. Streetman, B.G. and Banerjee, S., *Solid State Electronic Devices,* Prentice Hall, Uppre Saddle River, New Jersey, 2000, pp 214-216.

High Performance InGaAsSb TPV Cells via Multi-Wafer OMVPE Growth

Zane A. Shellenbarger, Gordon C. Taylor, Ronald K. Smeltzer, Jizhong Li, Ramon U. Martinelli, and Kalipada Palit

Sarnoff Corporation
Princeton, New Jersey 08543

Abstract. The fabrication and performance of InGaAsSb thermophotovoltaic cells are described. The InGaAsSb layers, grown by organometallic vapor-phase epitaxy in a multi-wafer reactor, with a 0.53 eV bandgap are lattice-matched to a GaSb substrate. Growth series with up to thirty 50 mm wafers have been done with good control of material composition and carrier transport properties. With improved materials and metallization and with a modification to the cell edges, fill factors near 70% and a greater than 60% peak external quantum efficiency are obtained. A two order-of-magnitude increase in shunt resistance with a consequent 15% improvement in fill factor was achieved with the improved edge structure. Series resistance, about 20 mΩ, is the remaining limitation to cell performance and is closely correlated with fill factor.

INTRODUCTION

Continued development of thermophotovoltaic (TPV) systems to exploit available thermal sources offers the potential for high conversion efficiency systems in a variety of applications. To achieve high efficiencies with many available sources, devices in low bandgap semiconductors are required. One low bandgap material system, the quaternary InGaAsSb, has the advantage of being lattice matched to available GaSb wafers, thus offering the possibility of a low concentration of crystalline defects with consequently good carrier lifetimes [1, 2]. In this paper, we report the recent advances in preparation of the quaternary material, the improvements in cell fabrication methods, and the cell performance characteristics achieved.

MATERIAL DEVELOPMENT

The active material, InGaAsSb, of the p-on-n cells is prepared by organometallic vapor-phase epitaxy (OMVPE) on a GaSb substrate. The target composition is $In_{0.16}Ga_{0.84}As_{0.14}Sb_{0.86}$ with a 0.53 eV bandgap. A low-pressure, vertical rotating-disk reactor with a capacity of three two-inch wafers is used. The system has a vertical gas flow, high-speed rotation, and a turbo-pumped low-pressure load-lock. Precursors used are triethylgallium (TEG), trimethylindium (TMI), trimethylantimony (TMSb), and 1000-ppm arsine (AsH_3) in hydrogen. Doping is done with 1000-ppm dimethylzinc (DMZ) in hydrogen for p-type material and 100 ppm hydrogen selenide

CP653, *Thermophotovoltaic Generation of Electricity: 5th Conference*
edited by T. J. Coutts, G. Guazzoni, and J. Luther

(H_2Se) in hydrogen for n-type material. Typical growth conditions are 60 torr pressure, 100 rpm wafer rotation, and 19 slpm H_2 carrier gas flow. Achievement of high material quality for this low bandgap composition depends on temperature and substrate orientation. The typical growth temperature is 500°C and typically a (100) substrate misoriented 6° towards the (111)B direction is used. The structure of the cells is schematically illustrated in Figure 1.

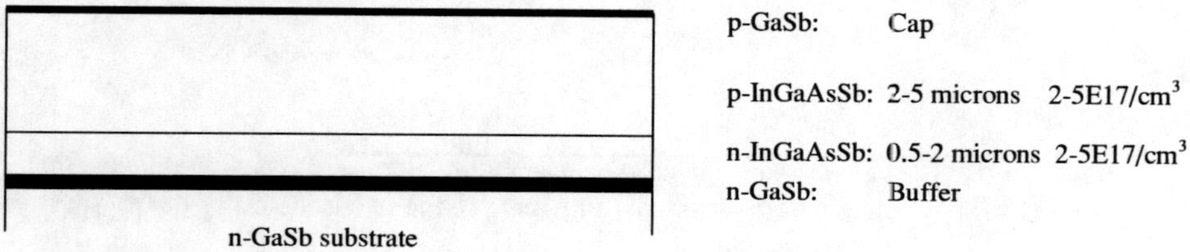

FIGURE 1. Schematic device layer structure.

Material uniformity is essential for a good yield of high efficiency cells. Properties of the layers are assessed with double-crystal X-ray diffraction, photoluminescence (PL), and Hall measurements. Photoluminescence, which is the key measure of uniformity, is recorded at the center of the wafer and at four points 5 mm from the edge of the wafer. Typically the variation of the PL peak had been ± 50 nm, but is now typically ± 5 nm. For three wafers from a single growth run, variation in the peak PL from the center of the wafers is typically within ± 3 nm. Variation in the peak PL from the center of the wafers from run to run is typically within ± 6 nm.

The improved uniformity in PL was achieved with a low, 100 rpm, susceptor rotation speed and with TEG instead of trimethylgallium (TMG) as the gallium source. Figure 2 shows the effect of rotation speed on peak PL across a wafer. It is speculated that the higher rotation speeds disturb the gas source profile, especially near the outer edge of the susceptor. With the lower rotation speed, improved surface morphology is obtained and the surface morphology is less sensitive to the V/III ratio. The use of TEG ensures more complete pyrolization at the growth temperature, thus reducing nonuniformity effects due to temperature variations across the susceptor. A third factor that may be important is adoption of a consistent surface preparation procedure to remove native oxide on the substrate prior to loading into the growth system.

Achievement of the target composition, with the 0.53 eV bandgap, increased the cutoff wavelength, defined at one-half of the peak value of the spectral quantum efficiency (SQE) characteristic, from an initial value of about 2125 nm to 2300 nm and reduced the variation among wafers from ± 45 nm to ± 5 nm. Thus, more of the input radiant energy can be utilized and cell-to-cell variability in performance is reduced.

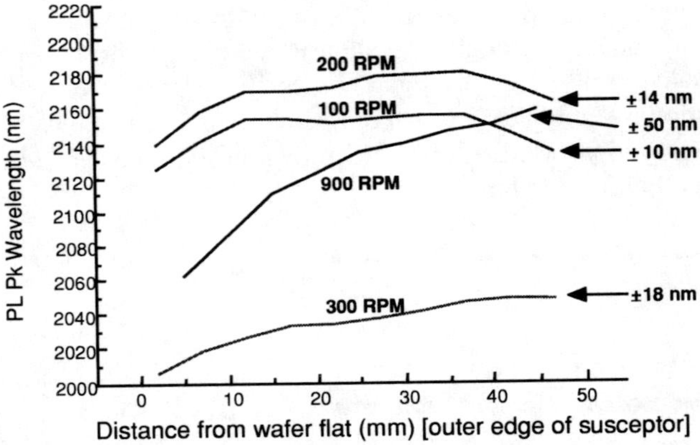

FIGURE 2. Peak PL across the wafer for various rotation speeds.

Data indicate almost identical properties for the three wafers from a growth run, not surprising since the wafers are symmetrically located. To demonstrate reproducibility, a growth series of 30 wafers was done. It was found important to keep the V/III ratio slightly higher than its lowest value, 3.4, to minimize run-to-run variations and formation of metal droplets. With the V/III ratio slightly higher, 3.6, a very small area at the edge of wafers was adversely affected, but a much larger area of good morphology was consistently obtained. Among the 30 wafers, the average peak PL at the center of the wafer was 2351 nm with a standard deviation of 9 nm and the average lattice mismatch was 0.074% with a standard deviation of 0.03%. These observations suggest that material can be prepared in reasonable volume for the fabrication of TPV cells.

Following fabrication of cells from the wafers, a standard set of measurements is taken from 5 mm x 10 mm cells, as many as 24, to assess uniformity and yield. The data indicate that good uniformity and yield from multiple wafers are achievable. Standard deviations less than about 3% for peak SQE and open circuit voltage are typical. Fill factor, which can be affected by cell processing, typically has a slightly larger variation. In a later section, fundamental cell characteristics related to the process improvements, which are discussed in the next section, are described in detail.

PROCESS OVERVIEW

Fabrication of the p-on-n cells begins with steps to remove contaminants and surface oxide. The p-contact, Ti/Au/Pt/Au (10nm/100nm/100nm/50nm), for the current collection grid is prepared by a lift-off process. Contact resistance and adhesion are improved by a 2 min sinter at 300 °C. The grid conductivity is increased with 5 μm thick plated gold, formed with a two layer photoresist process. The first resist layer is patterned with lines slightly narrower than the grid, and a seed layer of Ti/Au/Ti (150nm/100nm/60nm) is deposited. A second photoresist layer is patterned to etch off the top titanium and expose the gold layer in the grid area. After plating and resist removal, the titanium and gold layers are removed. Finally, the photoresist that protected the semiconductor during plating and etching is removed with acetone and a plasma clean-up.

Contact to the n-GaSb substrate is made with Ni/Au/Ge/Ni/Au (5nm/45nm/20nm/15nm /200nm). A 60 sec sinter at 250 °C is done. A diffusion barrier of Ti/Pt/Au (50nm/100nm/ 200nm) with a gold overcoat is added to facilitate solder mounting of the cells for testing.

The process is completed by dicing the wafer with a protective layer of photoresist on the surface. The mechanical damage from sawing is removed by etching the diced wafer in a 10:1 mixture of 50 wt % citric acid and hydrogen peroxide. Removal of approximately 1 μm of material from the device edges is adequate to obtain a high shunt resistance, as described later.

The specific contact resistances were evaluated with GaSb wafers metallized with a Cox and Strack [3] pattern. Contact resistances for the sintered Ti/Au/Pt/Au p-type contact are in the low 10^{-7} Ωcm^2 range. For the n-type contact, $2.5x10^{-5}$ Ωcm^2 was measured. A metal system for p-type contacts composed of Ti/Pt/Ag/Pt/Au (30nm/100nm/ 5000nm/100nm/200nm) and patterned by lift-off yielded $6x10^{-6}$ Ωcm^2 without sintering. Higher conductivity and lower cost with respect to gold are advantages of the silver-based p contact.

Two sizes of cells, 4 mm x 4 mm and 5 mm x 10 mm, are produced with one photomask. Although the layouts for the current collection grids are slightly different, basically the grids of the two cells consist of 10 μm wide lines on a 100 μm spacing with one bus bar.

CELL PERFORMANCE

A variety of measurements are done to assess material and process variants. Dark current-voltage characteristics are used to extract the zero-bias slope as the shunt resistance. As noted later, this definition of shunt resistance yields a good fit to a model calculation. Dark forward-bias, current-voltage characteristics are employed to extract the diode ideality factor, n, and the dark current density, Jo. Short circuit current versus open circuit voltage plots for different light intensities also yield the diode ideality factor and the dark current density. Output characteristics with illumination to a short circuit current density 2 A/cm^2 are used to determine fill factor and to monitor open circuit voltage. For series resistance, output characteristics are taken at different light intensities and the method described by Handy [4, 5] is used. The merit of the Handy method is that series resistance is determined from operating characteristics with no need to assume a value for any device parameter, to assume an ideal characteristic, or to have a high shunt resistance. The Handy method yields physically reasonable values of series resistance that correlate well with fill factor. Finally, SQE of the cells is determined from an absolute measurement at 1.55 μm and from the relative SQE in the range from 1.0 to 2.6 μm.

For the measurements discussed here, the cells were packaged by soldering them onto a brass pallet with a lead attached along the main bus bar to minimize external series resistance. The pallet serves as a heat sink and is attached to a water-cooled copper block for measurements. Open circuit voltage measurements indicate that there is no appreciable temperature rise during the measurements. In the remainder of this section, the important test results are described.

Low shunt resistance had been a substantial limitation to cell performance until addition of the edge etch to the process flow. To illustrate, Figure 3 shows diode characteristics before and after etching the edges of a cell. As measured by the zero-bias slope, edge etching increased shunt resistance from 13 to 860 Ω, with a resulting fill factor of 70% obtained from the operating characteristic. The fill factor prior to etching the edges of this cell was not measured, but fill factors near 60% were typical prior to addition of the edge etch to the process.

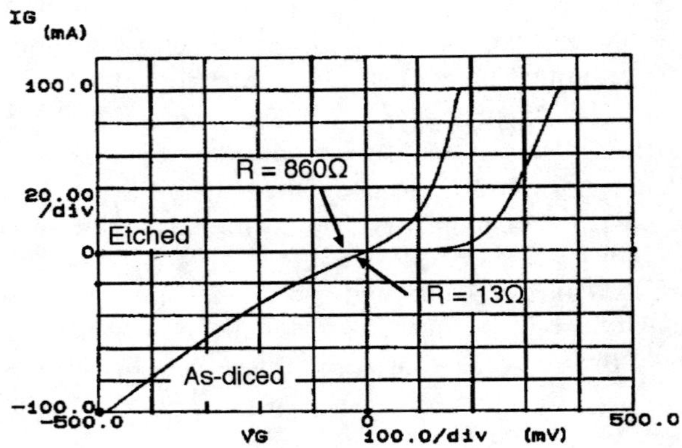

FIGURE 3. Pre- and post-edge-etch diode characteristics.

To illustrate more clearly the shunt resistance measurement and its effect on device characteristics, cells were characterized pre- and post-edge-etch. For the same cell, Figure 4 illustrates dark forward current-voltage characteristics measured pre- and post-edge-etch and calculated from the post-etch characteristic shunted by the initial measured shunt resistance. The effect of the edge etch on this cell was to increase the shunt resistance from 10.4 Ω to 350 Ω. The calculation of the shunted device characteristic agrees very closely with the pre-etch measurement, thus validating the zero-bias slope as a useful method to determine shunt resistance. The ideality factor of the diode after etching is close to one over a wide range of voltages, as indicated by the line for n=1 in Figure 4. The as-diced characteristic has very large excess currents over much of the voltage range, indicating that conduction through the shunt resistance path dominated the device behavior before the edge was etched. The fill factor for this particular cell increased from 61 to 68% after the edges were etched. Elimination of the large excess shunt current is also evident in the plot of short circuit current versus open circuit voltage, Figure 5, which yields a diode ideality factor near one and a dark current density of 3.6E-5 A/cm^2.

Under very low illumination conditions shunt resistance can be approximated [6] as the quotient of open circuit voltage to short circuit current. One comparison yielded a shunt resistance value from the low-illumination approximation within a few percent of that from the slope measurement, which however is a much more convenient procedure.

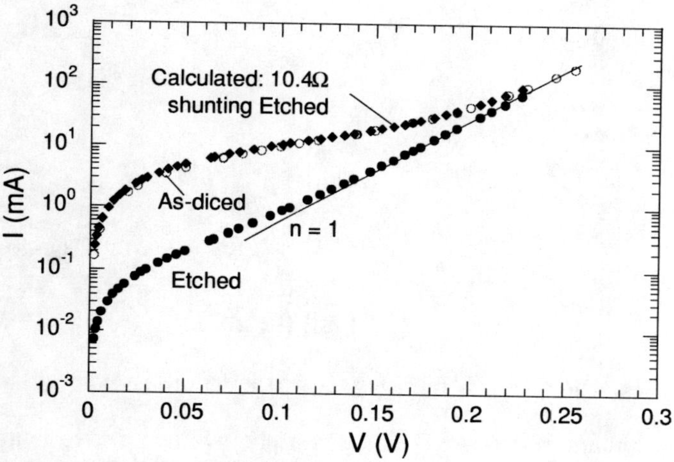

FIGURE 4. Dark forward characteristics: as-diced (open circles), after edge etching (filled circles), and calculated from etched characteristic shunted by the measured shunt resistance (filled diamonds) to illustrate the validity of the shunt resistance model.

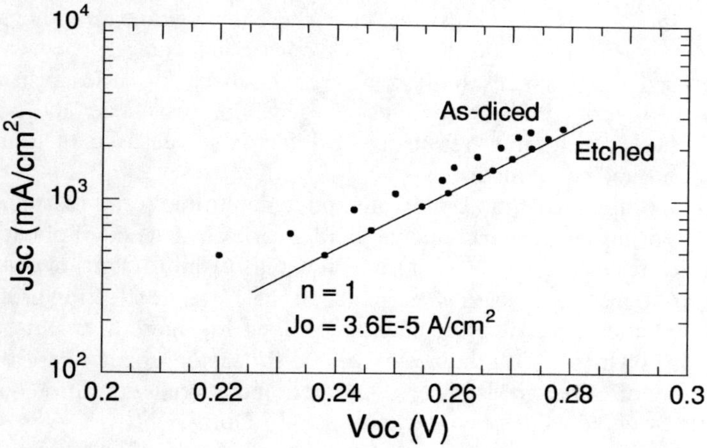

FIGURE 5. Short circuit current density versus open circuit voltage before and after edge etch.

To quantify the effect of shunt resistance and to determine the range of shunt resistance values for which the fill factor is independent of shunt resistance, the basic diode equation was solved, Figure 6, without series resistance for typical values of dark current. Converting from Ωcm^2 for the two cell sizes, one finds about 190 Ω for the small cells and about 60 Ω for the large cells yield fill factors independent of shunt resistance. Most cells now meet the respective criterion for shunt resistance. Presumably, series resistance is the remaining limitation to fill factor, now typically close to 70%, which is about three units from the theoretical limit shown in Figure 6.

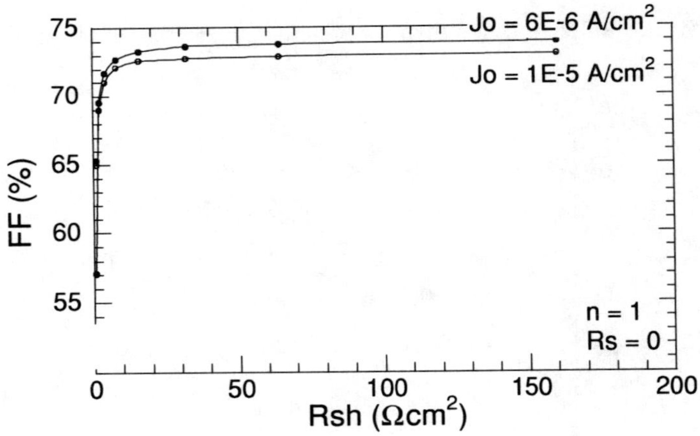

FIGURE 6. Calculated fill factors for ideal diodes with no series resistance.

The measurement of series resistance in a photovoltaic device is not trivial. The method of Handy was adopted because it uses only measured cell characteristics. In the Handy method, output characteristics are taken at various illumination levels, and an increment of current from the short circuit value is marked-off on each characteristic. Were there no series resistance, a line drawn through the points would be parallel to the current axis; in the real case, there is a voltage drop due to the series resistance, and the slope of the line is the series conductance. Because packaged devices are used for the measurement, a large database for series resistance is not available. However, with a sufficiently large shunt resistance and a correlation between fill factor and series resistance, fill factor serves as a measure of series resistance for a fixed cell structure.

An example of the use of the Handy method to determine series resistance is shown in Figure 7. Various current increments yield a series resistance of about 18 mΩ. A value for series resistance on the order of a few tens of milli-Ohms is consistent with an estimate of 10 mΩ for the series resistance of the current collection grid. Measured fill factor and series resistance for cells are plotted in Figure 8 to illustrate that the relationship is consistent with the best expected fill factor, about 73% from Figure 6, in the absence of series resistance. Another operational definition [7] of series resistance also yielded values of a few tens of milli-Ohms.

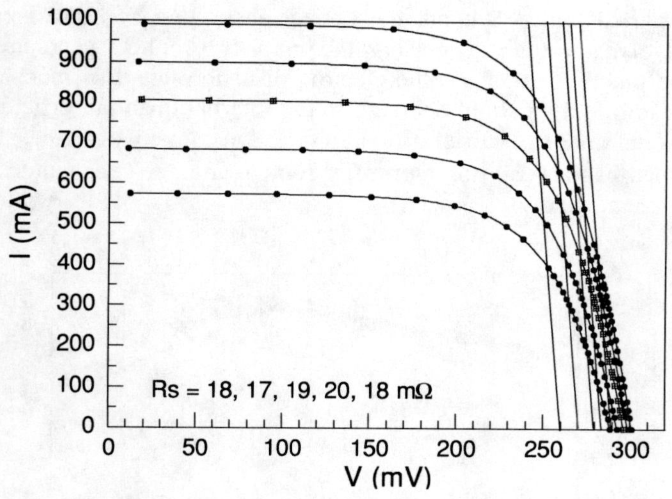

FIGURE 7. Series resistance determination from a family of operating characteristics for a large cell.

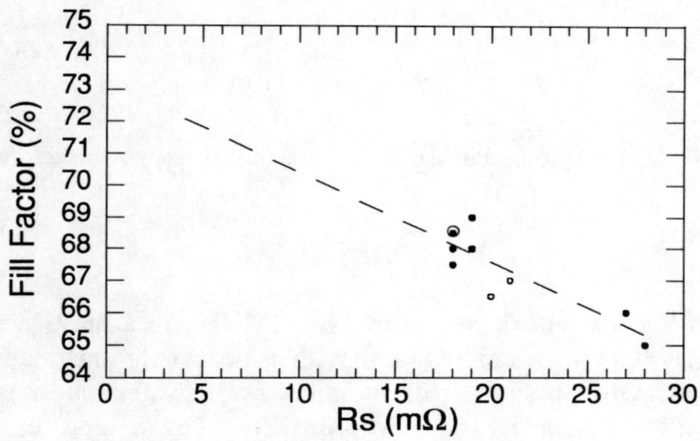

FIGURE 8. Fill factor vs. series resistance from Handy method measurements.

The fundamental optical performance of the cells is assessed by the measurement of external SQE. Figure 9 shows the measured SQE over the wavelength interval from 1 to 2.6 µm of four cells from two different wafers. The solid and the open data points identify data associated with each wafer. The data were taken at very low illumination levels, under short-circuit-current conditions of less than 1 µA.

The four cells have similar spectral responses. The long-wavelength cut-off is about 2.4 µm, indicating good control and uniformity of the composition both within a wafer and wafer-to-wafer. The SQE curves show peak values of 61-65% at 2 µm. Since the reflectivity of the cells is about 33% at 2 µm, the internal quantum efficiency is about 94%. The decrease in the SQE curves to 50% at 1 µm has two causes. First and most importantly, an increase in the reflectivity to 37% at 1 µm decreases the fraction of absorbed light. Second, the optical absorption constant increases with decreasing wavelength, thus decreasing the number of photo-generated electrons collected at the junction. Because the electron diffusion length is much longer than the p-emitter width, only a slight decrease in the internal quantum efficiency results. A one-dimensional minority-carrier diffusion model accurately describes these effects [8]. With a broad-band, 1-2.4 μm, anti-reflection coating, the cells would be efficient converters of radiant energy.

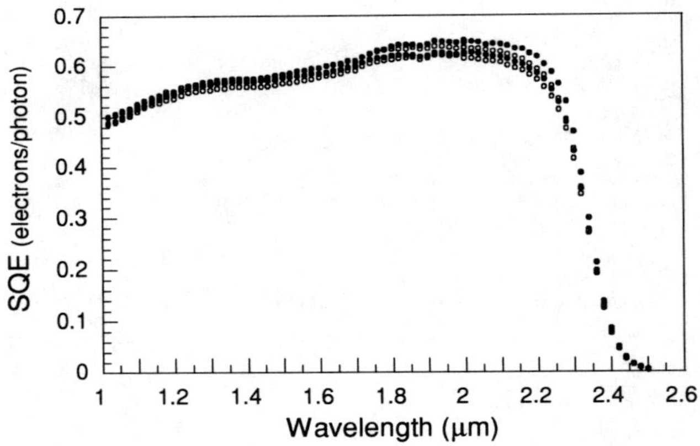

FIGURE 9. Typical external spectral quantum efficiency characteristics.

SUMMARY

InGaAsSb TPV cells with layers grown by OMVPE in a multi-wafer reactor are produced with fill factors near 70% and with a peak external quantum efficiency greater than 60%. Growth series with as many as 30 wafers show good control of material composition and carrier transport properties. Key to good cell performance is the two order-of-magnitude increase in shunt resistance achieved with an improved process. Dark forward diode characteristics without the process improvement are modeled closely as the improved diode characteristic shunted by a measured value for initial shunt resistance. Series resistance, typically about 20 mΩ, is the remaining limitation to cell performance for the current device architecture and is closely correlated with fill factor. Sarnoff TPV diode technology when combined with module technology by Sarnoff [9] and spectral control technologies provided by other organizations has led to TPV conversion efficiencies in excess of 17% [10].

REFERENCES

1. C. A. Wang, H. K. Choi, G. W. Turner, D. L. Spears, and M. J. Manfra, "Lattice-Matched Epitaxial GaInAsSb/GaSb Thermophotovoltaic Devices," in Thermophotovoltaic Generation of Electricity, edited by T. J. Coutts et al., *AIP Conference Proceedings* **401**, 75-87 (1997).

2. G. W. Charache, P. F. Baldasaro, L. R. Danielson, D. M. DePoy, M. J. Freeman, C. A. Wang, H. K. Choi, D. Z. Garbuzov, R. U. Martinelli, V. Khalfin, S. Saroop, J. M. Borrego, and R. J. Gutmann, "InGaAsSb thermophotovoltaic diode: Physics evaluation," *J. Appl. Phys.*, **85**, 2247-2252 (1999).
3. R. H. Cox and H. Strack, "Ohmic Contacts for GaAs Devices," *Solid-State Electronics*, **10**, 1213-1218 (1967).
4. R. J. Handy, "Theoretical Analysis of the Series Resistance of a Solar Cell," *Solid-State Electronics*, **10**, 765-775 (1967).
5. D. K. Schroder, *Semiconductor Material and Device Characterization*, second edition, John Wiley & Sons, New York, 1998, p. 210-211.
6. D. S. H. Chan and J. C. H. Phang, "A Method for the Direct Measurement of Solar Cell Shunt Resistance," *IEEE Trans. Electron Devices*, **ED-31**, 381-383 (1984).
7. K. Rajkanan and J. Shewchun, "A Better Approach to the Evaluation of the Series Resistance of Solar Cells," *Solid-State Electronics*, **22**, 193-197 (1979).
8. R. U. Martinelli, D. Z. Garbuzov, H. Lee, N. Morris, T. Odubanjo, G. C. Taylor, and J. C. Connolly, "Minority-Carrier Transport in InGaAsSb Thermophotovoltaic Diodes," in Thermophotovoltaic Generation of Electricity, edited by T. J. Coutts et al., *AIP Conference Proceedings* **401**, 389-395 (1997).
9. Y. Z. Yu, R. U. Martinelli, G. C. Taylor, Z. Shellenbarger, R. K. Smeltzer, K. Palit, and D. Channin, "High-Efficiency Multi-Cell TPV Module Fabrication and Performance," Fifth Conference on Thermophotovoltaic Generation of Electricity, October 2001.
10. P. F. Baldasaro, "Summary Analysis of Thermophotovoltaic Efficiency and Power Density Issues," Fifth Conference on Thermophotovoltaic Generation of Electricity, October 2001.

Lattice-Matched GaInAsSb/AlGaAsSb/GaSb Materials for Thermophotovoltaic Devices

C.A. Wang, C.J. Vineis*, H.K. Choi#, M.K. Connors, R.K. Huang

Lincoln Laboratory, Massachusetts Institute of Technology, Lexington, MA 02420-9108
**now at AmberWave Systems Corporation, Salem, NH 03079*
#now at Kopin Corporation, Taunton, MA 02780

L.R. Danielson, G. Nichols, G.W. Charache+

Lockheed Martin, Inc., Schenectady, NY 1230
+now at Princeton Lightwave, Cranberry, NJ 08512

D. Donetsky, S. Anikeev, G. Belenky

State University of New York, Stony Brook, NY 11794-2350

Abstract. High-performance GaInAsSb/AlGaAsSb/GaSb thermophotovoltaic (TPV) devices with quantum efficiency and fill factor near theoretical limits and open-circuit voltage within about 15% of the limit are reported. This paper discusses detailed studies of GaInAsSb epitaxial growth, the microstructure, and minority carrier lifetime that have led to these results. For further improvements in TPV cell performance, device structures with either a distributed Bragg reflector or a back-surface reflector are described.

INTRODUCTION

Recently, a thermodynamic analysis of thermophotovoltaic (TPV) system performance reported the importance of below-bandgap photon recuperation in determining tradeoffs between system efficiency and power density [1]. It was shown that the efficiency depends not only on the bandgap E_g of TPV cells but also on the reflectivity of the below-bandgap filter. The E_g where efficiency is a maximum decreases as filter reflectivity decreases. Thus, for TPV systems with radiator temperatures of approximately 1000 °C and a 95% reflective filter, a maximum efficiency was predicted for TPV devices with E_g of 0.5 eV [1].

GaInAsSb quaternary alloys have E_g that can be tailored from about 0.3 to 0.7 eV when lattice matched to GaSb substrates. Consequently, materials and device development for GaInAsSb TPV cells has been ongoing for several years [2-13], and devices have been demonstrated using structures prepared by all the major

CP653, *Thermophotovoltaic Generation of Electricity: 5th Conference*
edited by T. J. Coutts, G. Guazzoni, and J. Luther

epitaxial technologies, including liquid phase epitaxy [4-6], molecular beam epitaxy [7-8], and organometallic vapor phase epitaxy (OMVPE) [8-11]. This paper reports the current status of OMVPE-grown GaInAsSb TPV cells and the detailed studies that were performed to optimize GaInAsSb epitaxial layers for high-performance TPV cells. In addition, enhancements in performance by incorporating either a distributed Bragg reflector (DBR) [14] or back-surface reflector (BSR) [15] are discussed.

STATUS OF TPV CELL PERFORMANCE

A performance factor used for TPV device evaluation [1] is the diode efficiency η_{diode}, which is given by $\eta_{diode} = QE*F_0*(qV_{oc}/E_g)*FF$, where QE is the average internal quantum efficiency, F_0 is the usable fraction of thermal radiation with energy greater than the diode bandgap, qV_{oc}/E_g is the voltage factor based on electronic charge q, open-circuit voltage V_{oc}, and E_g of the diode material, and FF is the fill factor. Considering parameters that pertain specifically to diode performance, it was reported that theoretical limits for QE, qV_{oc}/E_g, and FF are 1, 0.7, and 0.7, respectively [2].

Figure 1 schematically shows a typical GaInAsSb/AlGaAsSb/GaSb TPV device structure that is being grown by OMVPE [8-11]. The GaInAsSb alloy composition corresponds to E_g of 0.52 to 0.55 eV. The structure consists of a thicker p-type emitter layer compared to the n-type base layer and an AlGaAsSb or GaSb window layer, and p-GaSb contact layer. This structure is used because shallow n-type contacts are difficult to prepare on GaSb-based materials and because of the larger minority carrier diffusion length in the p-type quaternary alloy compared to that in the n-type alloy. Surface recombination is detrimental to device performance [8], and both AlGaAsSb and GaSb window layers were shown to be effective in reducing interfacial recombination velocity [11].

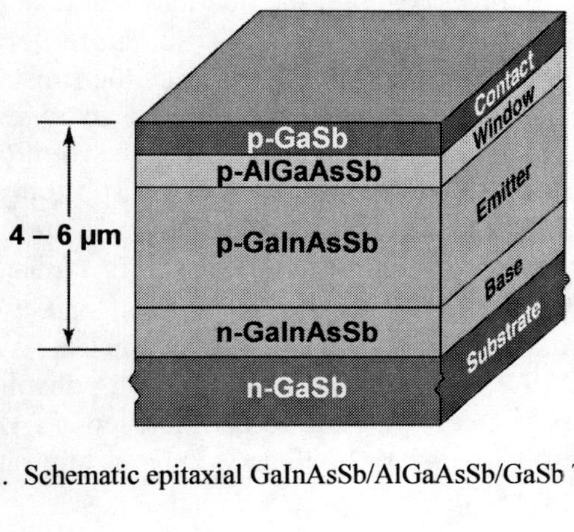

FIGURE 1. Schematic epitaxial GaInAsSb/AlGaAsSb/GaSb TPV device structure.

The spectral response for an uncoated TPV cell is shown in Figure 2. The data takes into account the measured surface reflectivity. Response is out to 2.5 μm, and QE is approaching the theoretical limit. Cell data of the peak value of EQE (uncorrected for surface reflection), V_{oc}, and FF for cells fabricated over several years are presented in Figures 3a, 3b, and 3c, respectively. The data are taken from uncoated cells in which the emitter layer thickness, emitter and base layer doping levels, window layers, and device fabrication techniques had been intentionally varied in an effort to study parameters influencing diode performance. The peak EQE and FF values currently approach theoretical limits, while the maximum voltage factor is 0.6.

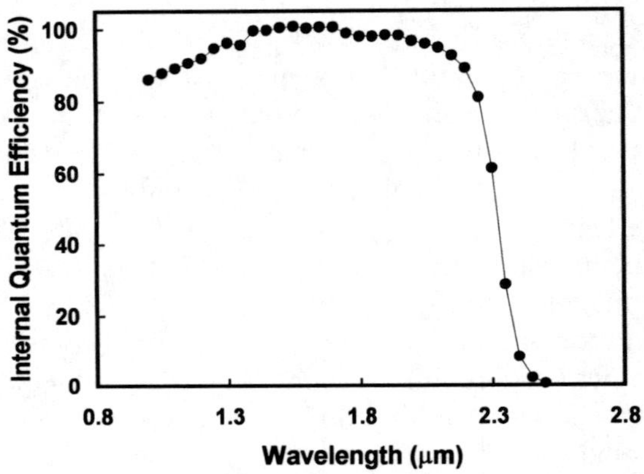

FIGURE 2. Spectral response of GaInAsSb/AlGaAsSb/GaSb TPV cell.

MATERIALS STUDIES

Factors that have a major impact on the quality of epitaxial GaInAsSb alloys, and thus TPV cell performance, include GaSb substrate quality, substrate preparation, and OMVPE growth parameters. Since the substrate is the template for epitaxial growth, its quality and preparation are extremely critical. GaSb substrates are less developed compared to the more conventional GaAs and InP III-V alloys, and the availability of high-quality GaSb substrates has been limited. GaSb is also a softer material, and thus, GaSb substrates are more susceptible to damage during substrate processing. Wafer bow is typically 10 μm and is considerably larger compared to only a few microns for GaAs. This bow can impact heat transfer during epitaxial growth and affect epilayer uniformity. In addition, a residue of SiO_2, which is used for substrate polishing, was sometimes observed on the GaSb surface by atomic force microscopy (AFM). This surface contamination will result in epitaxial defects. Thus, routine inspection of GaSb substrates quality is important.

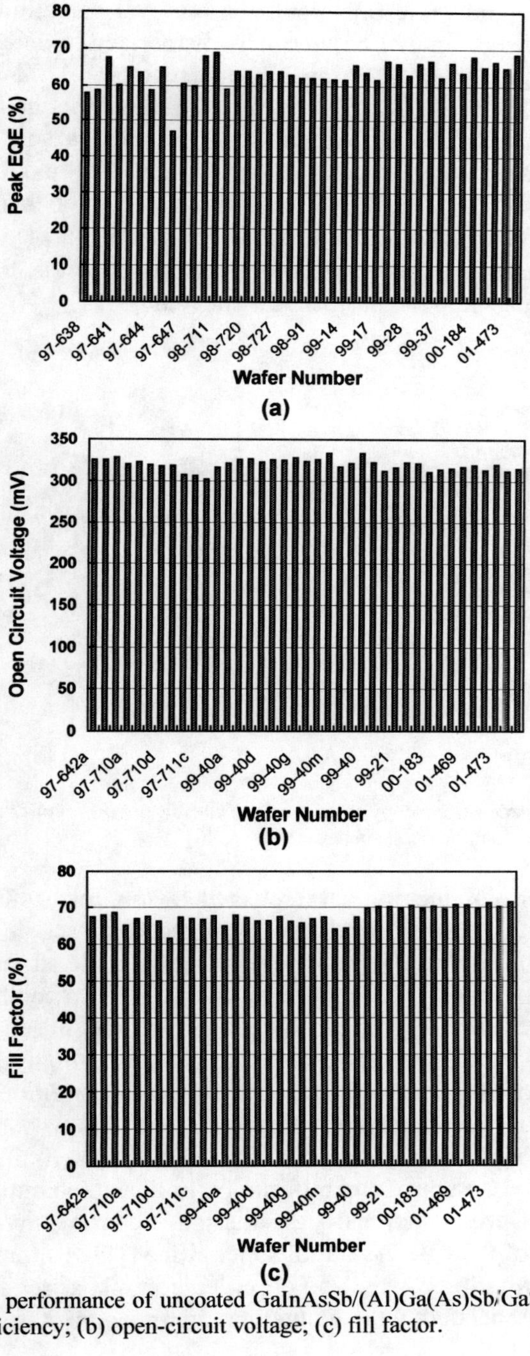

FIGURE 3. Device performance of uncoated GaInAsSb/(Al)Ga(As)Sb/GaSb TPV cells: (a) peak external quantum efficiency; (b) open-circuit voltage; (c) fill factor.

A critical step in substrate preparation is removal of gallium and antimony native oxides. Oxides can either be thermally evaporated in the growth reactor or chemically etched before introduction into the reactor. AFM images of GaSb grown either on an epi-ready substrate, where the oxide thermally evaporated, or on a chemically etched substrate are shown in Figures 4a and 4b, respectively. The morphology is rougher for the sample where the oxide was thermally evaporated. This is due to a thermally activated reaction between GaSb and antimony oxide, which consumes and thus roughens the surface [15]. When the oxide is chemically removed prior to loading into the reactor, there is negligible oxide and therefore, a smoother surface is achieved.

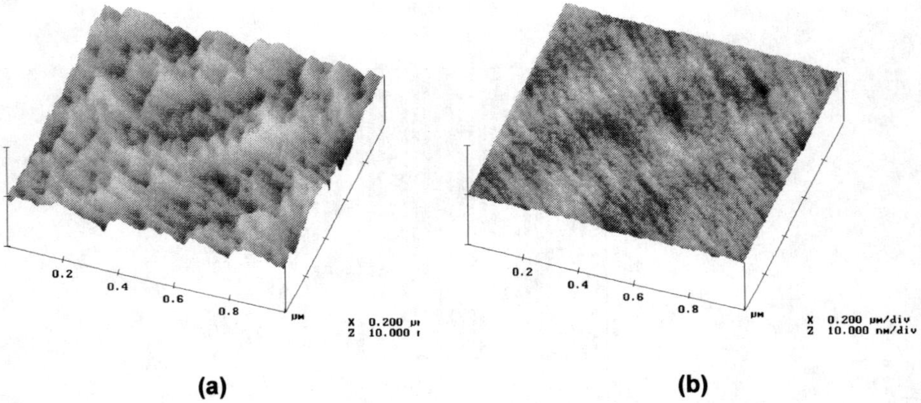

FIGURE 4. Atomic force microscopy images of (a) GaSb grown on epi-ready GaSb substrate; (b) GaSb grown on chemically etched substrate.

OMVPE growth parameters were selected to minimize phase separation in GaInAsSb. Phase separation results in localized microscopic regions that are GaAs- and InSb-rich, and this inhomogeneity can degrade structural and electro-optical quality [11], and thus reduce minority-carrier recombination lifetime. Previously reported growth studies indicated that phase separation could be effectively limited by using low growth temperatures, high growth rates, and specific substrate orientations [11,17]. Such growth conditions reduce adatom surface diffusion during epitaxial growth.

Phase separation, however, cannot be completely eliminated because of thermodynamics, and an unusual manifestation of phase separation was observed. Figure 5a shows a cross-sectional transmission electron microscopy image of GaInAsSb with E_g of 0.5 eV. A natural superlattice (NSL) spontaneously formed at the onset of GaInAsSb growth. The observed contrast arises from a very small degree of strain associated with a modulation in composition in the vertical direction. Figure 5b shows PL measurements at 4 and 300 K. Composition modulation is expected to increase the PL full width at half-maximum (FWHM) because of either layer thickness variations in the NSL or local variations in alloy composition. However, the PL data indicate FWHM of 9.5 meV, which is narrow for this alloy composition. The dependence of PL peak energy on temperature is similar to that observed for type-II band alignment [18]. Type-II alignment can enhance minority-carrier lifetime since electrons and holes are spatially separated [19]. However, if the alloy modulation results in material with E_g that is lower than the average alloy, the dark current would be higher.

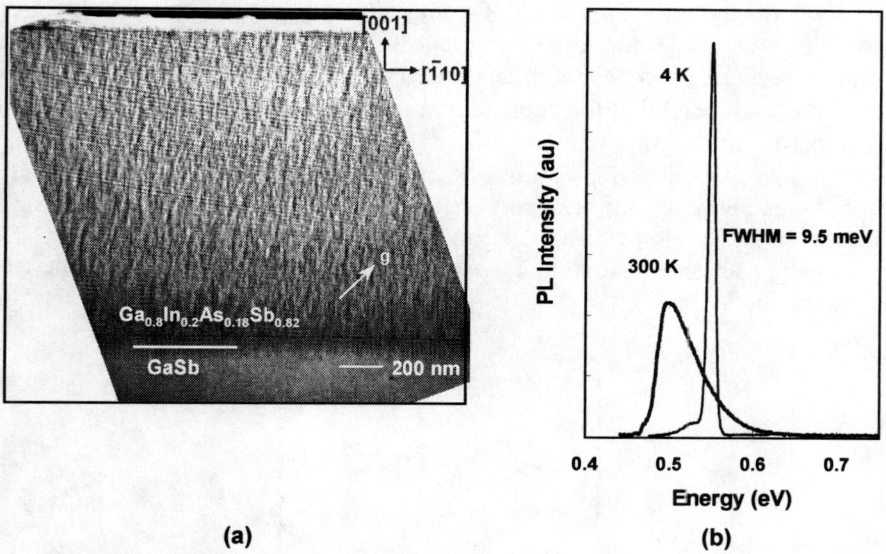

FIGURE 5. Natural superlattice in GaInAsSb: (a) cross-sectional transmission electron micrograph and (b) photoluminescence at 4 and 300 K.

Previously, it was suggested that the dominant bulk recombination mechanisms in GaInAsSb are radiative and Auger recombination and not Shockley-Read-Hall recombination [3,13]. The electron lifetime of p-GaInAsSb was measured by time-resolved PL [20]. Double-heterostructures with GaInAsSb doped at $2x10^{17}$ cm^{-3} were grown with different GaInAsSb thicknesses in order to separate the contributions of bulk and interfacial recombination processes according to the equation $1/\tau_{PL} = 1/\tau_{BLK} + 2S/W$, where τ_{PL} is the PL decay, τ_{BLK} is a bulk lifetime, S is the surface recombination velocity which is assumed to be equal at the front and back heterointerfaces, and W is the active layer thickness. In this approximation, S is assumed to be relatively small compared to the ratio of minority carrier diffusion constant D to W (S<D/W), and photon recycling is negligible.

Three sets of samples were grown, each with different capping layers: nominally undoped p-GaSb ($1x10^{16}$ cm^{-3}), p-GaSb ($2x10^{18}$ cm^{-3}), and p-AlGaAsSb ($2x10^{17}$ cm^{-3}). The results are shown in Figure 6. All three plots demonstrate a linear dependence consistent with the equation. The structures with undoped p-GaSb cap layers resulted in the highest S of 3100 cm/s. In structures with p-GaSb cap layers doped to $2x10^{18}$ cm^{-3}, S was substantially smaller at 1140 cm/s. Thus, the use of a heavily doped cap layer makes it possible to suppress interfacial recombination. S was further reduced to 720 cm/s for structures with AlGaAsSb cap layers. Higher S values are due to accumulation of electrons at the GaInAsSb/GaSb type-II interface. Doping the cap layers reduces electron accumulation, while the AlGaAsSb cap layers results in no valence band offset [8]. Photon recycling can be significant and it can contribute to the slope and the offset. It was estimated that the contribution to S is about 200 cm/s for structures doped at $2x10^{17}$ cm^{-3}.

For nominally undoped p-GaInAsSb, photon recycling is negligible. The smallest S was 380 cm/s for undoped GaInAsSb ($\sim 1x10^{16}$ cm^{-3}) with AlGaAsSb cap layers. In another set of samples, undoped GaInAsSb with undoped p-GaSb capping layers was also evaluated. Here a lower estimate of the bulk lifetime was determined to be 900 ns.

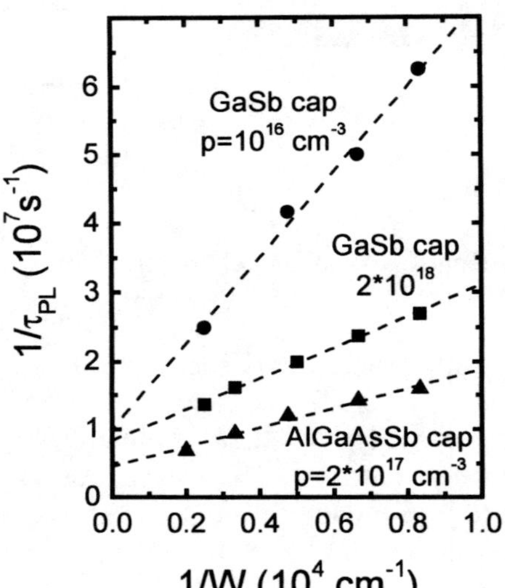

FIGURE 6. Inverse PL decay as a function of inverse thickness for three sets of GaInAsSb structures with different cap layers.

REFLECTORS FOR GaSb-BASED TPV DEVICES

Reflectors can enhance TPV performance by providing double-pass absorption for photons that are not absorbed in the first pass, especially near the band-edge. This allows for a thinner emitter layer and thus lower dark current. Furthermore, a reflector can enhance photon recycling and thus increase minority-carrier lifetime and V_{oc}. However, substrate absorption can reduce the effectiveness of a BSR [15], while increased series resistance can be problematic with DBR structures [21]. Preliminary results for both approaches are presented.

For either approach, low series resistance contacts and high shunt resistance R_{SH} are critical to device performance. Ohmic contacts to n- and p-GaSb were formed by depositing Au/Sn/Ti/Pt/Au and Ti/Pt/Ag/Pt/Au, respectively, and alloying the n-metal at 300 °C. The p-type contacts were not alloyed. The specific contact resistivity ρ_c was measured by the Cox-Strack method [22]. It was found that ρ_c for n- and p-contacts was $2.3x10^{-6}$ and $1.6x10^{-6}$ ohms-cm^2, respectively. It is slightly lower for the p-contacts than for the n-contacts because the Fermi level is pinned close to the valence band edge.

Large-area (0.5 cm^2) TPV cells were fabricated using a conventional photolithographic process. A single 0.15-mm-wide busbar connected to 10-µm-wide grid lines spaced 100 µm apart was used to make electrical contact to the front surface. TPV cells were defined by wafer sawing. The cutting introduces significant damage to the side walls and results in very low R_{SH} and degradation of FF. Saw-cut damage was removed with chemical etching, and it was found that the R_{SH} increased from a few ohms to hundreds of ohms, while the values for FF increased from below 60% to 70%.

GaInAsSb/GaSb TPV structures were grown with and without a five-period AlGaAsSb/GaSb DBR tuned to have reflectivity centered at 2.4 µm. The 300 K PL spectra for the TPV control structure without a DBR and the DBR/TPV structure are shown in Figure 7a. The PL intensity is enhanced by about 40%. The EQE of uncoated TPV and DBR/TPV devices is shown in Figure 7b. No enhancement in EQE near the band-edge was observed, which may be due to the narrow bandwidth (~400 nm) of the DBR centered at 2.4 µm. It is expected that increased EQE may be possible for TPV devices with a DBR centered at 2.2 µm. V_{oc} for the device with the DBR increased to 306 mV compared to 285 mV for the device without the DBR, or about 7%. Assuming a change only in recombination lifetime, a photon-recycling factor was estimated to be about 2.5. The series resistance for the cell with the DBR is slightly higher at $1.2x10^{-2}$ Ω, compared to $0.8x10^{-2}$ Ω without the DBR. Grading and doping the interfaces in the DBR structure should reduce the series resistance [21].

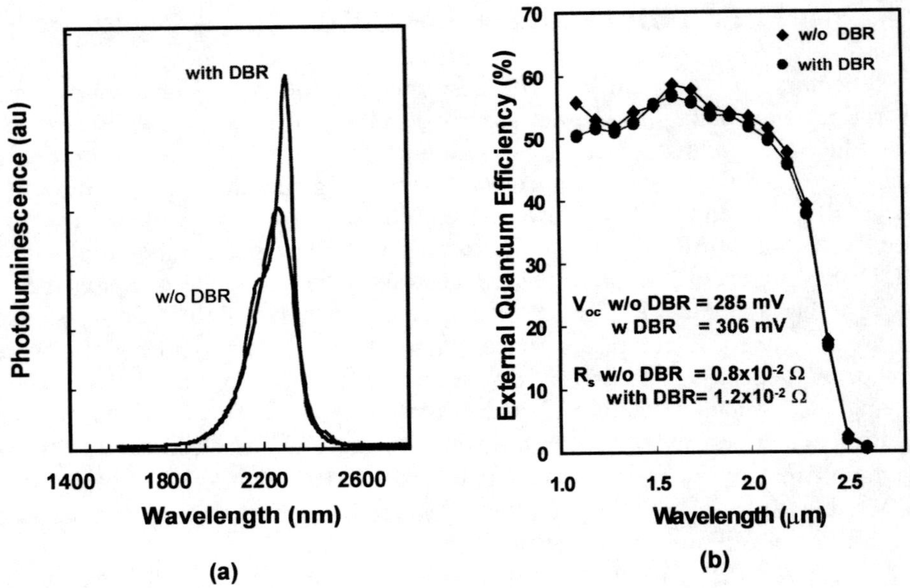

FIGURE 7. Comparison of GaInAsSb/GaSb TPV devices with and without distributed Bragg reflector: (a) 300 K PL spectra; (b) external quantum efficiency.

GaInAsSb/GaSb TPV devices were also fabricated with a BSR. Free-carrier absorption in the n-GaSb substrate can be reduced by thinning the substrate [15], and a reflectivity of nearly 90% through the GaSb substrate was measured for a substrate thinned to 100 μm thickness. A hybrid ohmic contact and BSR was constructed on the backside of a thinned and polished TPV cell. The EQE of an uncoated BSR-TPV device with substrate thickness of about 120 μm is shown in Figure 8. For comparison, the EQE of a conventionally processed TPV cell from the same wafer is also shown. The reflection limit of GaSb is also shown in the dotted-dashed line. In the wavelength regime from about 1.1 to 1.7 μm, the BSR-TPV cell and the reference cell have approximately the same EQE since the GaSb substrate is absorbing below 1.7 mm. From about 1.7 to 2.5-μm, the EQE of the BSR-TPV cell exceeds that of the reference cell, and indicates the enhancement due to the double-pass absorption. The peak difference in EQE is about 8% at a wavelength of 2.35 μm, which corresponds to a fractional increase in EQE of about 20% at this wavelength.

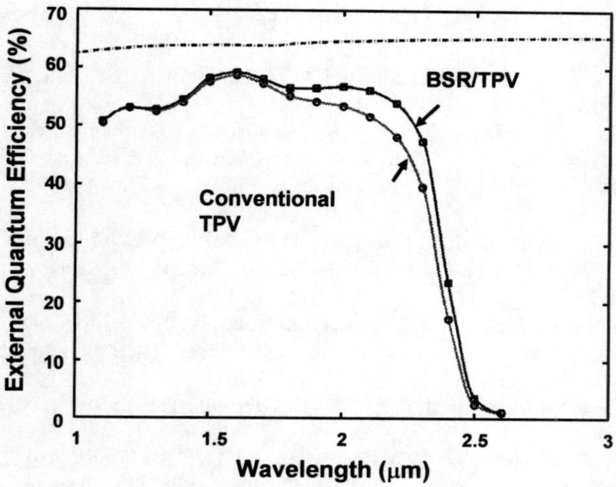

FIGURE 8. External quantum efficiency of GaInAsSb/GaSb TPV cells with and without back-surface reflector.

CONCLUSIONS

The optimization of GaInAsSb/(Al)Ga(As)Sb/GaSb TPV devices requires extensive effort in materials growth and characterization, device design, fabrication, and testing. High-performance TPV cells have been fabricated, indicating good control and reproducibility of both epitaxial growth and device processing. Values of EQE and FF are approaching theoretical limits, while the voltage factor is within 15% of the limit. Further improvements in the voltage factor should be forthcoming with increased understanding of recombination mechanisms in GaInAsSb. Preliminary device results on cells with a DBR structure indicate an increase in V_{oc}. Improvements in EQE was measured for TPV cells with a BSR.

ACKNOWLEDGMENTS

The authors gratefully acknowledge D.R. Calawa, J.W. Chludzinski, S. Hoyt, P.M. Nitishin, D.C. Oakley, and V.A. Todman for technical assistance. This work was sponsored by the Department of Energy under Air Force Contract No. F19628-00-C-0002. The opinions, interpretations, conclusions, and recommendations are those of the author and are not necessarily endorsed by the United States Government.

REFERENCES

1. P.F. Baldasaro, J.E. Raynolds, G.W. Charache, D.M. DePoy, C.T. Ballinger, T. Donovan, J.M. Borrego, J. Appl. Phys. **89**, 3319-3327 (2001).
2. G.W. Charache, J.L. Egley, D.M. DePoy, L.R. Danielson, M.J. Freeman, R.J. Dziendziel, J.F. Moynihan, P.F. Baldasaro, B.C. Campbell, C.A. Wang, H.K. Choi, G.W. Turner, S.J. Wojtczuk, P. Colter, P. Sharps, M. Timmons, R.E. Fahey, K. Zhang, J. Electron. Mater. **27**, 1038-1042 (1998).
3. G.W. Charache, P.F. Baldasaro, L.R. Danielson, D.M. DePoy, M.J. Freeman, C.A. Wang, H.K. Choi, D.Z. Garbuzov, R.U. Martinelli, V. Khalfin, S. Saroop, J.M. Borrego, R.J. Gutman, J. Appl. Phys. **85**, 2247-2252 (1999).
4. Z.A. Shellenbarger, M.G. Mauk, J.A. Cox, M.I. Gottfried, P.E. Sims, J.D. Lesko, J.B. McNeely, L.C. DiNetta, R.L. Mueller, 'Improvements in GaSb-Based Thermophotovoltaic Cells,' AIP Conf. Proc. 401, 117-128 (1997).
5. O.V. Sulima, R. Beckert, A.W. Bett, J.A. Cox, M.G. Mauk, IEE Proc.-Optoelectron. **147**, 199-204 (2000).
6. V.M. Andreev, V.P. Khvostikov, V.D. Rumyantsev, S.V. Sorokina, M.Z. Shvarts, 'Single-junction GaSb and tandem GaSb/InGaAsSb and AlGaAsSb/GaSb thermophotovoltaic cells,' Proc. 28th IEEE Photovoltaics Spec. Conf., 1265-1268 (2000).
7. V.B. Khalfin, D.Z. Garbuzov, H. Lee, G.C. Taylor, N. Morris, R.U. Martinelli, J.C. Connolly, 'Interfacial Recombination in In(Al)GaAsSb/GaSb Thermophotovoltaic Cells,' AIP Conf. Proc. 460, 247-255 (1999).
8. H.K. Choi, C.A. Wang, G.W. Turner, M.J. Manfra, D.L. Spears, G.W. Charache, L.R. Danielson, D.M. DePoy, Appl. Phys. Lett. **71**, 3758-3760 (1997).
9. C.A. Wang, H.K. Choi, D.C. Oakley, G.W. Charache, J. Cryst. Growth **195**, 346-355 (1998).
10. C.A. Wang, H.K. Choi, S.L. Ransom, G.W. Charache, L.R. Danielson, D.M. DePoy, Appl. Phys. Lett. **75**, 1305-1307 (1999).
11. C.A. Wang, H.K. Choi, G.W. Charache, IEE Proc.-Optoelectron. **147**, 193-198 (2000).
12. J.M. Borrego, S. Saroop, R.J. Gutman, G.W. Charache, T. Donovan, P.F. Baldasaro, C.A. Wang, J. Appl. Phys. **89**, 3753-3759 (2001).
13. S. Saroop, J.M. Borrego, R.J. Gutman, G.W. Charache, C.A. Wang, J. Appl. Phys. **86**, 1527-1534 (1999).
14. S.P. Tobin, S.M. Vernon, M.M. Sanfacon, A. Mastrovito, 'Enhanced Light Absorption in GaAs Solar Cells with Internal Bragg Reflectors,' Proc. 22nd IEEE Photovoltaics Spec. Conf., 147-152 (1991).
15. G.W. Charache, D.M. DePoy, P.F. Baldasaro, B.C. Campbell, 'Thermophotovoltaic Devices Utilizing a Back Surface Reflector for Spectral Control,' AIP Conf. Proc. 321, 339-350 (1996).
16. C.J. Vineis, C.A. Wang, K.F. Jensen, J. Cryst. Growth **225**, 420-425 (2001).
17. C.A. Wang, D.R. Calawa, C.J. Vineis, J. Cryst. Growth **225**, 377-383 (2001).
18. N.N. Ledentsov, J. Bohrer, M. Beer, F. Heinrichsdorff, M. Grundmann, D. Bimberg, S.V. Ivanov, B.Ya. Meltser, S.V. Shaposhnikov, I.N. Yassievich, N.N. Faleev, P.S. Kop'ev, Zh.I. Alferov, Phys. Rev. **B 52**, 14058-14066 (1995).
19. R.H. Ahrenkiel, S.P. Ahrenkiel, D.J. Arent, J.M. Olson, Appl. Phys. Lett. **70**, 756-758 (1997).
20. D. Donetsky, C.A. Wang, S. Anikeev, G. Belenky, S. Luryi, G. Nichols, 'Reduction of the Interfacial Recombination Rate in GaInAsSb/GaSb Heterostructures,' 44th Electron. Mater. Conf., 26-28 June 2002, Santa Barbara, CA.
21. K.Tai, L. Yan, Y.H. Wang, J.D. Wynn, A.Y. Cho, Appl. Phys. Lett. **56**, 2496-2498 (1990).
22. R.H. Cox, H.Strack, Solid-State Electron. **10**, 1213-1218 (1967).

High-Efficiency Multi-Cell TPV Module Fabrication and Performance

Y. Z. Yu*, R. U. Martinelli*, G. C. Taylor*, Z. Shellenbarger*, R. K. Smeltzer*, J. Li, K. Palit*, S. R. Burger†, R. P. Cardines†, and L. R. Danielson†, C. A. Wang†† and M. K. Conners††

*Sarnoff Corporation, CN5300, Princeton, New Jersey 08543-5300
†Lockheed Martin, Inc., Schenectady, New York 12301-1072
††Lincoln Laboratory, Massachusetts Institute of Technology, Lexington, Massachusetts 02173-9108

ABSTRACT

Sarnoff Corporation has developed thermophotovoltaic (TPV) modules using either Sarnoff grown and processed InGaAsSb/GaSb TPV cells or cells fabricated by our co-workers. The TPV module fabrication includes substrate design, fabrication and module assembly. The substrate comprises an AlN base plate and Al_2O_3 septa. The septa separate the cells and support metal layers to electrically connect the bottom of a cell to the top of the adjacent septum. A welded gold ribbon connects the septum to the topside of the adjacent cell. The detailed structure of the substrate, as well as the module assembly process, is discussed. TPV modules of 1 cm x 1 cm area with two 1 cm x 0.5 cm cells and 2 cm x 2 cm area with eight cells are fabricated routinely. Alternative approaches to the Al_2O_3/AlN substrate and ribbon connection have been explored for low-cost, large-scale production. Silicon based substrates show promising results. KOH wet etching of silicon produces septa with straight walls having perfect 90-degree angles at the bottom without fillets. Plated gold forming an "air bridge" to connect the cell busbar and septum has the potential to replace the welded ribbon. Packaging p-on-n and n-on-p cells alternatively can drastically reduce the complexity of the module structure by facilitating all connections on a planar surface without septa. Sarnoff module technology, when combined with Sarnoff TPV-cell technology and spectral-control technologies from other organizations, has led to TPV power-conversion efficiencies exceeding 17 % [1].

INTRODUCTION

A key process for the development of thermophotovoltaic (TPV) systems is cell packaging. In addition to InGaAsSb cell growth and fabrication [2-5], Sarnoff's TPV module development effort includes substrate design, fabrication and module assembly. Activities were split into two aspects. The first was to provide modules in small quantities for performance evaluation in proof-of-principle studies to obtain high

CP653, *Thermophotovoltaic Generation of Electricity: 5th Conference*
edited by T. J. Coutts, G. Guazzoni, and J. Luther

TPV power-conversion efficiency. At the same time, based on what was achieved from test modules, the second effort was to find better approaches in connection with the long-term goal of mass production. This paper reports our current status in TPV module development.

The goal of the packaging project is to provide a substrate on which 1 cm x 0.5 cm cells are affixed and electrically connected, either in parallel or in series, with minimum connection resistance. Figure 1 shows the front surface of a typical cell. Parallel metal fingers, 10-μm-wide, 0.5-cm-long, and spaced on 100-μm centers, collect the photogenerated currents from the cell and conduct them to the busbar. The 125-μm-wide, 1-cm-long busbar connects to the fingers at the edge of the cell. Since all cells in a module are p-on-n diodes, the backside metal of one cell must be connected to the front side busbar of the adjacent cell in series with it. The challenge is to minimize the dead space between cells and to reflect the photons hitting the dead space back into the radiator for photon recycling, which boosts overall module efficiency. Also, to mount an InP-based multi-layer dielectric filter on top of the module for spectral control, the top of the assembled module must be as flat as possible to minimize the thickness of the optical adhesive used to hold the spectral-control filter in place. In addition, the substrate must have a thermal conductivity such that with a 5-W/cm^2 power density input, the increase of the cell temperature is no more than 1 K.

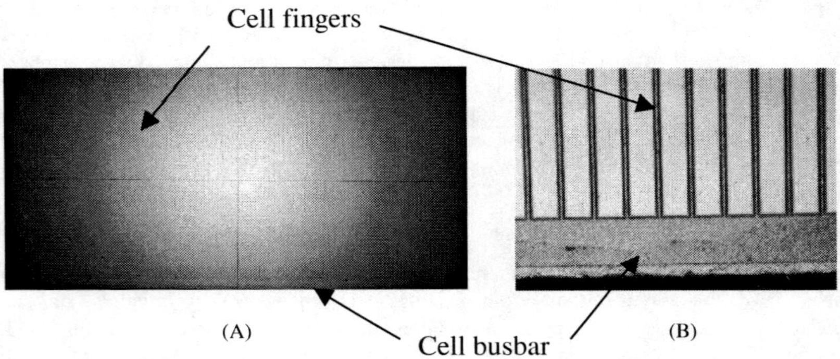

FIGURE 1. Photograph of a 1 x ½ cm^2 TPV cells.
(A): Entire cell (B): Magnified area near the busbar.

Al_2O_3/AlN SUBSTRATE DESIGN ARCHITECTURE

A schematic diagram of the substrate used for small-quantity module-performance evaluation is shown in Figure 2. Fabrication starts with grooves cut into the 0.5-mm-thick AlN sheet. Then Ti/Pt/Au seed metal is deposited in a directional metal evaporator, so that only right-hand sidewall of the grooves is coated by the metal. About 12 μm of copper and 5 μm of lead/tin are plated on top of the seed metals. The copper layer enhances the electrical conductivity, and the lead/tin protects the copper and facilitates soldering. The septa are prepared by dicing 125-μm-thick alumina into strips of a width determined by the 0.5-mm-thick TPV cells. The same seed metals are deposited on only the top and one sidewall of the septa. Copper and gold layers are plated on top of the seed metal coverage, so the copper is protected from oxidation by the gold. The septa are then inserted into the grooves in the AlN with spacers in between to hold them perpendicular to the substrate surface. The assembly is heated to 200 °C to solder the septa in place. After protecting the gold with photoresist, the substrate floor is plated with an additional 5 μm of indium.

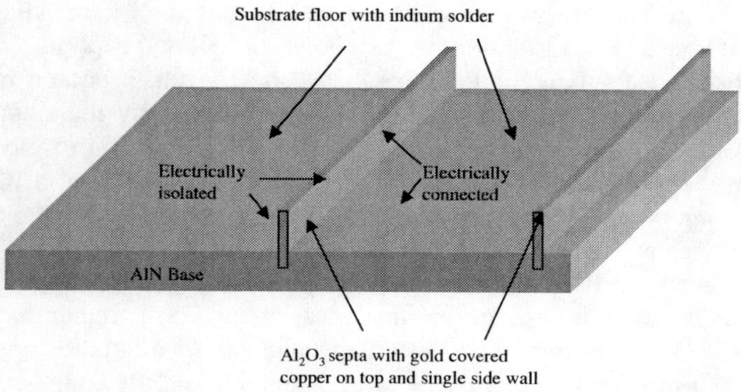

FIGURE 2. Schematic diagram of the Al_2O_3/AlN substrate.

The substrate is diced into 1 cm x 1 cm and 2 cm x 2 cm sizes to host two and eight TPV cells, respectively. The finished substrate has septa about 500 μm high and 125 μm wide with the spacing between septa of slightly more than 0.5 cm. The cells fit into the substrates with a 25-30 μm lateral and a 5-30 μm thickness tolerance.

A sample was prepared to evaluate the thermal resistance of the module assembly. It comprised a 0.5-mm-thick AlN substrate with seed metal, copper, lead/tin, and indium metallization applied. A 0.5-mm-thick GaSb chip was soldered to the AlN with Au/Sn. This sample simulates the TPV cell mounted in the substrate. The specific thermal resistance of the solder joint was measured to be 0.093 Kcm^2/W. With a 0.5-mm-thick GaSb and AlN samples and with thermal resistivities of 2.86 and 0.56 Kcm/W for GaSb and AlN, respectively, one obtains a specific thermal resistance of 0.143 Kcm^2/W for GaSb and 0.028 Kcm^2/W for AlN. The calculated specific thermal resistance for the assembled module is 0.264 Kcm^2/W, which is equivalent to a 1.32 K temperature rise with a 5-W/cm^2 power density input. This is close to the targeted value of 1K. The major contribution of thermal resistance is from the GaSb chip. If the GaSb cells were thinned to 0.25 mm, the temperature rise would be reduced to less than 1 K.

Module Assembly

In the module assembly process, one first places cells in the proper orientation and inserts precut 25- to 50-μm-thick Teflon tape as shims between a cell and its adjacent septum to protect the cell junction. Since the unprotected p-n junction is only a few micrometers below the cell surface, which is flush with the septum top, cells too close to the septum may suffer damage in the later gold-ribbon welding process. The substrate populated with cells is then placed on a hot plate at 180 °C for 5-10 s to reflow the plated indium solder. After the cells are soldered to the substrate, pieces of 12-μm-thick gold ribbon are welded onto the cell busbars and their adjacent septa. The ribbon width is selected to cover the gap between cell and septum. The welded ribbon not only connects the cells in series, but also serves as a photon reflector, so that some of the photons hitting the inactive areas covered by the ribbon may be reflected back into the radiator for recycling. It requires precision to weld the ribbon to the 125-μm-wide busbar, because the welder tip may hit the edge of the cell and damage the p-n junction. From a series of tests, welding parameters were determined that assured sufficient mechanical strength and electrical conductivity of the weld joints without causing cell structural damage or performance degradation. For the 1 cm x 1 cm module the module series resistance was about 1 mΩ greater that of the two cells in series. With micrometer-adjusted mechanical manipulation stages and a custom-made 75-μm-wide welder tip, ribbons can be successfully welded without cell damage.

Figure 3 shows a schematic diagram of an assembled module. .After assembling the module, two power-take-off tabs made of gold-plated copper are welded to both ends of the module. The assembly is then mounted to the top of a gold-plated copper pedestal to measure the power-conversion efficiency. Power-conversion efficiency is the ratio of the electrical power output to the radiant power absorbedIn addition to 1 cm x 1 cm modules with two cells, 2 cm x 2 cm modules that hold eight cells were also fabricated. Shown in Figure 4 is a picture of the larger size module.

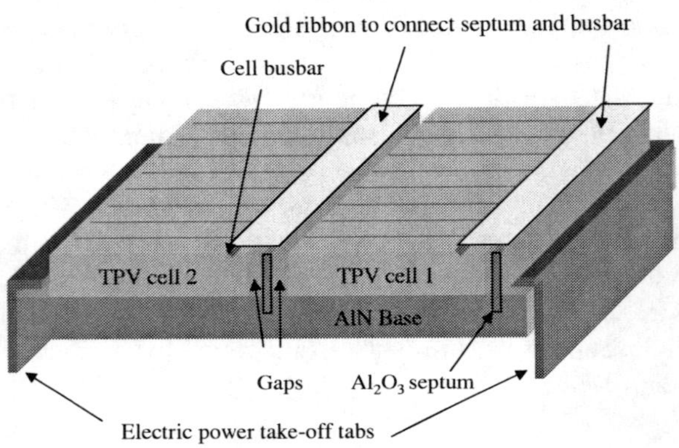

FIGURE 3. Schematic diagram of a two-cell, 1 cm^2 module.

FIGURE 4. Photograph of a 2 cm x 2 cm, 8-cell TPV module.

Module Test

The open-circuit voltage (V_{oc}), the short-circuit current (I_{sc}), and the fill factor (FF) of the modules were measured under the illuminations from a high-intensity flash lamp and from a 1150 °C blackbody radiator. Figure 5 shows the I-V characteristics of a typical module for each case. The flash lamp generates much more electrical power.

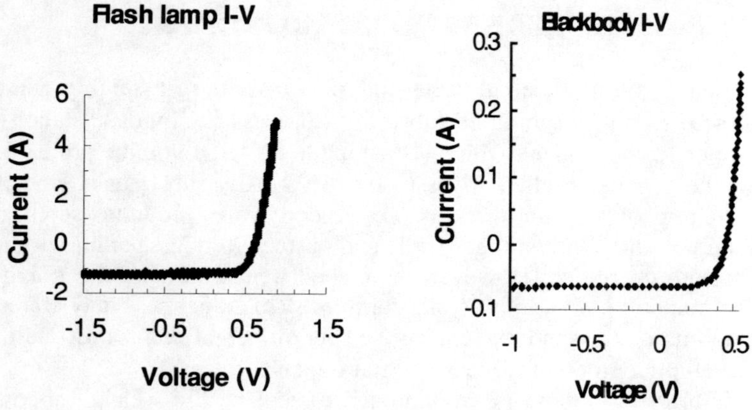

FIGURE 5. Typical I-V curves of a 1 cm^2 TPV module under Flash Lamp and Blackbody Illumination.

In Table 1 the module performance data extracted from the flash-lamp I-V curves shown in Figure 5 are compared with the individual cell data. The two p-on-n cells used in this module were grown at Sarnoff and metallized at Lincoln Laboratory. We evaluate the quality of module fabrication process by comparing module short-circuit current, I_{sc}, open circuit voltage, V_{oc}, and fill factor, FF, with the corresponding parameters of the individual cells taken under identical illumination before the cells are assembled into the module. The cells have the same I_{sc} values of 1.22 A, while the module generates I_{sc} = 1.18 A. Open-circuit voltages of the cells are nearly identical at 0.299 and 0.296 V, and V_{oc} of the module is slightly larger than the sum of the two at 0.600 V. The FFs are all above 69 %, and that of the module is slightly lower, leading to a maximum power output P_{max} = 0.489 W for the module. The module P_{max} is 3 % less than the sum of the cells' P_{max} values. This shows that the process of assembling the module reduces slightly the expected performance of the two-cell series combination.

TABLE 1. Performance data of a typical 1 cm^2 module compared with that of its constituent cells

	I_{sc} (A)	V_{oc} (V)	FF (%)	P_{max} (W)
Cell-1	1.22	0.299	69.5	0.254
Cell-2	1.22	0.296	69.8	0.252
Module	1.18	0.600	69.1	0.489

When coupled with spectral-control filters from Lockheed Martin, Inc., modules of this type exhibited TPV power-conversion efficiencies exceeding 17 % [1].

ALTERNATIVE APPROACHES

Although the multi-piece substrate and ribbon-welding assembly achieved the proof-of-principle goal, the module fabrication process is obviously labor-intensive and hence, operator-dependent, which is unsuitable for large-volume production.

One of the promising alternatives to the Al_2O_3/AlN substrate is etched silicon. KOH wet etching of a 1-mm-thick <110> silicon wafer produces septa along the <111> crystal direction with straight walls and perfect 90-degree angles at the bottom of the septa with no fillets. However, the etched trench surfaces are rough because they are a combination of various crystal surfaces. This roughness may affect the cell-substrate bonding quality and make accurate septum height control difficult. We are currently developing protocols to produce flat trench surfaces.

The "air bridge" method is an approach to replace the welded ribbon for cell connections. The idea is to fill gaps between busbars and septa with photoresist and to expose the area where the busbars and septa are located. Plating gold on the photoresist to connect the exposed busbars and septa covers the 25-μm-wide gaps. Removal of the residual photoresist makes the plated gold an "air bridge". The air bridge may relieve the stress due to thermal expansion when the module undergoes temperature cycling, and therefore, may mitigate a possible limitation to operational life. An air bridge was successfully formed in our preliminary studies, although issues such as air bubbles in the photoresist and imperfection of the bridge remain to be solved.

One way to simplify drastically the substrate structure and module assembly process is to package rows of p-on-n and n-on-p cells alternately. In this way, one only needs to connect the backside-to-backside and the front-side-to-front-side of cells in adjacent rows. One can eliminate the septa, which reduces the inactive area. A special 2 cm x 2 cm substrate was fabricated for the proof-of-principle study. One module was assembled with four 1 cm x 1 cm cells, two of each type, as shown in Figure 6. Two n-on-p cells were mounted in parallel on one side of the gap, and two p-on-n cells were mounted in parallel on the other side of the gap. Two straps along the central busbars connected the two rows in series.

The I_{sc}-V_{oc} characteristic of the module is presented in Figure 7. At a photogenerated current of 0.9 A (0.225 A/cm^2) the module had an open-circuit voltage of 480 mV. A Sarnoff 4 mm x 4 mm TPV test cell from the wafer that produced the p-on-n cells generated an open-circuit voltage of 320 mV at a short-circuit current density of 4.3 A/cm^2 (690 mA). The characteristics of the n-on-p cells from our co-workers were unknown to us. To evaluate the module performance, we assumed that these n-on-p cells have the same characteristics as those of the Sarnoff cells. The single-cell I_{sc}-V_{oc} curve shown in Figure 7 was generated using half of the module short-circuit current and half of the module open-circuit voltage, because the module has a pair of parallel-connected cells, each of which comprises an n-on-p and a p-on-n cell connected in series.

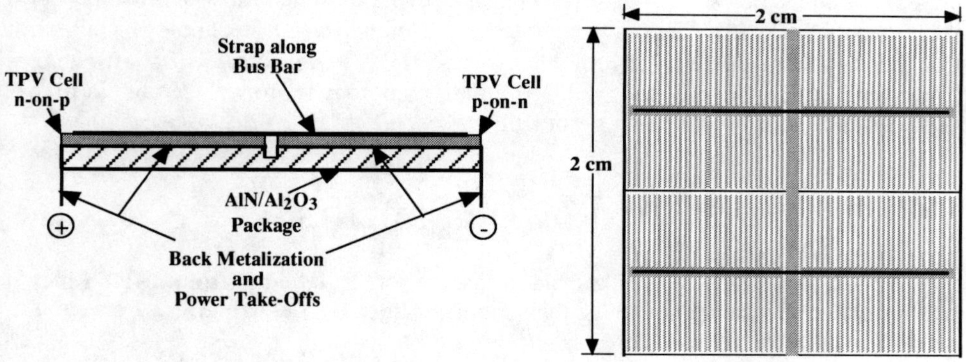

FIGURE 6. Schematic diagram of the package for alternating n-on-p and p-on-n cell arrangement.

Extrapolation of the "single-cell curve" in Figure 7 to the current of 4.3 A (same as current density of 4.3 A/cm^2 because each cell has an area of 1 cm^2) corresponds to an open-circuit voltage very of 310 mV. This is very close to V_{oc} = 320 mV measured with the 4 mm x 4 mm p-on-n test cells at the same current density. We conclude, therefore, that the module performs as well as the p-on-n single cell and that arrays of cells with alternating polarities is a possible approach to large-scale manufacturing.

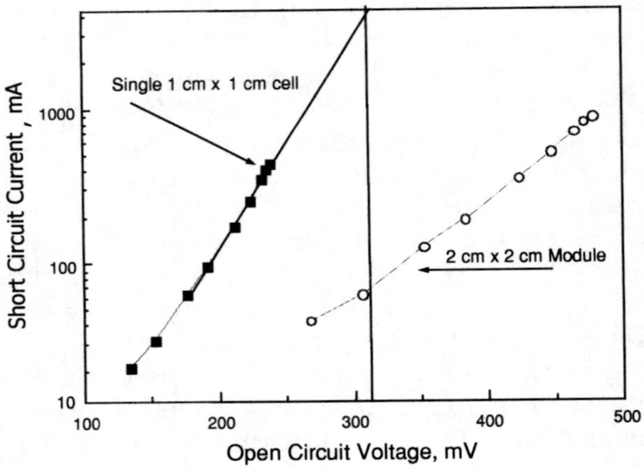

FIGURE 7. Performance of the alternating n-on-p and p-on-n cell 2 cm x 2 cm module (open circles) and calculated single cell data (solid squares).

SUMMARY

Sarnoff has designed, fabricated and characterized multi-cell TPV modules in 1 cm x 1cm and 2 cm x 2 cm sizes. When Sarnoff module and TPV-cell technologies were combined with Lincoln Laboratory cell-metallization technology and with Lockheed Martin spectral-control technology, TPV power-conversion efficiencies exceeding 17 % [1] were achieved. Alternative approaches to reduce cost in higher volume production have shown promising results.

ACKNOWLEDGEMENTS

We acknowledge the technical assistance of R. Huang from MIT Lincoln Laboratory. We also acknowledge the support of Lockheed Martin, Inc.

REFERENCES

1. J. Beausang, "*Thermodynamic Analysis of Thermophotovoltaic Efficiency and Power Density Tradeoffs*", Fifth Conference on Thermophotovoltaic Generation of Electricity", Rome, Italy, 15 – 19 September, 2002
2. V. B. Khalfin , D. Z. Garbuzov, H. Lee, G. C. Taylor, N. Morris, R. U. Martinelli, J. C. Connolly, "*Interfacial Recombination in In(Al)GaAsSb/GaSb Thermophotovoltaic Cells*", in

Thermophotovoltaic Generation of Electricity edited by T. J. Coutts et al, *AIP Conference Proceedings* **460**, 247-255 (1998).

3. R. U. Martinelli, D. Z. Garbuzov, H. Lee, N. Morris, T. Odubanjo, G. C. Taylor, and J. C. Connolly, "*The Quantum Efficiency of InGaAsSb Thermophotovoltaic Diodes*", in Space Technology and Applications International Forum-1998. edited by M. S. El-Genk *AIP Conference Proceedings* **420**, 1394-1399 (1998).
4. D. Z. Garbuzov, R. U. Martinelli, V. B. Khalfin, H. Lee, N. Morris, T. Odubanjo, G. C. Taylor, J. C. Connolly, G. W. Charache, and D. M. Dupoy, "*A Novel Approach for the Improvement of Open Circuit Voltage and Fill Factor of InGaAsSb/GaSb Thermophotovoltaic Cells*", in Space Technology and Applications International Forum-1998. edited by M. S. El-Genk *AIP Conference Proceedings* **420**, 1400-1409 (1998).
5. R. U. Martinelli, D. Z. Garbuzov, H. Lee, N. Morris, T. Odubanjo, G. C. Taylor, and J. C. Connolly, "*Minority-Carrier Transport in InGaAsSb Thermophotovoltaic Diodes,*" in Thermophotovoltaic Generation of Electricity, edited by T. J. Coutts et al., *AIP Conference Proceedings* **401**, 389-395 (1997).

Strain-Compensated InGaAs/InGaAs Quantum Well Cell With 2 μm Band-Edge

Carsten Rohr*, Paul Abbott*, Ian Ballard*, James P. Connolly*, Keith W. J. Barnham*, Lucia Nasi†, Claudio Ferrari†, Laura Lazzarini†, Massimo Mazzer** and John Roberts‡

*Experimental Solid State Physics, Blackett Laboratory, Imperial College London, SW7 2BW, U.K.
†CNR-IMEM, Parco Area delle Scienze 37/A, 43010 Fontanini-Parma, Italy
**CNR-IMM, University of Lecce, 73100 Lecce, Italy
‡EPSRC III-V Facility, University of Sheffield, Sheffield S1 3JD, U.K.

Abstract. Strain-compensated Quantum Well Cells (QWCs) have been shown to extend the absorption to longer wavelengths than attainable with lattice-matched material, while retaining a similar or better dark current. This is of particular interest for thermophotovoltaic (TPV) applications with low temperature sources where low energy bandgaps are required, for example in combination with a Holmia emitter of 1.95 μm peak emission. Here we report on a two quantum well $In_{0.74}Ga_{0.26}As/In_{0.36}Ga_{0.64}As$ device which absorbs out to 2040 nm, as observed by cathodoluminescence. Our modelling of the spectral response, including quantum and strain effects, is also consistent with this result. We show that the material quality measured by transmission electron microscopy (TEM) is excellent, exhibiting sharp interfaces. However, the electrical properties under illumination are less encouraging: At room temperature not all carriers are collected in forward bias. We present results on the field and temperature dependence of carrier escape and collection.

INTRODUCTION

In thermophotovoltaics (TPV) low-bandgap semiconductors are used to convert heat into electricity. In order to achieve efficient conversion, a selective emitter is heated by a heat source and emits (near) infrared radiation in a narrow band, which should be matched to the bandgap of the photovoltaic device. For temperatures of about 1500 K, a typical combustion temperature for example, the rare-earth selective emitter Holmia with an emission peak of about 1950 nm is very suitable. Hence the photovoltaic device should absorb this emission band, including wavelengths beyond 2 μm. Standard TPV semiconductor materials such as GaSb and InGaAs lattice-matched to InP only absorb as far as about 1.75 and 1.7 μm respectively, and are therefore not suitable for this application.

In quantum well cells (QWCs) it is possible to use strain-compensation techniques to overcome these material limitations [1]. Compressively strained quantum wells (QWs) with a lower bandgap than the lattice-matched material are compensated with tensile strained barriers so that the shear forces at the interfaces are minimised [2, 3]. For highly strained layers the difference in their elastic properties has to be taken into account [4] rather than using the simpler strain-thickness approach. Strain-compensation avoids deterioration of the material quality due to the formation of misfit dislocations which

CP653, *Thermophotovoltaic Generation of Electricity: 5th Conference*
edited by T. J. Coutts, G. Guazzoni, and J. Luther

TABLE 1. Sample description of MR1674 (nominal).

Layer	Thickness (Å)	Material	Doping	Conc. (cm^{-3})
Cap	1000	InGaAs	p	1E+19
Emitter	2000	InP	p	2E+18
Spacer	2000	InP	i	
$2 \times \frac{1}{2}$ Barrier	50	$In_{0.36}Ga_{0.64}As$	i	
$2 \times$ Quantum well	100	$In_{0.74}Ga_{0.26}As$	i	
$2 \times \frac{1}{2}$ Barrier	50	$In_{0.36}Ga_{0.64}As$	i	
lattice-matched	200	$In_{0.53}Ga_{0.47}As$	i	
Base / buffer	5000	InP	n	1E+18
Substrate		InP	n	

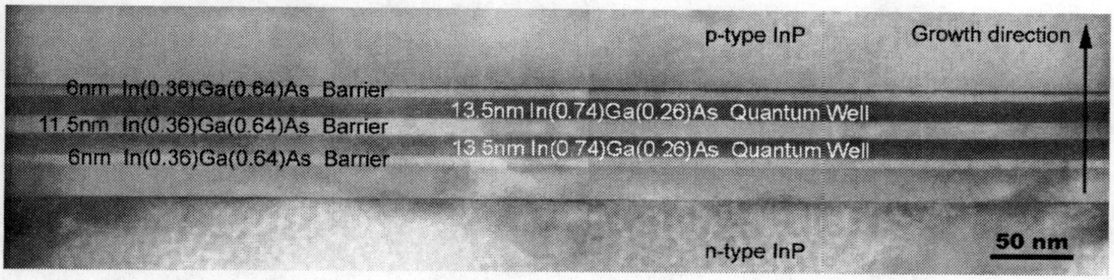

FIGURE 1. Cross-sectional TEM of the two quantum well sample.

is otherwise typical for strained (or strain-relaxed) materials exceeding their critical thickness. Strain-compensation therefore allows engineering of the bandgap beyond what is possible with lattice-matched material while retaining good material quality [3, 5, 6]. The individual layers have to remain below their respective critical thickness, and strain-compensation is also limited by the growth conditions when depositing highly strained material.

In this paper we report a strain-compensated QWC (MR1674) where we have achieved absorption at wavelengths greater than $2\,\mu$m. This two QW $In_{0.74}Ga_{0.26}As/In_{0.36}Ga_{0.64}As$ sample was grown by metal-organic vapour phase epitaxy (MOVPE) and the sample description is given in Table 1.

MATERIAL CHARACTERISATION

Cross-sectional transmission electron microscopy (TEM) is shown in Fig. 1. One can see good material quality with sharp interfaces. The layer thicknesses, however, are significantly larger than nominal (compare with Table 1) but still below their respective critical thicknesses. Although the structure is highly strained, +1.44% for the QWs and -1.18% for the barriers, we do not observe any three-dimensional growth (see Fig. 1).

The atomic force microscopy (AFM) image in Fig. 2 shows quite good surface morphology. The surface is flat with a few pits of less than 0.5 μm diameter.

The cathodo-luminescence (CL) spectrum in Fig. 3 clearly shows a peak at 2040 nm,

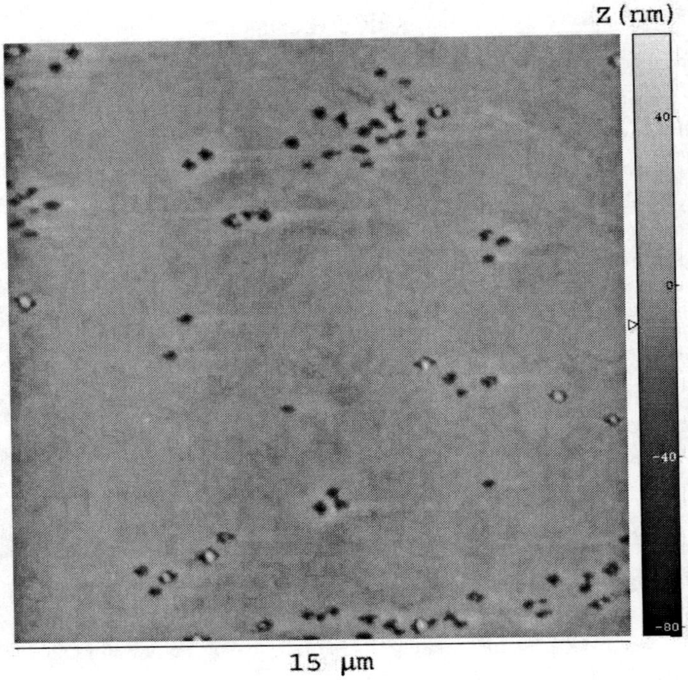

FIGURE 2. Atomic Force Microscopy (AFM) of MR1674.

although it is quite broad. There are additional 'humps' at about 1800-1850 nm and around 1650 nm which could be due to other energy levels in the quantum wells. It is very encouraging to see the luminescence peak at more than 2 μm with such a good material quality.

According to our model SOL, developed to calculate quantum efficiencies (QE) including strained quantum wells [7], an absorption edge of about 2040 nm with a quantum well width of 135 Å is consistent with an In composition of about 74-76% in the QW. The luminescence peak can be calculated from the absorption using the detailed balance model and is consistent with the measured CL peak. As shown in Fig. 4, SOL is within errors in agreement with TEM corrected dimensions and for nominal compositions (the QE data is discussed later).

If the barrier composition is the nominal 36% In, and for a barrier width of 115 Å, the strain can be compensated for at least 10 periods, using Matthews & Blakeslee [8] to calculate the critical thickness of the multi quantum well stack. Using a perhaps more realistic calculation of the critical thickness [9, 10] suggests that a maximum of 21 repeats would be possible in this structure. With nominal dimensions, in theory, we would be able to accommodate up to 61 or 421 periods, depending on how the critical thickness is calculated. But this practically unlimited number of repeats would require very high growth accuracy of barrier and QW dimensions and material compositions.

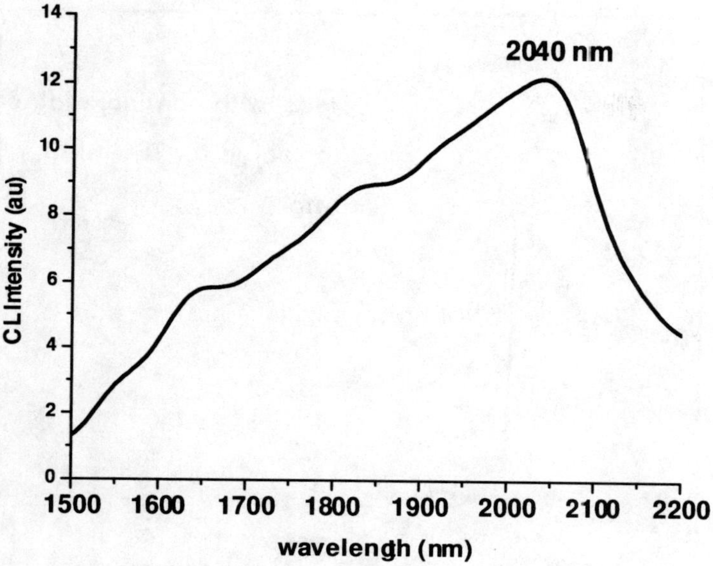

FIGURE 3. Cathodo-luminescence spectrum of MR1674.

ELECTRICAL AND OPTICAL CHARACTERISATION

The material was processed into our standard test devices of 1mm mesas with 200 μm wide ring contacts leaving a 600 μm diameter optical window.

Photocurrent

The photocurrent is measured using a monochromator and a lock-in technique, and hence the dc dark current is not detected leaving only the photo-generated current or photocurrent.

Spectral Response

In Fig. 4 the spectral response data of MR1674 with no anti-reflection (AR) coating and the InGaAs cap left on is shown, measured in reverse bias (-1V) to ensure full carrier collection. The absolute level of the spectral response beyond the bandgap of InP (about 920 nm) is expected to be low due to the presence of only two QWs but the response has been extended beyond 2 μm. However, due to limitations of the current experimental set-up we cannot measure the photocurrent accurately much beyond 2 μm.

The modelled spectral response, also shown in Fig. 4, fits the data quite well, particularly for the correct dimensions taken from TEM. As we have observed in other

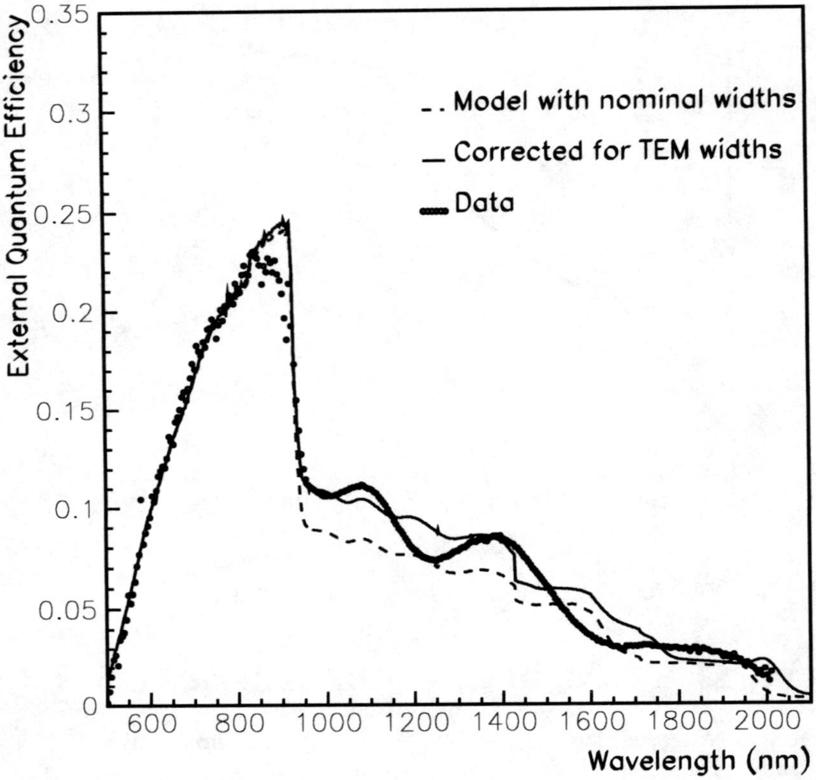

FIGURE 4. Modelled spectral response of MR1674 (with no anti-reflection coating, the InGaAs cap left on, in reverse bias and with a back surface reflectivity of 65%) for nominal dimensions and corrected for TEM widths.

samples [6], we assume a reflectivity of 65% from the back contact which boosts the QW response significantly because of successive internal reflections. The poor response particularly of the bulk material (below ~920 nm) is largely due to high surface recombination (10^5 m/s) of and absorption in the InGaAs cap layer which would be removed from the optical window area of a device before AR coating. In any case, for TPV applications with a selective emitter, these wavelengths are not important as a narrow band illuminating spectrum is expected to be mainly in the wavelength range of the QW response.

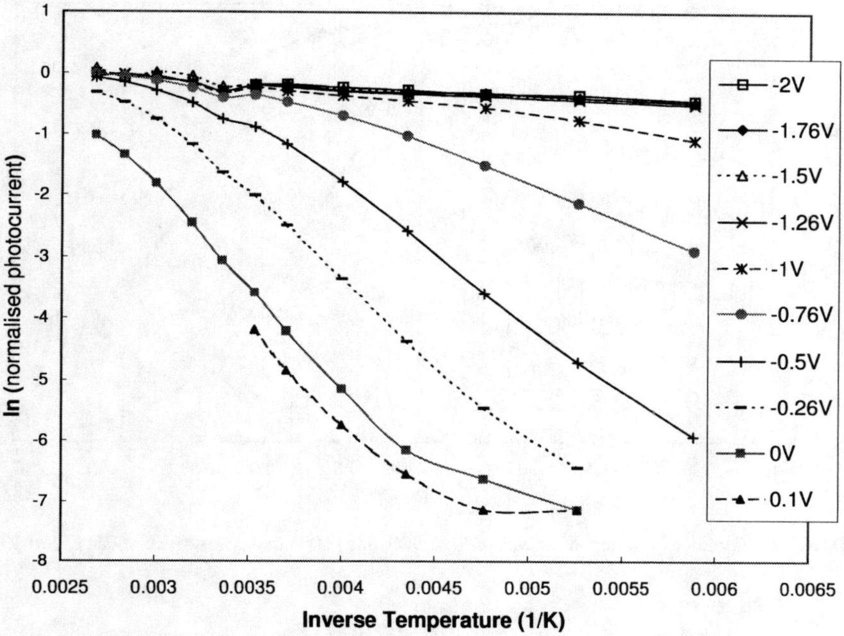

FIGURE 5. Arrhenius plot of the logarithm of the photocurrent vs. inverse temperature from monochromatic I-V at 1800 nm for various temperatures of sample MR1674.

Carrier Escape

The spectral response measurements of MR1674 were performed at 25°C and at 100°C and at 0 V and -2.5 V bias. At elevated temperatures such as 100°C the QW response is significantly higher at zero and forward bias which indicates thermal excitation out of the wells. At -2.5 V reverse bias the QW response is higher still but virtually the same for 25°C and 100°C. This suggests that saturation is occurring and all carriers are collected by field assisted escape at that bias.

In Fig. 5 we show the Arrhenius plot of the logarithm of the photocurrent vs. inverse temperature. The data is taken from bias-dependent photocurrent measurement at a monochromatic wavelength of 1800 nm, solely generating carriers deep in the QWs, for temperatures between 170 and 373 K. For temperatures below room temperature a cryostat was used and for higher temperatures a temperature stage with a Peltier plate was employed. In reverse bias the carrier escape rate increases as the wells and barriers increasingly become trapezoidal and the effective barrier height is reduced. Hence the photocurrent saturates at high reverse bias and in principle also at high temperatures [11]. The saturation for high reverse bias is clearly visible in Fig. 5 for reverse biases below -1.26 V across the temperature range. However, saturation at high temperatures due to thermionic carrier escape and thermally assisted tunnelling requires temperatures above 100°C for biases greater than about -0.5 V.

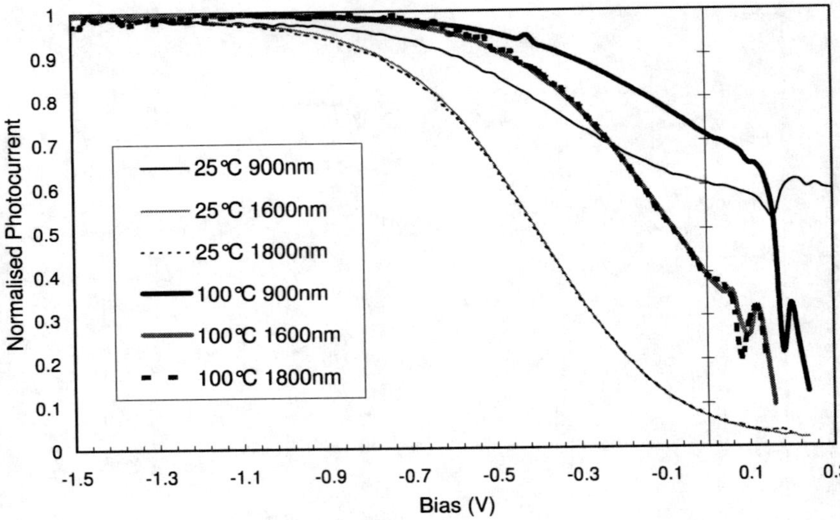

FIGURE 6. Monochromatic I-V of MR1674 at 900 nm, 1600 nm and 1800 nm at 25°C and 100°C.

Monochromatic I-V

Monochromatic I-V measurements are photocurrent measurements across a range of biases at a particular wavelength. In Fig. 6 monochromatic I-V measurements at 25°C and 100°C are shown for various wavelengths, which were chosen to excite carriers in different parts of the cell. 900 nm is above the bandgap of the InP, 1600 nm excites into the lattice-matched InGaAs layer and 1800 nm excites solely into the QWs (see Table 1 and Fig. 7). Again it is clear from Fig. 6 that an elevated temperature is aiding carrier escape but that collection has not saturated, even at 100°C. The reduction in photocurrent as we move from far reverse bias towards forward bias, even for above-bandgap photons at 900 nm, suggests problems with carrier collection, possibly due to background doping in the i-region.

Dark Current

The dark currents of these devices have only a small spread, barring a little parallel resistance, indicating good processing. Fig. 8 shows that the dark current of MR1674 is at a similar level to previous multi-quantum-well structures such as a 30 QW strain-compensated structure with an absorption edge of about 1.77 μm [5]. MR1674 has deeper QWs, which should increase the dark current, but on the other hand it has only 2 QWs (compared with 30 QWs in MR1288), which should lower the dark current. In Fig. 8, the dark current of MR1674 is also compared with bulk InGaAs control cells. It has a better dark current than the InGaAs p-n E004, a homo-structure, but worse than the InGaAs p-i-n double hetero-structure MR1317.

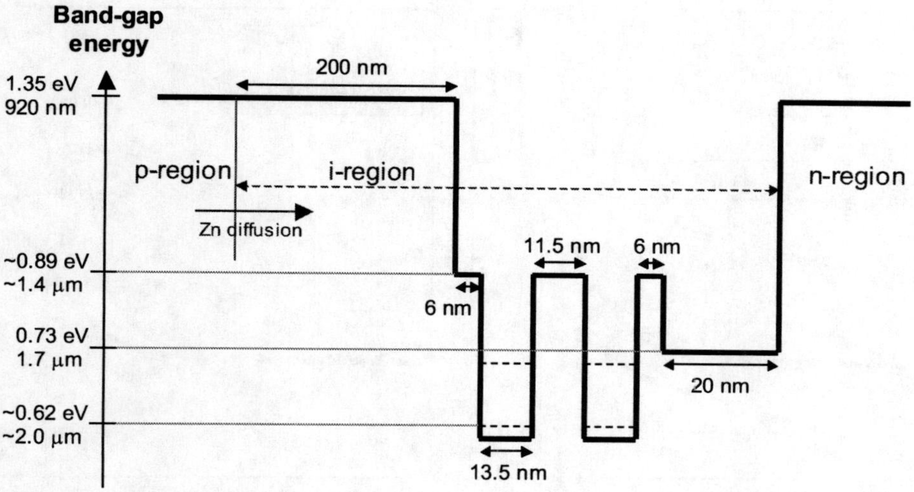

FIGURE 7. Schematic diagram of the Conduction Band of MR1674 in flat band condition (not to scale).

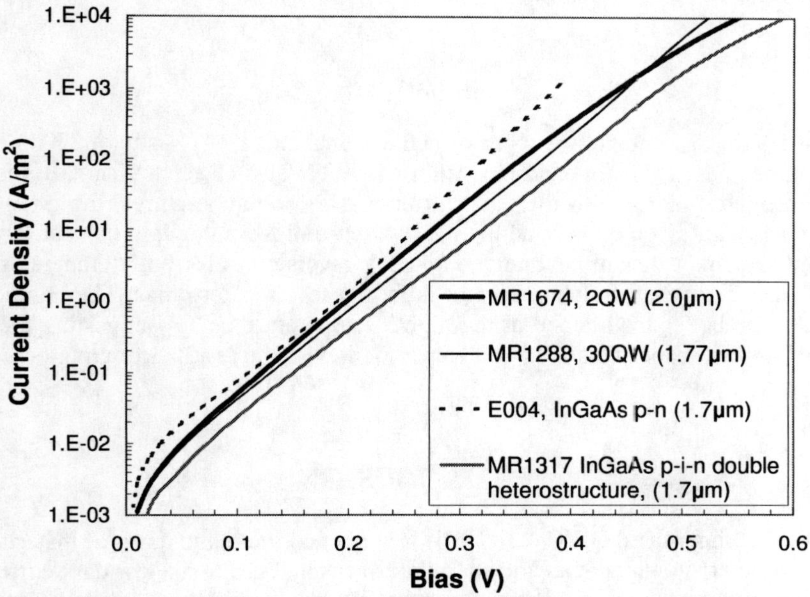

FIGURE 8. Dark current density of MR1674 compared with other samples. The absorption edge is indicated in brackets.

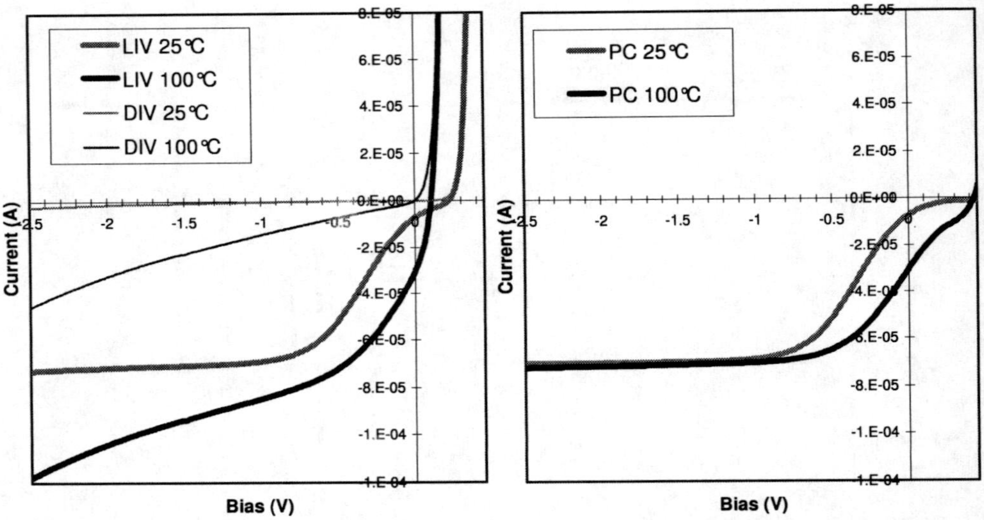

FIGURE 9. Sample MR1674. **(a)** Light I-V (LIV) and dark current (DIV) at 25°C and 100°C. **(b)** Photocurrent (PC) at 25°C and 100°C.

Light I-V

Light I-V measurements were performed at 25°C and 100°C using a 3200 K black-body source and a 1500 nm band-pass filter with a FWHM of about 90 nm. In Fig. 9(a), these are plotted along with the dark currents at those temperatures. Fig. 9(b) shows the photocurrent, given by subtracting the dark current from the light current. The photocurrent remains constant beyond about -1.5 V reverse bias for both temperatures, but at 25°C it is clear that the photocurrent is falling towards zero bias. The photocurrent appears to be better maintained at the higher temperature, again suggesting thermally assisted carrier escape. This is in good agreement with the results from monochromatic I-V measurements.

CONCLUSION

This strain-compensated QWC (MR1674) is very encouraging in terms of material quality, in particular showing clean and sharp interfaces in TEM and a low dark current. We were able to demonstrate that strain-compensation can be used to extend the absorption edge of this material system beyond 2 μm.

However, it seems evident that the incomplete carrier escape contributes to the poor electrical performance of these devices under illumination, in particular at room temperature. The carrier collection improves with increasing temperature, although it is not complete at 100°C yet. Transport problems could also be partly responsible.

In another strain-compensated QWC with 40 QWs and an absorption edge of

1.95 μm, we have observed full carrier escape at room temperature and around zero and modest forward bias [12].

REFERENCES

1. Rohr, C., Barnham, K. W. J., Connolly, J. P., Nelson, J., Button, C., and Clark, J., "Potential of InGaAsP quantum well cells for thermophotovoltaics," in *Proc. 26th International Symposium on Compound Semiconductors*, Institute of Physics Conference Series 166, Institute of Physics Publishing, Bristol and Philadelphia, 2000, pp. 423–426.
2. Ekins-Daukes, N., Barnham, K., Connolly, J., Roberts, J., Clark, J., Hill, G., and Mazzer, M., *Appl.Phys.Lett*, **75**, 4195–4197 (1999).
3. Rohr, C., Connolly, J. P., Ekins-Daukes, N., Abbott, P., Ballard, I., Barnham, K. W., Mazzer, M., and Button, C., *Physica E: Low-dimensional Systems and Nanostructures*, **14**, 158–161 (2002).
4. Ekins-Daukes, N., Kawaguchi, K., and Zhang, J., *Crystal Growth & Design*, **2**, 287–292 (2002).
5. Rohr, C., Connolly, J., Barnham, K. W., Mazzer, M., Button, C., and Clark, J., "Strain-balanced $In_{0.62}Ga_{0.38}As/In_{0.47}Ga_{0.53}As$ (InP) Quantum Well Cell for Thermophotovoltaics.," in *Proc. 28th IEEE Photovoltaic Specialists Conference*, IEEE, USA, 2000, pp. 1234–1237.
6. Abbott, P., Rohr, C., Connolly, J. P., Ballard, I., Barnham, K., Ginige, R., Corbett, B., Clarke, G., Bland, S. W., and Mazzer, M., "A Comparative Study of Bulk InGaAs and InGaAs/InGaAs Strain-Compensated Quantum Well Cells for Thermophotovoltaic Applications," in *Proc. 29th IEEE PV specialists conf.*, IEEE, USA, 2002.
7. Connolly, J. P., Nelson, J., Ballard, I., Barnham, K. W. J., Rohr, C., Button, C., Roberts, J., and Foxon, T., "Modelling Multi Quantum Well Solar Cell Efficiency," in *Proc. 17th European Conference and Exhibition on PV Solar Energy*, WIP, Munich, 2001.
8. Matthews, J. W., and Blakeslee, A. E., *J.Crystal Growth*, **27**, 118–125 (1974).
9. Drigo, A. V., Aydinli, A., Carnera, A., Genova, F., Rigo, C., Ferrari, C., Franzosi, P., and Salviati, G., *J.Appl.Phys*, **66**, 1975–1983 (1989).
10. Romanato, F., Napolitani, E., Carnera, A., Drigo, A., Lazzarini, L., Salviati, G., Ferrari, C., Bosacchi, A., and Franchi, S., *J. Appl. Phys.*, **86**, 4748–4754 (1999).
11. Zachariou, A., Barnes, J., Barnham, K. W. J., Nelson, J., Tsui, E. S. M., Epler, J., and Pate, M., *J.Appl.Phys.*, **83**, 877–881 (1998).
12. Abbott, P., Rohr, C., Connolly, J., Ballard, I., Barnham, K., Ginige, R., Clarke, G., and Mazzer, M., "Characterisation of Strain-Compensated InGaAs/InGaAs Quantum Well Cells For TPV Applications," in *The Fifth Conference on Thermophotovoltaic Generation of Electricity*, edited by T. J. Coutts, G. Guazzoni, and J. Luther, AIP Conference Proceedings, American Institute of Physics, Woodbury, New York, 2002, in this volume.

The Design, Fabrication and Evaluation of InGaAs/InP TPV Cells for Commercial Applications

Ravin Ginige*[+], Carmel Kelleher*, Brian Corbett*
Just Hilgarth** and Graham Clarke***

* NMRC, University College Cork, Cork, Ireland
** RWE Solar GmbH, Germany
*** IQE, Cardiff, UK
+ Corresponding Author: E-mail: rginige@nmrc.ie
Tel: + 353 21 490 4267/4177

Abstract. We report on the design, fabrication and evaluation of InGaAs/InP TPV cells. The fabrication process was developed for single wafer laboratory processing of lattice matched $In_{0.53}Ga_{0.47}As$ cells and subsequently adapted for batch processing in industry. The 1 x 1cm cells and 1mm diameter mesa structures were evaluated using Dark IV and the PVIV techniques. Of the batch (volume) processed 1 x 1 cm cells, the best measured efficiency under AM0 conditions (Si standard) was 10.9% (batch efficiency was 10.4%), with a short circuit current density (I_{SC}) of 54.6 mA/cm^2, open-circuit voltage (V_{OC}) of 363 mV and a fill factor (FF) of 74%. Using an IR lamp, a current density of 1442.8 mA/cm^2, open-circuit voltage of 418.5 mV and a fill factor of 65.7% were recorded. We also point out the salient design and fabrication features employed in realising the above quoted results, which we believe, are the first reported results from an industrial volume production. Problems associated with volume production are also discussed with solutions.

INTRODUCTION

Thermophotovoltaic cells are used to convert thermal radiation to electrical energy from hot emitters (radiators) operating around 1000-2000°K. The emitted radiation is mainly in the long wavelength (peak emission from a Holmium emitter at 1500°K is at 2.0μm and for an Erbium emitter at 1870°K is at 1.6μm [1]) and as such narrow bandgap cells such as $In_{(0.53)}Ga_{(0.47)}As$ cells with a bandgap of 0.72eV (1.7μm) are well suited for this purpose.

The thermophotovoltaic (TPV) system that such cells are designed for, comprises of a fuel burner, a selective emitter and high efficiency PV cells that have been tailored to absorb the radiative emission from the selective emitter and filter combination [2]. The filters reflect the infrared emission that the cell will not be able to convert to

CP653, *Thermophotovoltaic Generation of Electricity: 5th Conference*
edited by T. J. Coutts, G. Guazzoni, and J. Luther

electricity. Unlike the III-V solar cells, which are in the main used for space applications under normal incidence and one-sun conditions, TPV cells are operated under many suns and oblique angles of incidence. Their application range from military to commercial, such as a range extender for an electric motor vehicle in our case [3]. Without exception, in a range of applications, TPV cells are operated at very high power densities (W/cm^2), many amperes per square centimetre and as such need to be designed to minimise resistive losses. Another distinction from the bulk of today's solar cells which are made from Silicon (Si) and Gallium Arsenide (GaAs) is the narrow bandgap of the TPV cells. These attributes make the design and fabrication of TPV cells different to GaAs and Si based solar cells.

We demonstrate that an industrial GaAs cell process can be adapted with minimal but critical modifications for processing 'state of the art' TPV cells and that it can be adapted from a single wafer laboratory process to an industrial batch process. Also, we present the characterisation tool, 'dark forward IV' measurements as an effective tool in determining if cells would perform well under AM0 or IR conditions.

TPV CELL STRUCTURE AND OPTIMISATION

The epitaxial layer structure used in the manufacture of the TPV cells consisted of a highly doped p^+-InGaAs 'Cap' layer, p-InP window, p-InGaAs emitter, n-InGaAs base layer, n-InP buffer and an n-InP substrate. The main elements of the cell structure is complete with back contact metallisation and the grid metallisation which minimises series resistance and maximises the cell area available for the absorption of the thermal irradiance, a conflicting set of requirements. To enhance the optical absorption and reduce the reflectance at the window layer, an anti-reflection layer is deposited. While such a structure, meets the general PV cell requirements, additional process steps (e.g. incorporation of a trench surrounding the TPV cell) are needed to realise high efficiency low band gap TPV cells.

The performance of the cell is governed by many factors such as doping, thickness of the active layers etc. Poor morphology i.e. textured, cloudy surface (indicative of growth related defects) and/or 'large' defects are known to degrade cell performance.

FABRICATION OF THE 'TRENCH TPV CELL': BASELINE MANUFACTURING PROCESS

The industrial GaAs solar cell fabrication process (GaAs/GaAs or GaAs/Ge) consists typically of five step; front contact metalisation, 'cap' etch, deposition of the anti-reflection coating, contact pad opening and the rear back metalisation. This process gives excellent results for GaAs cells, with low dark current. GaAs being a large band gap material (1.42eV) the 'edge' leakage is not a large component of the dark current. However with narrow bandgap material such as $In_{0.53}Ga_{0.47}As$ (0.7eV) the 'edge' leakage is a major contributor to the dark current and as such edge isolation needs to be an integral component in the fabrication process. This is accomplished by enclosing the cell in a trench (with the side-walls protected by an insulation i.e. anti-reflection coating). In figure 1 below, we compare the effect of the forward current with and without a trench. The 'No-Trench' curve is clearly seen to be about two orders of magnitude higher than the 'After Trench Etch' curve, illustrating the importance of this process step. Another crucial step for reproducible and high efficiency PV cells is the back alloy step. Measurements after the deposition of n-metal shows the presence of series resistance, indicated by the tendency for the I-V characteristics (curve 'After N-Metal' in figure 1) to 'flatten' out. On alloying the back contact the 'high' series resistance is removed as seen from the absence of the 'flattening' of the IV curve (see curve 'After Alloy'). Note that despite the rear side etching to remove epitaxial growth contamination prior to the back metal evaporation and the large area, the alloy step is a critical process step, as is the trench isola tion step.

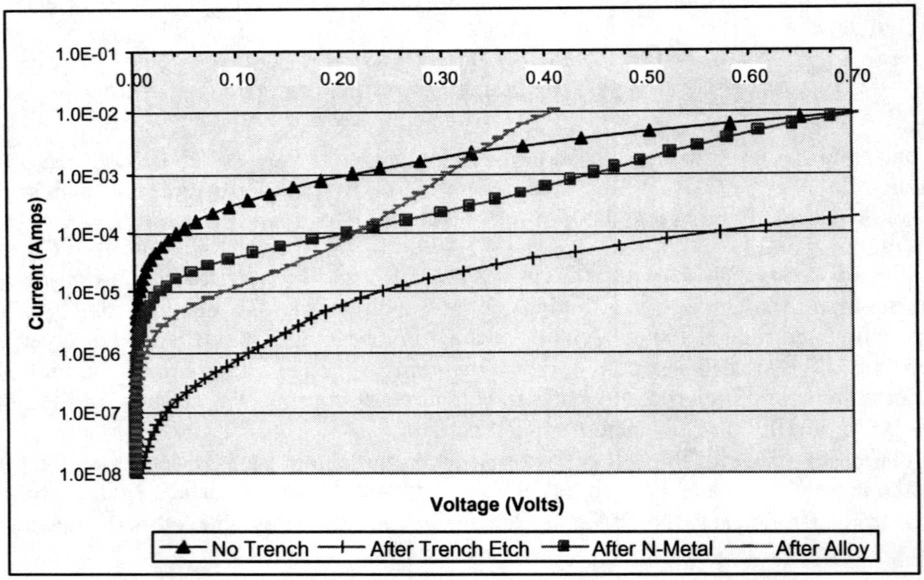

FIGURE 1. The dark IV characteristics of a typical 1 x 1 cm^2 InGaAs cell before and after trench and alloy steps. The two order magnitude lowering of the current after the trench isolation step illustrates the importance of the trench process in low bandgap material such as InGaAs. The effect of the back alloy step is shown by the absence of the 'flat' region in the post alloy curve, signifying the absence of the series resistance.

We have already outlined the basic five step industrial process that is used by volume manufacturers of GaAs PV cells. For low bandgap TPV cells, we have demonstrated with the use of forward dark IV characterisation the additional process steps that need to be incorporated to the well tried out GaAs cell process, for realising low leakage, low resistance cells. This baseline process is given in figure 2.

The baseline process consists of a front metallisation process, followed by the cap etch (using the metal as a mask) to expose the areas onto which the thermal radiations falls. This is followed by the deposition of the two layer anti-reflection coating. This overlayer covers the entire cell and thus is followed by the photomasking for the openings of the contact pad area. The next vital process is the trench isolation that surrounds the cell. Note, that this step cannot precede the anti-reflection coating, as the TiO_2/Al_2O_3 due to its metallic components will short the junction. Also note, this step leaves the side-walls of the trench exposed to the environment with no side passivation. Finally the back metallisation, alloying and sawing (dicing) steps completes the fabrication sequence. As already mentioned the trench and alloying is crucial for good and reproducible results.

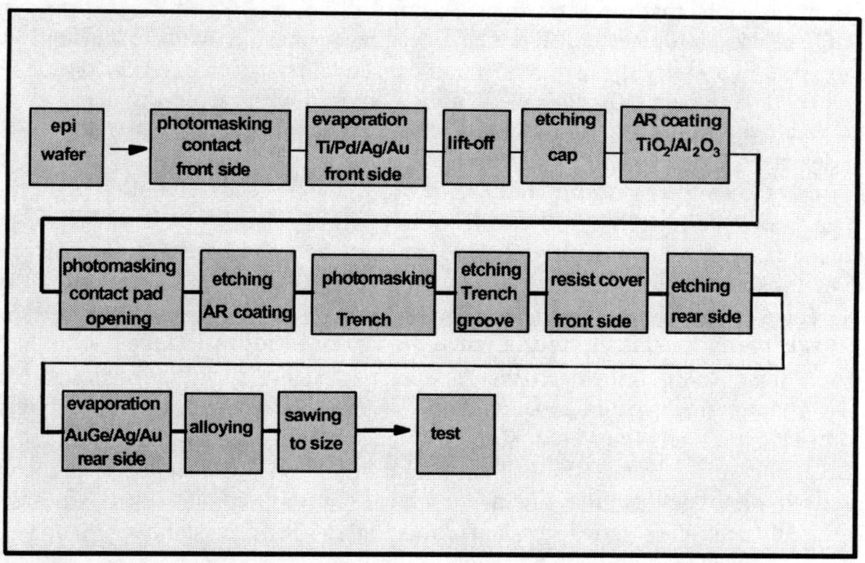

FIGURE 2. The Baseline Process Flow for the low bandgap InGaAs TPV Cells

The other significant process parameter is the choice of etchants. Highly selective etchants are used throughout. The InGaAs etch is H_2O_2: H_3PO_4: H_2O and the InP etch is H_3PO_4:HCl. The front metallisation consisted of Ti/Pt/Ag/Au metal and AuGe/Ag/Au for the back metal. The front metallisation scheme employed gave a specific contact resistance, which was in the low 10^{-7} ohm.cm^2. The anti-reflection coating is a dual layer coating of TiO_2/Al_2O_3 optimised to 1500nm with a reflectance less than 0.5%.

The basic process outlined above was used to process 20, 2" InGaAs on InP wafers. Each wafer contained 12, 1 x 1 cm^2 cells. The wafers/cells were processed as a batch process. The new process with the additional steps and etchants was designed to be totally compatible with the volume production system for GaAs space solar cells employed by one of the authors.

MATERIAL AND CELL CHARACTERISATION

The quality of the epitaxial material is an important factor in the performance of these TPV cells. In this study we have looked at material with differing number of defects and morphology. The main conclusions are, that poorer the morphology and greater the defect density, the worse is the IV characteristics. In addition to these, the presence of 'hillock' type of defects (even if the number of defects was considered to be low) was found to lead to very poor cells.

Figure 3 is a plot of Photo-IV (IR) and the dark IV characteristics of three typical 1x1cm^2 cells of different defect densities ranging from 1.3 to 10.5 defects/cm^2. It is clear from the plots, that the performance degradation in P_{MAX} and V_{OC} is least for the cell with the lowest defect density. The lower P_{MAX} and V_{OC} with increasing defect densities lead to a lower fill factor (FF) and consequently efficiency. The dark current too, as expected is seen to have considerably increased with increasing defect density. It was concluded from this work that a desired specification for epi-grown wafers is a defect density < 2 defects/cm^2, a defect size < 15μm^2 and no 'Hillock' type of defects.

Figure 4 shows the measured and calculated photo I-V characteristics of a typical 1 x 1 cm^2 solar cell under AM0 and IR illumination. The AM0 curves show good agreement between the measured (curve 1) and the calculated results (curves 2), however for the IR spectrum the calculated IR results (curve 4), indicates a higher open-circuit voltage and fill factor in comparison with the measured IR curve (curve 3). A likely cause for this discrepancy may be that of series resistance due to the grid design or metallisation. However, this has been ruled out as we have measured the dark I-V characteristics of an array made up of such 1x1cm^2 cells and have found no series resistance effects (curve not shown) to currents up to 2 amperes. The presence of a 'resistance effect' under high illumination conditions (1.4 amperes) is therefore believed to arise due to the lateral resistance in the emitter layer. In PV cell measurements, the dark series resistance and 'light' series resistance may differ due this effect arising from lateral conduction in the layer where much of the photovoltaic current is generated (In GaAs and in InGaAs latticed matched p/n cells much of the photovoltaic current is generated in the emitter).

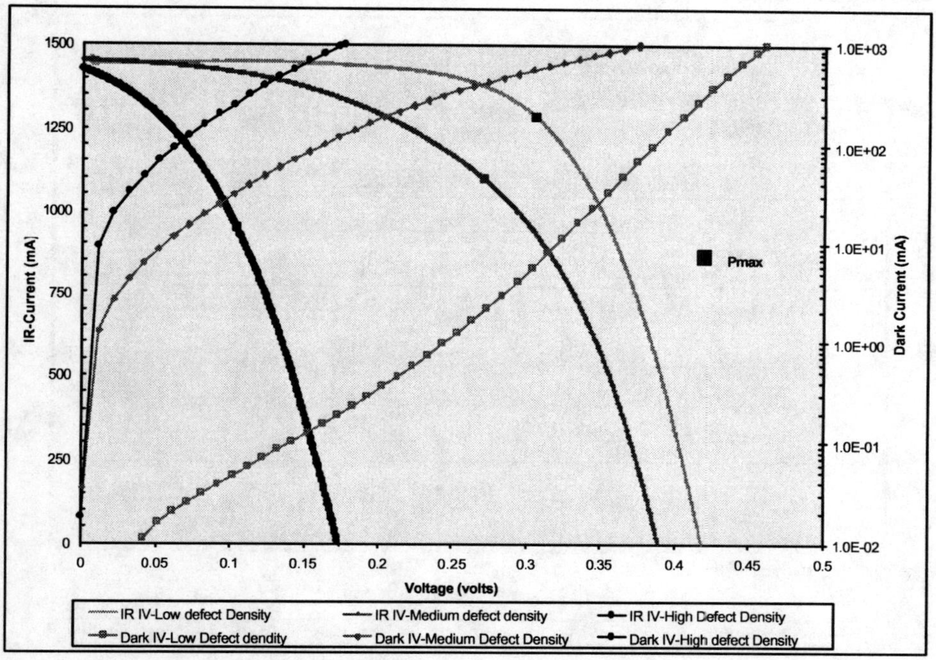

FIGURE 3. A plot of the IR-PVIV (IR current is shown on the Y1 axis) and dark IV (dark current on the Y2 axis) of three cells with different defect density (1.3 -10.5 defects/cm^2), showing that performance decreases with increasing defect density.

From the first batch of 92, 1 x 1cm^2 p/n InGaAs cells fabricated and measured under AM0 conditions, the best cell had an efficiency of 10.9% (with a batch efficiency of 10.4%). The short circuit current density of this cell was 54.6mA/cm^2 and the open-circuit voltage was 363mV, with a fill factor of 74%. This cell, measured under IR illumination had a current density of 1442.8 mA/cm^2, open-circuit voltage of 418.5 mV and a fill factor of 65.7%. The lower fill factor is believed to be for reasons already stated i.e. resistance due to lateral current spreading in the emitter.

The cells fabricated using the process outlined in figure 2, suffer from two problems. The first being a degradation due to sawing (dicing) of the cells and the second a time related degradation. The latter is believed to be due to exposure of the side-walls of the cells. In figure 5, four plots are shown, illustrating the more typical and the worst case degradation after sawing. It is evident from figure 5 that the saw action degrades the cells at the low current/low voltage end. However at the high current end i.e. around the operating point of the TPV cells (~0.4 volts) no measurable degradation is evident.

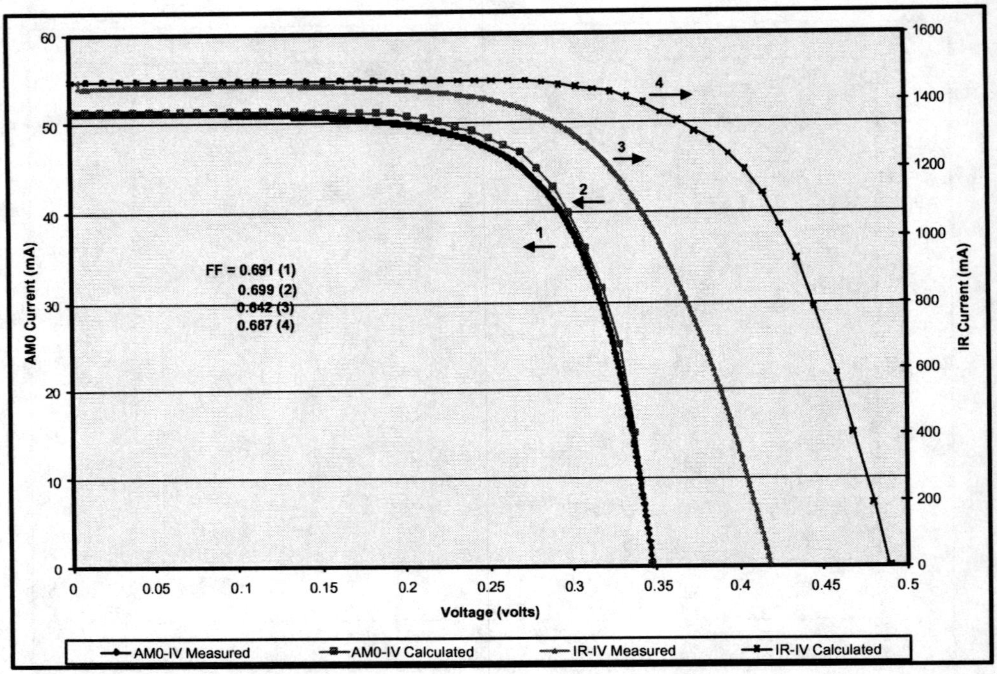

FIGURE 4. The measured AM0 IV characteristics (curves 1 and 2) shows good agreement, while the IR characteristics (curves 3 and 4) shows poor agreement. The poor IR agreement is thought to be due to lateral current spreading in the emitter layer.

A related problem is the degradation in the dark IV characteristics observed with time (figure not shown). Three cells (good, medium and poor) were measured over three consecutive weeks. It was found that the dark IV characteristics degraded the greatest for the best (low) dark current cell and least with no apparent change for the worst cell and in between for a mediocre cell. Again there was no measurable degradation at the high operating current end.

We are confident that these problems can be resolved by a simple modification to the process, to include removal of saw damage by etching and side-wall passivation. This modified process is currently being evaluated and will be reported in a separate publication.

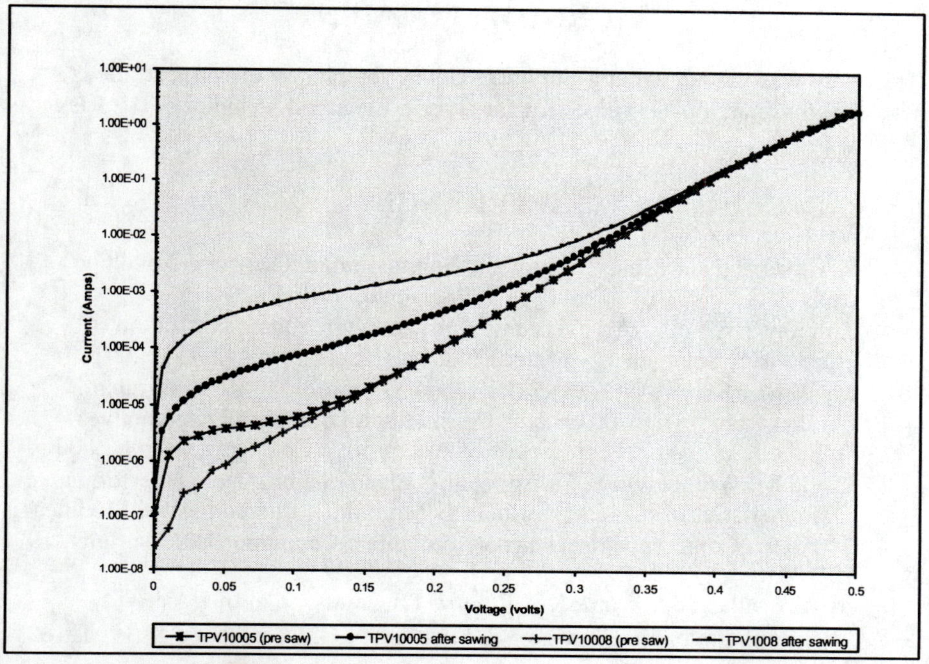

FIGURE 5. The Dark IV characteristics of two 1x1cm^2 cells, pre and post sawing. The plots, clearly indicates the degradation at the low voltage end on dicing the cells.

CONCLUSIONS

We have designed, fabricated and analysed latticed matched $In_{(0.53)}Ga_{(0.47)}As$ on InP TPV cells. Above all, a laboratory process has successfully been adapted to an industrial batch process. Many of the problems associated with processing low bandgap material have been identified and resolved. The time related degradation and the saw induced degradation have both been remedied. An epitaxial wafer specification has been arrived at, thereby increasing the process yield. We have also shown that the dark current-voltage (IV) measurement is a quick and reliable indicator of the quality of the cells or material. Cell efficiencies of 10.4% at 1 sun, AM0 for 1 x 1cm^2 have been achieved in batch processing, with a best efficiency of 10.9%. The efficiency measured is comparable to those reported in the literature [3,4].

ACKNOWLEDGMENT

This work was funded under a European Union contract on the project titled 'A Thermophotovoltaic Power Generator for Hybrid Electrical Vehicles (THE REV)'.

REFERENCES

1. P.A. Iles and C.L.Chu, "Design and Fabrication of Thermophotovoltaic Cells", First WCPEC, Dec. 5-9, 1994, Hawii, pp 1750-53
2. Timothy J. Coutts, "An overview of thermophotovoltaic generation of electricity", Solar Energy Materials & Solar Cells, 66, 443-452, 2001
3. Massimo Masser, Auturo De Risi, Domenico Laforgia, Keith Barnham, Carsten Rohr, "High Efficiency Thermophotovoltaics for Automotive Applications", World conference of Automotive Engineers, March 2000
4. David M. Wilt, Navid S. Fatemi et.al, " Electrical and Optical Performance Characteristics of 0.74 eV p/n InGaAs Monolithic Interconnected Modules", 3rd NREL Conference on Thermophotovoltaic Generation of Electricity, 1997, 237-247
5. M.W.Wanlass, J.S.Ward, K.A.Emery, T.J.Coutts, "$Ga_xIn_{1-x}As$ Thermophotovoltaic Converters", First WCPEC, December 5-9, 1994, 1685-1691

DESIGN AND OPTIMIZATION OF THERMOPHOTOVOLTAIC DEVICES USING A NEW MODELING APPROACH

Sherif Michael and LT. Panayiotis Michalopoulos

Naval Postgraduate School, Space Systems Academic Group
699 Dyer Rd, Monterey, CA 93943-5000, michael@nps.navy.mil

Abstract. A new method for developing a realistic model of any type of solar cell, with application to Thermophotovoltaic devices, is presented in this paper. Taking into account the high cost of research and experimentation involved with the development of advanced cells, we present here this novel methodology. An example model of an InGaP/GaAs/Ge multi-junction cell is prepared and is fully simulated. The major stages of the process will be explained and the simulation results are compared to published experimental data. An example of cell parameters optimization is also presented. The flexibility of the proposed methodology is demonstrated and example results are shown throughout the whole process.

INTRODUCTION

In our research of the existing literature, we have found a number of significant publications that fully describe various aspects of device characteristics and modeling. However, all of them were focused on very specific issues lacking the breadth of a complete simulation tool.

The Silvaco Software Package [1] is a simulation software tool targeting the area of electronics and VLSI design. One of its major products is the Virtual Wafer Fabrication package (VWF). This is a large suite of highly sophisticated tools - among which is ATLAS - that aid in the design and development of all types of semiconductor and VLSI devices. Such types range from simple bipolar transistors to EEPROMs. The phenomena modeled start from simple electrical conductivity and extend to thermal analysis, radiation, laser effects etc. A big variety of detailed layer growth processes and material properties (mobilities, recombination parameters, optical parameters, ionization coefficients, etc) add to the accuracy of the simulation. However, we have not seen any effort, from researchers or from solar cell manufacturers, to utilize this powerful tool for the modeling of advanced solar cells.

Even though any type of solar spectrum with various concentrations can be easily specified and utilized, the standard 1 sun AM0 was used throughout our research and in this paper as an example source.

CP653, *Thermophotovoltaic Generation of Electricity: 5th Conference*
edited by T. J. Coutts, G. Guazzoni, and J. Luther
2003 American Institute of Physics 0-7354-0113-6

THE VWF MODELING PROCESS

The Example Multi-Junction Cell Modeled

The multi-junction cell, selected as an example to be virtually fabricated in this paper, is constructed of three stacked (Ge, GaAs and InGaP) cells. A 3-D illustration of this cell is shown in Fig.1 below.

These cells were first developed, modeled and evaluated individually. This way, their efficiency can be optimized and their performance characteristics can be adjusted to the required range. In the beginning, only the p-n junction was built. The back surface field (BSF) layer was then added and the improvement on the device performance was observed. Substrate and window layers are the final additions to the individual cell models. Short circuit currents and open circuit voltages were obtained. I-V characteristics and spectral responses are also plotted. Producing snapshots of the structure and how various parameters were distributed throughout the different layers provided a very valuable insight. Such parameters include; electron and hole concentrations and densities, photogeneration rates voltage potential, and electrostatic fields. Figs 2 and 3 show electron and hole concentrations profile for an example single junction cell, while the photogeneration rates are shown in Fig. 4 for the same cell. The potential fields in the depletion region with BSF, are shown in Figs 5 and 6 respectively, of the InGaP cell.

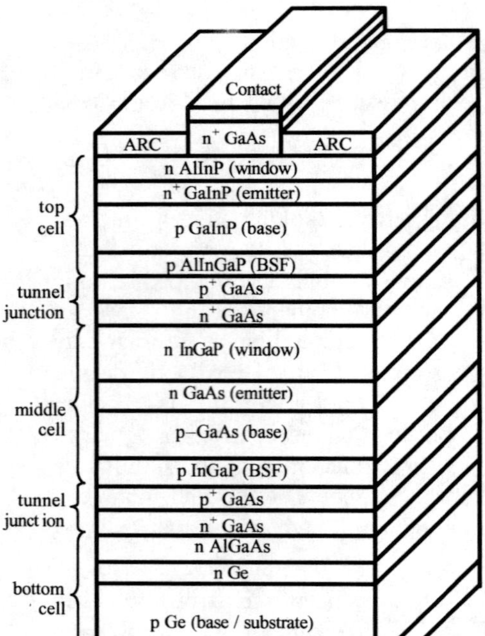

Fig 1: 3D layout of multi-junction cell

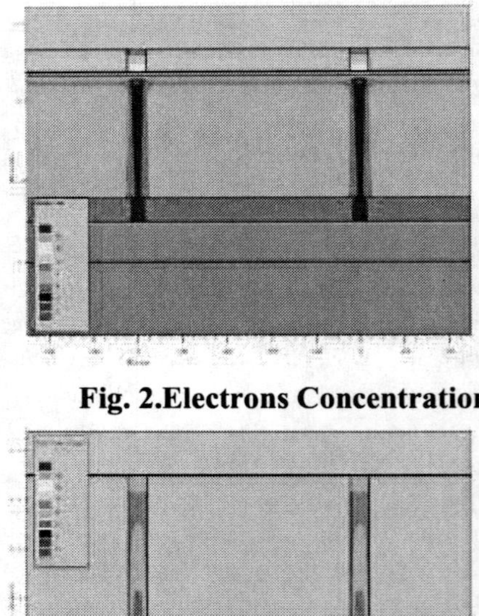

Fig. 2.Electrons Concentration

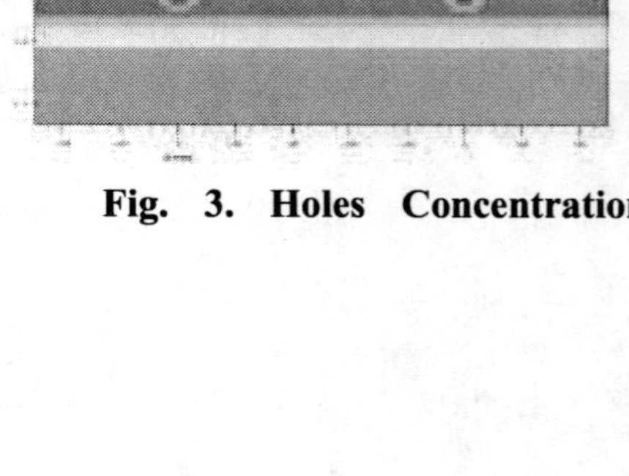

Fig. 3. Holes Concentration.

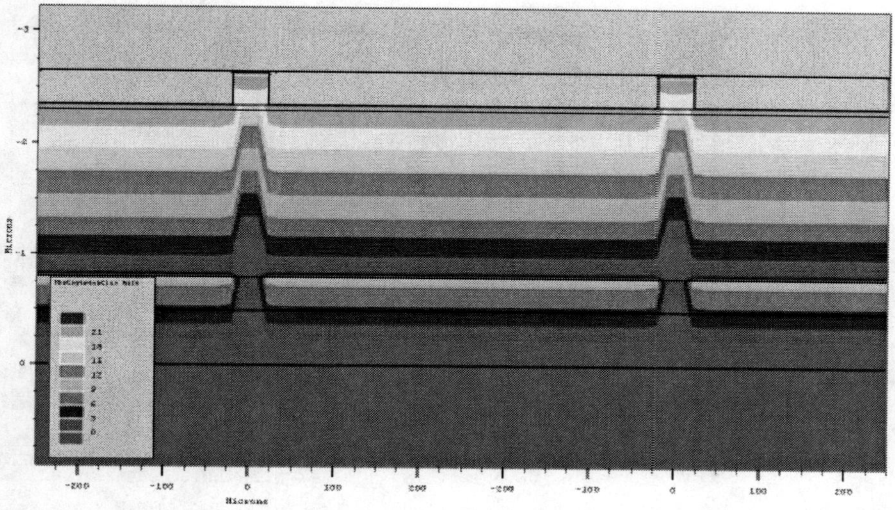

Fig. 4 Photogeneration under AM0.

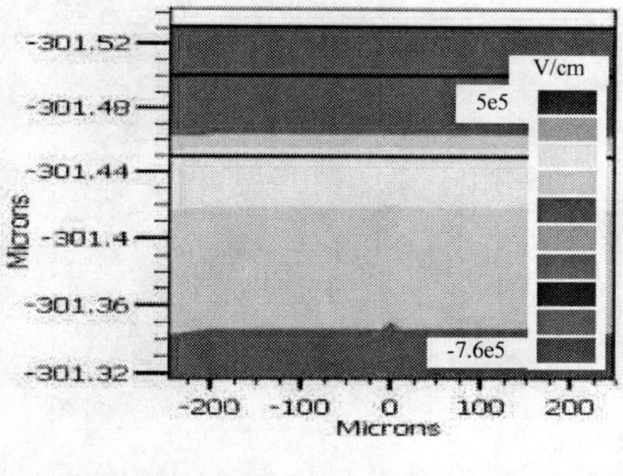

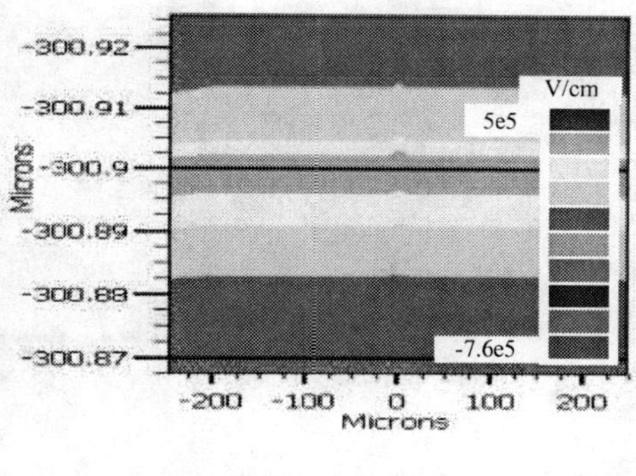

Fig 5: Depletion region field in InGaP cell cell

Fig 6: BSF layer field in InGaP

Mechanically stacked and multi-junction cell

The three cells were next placed in a mechanically stacked configuration. The spectral responses of the individual cells and those of the stack are shown in Fig 7, while Fig 8 shows the generated I-V characteristics of each cell, which compared favorably with corresponding experimental data for similar devices using the same parameters. The shadowing of the lower (GaAs, Ge) cells can be observed as depicted in Fig 9, while the potential field build-up is shown in Fig 10.

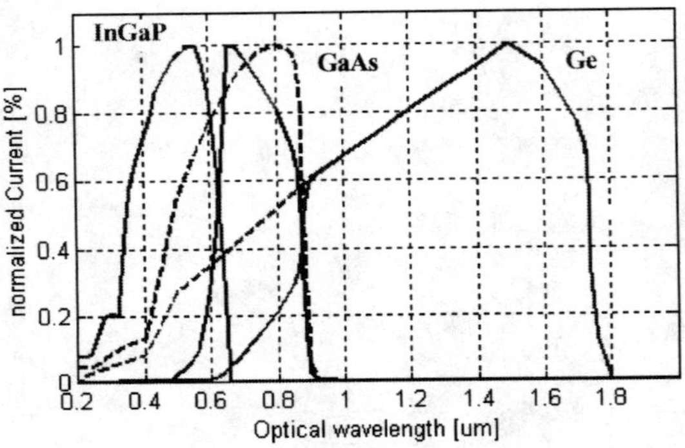

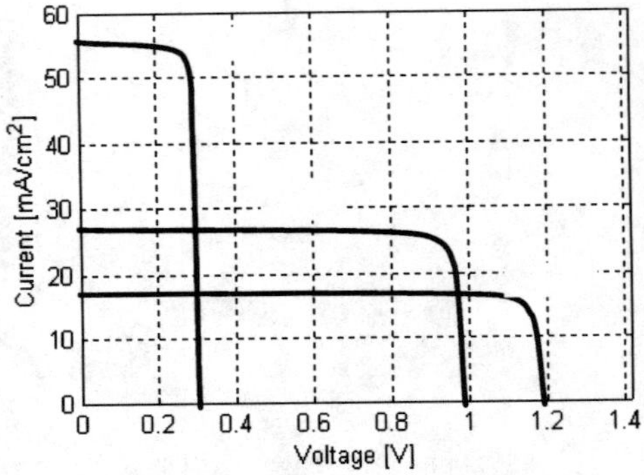

Fig 7: Spectral response of the individual cells (dashed) and of the stacked cell (continuous).

Fig 8: I–V characteristics of individual cells

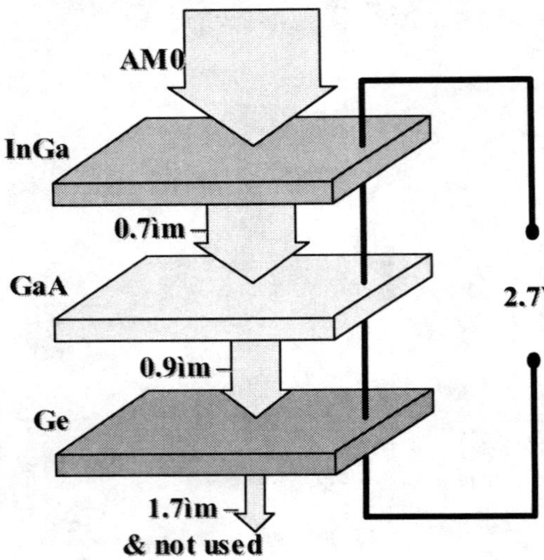

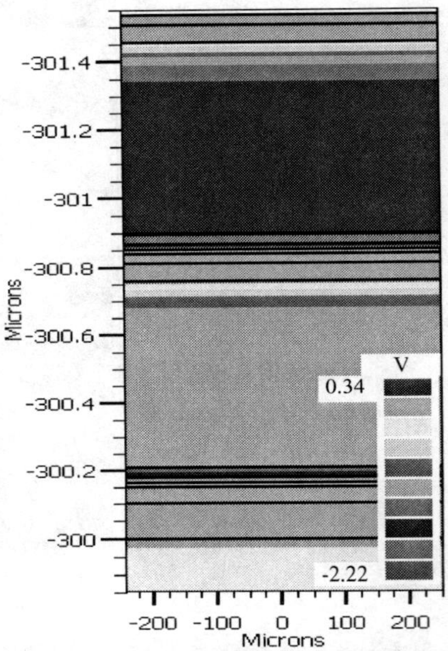

Fig. 9: Shadowing of photons in three mechanically stacked cells.

Fig 10: Potential build-up. (expanded view of the cells)

Two Tunnel junctions were also implemented and placed between each of the three cells. This resulted in the actual virtual fabrication of the entire multi-junction configuration, that was previously illustrated in Fig.1. The photogeneration rate for different wavelengths of the entire structure is presented in Fig 11. Spectral response of the overall multi-junction structure is displayed in Fig.12.

As a final step, anti-reflective coating (ARC) and golden contacts were added. The bottom contact was polished to reflect photons back in the cell, creating a back surface reflector (BSR). The expected generated I-V characteristics can be shown in Fig 13, which demonstrate the expected improvement in output power.

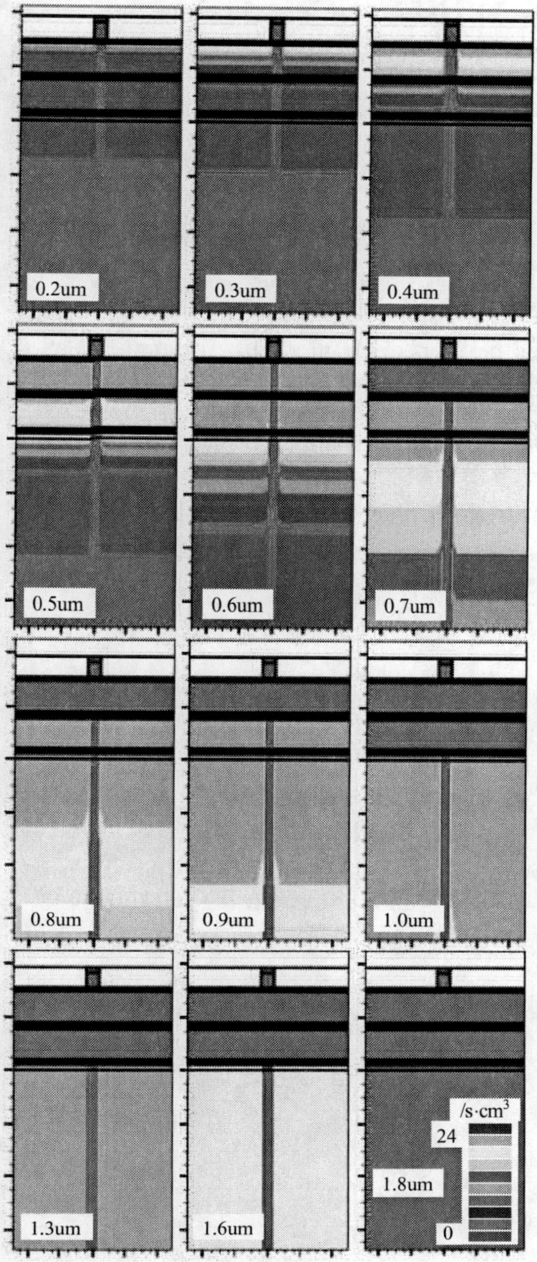

Fig 11: Photogeneration rate of the MJ cell model For different wavelengths of the AM0 spectrum.

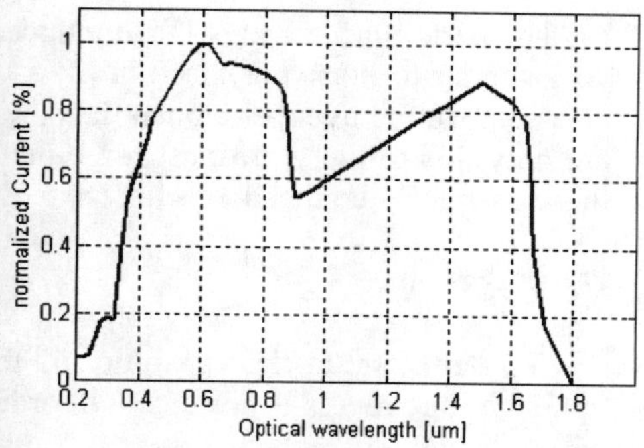

Fig 12: Spectral response

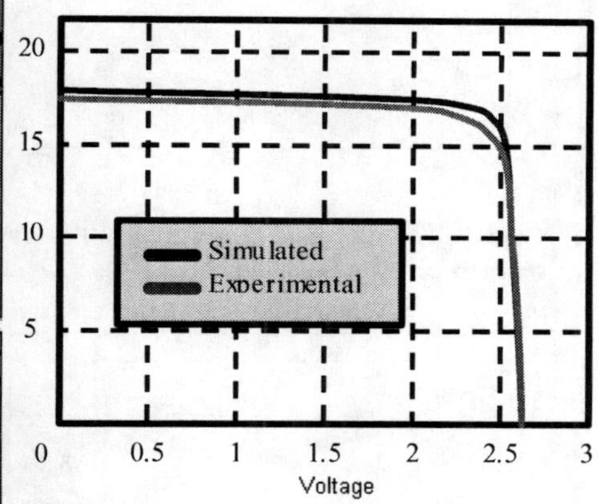

Fig 13: The overall MJ cell Characteristics.

COMPARISON OF RESULTS

The performance and results produced from the virtually fabricated MJ model, are very similar to those published in [2] as well as in many other papers. The results are also comparable to experimental data of similar manufactured cells. Minor differences are only due to the variations in the material and optical parameters used. If needed, those can easily be tuned to reflect experimental results more accurately.

Optimization

To demonstrate the capability of this developed MJ model, an example of an optimization process is presented. In order to maximize cell efficiency, a search for the optimum cell layers thicknesses was conducted. As expected, the results demonstrated in Figs. 14-17, verified some of the parameters and data of current MJ cell designs, that were previously obtained through years of research, experimentation and fabrication.

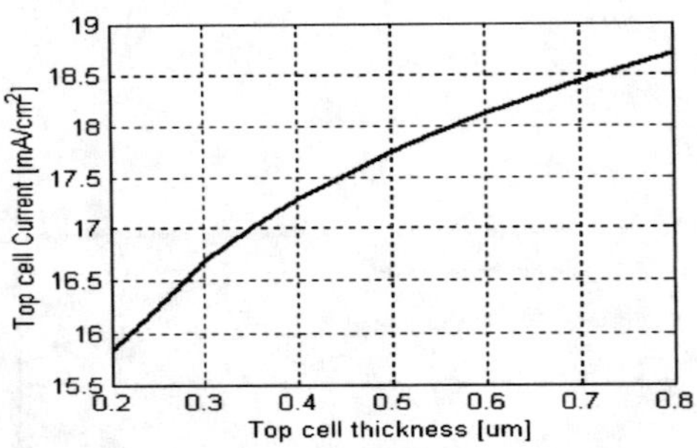

Fig 14: Top cell short-circuit current vs. thickness.

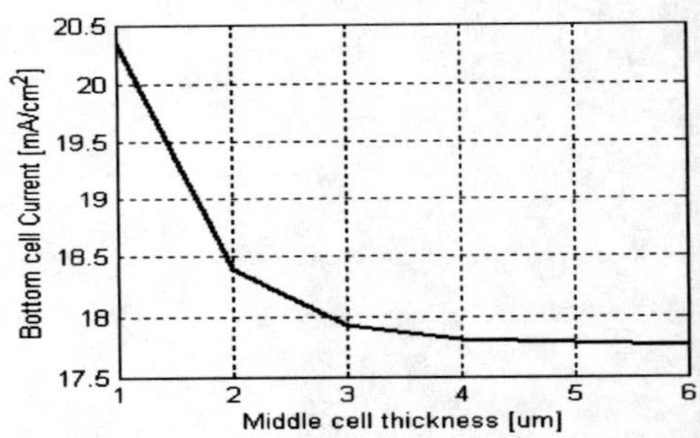

Fig 15: Bottom cell short-circuit current vs.middle cell thickness

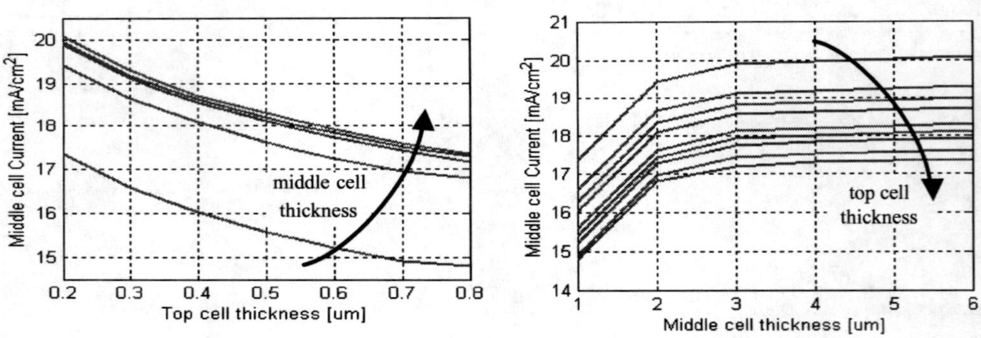

Fig 16: Middle cell short-circuit current vs. top and middle cell thickness

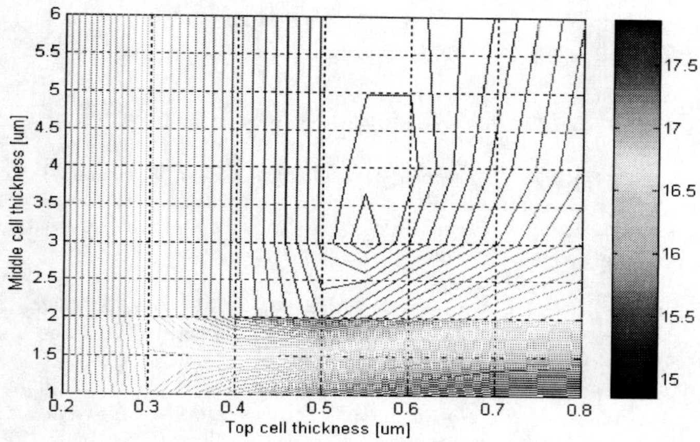

Fig 17: Total short-circuit current vs. top and middle cell thicknesses

CONCLUSION

A new technique has been presented for modeling and developing of multi-junction solar cells. This is the first time the SILVACO/ATLAS software package has been utilized in this research field. The developed models have demonstrated various material parameter effects on many of the cell characteristics, e.g. doping levels of different materials, thickness of different layers, etc. Simulation and modeling of wide bandgap material, e.g. Sb-based compounds, can also be incorporated, and currently underway. In our opinion, the introduction of this modeling technique to the thermophotovoltaic community will prove to be of great importance in aiding in the design, developing and optimization of advanced thermophotovoltaic cells.

REFERENCES

[1] *Silvaco ATLAS (vol I&II) user's manual*, Silvaco International.

[2] R.R. King, N.H. Karam, J.H. Ermer, N. Haddad, P. Colter, T. Isshiki, H. Yoon, H.L. Cotal, D.E. Joslin, D.D. Krut, R. Sudharsanan, K. Edmondson, B.T. Cavicchi, D.R. Lillington, "*Next-generation, high-efficiency III-V multijunction solar cells*", Photovoltaic Specialists Conference, 2000. Conference Record of the Twenty-Eighth IEEE , 2000, pp. 998 -1001

[3] D. Lillington, H. Cotal, J. Ermer, D. Friedman, T. Moriarty, A. Duda, "*32.3% efficient triple junction GaInP/sub 2//GaAs/Ge concentrator solar cells*", Energy Conversion Engineering Conference and Exhibit, 2000. (IECEC) 35th Intersociety , Volume: 1 , 2000, pp. 516 -521

[4] N.H. Karam, et al, "*Development and characterization of high-efficiency Ga/sub 0.5/In/sub 0.5/P/GaAs/Ge dual- and triple-junction solar cells*", Electron Devices, IEEE Transactions on , Volume: 46 Issue: 10 , Oct. 1999, pp. 2116 –2125

[5] K.C. Reinhardt, C.S. Mayberry, B.P. Lewis, T.L. Kreifels, "*Multijunction solar cell iso-junction dark current study*", Photovoltaic Specialists Conference, 2000, Conference Record of the Twenty-Eighth IEEE, pp 1118-1121, 2000

[6] P.K. Chiang, J.H. Ermer, W.T. Nishikawa, D.D. Krut, D.E. Joslin, J.W. Eldredge, B.T. Cavicchi, J.M. Olson, "*Experimental results of GaInP/sub 2//GaAs/Ge triple junction cell development for space power systems*", IEEE 25th Photovoltaic Specialists Conference Proceedings 1996, pp. 183 -186

[7] E. Van Kerschaver, J. Nijs, R. Mertens, M. Ghannam, "*Twodimensional solar cell simulations by means of circuit modelling*", Photovoltaic Specialists Conference, 1997., Conference Record of the Twenty-Sixth IEEE, pp 175-178, 1997

[8] R.K. Jain, D.J. Flood, "*Simulation of high-efficiency n/sup +/p indium phosphide solar cell results and future improvements*", Electron Devices, IEEE Transactions on, Volume: **41** Issue: 12, pp 2473-2475, Dec. 1994

[9] S. Kurtz, J.F. Geisz, D.J. Friedman, J.M. Olson, A. Duda, N.H. Karam, R.R. King, J.H. Ermer, D.E. Joslin, "*Modeling of electron diffusion length in GaInAsN solar cells*", IEEE 28th Photovoltaic Specialists Conference Proc. pp 1210-1213, 2000.

[10] R.J. Walters, G.P. Summers, S.R. Messenger, "*Analysis and modeling of the radiation response of multijunction space solar cells*", IEEE 28th Photovoltaic Specialists Conference Proc. pp 1092 -1097, 2000

[11] P. Panse, et al, "*Evaluation and modeling of junction parameters in Cu(In,Ga)Se/sub 2/ solar cells*", IEEE 28th Photovoltaic Specialists Conference Proc. pp 599 -602, 2000.

[12] V. Andreev, V. Khvostikov, et al." *Single-Junction GaSb and Tandem GaSb/InGaAsSb & AlGaAsSb/GaSb Thermophotovoltaic Cells,*" IEEE 28th Photovoltaic Specialists Conference Proc. , Anchorage, Alaska, 2000.

[13] Sherif Michael, Panayiotis Michalopoulos, "*Application of the SILVACO / ATLAS Software Package in Modeling and Optimization of State-of-the-Art Photovoltaic Devices,*"Proceedings of the 45 Midwest Symposium on Circuits and Systems, Tulsa Oklahoma, Aug 5-7, 2002.

SESSION 6

Cell Technology

Chair: V. M. Andreev

Zinc(P) diffusion in $In_{0.53}Ga_{0.47}As$ and GaSb for TPV devices

L.B.Karlina, B.Ya.Ber, P.A.Blagnov, M.M.Kulagina, A.S.Vlasov

IOFFE Physical-Technical Institute, 26 Polytechnicheskaya, St.-Petersburg, 194021, Russia

Abstract. The simultaneous diffusion of Zn and P into InGaAs and GaSb from a local vapour phase source was investigated. Zn diffusion allowed to form a TPV cells with *p*-layer (emitter) of 0.2-0.6 μm thickness without any additional post-diffusion operations. The penetration depth of Zn was determined by temperature and time of diffusion and was found to depend on defects and impurities at the InP/InGaAs heterointerfaces. This interface acts as a source of nonequilibrium neutral interstitials. We assume that the growth of thin undoped InP before the growth of InGaAs base region can decrease induced defects. This is recommended for an efficient suppression of the anomalous fast Zn diffusion. Doping profiles of the *p*-layers were obtained from the Raman spectra of the cell structure and by the Secondary Ion Mass Spectroscopy method. Good performance characteristics were demonstrated for InGaAs and GaSb TPV cells based on this technique.

INTRODUCTION

Different post-grown Zn diffusion technologies were proposed [1-5] during last years: application of Zn- containing polymer/glass spin-on films, sealed quartz ampoule diffusion, MOVPE based diffusion, a "pseudo-closed box" technique. The results of these investigations indicate the considerable potential of the diffusion technology. The main benefit of this technique is its low cost in obtaining good parameters of TPV devices.

The lattice-matched and mismatched InGaAs/InP structures for TPV devices have been grown by Metal-Organic Chemical Vapor Deposition (MOCVD) [6, 7].

The devices developed in this work were fabricated by the simultaneous diffusion of Zn and P into undoped layers of InGaAs grown by LPE on -InP (100) substrates. The method of diffusion from a local source applied to n-InGaAs has proved to be versatile, therefore it was also used for diffusion of Zn (P) into n-GaSb.

The choice of phosphorous as isovalent impurity was caused by several reasons. First, the treatment of an A^3B^5 material surface with phosphorous results in the passivation of surface defects and in the decrease of the surface recombination [8].

CP653, *Thermophotovoltaic Generation of Electricity: 5th Conference*
edited by T. J. Coutts, G. Guazzoni, and J. Luther

Second, phosphorous as it has been shown by us before [9] improves the radiation tolerance of InGaAs.

EXPERIMENTAL

Simultaneous diffusion of Zn and P from a local source was performed into heteroepitaxial InGaAs layers grown by liquid phase epitaxy. The initial concentration of carriers in these layers was $8*10^{16}$ - $1*10^{17}$ cm^{-3}. The diffusion experiments were carried out in an open quartz tube through which purified hydrogen flowed. Figure 1 shows the scheme of this process.

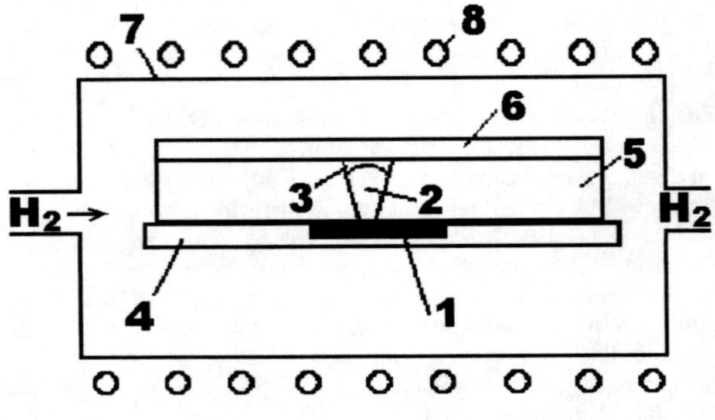

FIGURE 1. Scheme of the diffusion process from a local source. 1 - InGaAs/InP heterostructure or GaSb substrate. 2 - Sn-InP-Zn melt as source of Zn and P vapors, 3 - a specially designed bast having slit in the bottom, 4 - slider to transport substrate, 5 - cassette, 6 - cover, 7 - quartz tube with Pd purified H_2 inside the furnace, 8 - the furnace.

The diffusion technique used does not require controlling the vapor pressure of the volatile components of the III-V compounds in the whole system volume. It provides an excess phosphorus and zinc vapour pressure in a small volume in the immediate vicinity of the substrate. The local Zn diffusion source consisted of an InP-saturated Sn solution containing dissolved Zn from 0.9 to 5 wt.%. The initial Sn melt containing dissolved Zn was saturated by InP saturation substrate. The use of InP saturation substrates in the preparation of melts allows the process to be exactly controlled via the amount of InP dissolved in the melt at this temperature. It is well known that for indium and tin solutions saturated with InP, phosphorous vapour pressure (P_2 and P_4) is greater above the tin solutions at all temperatures in the range of 500-800^{0}C. In and Sn vapour pressures are negligible at these temperatures [10].

All samples were examined for traces of morphological damage after the diffusion treatment. The surfaces remained mirror smooth throughout. Zn(P) diffusion from Zn-P vapor phase was performed into n-GaSb(Te) (100) substrates. The depth distribution of Zn and P was investigated by the secondary ion mass spectrometry (SIMS) monitoring CsX^+ ions (X = As, P, Zn). The carrier concentration was determined by the Raman spectra measurements. The Raman scattering, spectra were excited with the 2.41eV argon laser beam and measured on a DFS-52 double monochromator. A (90°, 20°) backscattering geometry was used.

Photoluminescence (PL) spectra of (Zn, P) doped InGaAs were measured at 77K using Ar^+ laser for excitation and a LN_2 cooled PbS detector. PL excitation power density was about $5W/cm^2$.

RESULTS AND DISCUSSION

InGaAs

The initial experiments on Zn(P) diffusion into InGaAs layers grown on heavily doped (higher than $10^{18}cm^{-3}$) InP substrates showed insufficient reproducibility of results, since in some cases the anomalously high diffusion rate was observed. The effect of the InP/InGaAs heterojunction was discussed also before in [11]. It has been shown that the dislocation density both in material used and on the heterointerface affects the diffusion depth. Our investigations have shown that growth of undoped InP buffer layer on the substrate-layer heterointerface removes these defects. The results presented in this work have been obtained for the structures containing such a layer.

Figures 2, 3 show the SIMS profiles of the Zn and P concentrations for InGaAs samples. The plateau depth can be controlled quite accurately by changing the diffusion time (the time under source action) and temperature.

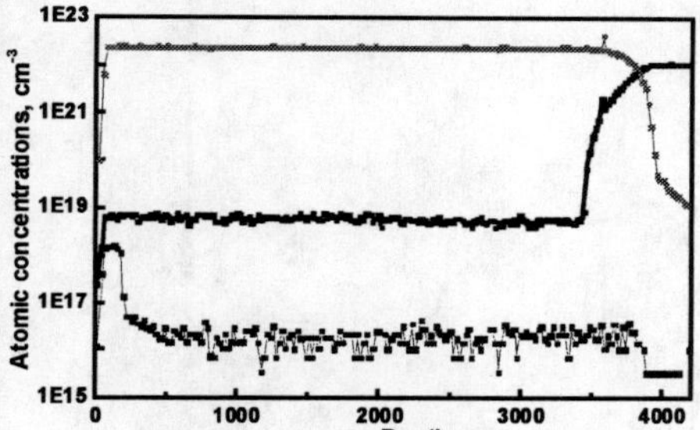

FIGURE 2. SIMS profiles of Zn and P concentrations in InGaAs cells at 500°C. Diffusion time: 10min.

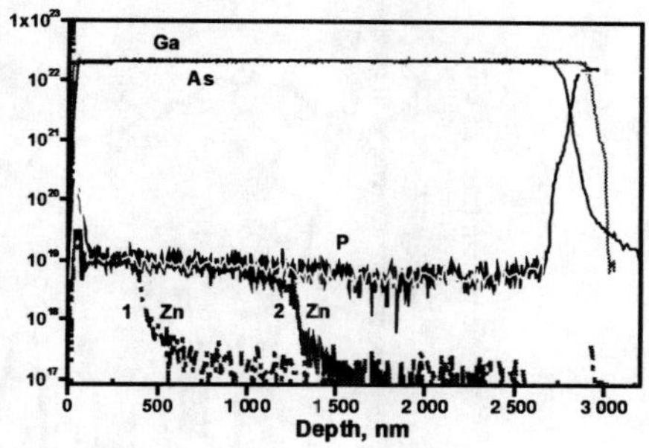

FIGURE 3. SIMS profiles of Zn and P concentrations in InGaAs cells at 600°C. Diffusion time: 1 ÷ 10min, 2 ÷ 30min.

The diffusion depths of the plateaus linearly depend on the square root of time, as would be expected from the diffusion theory. The diffusion coefficient calculated from the data on Fig.3 was $10^{-14}\,cm^2s^{-1}$. The solubility of zinc at the plateaus shown in Fig.3 was measured as $1*10^{19}cm^{-3}$, which is somewhat lower than reported earlier for zinc spikes grown from a gaseous phase by a closed ampoule technique [12]. We assume that this behavior is attributed to the thermodynamical conditions realized by the mixed Sn-P-Zn source. Obviously, the P vapour pressure over our source is greater than the As vapor pressure over Zn_3As_2. Probably, at simultaneous diffusion of zinc and phosphorus, which is isovalent impurity for InGaAs, amount of intrinsic point defects appearing during the diffusion process decreases. This results in the fact that the zinc diffusion slows down. We believe that the simultaneous diffusion of Zn and P into InGaAs effectively reduces diffusion loss through out. The corresponding TEM micrograph did not show the $Zn_3\,P_2$ and Zn precipitates.

The incorporation of isoelectronic atoms of P at the level of $1*10^{19}cm^{-3}$ into InGaAs was almost constant for layers of 3 μm thickness (Fig.2, 3).

GaSb

Fig. 4 shows the SIMS profile of the Zn and P concentrations for the co-doped (Zn, P) GaSb. A characteristic feature, which was consistently observed with Zn profile is the drop of the dopant concentration at the surface to about for $p = 2*10^{19}\ cm^{-3}$. The reason for this effect is, however, not yet clarified. It is the subject of further investigations.

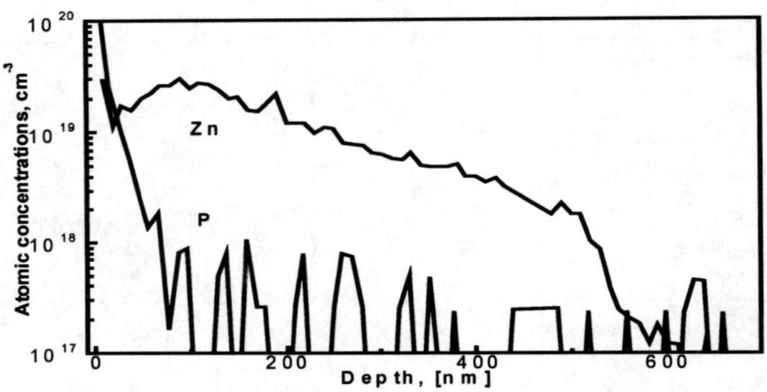

FIGURE 4. SIMS data for Zn and P diffusion into GaSb from Sn-InP-Zn source. Diffusion conditions: 10 minutes at 500^0C. (3472).

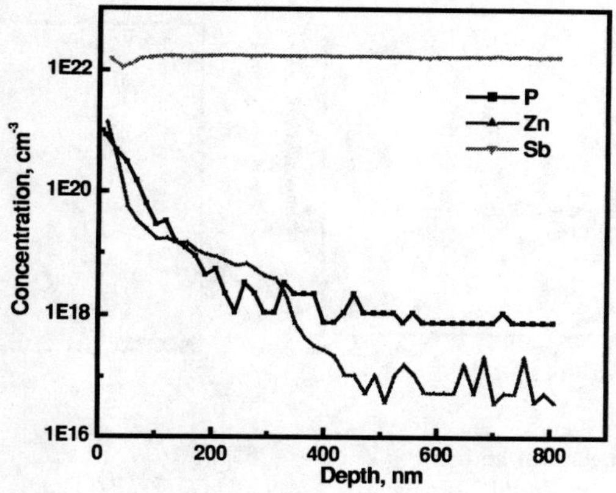

FIGURE 5. SIMS data for Zn and P diffusion into GaSb from Sn-InP-Zn source. Diffusion conditions: 10 minutes at 600° C.(BS 276).

In GaSb the diffusion has also been shown to depend on group V vapor pressure [2]. As has been mentioned above, the Zn and P diffusion from a local source allows to change easily the V group element pressure by varying the process temperature.

Fig. 5 shows the SIMS data for the 10 min diffusion time at 600 C. It is seen that diffusion of Zn decreases to values comparable with the data shown in Fig. 4. These results agree with the model of the Zn diffusion into GaSb considered in detail in [2].

In this case, the Zn diffusion profile has a strong Zn concentration gradient. Strong near – surface built-in electric fields decrease the influence of the surface recombination on the device characteristic [5].

Our studies have shown that the use of such structures for fabrication of TPV devices is possible without any additional post-diffusion treatment of the surface. In addition to SIMS measurements the InGaAs layers were characterized by optical methods. Raman scattering was used for estimation of the free carrier concentration in the subsurface (<50 nm) layer. In comparison with SIMS this method allows to estimate the concentration of free charge carriers independently of the impurity concentration. Our studies have shown that after diffusion the electron concentration in the "as grown" layers appeared to be $1*10^{17}cm^{-3}$ and the hole concentration - $1*10^{19}cm^{-3}$ after diffusion. The corresponding spectra measured in the {001} backscattering z(xy)z configuration are shown in Fig. 6, where curve 1 corresponds to "as grown" and curve 2- to after diffusion material. The values of free carrier concentration are in a good agreement with the SIMS measurements. This fact allows to conclude that after diffusion almost all Zn atoms reside on the proper lattice sites and are electrically active, and the obtained structure does not need any additional processing.

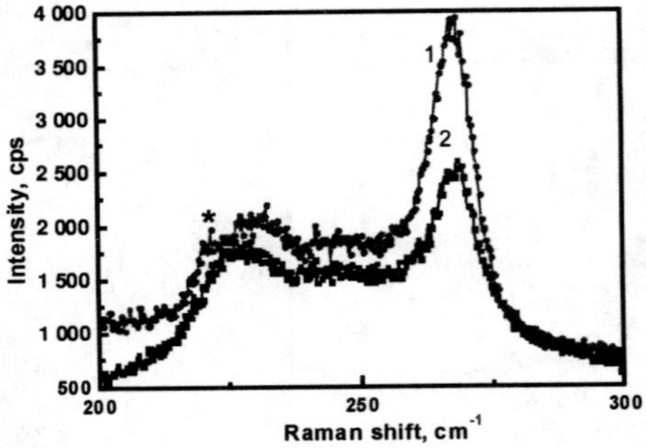

FIGURE 6. Raman spectra of an InGaAs sample before (curve 1) and after Zn diffusion (curve 2). The concentration are estimated $n=2*10^{17}cm^{-3}$ and $p=1*10^{19}cm^{-3}$. The asterisk * shows the Ar plasma scattered light.

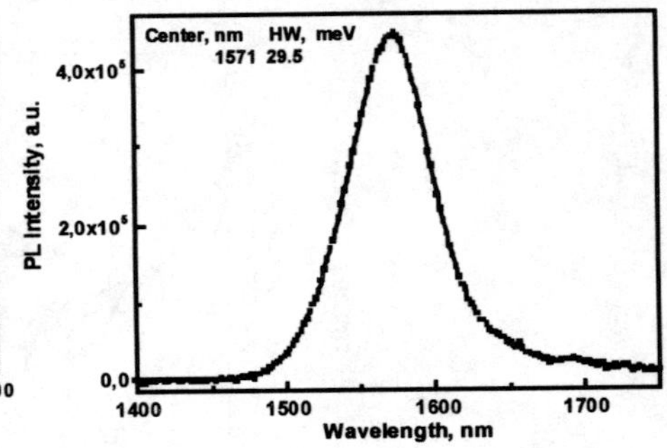

FIGURE 7. Photoluminescence spectrum of InGaAs(Zn,P) sample obtained at 77K. PL excitation power density is $5W/cm^2$.

Another method of characterization of layer and surface properties used was PL. The PL spectrum of an InGaAs (Zn, P) layer at 77K is shown in Fig.7. The observed 1.57μm line is attributed to the donor-acceptor pair recombination and its position and halfwidth (29,5 nm) indicating good layer properties.

FABRICATION OF InGaAs AND GaSb TPV CELLS

Cells with three different areas were fabricated by the procedure described above: 1, 0.16 and 0.04 cm^2. In some cases a precise thinning of diffused InGaAs and GaSb layers was performed through selective chemical etching.

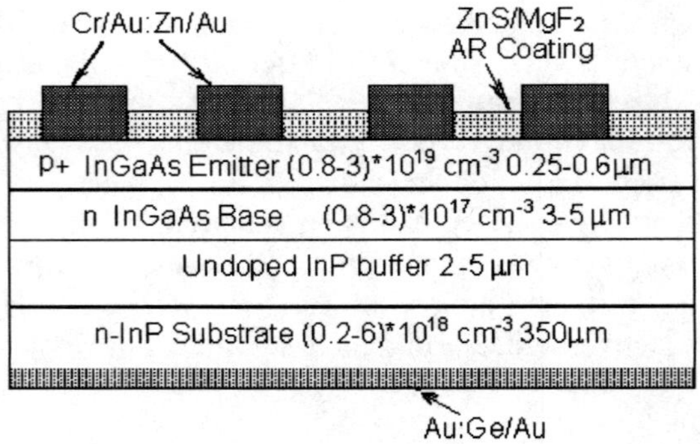

FIGURE 8. Schematic diagram of the p^+/n $In_{0.53}Ga_{0.47}$ TPV cells used in this study. Shown are the front and back contacts as well as the antireflection coating (ARC).

The GaSb cells consisted of an n-GaSb base (substrate) and a diffused p-GaSb emitter. The ARC and contacts were similar to those used in InGaAs cells (see Fig. 8).

INVESTIGATION OF InGaAs AND GaSb TPV CELLS.

External Quantum Efficiency Data.

InGaAs

The external quantum efficiency (QE) data for both InGaAs and GaSb structures showed very good collection efficiency.

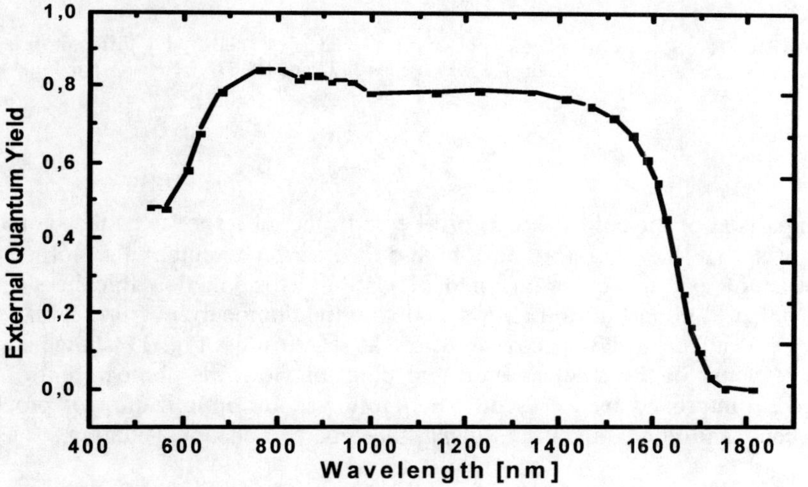

FIGURE 9. Spectral response of InGaAs TPV cell. Photocurrent density is 50.4 mA/cm^2 (under AMØ and 1 Sun conditions).

It should be noted that the external quantum yield of our TPV cells in the long wavelength region (1300-1700nm) is comparable with the value of the quantum yield of the best InGaAs cells obtained by the MOVPE [13] technique with an InP wide-band window. The photocurrent density of 26-28.8 mA/cm^2 was estimated for these cells, when the AMØ illumination spectrum is limited by the spectral interval of 900-1820 nm.

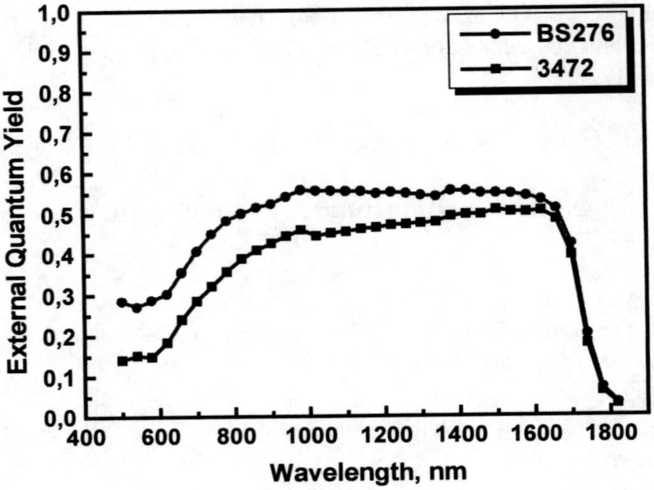

FIGURE 10. Spectra of photoresponses for GaSb TPV cells with different **p**-layer thicknesses (without ARC). SIMS data for the samples 3472 and BS 276 were shown in Fig.4 and Fig.5.

GaSb

Comparison of the SIMS (Fig.5) profile and spectral response of the sample (BS 276) shows that the thin (50 nm) high-doped surface region did not result in degradation of external quantum yield of GaSb. This junction thickness and the profiles of Zn(P) concentration allows to obtain the photocurrent from 26mA/cm^2 to 28.6 mA/cm^2 under AMO spectrum cut-off at < 900 nm. Fig. 11 shows a typical spectral response of these cells. From our point of view, the photocurrent of GaSb cells can be improved by Zn(P) diffusion followed by optimization of procedure, without using additional operation, such as the anodic or chemical etching.

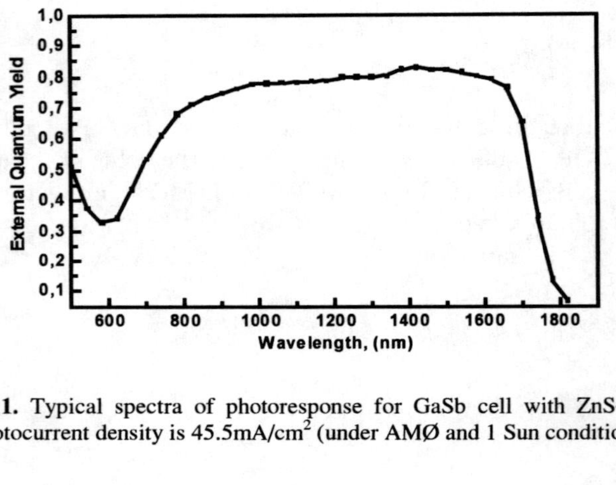

FIGURE 11. Typical spectra of photoresponse for GaSb cell with ZnS+MgF_2 antireflection coating. Photocurrent density is 45.5mA/cm^2 (under AMØ and 1 Sun conditions).

The open circuit voltage versus the short circuit current dependences.

InGaAs and GaSb TPV cells with different areas were tested under high-intensity illumination, using a flash lamp. The open-circuit voltage versus the short-circuit current dependences are shown in Fig. 12. InGaAs 0.16 cm^2 cells yielded an open-circuit voltage of 0.45V and a fill factor of 68-70% at a short-circuit current of 1.8A/cm^2. Thus, the resistive losses are relatively low in this case.

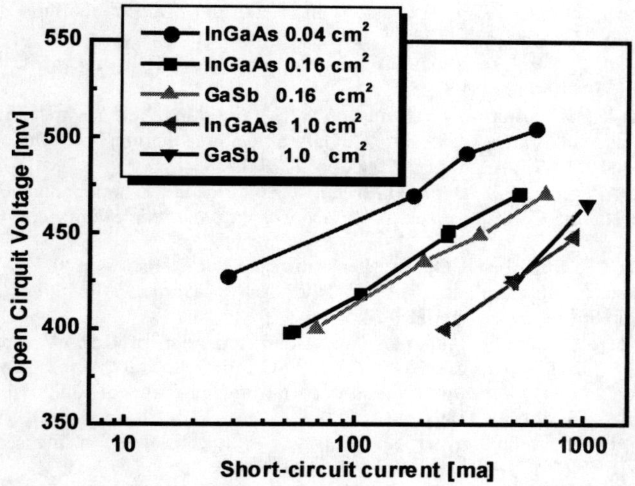

FIGURE 12. The open circuit voltage versus short-circuit current for InGaAs and GaSb TPV cells with different areas.

The results for InGaAs I-V data presented show that the undoped InP buffer used in our structure was effective in preventing the defects created the InP at interface InP-InGaAs. The dark current density is lower than in a lattice – matched InGaAs-InP cell with a high doped interface. In this case, the V_{oc} value is determinated by the injection component of the dark current and perimeter recombination slightly affects V_{oc}.

SUMMARY

The Zn(P) diffusion from local source applied to InGaAs and GaSb allows formation of the sharp Zn concentration gradient in the thin surface region and simultaneously eliminate "dead" layers on the surface. The use of the one-stage diffusion technology reduces evidently the TPV cell cost.

ACKNOWLEDGEMENTS

The authors thank V.M.Andreev for fruitful discussions, N.Kh.Timoshina for technical assistance.

REFERENCES

1. B.Ya.Ber et al. "Diffusion from polymer spin-on films: Measurements and simulations", *Vacuum Science and Technology* B **16**, 426-429 (1998).

2. G.J. Conibeer et al." Zinc Diffusion in Tellurium Doped Gallium Antimonide", *J. of Electron Materials* **25**, 1108-1112 (1996).

3. D.Franke et al. " Post–grown Zn diffusion into InGaAs/InP in a LP-MOVPE reactor", *J. of Crystal Growth* **195**, 112-116 (1998).

4. V.M.Andreev, V.P.Khvostikov, V.D.Rumyantsev, S.V.Sorokina, M.Z.Shvarts, "Single-junction GaSb and tandem GaSb/InGaAsSb & AlGaAsSb/GaSb thermophotovoltaic cells" in *Proceedings of 28th IEEE PVSC,* Anchorage, 2000, pp. 1265-1268.

5. O.V.Suluma, A.W.Bett, P.S. Dutta, H.Ehsani, R.J. Gurmann " Thermophotovoltaic cells on zinc diffused polycrystalline GaSb" in *Proc. 16th European Photovoltaic Solar Energy Conference*, Glasgow, 2000.

6. D.Wilt et al., "Electrical and Optical Performance Characteristics of 0.74eV p/n InGaAs Monolithic Interconnected Modules" in *Proc. of Third NREL Thermophotovoltaic Generation of Conference*", Colorado Springs, 1997, pp. 237-247.

7. C.S.Murray et al. "Growth, processing and Characterization of 0.55-eV n/p/n monolithic interconnected modules" in *Proceedings of 28th IEEE PVSC*, Anchorage, 2000, pp. 1238-1241.

8. G.Wang et al. "PH_3/H_2 plasma passivation of metal-organic chemical vapor deposition grown GaAs on Si", *J. of Appl. Physics* **88**, 3689-3694 (2000).

9. L.B.Karlina et al. " Radiation hard $In_{0.53}GaAs_{0.47}$ solar cells with Zn diffused emitters" in *Proceedings of 26th IEEE PVSC*, Anaheim, 1997, pp. 1007-1010.

10. E.Buehler et al. "Solubility of InP and CdS in Cd,Sn.In, Bi and Pb.", *J. of Crystal Growth* **35**, 60-64 (1976).

10. K.Kurishima et al. "Control of Zn diffusion in InP/ InGaAs heterojunction bipolar transistor structures grown by metalorganic vapor phase epitaxy", *J. of Appl. Phys.* **79,** 4017- 4023 (1996).

11. U.Wielsch et al. *Semicond. Sci. Technol.* **5**, 923-927 (1990).

12. M.A.Smith et al. "Solar cells for NASA RAINBOW concentrator" in *Proceedings of 28th IEEE PVSC*, Anchorage, 2000, pp. 1139-1141.

Low-Bandgap Ge and InAsSbP/InAs-Based TPV Cells

V.M.Andreev, V.P.Khvostikov, O.V.Khvostikova, E.V.Oliva, V.D.Rumyantsev, M.Z.Shvarts

Ioffe Physico-Technical Institute, 26 Polytechnicheskaya, St.Petersburg, 194021, Russia

Abstract. *P*-GaAs/*p*-Ge/*n*-Ge heterostructures with a submicron (0.1μm) GaAs window for the thermophotovoltaic (TPV) application have been fabricated using the combination of LPE and Zn–diffusion techniques. Due to the increase of the open circuit voltage by growth of a GaAs wide bandgap window, an increase of efficiency of Ge based TPV cells from 2.6 – 4.8% up to 3.1 – 6.1 % takes place for the emitter temperature in the range of 1400 – 1900 K. The efficiencies higher than 13% under the cut-off at $\lambda<900$ nm AM0 solar spectrum have been achieved in photovoltaic (PV) concentrator cells with a GaAs window layer at photocurrent densities of 3-25 A/cm^2.
By means of LPE growth and Zn diffusion techniques, TPV cells based on *p*-InAsSbP/*n*-InAsSbP/*n*-InAs structures were fabricated sensitive in the infrared range of 2.5-3.4 μm. Lattice-matched InAsSbP (Eg=0.45-0.48eV) quaternary alloy layers were grown on the (100) InAs substrates at 850 K by the step-cooling LPE technique. Low-temperature (600-630 K) pseudo-closed box Zn diffusion was used to produce *p-n* junctions in InAsSbP and InAs. Developed InAsSbP-based TPV cells demonstrate rather good perfomance, which can be improved by the structure optimisation.

INTRODUCTION

Germanium was the first semiconductor, which was suggested in the 1960s for the thermophotovoltaic application. However, advantages of the first TPV systems based on germanium have not been realized, such as low cost and their commercial availability, because of a poor performance obtained in Ge-based cells. During the last decade intensive investigations of Ge-based PV cells were carried out, stemed from the fact that Ge had become the basic material (substrate and bottom cell) widely used already for fabrication of high efficiency InGaP/GaAs/Ge concentrator solar cells [1]. Therefore characteristics of the cells are interesting not only for blackbody emission (TPV application), but for solar radiation as well. The characteristics of Ge-based TPV cells can be improved by growth of a GaAs wide bandgap window on the Ge wafer. This increases the open circuit voltage and as a result gives higher efficiency of the TPV system. This paper presents the results of investigations of Ge based TPV cells with and without GaAs "window" layers. Optimization of the TPV system may imply a choice of the emitter spectrum and a possibility to return the nonused part of radiation from the receiver back to the emitter surface supplying it by an "additional" power. The way to improve the TPV efficiency is to utilize the non-absorbed in the cell part of the radiation in the emitter spectrum region returning it back to the emitter surface where these low-energy photons can be transformed again into photons of the whole blackbody (graybody) spectrum. The investigation results of the TPV cells with back-surface mirrors are presented as well.
Efficiency of TPV cells when used with lower temperature heat source (1250-1500 K) could be essentially improved by extending the cut-off wavelength behind 2.0 μm [2-6]. With this point of view III-V solid solutions represent an important class of semiconductor materials, which are

CP653, *Thermophotovoltaic Generation of Electricity: 5th Conference*
edited by T. J. Coutts, G. Guazzoni, and J. Luther

useful for the fabrication of such devices. Results of InAsSbP and InAs TPV cell developments are discussed in this paper.

Ge TPV CELLS

The pseudo-closed box Zn diffusion from vapour phase procedure was applied for formation a *p*/*n* junction in *n*-Ge wafers. Patterned silicone nitride or SiO_2 dielectrics on the top surface were used as a diffusion mask so that the junction was formed only in the designated illumination area. It permitted to reduce current leakages. A busbar was deposited on this dielectric layer and only ohmic contact grid fingers touched the junction region. A typical spacing between fingers was within 100-200 μm with shadowing of 5-10%. Fabricated cells had an optimized MgF_2/ZnS ARC. The cross section of a Ge-based device is shown in Fig.1.

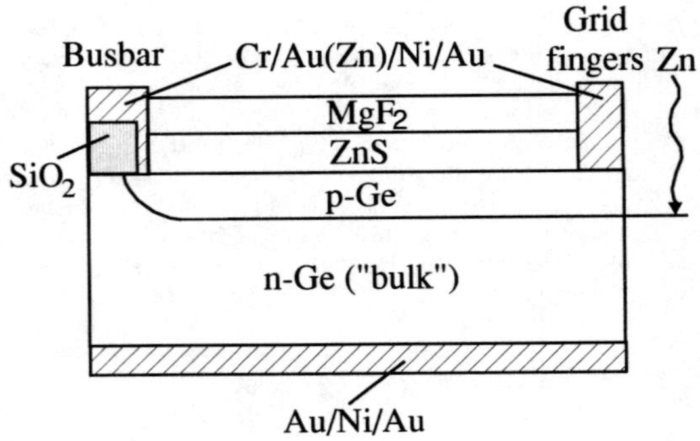

FIGURE 1. Schematic cross section of a *p*-Ge / *n*-Ge TPV cell fabricated by the Zn-diffusion

Such cells demonstrate high photosensitivity in the IR part of spectrum. The external quantum yield as high as 0.9 (Fig. 2, curve 1) and a high short circuit current of 29.1 mA/cm^2 were measured in this Ge cells, illuminated by sunlight with the cut-off at $\lambda < 900$ nm AMO spectrum. $I_{sc} = 50.3$ mA/cm^2 was obtained under sunlight with the cut-off at $\lambda < 580$ nm AMO spectrum. The calculated photocurrent density was 3.7 A/cm^2 assuming the emitter temperature of 1473 K. The open circuit voltage of Ge TPV cells appeared to be worse than that of previously developed GaSb based TPV cells [7-8]. In order to improve this parameter the technology of LPE growth of a thin (0.1μm) GaAs window on a Ge wafer has been developed.

GaAs/Ge TPV cells

TPV and PV cells based on the *p*-GaAs/*p*-Ge/*n*-Ge heterostructure with a thin (0.1 μm) GaAs layer were fabricated by the low-temperature liquid-phase epitaxy and by an additional Zn-diffusion process into the *p*-GaAs-*n*-Ge heterostructure. The advantage of this technology compared to MOCVD, widly used now for fabrication of GaInP/GaAs/Ge triple-junction solar cells on Ge substrates, is avoiding the undesirable diffusion of Ga and As atoms from vapour phase owing to much lower growth temperature.

The growth was carried out at 380 ^{0}C from a supercooled liquid phase with a relatively fast cooling rate of 2 ^{0}C/sec in order to avoid dissolution of the Ge substrate. A conventional graphite sliding boat in a Pd-diffused hydrogen atmosphere was used for LPE growths. The epitaxy was carried out using Pb-rich melts due to very low solubility of germanium and due to the layers of GaAs did not contain Pb atoms. We used the phase diagram data from the work [9] as a starting point. It shoud to be noted that even at high cooling rate the growth rate is relatively low (~0.05μm/s) allowing an accurate thickness control. The Zn diffusion procedure from the vapor phase was applied to form *p-n*-junctions in GaAs/Ge heterostructures.

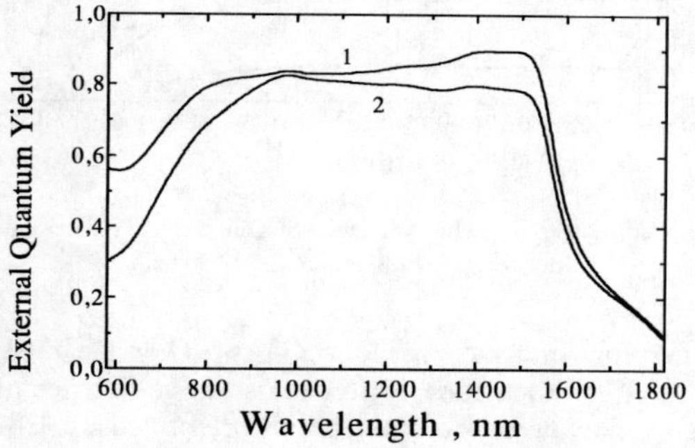

FIGURE 2. Spectral response of *p*-Ge/*n*-Ge (curve 1) and *p*-GaAs/*p*-Ge/*n*-Ge (curve 2) TPV cells (8% shadowing by contact grid).

Figure 3 shows the cross section of the GaAs/Ge cell structure. A high external quantum yield (Fig.2, curve 2) in this cell demonstrates low densities of recombination centers in the *p*-GaAs/*p*-Ge interface. The short circuit current of 27.7 mA/cm^2 was measured under sunlight with the cut-off at $\lambda < 900$ nm AMO spectrum. The photocurrent density of 3.3 A/cm^2 has been calculated for the emitter temperature of 1473 K.

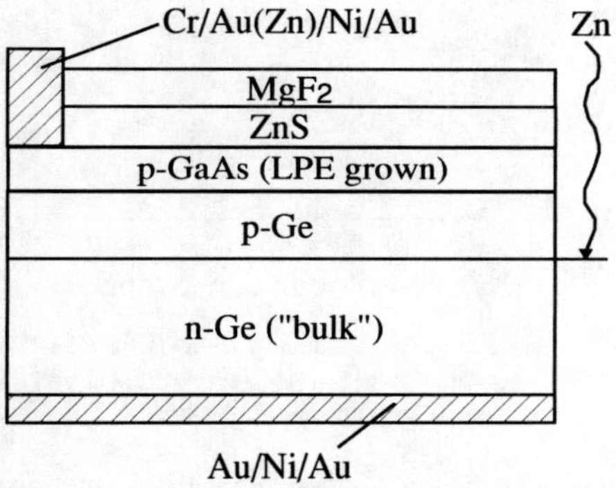

FIGURE 3. Cross section of a p-GaAs/p-Ge/n-Ge TPV cell, fabricated by LPE growth of GaAs and Zn-diffusion.

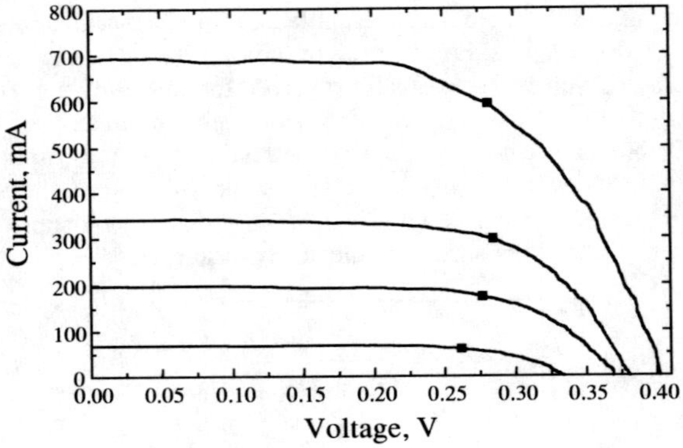

FIGURE 4. Flash illuminated current-voltage characteristics of a Ge based TPV cells with a GaAs wide bandgap window layer. Cell area of 2.5×2.5 mm^2.

Fig.4 represents a set of current-voltage curves for *p*-GaAs/*p*-Ge/*n*-Ge TPV cells recorded under flash illumination at various light intensities. Fig.5 demostrates efficiency of Ge based TPV cells with and without a GaAs window layer versus the photocurrent density. Efficiencies higher than 13% under the cut-off ($\lambda < 900$ nm) solar AM0 spectrum have been achieved in PV cells with a GaAs window layer at the photocurrent densities of 3-25 A/cm^2 (curve 1). This efficiency seems to be the highest among those published for Ge-based cells measured in these conditions.

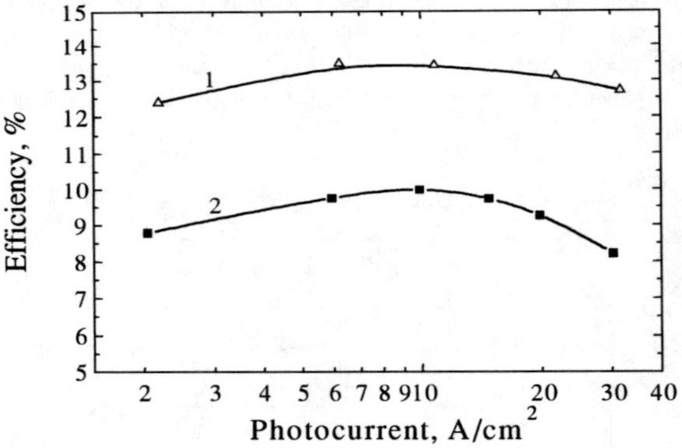

FIGURE 5. Efficiency as a function of the photocurrent density for a Ge-based TPV cell. Efficiency was calculated for illumination by sunlight with the cut-off at $\lambda < 900$ nm AM0 spectrum. Curve 1 - with a GaAs wide gap window layer; 2 - without a GaAs layer.

Fig. 6 and 7 show the *FF*, V_{oc} and efficiency dependencies as a function of the photocurrent density generated in *p*-GaAs/*p*-Ge/*n*-Ge based TPV cells in comparison with that for *p*-Ge/*n*-Ge cells for different blackbody emitter temperatures. It is seen that due to improvement of the open circuit voltage an increase of efficiency for a Ge based cell from 2.6 – 4.8% (Fig.7, curve 2) up to 3.1 – 6.1 % (Fig.7, curve 1) takes place for photocurrent densities of 3-20 A/cm^2. Such efficiency values have been calculated for the blackbody emitter temperature in the range of 1400 – 1900 K on the assumption that the TPV cell is placed close to the source (view factor is 100%). But the TPV cell efficiency degraded when operated at elevated temperature (especially when placed very close to the heat source). To reduce the cell operating temperature and to rise the overall system efficiency the unabsorbed radiation at the wavelength beyond the cut-off wavelength of a specified TPV cell could be reflected back to the emitter from the built-in back surface reflector.

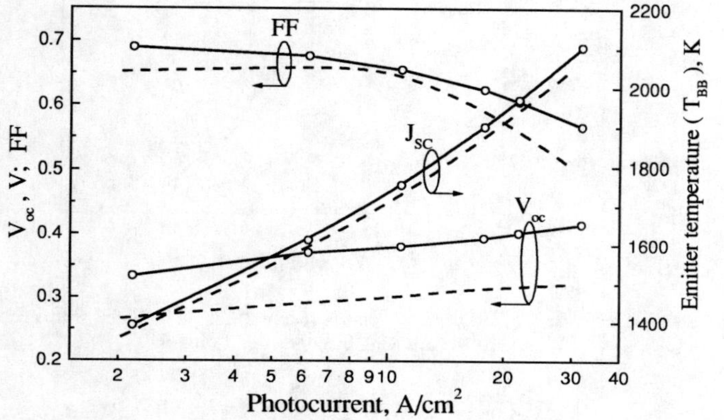

FIGURE 6. Generated photocurrent vs blackbody emitter temperature and fill factor and open circuit voltage as a function of generated photocurrent density for Ge based TPV cells: solid lines with open dots - with a GaAs wide gap window layer; dash lines - without a GaAs layer.

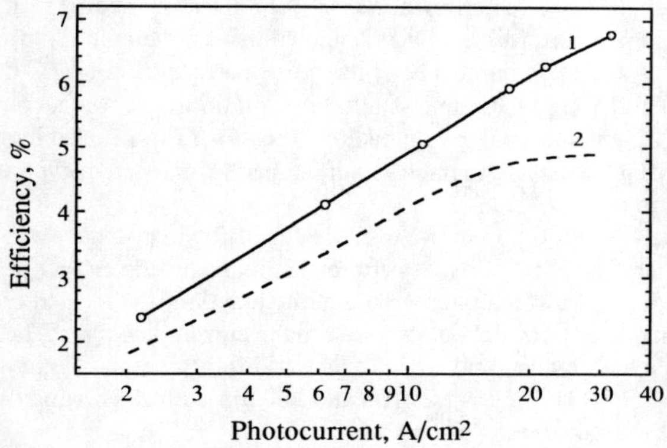

FIGURE 7. Efficiencies of the Ge TPV cells under blackbody (1400-1900 K) radiation as a function of the generated photocurrent: 1 - with a GaAs window layer; 2 - without a GaAs layer.

A built-in MgF_2/Au back surface mirror was developed for TPV cells to return non-absorbed sub-bandgap photons back to the emitter supplying it by an additional power. As a result of front ARC and back surface mirror optimization the total reflectance at wavelengths longer than 1.8 μm was increased up to 82-87% for Ge cells (Fig. 8). In this case the efficiency of the TPV system could be increased up to 9.3% for 1473 K emitter temperature and up to 11.5% for 1673 K emitter temperature assuming 85% return efficiency for photons with $\lambda > 1.8$ μm.

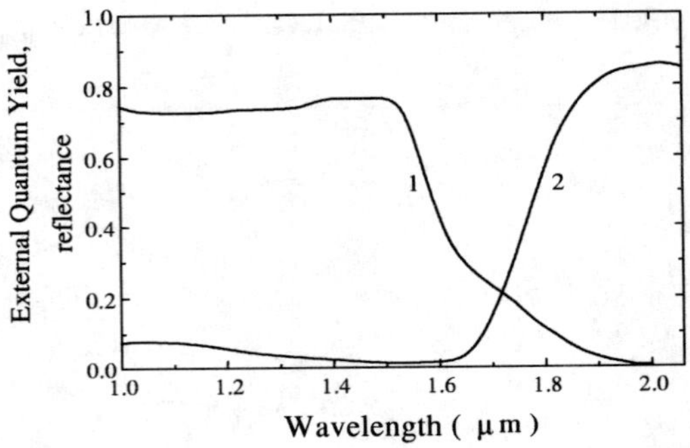

FIGURE 8. Spectral response (curve 1) and reflectance (curve 2) for Ge TPV cell with a back surface mirror at 10% shadowing by contact grid fingers.

TPV CELLS BASED ON InAsSbP/InAs

Narrow gap epitaxial InAsSbP (0.45-0.48eV) cells were fabricated from *p*-InAsSbP/*n*-InAsSbP/*n*-InAs heterostructures grown on (100) *n*-InAs substrates. Epitaxial growth of *n*-InAsSbP quaternary layers lattice matched to InAs was carried out by liquid phase epitaxy from a Sb-rich melt. The lattice-mismatch ratio $\Delta a / a$ =0.15% was estimated from X-ray diffraction pattern. Liquid-solid phase equilibrium was obtained using a "simple" solution model. We used calculation results as a starting point. The epitaxy was performed at 570^{0}C during 10 min (step-cooling technique) under 10-11 ^{0}C super-saturation condition to avoid a composition gradient in the epitaxial layer. A conventional graphite sliding boat in a Pd-diffused hydrogen atmosphere was used in the LPE growth. The unintentionally doped layers were n-type with a background carrier concentration of n~10^{17} cm^{-3}.
Low-temperature (340-350 °C) pseudo-closed box Zn diffusion process was applied to form *p-n* junction in quaternary InAsSbP alloy. A patterned silicone nitride or SiO_2 dielectric layers on the top surface were used as a diffusion mask so that the junction was formed only within the designated illumination area, which allowed to reduce current leakages. The busbar was deposited on this dielectric layer and only contact grid fingers contacted to the junction region. The typical spacing between fingers was 100 and 200 μm with shadowing of 5-10%. The cross section of InAsSbP-based devices is shown in Fig.9.
Fig.10 represents the spectral response of *p*-InAsSbP/*n*-InAsSbP/*n*-InAs TPV cells. A not sufficient diffusion length for photogenerated holes in the epitaxial n-InAsSbP base layer probably causes reduction of the photosensitivity in the long wavelength region. V_{oc} value was 0.15V in InAsSbP cells with E_g=0.45 eV at photocurrent density of 3 A/cm^2.

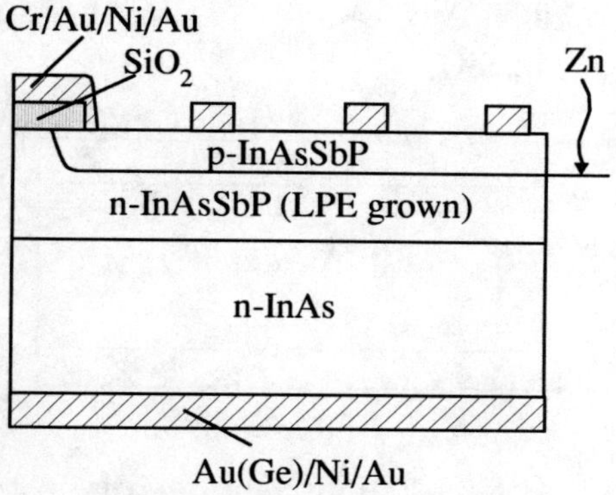

FIGURE 9. Cross section of a TPV cell based on the *p*-InAsSbP/*n*-InAsSbP/*n*-InAs heterostructure.

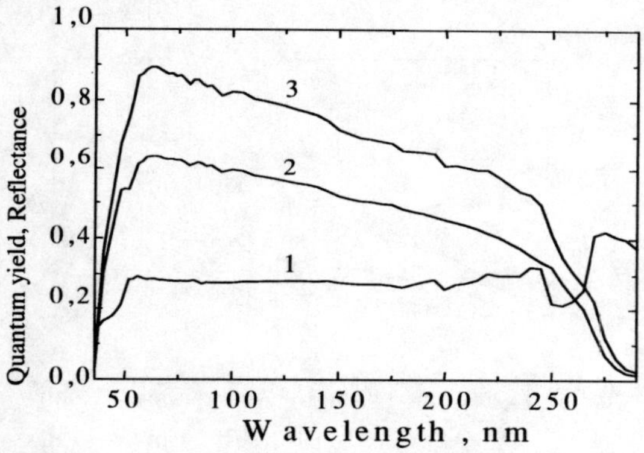

FIGURE 10. Reflectance (1) and spectral responses (curve 2 - external and curve 3 - internal quantum yields) of a cell based on the *p*-InAsSbP/*n*-InAsSbP/*n*-InAs heterostructure.

TPV CELLS FABRICATED FROM InAs "BULK" MATERIAL

The InAs-based TPV cells were fabricated by using wafers prepared from ingots grown by the Czochralski technique, Te-doped. The InAs wafers were subjected to the zinc diffusion procedure by means of the "pseudo-closed box" technique to form *p-n* junction in the photoactive area of the cells from a pure zinc source. The cross section of InAs-based devices is shown in Fig.11. Fig.12 represents the spectral response of a *p*-InAs/*n*-InAs TPV cell, which demonstrates rather high photosensitivity up to 3.5 μm at room temperature. Improvement of the photosensitivity in the short-wavelength (λ<2.5μm) part of the spectrum can be obtained by growth of a wide-bandgap lattice-matched window InAsSbP layer.

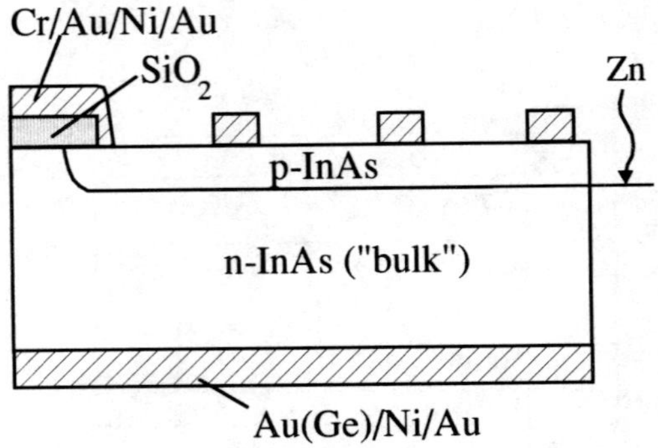

FIGURE 11. Cross section of a TPV cell based on the *p*-InAs/*n*-InAs structure.

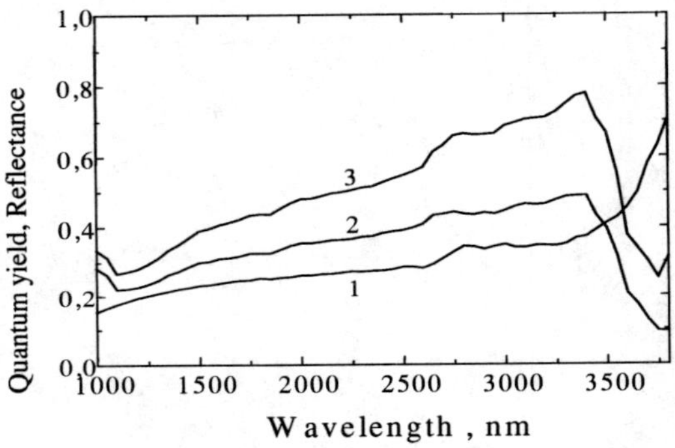

FIGURE 12. Reflectance (1) and spectral responses (curve 2 - external and curve 3 - internal quantum yields) of *p-n* InAs TPV cells fabricated from a "bulk" material by the Zn diffusion process.

CONCLUSION

TPV cells based on the *p*-GaAs/*p*-Ge/*n*-Ge heterostructure with thin (0.1 μm) GaAs top layers have been fabricated by the LPE with the additional Zn-diffusion process before growth. The short circuit current of 29.1 mA/cm^2 and efficiency higher than 13% under sunlight with the cut-off at $\lambda < 900$ nm AMO spectrum have been achieved in these *p*-GaAs/*p*-Ge-*n*-Ge PV cells at the photocurrent densities of 3-25 A/cm^2. Due to the increase of the open circuit voltage an increase of efficiency of the Ge based cell from 2.6–4.8% up to 3.1–6.1 % takes place for the photocurrent densities of 3-20 A/cm^2 generated by radiation from the blackbody with emitter temperature in the range of 1400 – 1900 K. Taking into account lower cost of Ge-based cells and the higher sub-bandgap photon reflection in the developed Ge-cells with a back-surface mirror the conclusion can be made that these Ge TPV cells are rather promising ones.

Photosensitivity of PV cells based on *p*-InAsSbP/*n*-InAsSbP/*n*-InAs structures in the infrared range was widened up to 2.5-3.4 micrometers by combination of LPE growth and Zn diffusion. It alows to make a conclusion that these PV cells are good for applications in the TPV generators with low-temperature (~1000 ^{0}C) emitters.

ACKNOWLEDGEMENTS

Authors would like to thank Zh.I.Alferov for continuous support and to express our gratitude to V.R.Larionov, A.I.Dement'eva and N.H.Timoshina for their contribution and assistance to this work.

REFERENCES

1. Cotal, H.L., Lillington, D.R., Ermer, et al., "Triple-Junction Solar Cells Efficiencies above 32%: The Promice and Challenges of Their Application in High-Concentration-Ratio PV Systems", in *Conference Record of the 28th IEEE PVSC*, Anchorage, 2000, pp. 955-960.
2. Andreev, V.M., Khvostikov, V.P., Larionov, V.R., et al., "Tandem GaSb/InGaAsSb Thermophotovoltaic cells", in *Conference Record 26th IEEE PVSC*, Anaheim, 1997, pp.935-939.
3. Mauk, M.G., Shellenbarger, Z.A., Cox J.A., et al., "Liquid-Phase Epitaxy of Low-Bandgap III-V Antimonides for Thermophotovoltaic Devices", *J. of Grystal Growth* **211**, 189-193 (2000).
4. Sulima, O.V., Beckert, R., Bett, A.W., et al., "InGaAsSb Photovoltaic Cells with Enhanced open-circuit voltage", in *Third International Conference Mid-Infrared Optoelectronics: Materials and Devicies*, Aachen, 1999.
5. Matveev, B., Aidaraliev, M., Zotova, N., et al, "Room Temperature InAs Photodiode – InGaAs LED Pairs for Methane Detection in Mid-IR", *Sensors and Actuators* **B 51**, 233-237 (1998).
6. Stoyanov, N.D., Mikhailova, M.P., Yakovlev, Yu.P., et al., "Type II Heterojunction Photodiodes in a GaSb/InGaAsSb system for 1.5–4.8 μm range", *Semiconductors* **35**, N4 (2001).
7. Andreev, V.M., Khvostikov, V.P., Khvostikova O.A., et al., "Thermophotovoltaic cells with sub-bandgap photon recirculation", in *17th European Photovoltaic Solar Energy Conference*, Munich, 2001, to be published.
8. Bett, A.W., Keser, S., Stollwerck, G., et al., "Large-Area GaSb Photovoltaic Cells", in *Third NREL Conference on Thermophotovoltaic Generation of Electricity*, AIP Conference Proceedings 401, New York, 1997, pp.41-53.
9. Immorlica, A., Ludington, Jr. and Burt, "Diffusion coefficient and solubility of Ge and GaAs in Pb and application to LPE growth of Ge on GaAs", *Journal of Crystal Growth* **51**, 131-139 (1981).

GaSb and $Ga_{1-x}In_xSb$ Thermophotovoltaic Cells using Diffused Junction Technology in Bulk Substrates

P.S. Dutta[1,2], J.M. Borrego[1,2], H. Ehsani[3], G. Rajagopalan[1], I.B. Bhat[1,2], R.J. Gutmann[1,2], G. Nichols[3], and P.F. Baldasaro[3]

[1]*Center for Integrated Electronics and Electronics Manufacturing*
[2]*Department of Electrical, Computer and Systems Engineering*
Rensselaer Polytechnic Institute, Troy, NY – 12180
[3]*Lockheed-Martin Inc., Schenectady, NY - 12301*

Abstract. This paper presents results of experimental and theoretical research on antimonide- based thermophotovoltaic (TPV) materials and cells. The topics discussed include: growth of large diameter ternary GaInSb bulk crystals, substrate preparation, diffused junction processes, cell fabrication and characterization, and, cell modeling. Ternary GaInSb boules up to 2 inches in diameter have been grown using the vertical Bridgman technique with a novel self solute feeding technique. A single step diffusion process followed by precise etching of the diffused layer has been developed to obtain a diffusion profile appropriate for high efficiency, p-n junction GaSb and GaInSb thermophotovoltaic cells. The optimum junction depth to obtain the highest quantum efficiency and open circuit voltage has been identified based on diffusion lengths (or minority carrier lifetimes), carrier mobility and experimental diffused impurity profiles. Theoretical assessment of the performance of ternary (GaInSb) and binary (GaSb) cells fabricated by Zn diffusion in bulk substrates has been performed using PC-1D one-dimensional computer simulations. Several factors affecting the cell performances such as the effects of emitter doping profile, emitter thickness and recombination mechanisms (Auger, radiative and Shockley-Read-Hall), the advantages of surface passivation and the impact of dark current due to the metallic grid will be discussed. The conditions needed for diffused junction cells on ternary and binary substrates to achieve similar performance to the epitaxially grown lattice- matched quaternary cells are identified.

INTRODUCTION

The availability of bulk substrates of III-V ternary or quaternary substrates with diffused p-n junctions seems to be the most promising option for low cost TPV cell technology. The present work is focused towards evaluating the feasibility of using diffused junctions of bulk GaInSb substrates for TPV cells. A comprehensive analysis of all the issues related to this technology, namely, crystal growth, wafer preparation, diffusion technology, device fabrication, processing and characterization, and, device modeling has been carried out. GaSb has been used as the model baseline material for both experimental and theoretical modeling efforts.

CRYSTAL GROWTH OF GaInSb

Bulk crystals of GaInSb are grown by the vertical Bridgman method from stoichiometric melts [1]. Polycrystalline compound formation was carried out from 6N pure gallium, indium and antimony in an open silica crucible in the presence of eutectic LiCl-KCl salt encapsulation and argon ambient. Synthesized polycrystalline charges were used in subsequent experiments to grow single crystals on <111>B or <100> oriented seeds of compositionally graded GaInSb starting from GaSb or InSb. Unlike binary

CP653, *Thermophotovoltaic Generation of Electricity: 5th Conference*
edited by T. J. Coutts, G. Guazzoni, and J. Luther

compounds such as GaSb or InSb, growth of ternary alloys still remains a challenge. The major problem in ternary or quaternary alloy growth arises from the wide separation between the liquidus and solidus in the phase diagram [1]. Crystals grown from melt inevitably exhibit mechanical cracks. Furthermore, the alloy concentration changes continuously along the growth direction due to segregation. Hence, no two wafers sliced from two different axial positions in the crystal will have the same composition. In this work, a self-solute feeding method has been developed to grow uniform ternary and quaternary alloys [1]. Crack-free, compositionally homogeneous GaInSb crystals in the band gap range of 0.4 – 0.7 eV have been grown by this method. To describe how the process works, an example of homogeneous $Ga_{1-x}In_xSb$ growth is presented below.

The growth experiment is initiated with the exact mole fraction of GaSb and InSb as desired in the final crystal. The growth temperature is kept constant at the solidus temperature corresponding to the desired alloy composition (of the crystal). At the growth temperature, the melt composition is decided by the liquidus of the ternary phase diagram. Hence only a fraction of the initial charge is in liquid phase. As solidfication proceeds, the liquid gets depleted in the higher melting binary component which is then replenished from the undissolved solute. In the case of GaInSb growth, the solvent is Ga-In-Sb psuedo-binary liquid and the undissolved solute is GaSb. The driving force for solute transport (and growth) is primarily diffusion controlled. This drastically reduces the growth rate of the alloys to ~ 0.1 mm/hr. To enhance the growth rate, a temperature gradient was imposed to increase the solute dissolution rate and the rate at which the solute arrives at the solid-liquid interface. With a temperature gradient of 30 °C/cm, maximum growth rate of 1.5 mm/hr for inclusion free crystals could be attained. Figure 1a shows a uniform 2 inch diameter polycrystal of $Ga_{0.75}In_{0.25}Sb$ grown using the self-solute feeding method. A single crystal and a bi-crystal GaInSb grown from InSb and GaSb single crystalline seeds respectively are shown in Figures 1b and 1c, respectively.

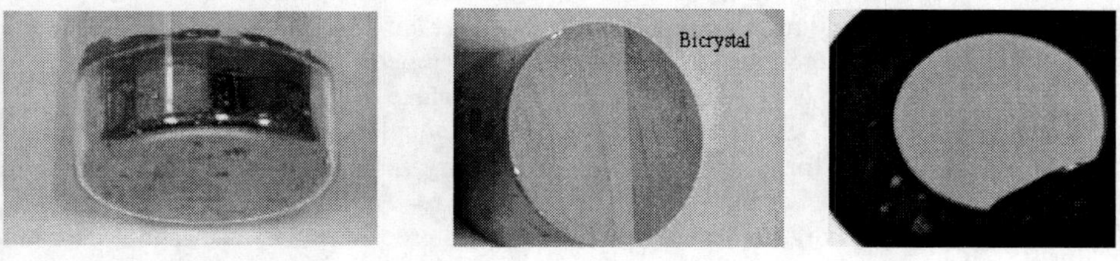

(a) (b) (c)

Figure 1 (a) Two inch diameter compositionally homogeneous polycrystal of $Ga_{0.75}In_{0.25}Sb$ grown using the self solute feeding method, (b) Bi- crystal and (c) Single crystal of compositionally graded GaInSb grown from GaSb and InSb single crystalline seeds, respectively.

SUBSTRATE PREPARATION of GaInSb

Ternary alloys are fragile and extreme care is needed when handling during the slicing and polishing of wafers. The as-grown GaInSb boules were sliced either using a single wire saw from Princeton Scientific Company, with a 50 μm thick tungsten wire and an abrasive slurry feeder, or a Southbay Technology diamond coated blade saw. For mechanical lapping of ternary substrates, the use of large diameter abrasives (> 5 μm) introduces permanent cracks in the wafers. Cracking can be avoided by using smaller abrasive particles (~ 1 μm) for lapping. However, this increases the lapping duration. Prolonged lapping and polishing cycles can also lead to mechanical damage of the wafers. In this work, a novel approach for preparing damage-free GaInSb wafers using the Baikalox CR- series of agglomerate- free alumina slurries has been developed [2]. Rapid removal of material has been evidenced with specific slurries containing abrasive particles of 1 μm and less. The high removal rate exhibited by these slurries is attributed to the agglomerate- free nature resulting from special slurry preparation procedures. The final polishing step was done using a soft velvet pad and a high pH (>10) silica slurry (Glanzox) from Southbay Technology that helps in removing the surface contaminants left behind by the alumina slurry. Wafers were finally cleaned in deionized water with scrub and sponges provided by Rippey Corporation. Surface evaluation of the wafers using a Nanospec atomic force microscope (AFM) revealed surface roughness of less than 0.5 nm (in a 10 μm x 10 μm scan area). This is comparable to the best available commercial GaSb wafer surfaces.

Zinc DIFFUSION IN GaSb AND GaInSb

The junction fabrication technology is based on zinc diffusion in n-GaSb and n-GaInSb substrates (Te- doped) [3-6]. The typical zinc profile after a single step diffusion process consists of a heavily doped near-surface region, followed by a plateau and a graded region near the metallurgical junction [3-6]. There are two requirements for obtaining high quantum efficiency and low dark current (and high fill factor). The light absorbing layer should have a low doped concentration for a high minority carrier diffusion length and hence be shallow in depth for high doping concentration (typically the case with a diffused junction emitter). On the other hand, for a good ohmic contact, a heavily doped (for low contact resistivity) and deep junction (to prevent metal spiking from shorting the cell) is required. A high near surface electric fields resulting from diffused impurity profile is advantageous for devices as it reduces the surface recombination velocity for minority carriers (necessary for high quantum efficiency and low dark currents).

In the present work, a single step diffusion and precise emitter etching process has been developed to take advantage of the high near surface electric fields inherent in diffused junctions. For diffusion in GaSb, <100> oriented n-GaSb commercial wafers from Wafer Tech. Inc. with carrier concentrations of $2 - 4 \times 10^{17}$ cm^{-3} were used. For diffusion in GaInSb, polycrystalline or un-oriented single crystalline wafers of n-GaInSb grown in-house with carrier concentrations in the range of 7×10^{17} - 2×10^{18} cm^{-3} were used. Prior to diffusion, the wafers were degreased in xylene, acetone and methanol, followed by etching the native oxide in HCl. The diffusion was carried out in a pseudo-closed graphite box with the wafer placed above the zinc pellet (approximately 1 cm away). Optimum diffusion temperatures and times have been established for GaSb and GaInSb through post diffusion surface analysis and feedback from device performances (QE, I-V). For GaSb, Zn-diffusion at 500 °C for 5 hours seems to be suitable, whereas for

GaInSb in the band gap range of 0.5 – 0.6 eV, 450 °C for 2 hours gave better results. With the above conditions, typical junction depth for GaSb was in the range of 0.6 – 0.8 μm and 0.3 – 0.6 μm for GaInSb depending on alloy composition.

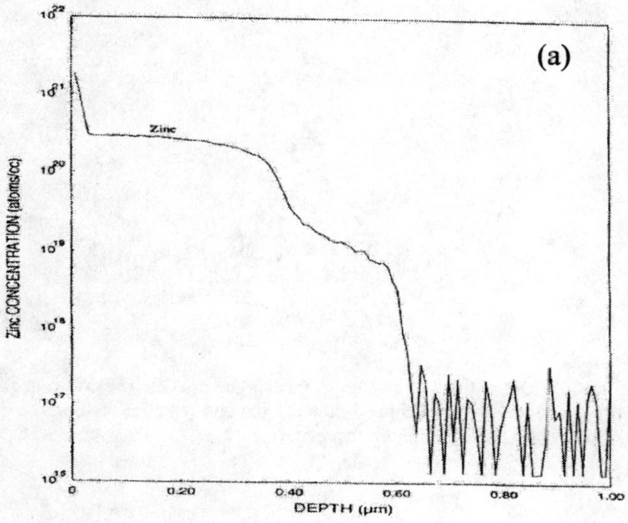

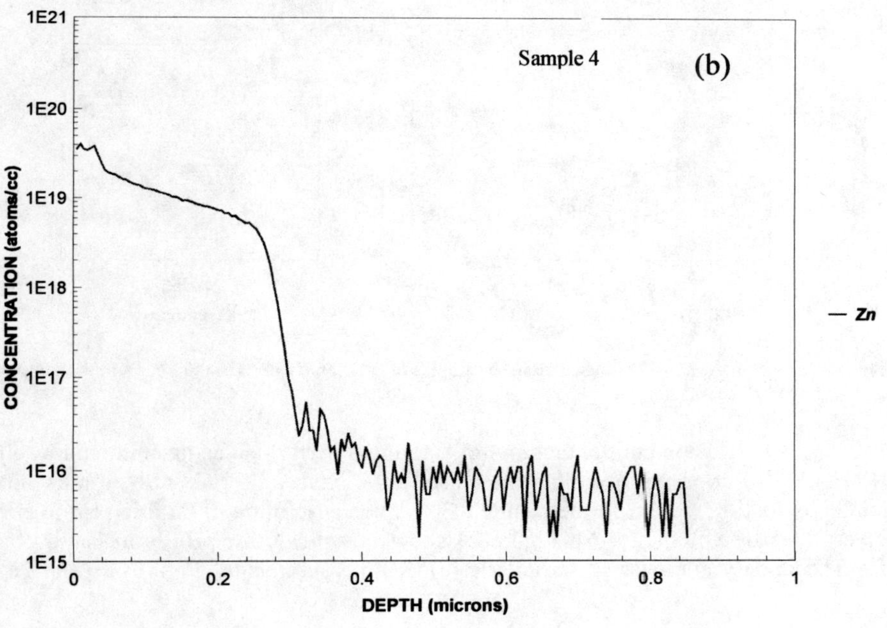

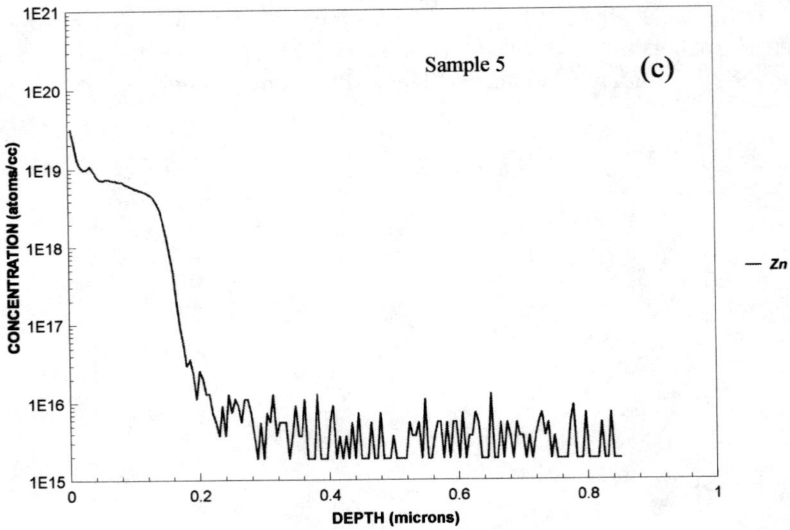

Figure 2. Zinc profiles in n-GaSb diffused at 500 °C for 5 hours (as determined using secondary ion mass spectrometric (SIMS) technique): (a) as-diffused sample, (b) and (c) after etching the diffused sample in ammonium sulphide for 29 min and 55 minutes, respectively.

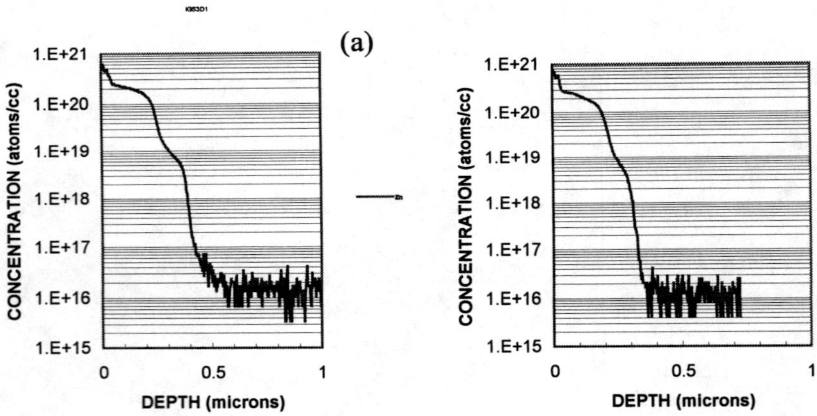

Figure 3. Zinc profiles in as-diffused GaInSb samples of (a) 0.56 eV and (b) 0.6 eV. Both the samples were diffused at 450 °C for 2 hours.

Post diffusion emitter etching for GaSb was performed using ammonium sulphide [$(NH_4)_2S_x$] solution with etching rate close to 0.2 nm/sec. For GaInSb emitter thinning, anodic oxidation was performed using a 3% aqueous solution of tartaric acid in ethylene glycol (1:2 by volume) [5,6]. Figure 2a shows a typical zinc profile in GaSb obtained from secondary ion mass spectrometric (SIMS) measurements. Figures 2b and 2c show the zinc profiles in the samples after etching in ammonium sulphide. Figures 3a and 3b show the zinc profiles in as-diffused GaInSb with band gaps of 0.56 and 0.6 eV, respectively clearly indicating that the zinc diffusion rate increases with the increase in indium concentration in GaInSb.

To monitor the junction depth during the post diffusion emitter etching, photovoltage measurements were performed using a two-point probe set-up [7]. The diffused sample was placed on a metal block and the top surface was contacted with a metal point probe. A tungsten lamp was used to illuminate the sample and the shift in I-V (the photovoltage) in the low current regime was measured using an oscilloscope. The photovoltage depends on the junction depth and the surface electric field (due to diffused impurity profile). A typical photovoltage profile as a function of cumulative etching times is shown in Figure 4. Initially the photovoltage increases slowly, followed by a rapid increase; beyond the maximum value, the photovoltage decreases drastically until the junction is reached. The maximum photovoltage corresponds to a thin emitter with surface zinc concentration (as in Fig. 2c) an order of magnitude lower than in the as-diffused sample (Fig. 2a). As will be described in the next section, the external QE increases with the decrease in emitter depth, whereas the open circuit voltage (V_{oc}) decreases. Hence it is essential to optimize the junction depth for achieving maximum power output instead of either higher QE or V_{oc}. By comparing the external QE and V_{oc} in the final cells with photovoltage of the etched junction, it has been ascertained that a photovoltage of 100 – 150 mV will result in the best GaSb TPV cell. For the best GaInSb TPV cells, the photovoltage ranged between 20 to 40 mV.

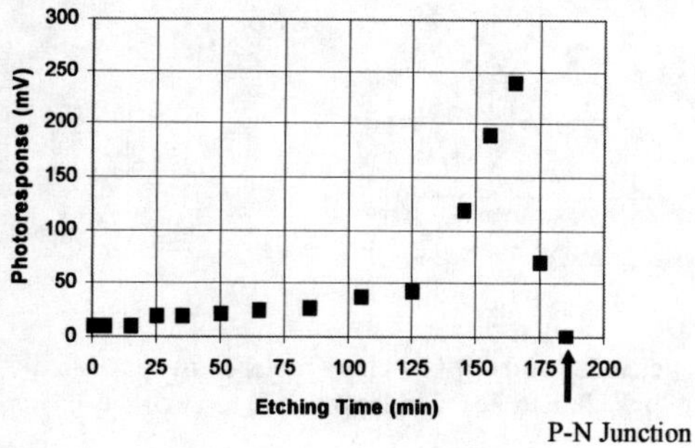

Figure 4. Photovoltage versus etching time for GaSb p-n junction during the emitter thinning process using ammonium sulphide solution

THERMOPHOTOVOLTAIC CELL FABRICATION AND CHARACTERIZATION

The TPV cells fabrication process (such as front and back side metallization, lift-off scheme, mesa etching) and cell characterization schemes were identical to that reported earlier [8]. The quantum efficiencies and I-V characteristics for the two GaSb cells fabricated using the single step diffusion and etching technique (with zinc profiles as shown in Figures 2b and 2c) are shown in Figures 5a and 5b, respectively. The values of quantum efficiencies and dark currents are amongst the best reported for GaSb TPV cells. It is important to note that there is a one-to-one correlation between the I-V and emitter depth. Even though the external QE are identical for the cells, a thinner junction results in an increase in leakage current (and hence a decrease in open circuit voltage). There is an optimum emitter depth to obtain the highest QE and lowest leakage current. If the emitter is thick, the external QE will decrease due to the high doping concentration in the emitter region which decreases the minority carrier lifetime (see Figure 6a). On the other hand, if it is too thin, the leakage current increases which decreases the V_{oc} (as shown in Figure 6b). The optimum emitter thickness to obtain the highest QE with lowest leakage current is dependent on the zinc profile. With the present single step diffusion profiles and etching technique, the optimum junction depth for the best GaSb cell performance is estimated to be approximately 0.4 μm.

Figures 7a and 7b show the external QE and I-V for three polycrystalline GaInSb TPV cells with un-optimized junction depths. Both the external QE and V_{oc} in these cells are comparable to those reported for epitaxial grown cells. These results are quite encouraging. Further improvements in performances may be possible by optimizing the junction depth and the base doping of the substrates.

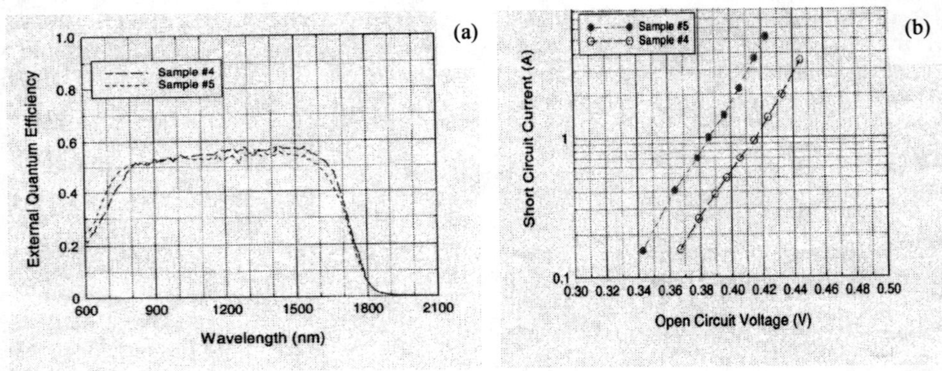

Figure 5. (a) Quantum efficiencies and (b) I-V characteristics of the zinc diffused GaSb TPV cells with diffusion profiles shown in Figs. 2b and 2c (samples 4 and 5, respectively).

MODELING OF Zn DIFFUSED GaSb AND GaInSb TPV CELLS

To complement the experimental part of this study, computer simulations of the Zn diffused GaSb and GaInSb TPV cells were performed. The PC-1D commercial program [9] was used with appropriate material parameters for GaSb and for the GaInSb alloys. Material constants such as energy band parameters, index of refraction, optical absorption coefficients, mobility as a function of the doping concentration are well known for GaSb and they can be extrapolated with a good degree of certainty for the GaInSb alloys [10]. Two parameters that are difficult to obtain accurately for the GaInSb alloys are the Auger recombination coefficient and the band gap narrowing. Values obtained from recent computer calculations taking into account the non-parabolic nature of the energy bands were used [11].

The Zn doping profile in the emitter, which was obtained from SIMS measurements in the diffused bulk samples, and the base doping which was obtained from Hall measurements on the bulk samples before diffusion were taken into account in the computer simulation. The value of the front surface recombination velocity was chosen so that the simulated external quantum efficiency at 600 nm corresponded to the measured value in the TPV cells with a low doped and thin emitter. The base length of the TPV cells was chosen to be 100 μm so the value of the back surface recombination velocity would not be significant. In addition the metallic grid, covering 10 % of the total cell area, was considered as a parasitic diode with a dark current determined by the metallic grid area, which had a surface recombination velocity equal to the thermal velocity.

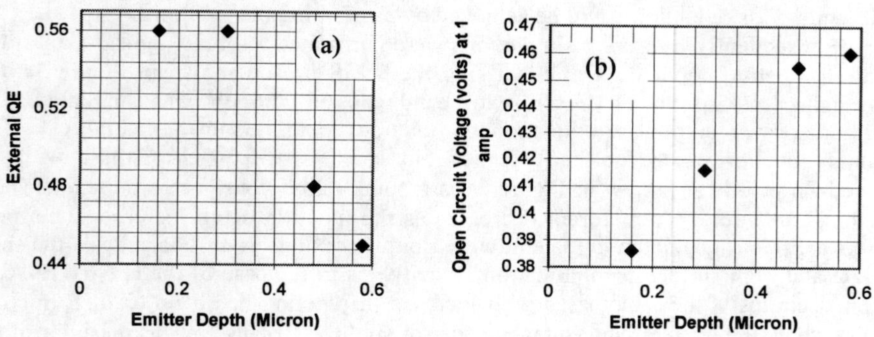

Figure 6. (a) External QE and (b) V_{oc} versus emitter depth for GaSb TPV cells

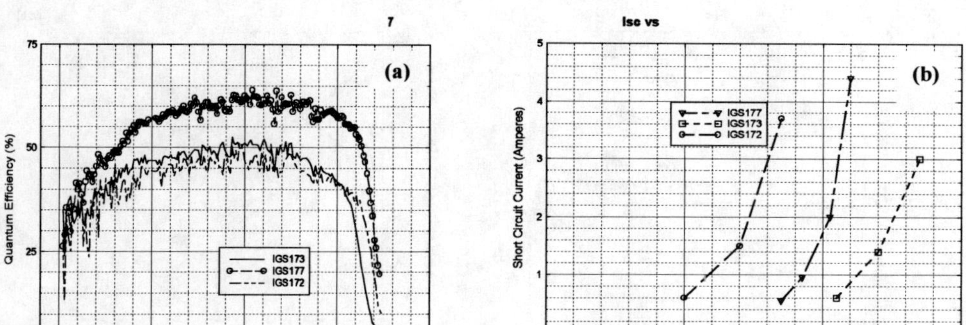

Figure 7. (a) External Quantum Efficiencies and (b) I-V characteristics of the zinc diffused GaInSb TPV cells

Using the SIMS profiles shown previously in Figures 2b and 2c, the performance of GaSb TPV cells with an N- doped base of $3x10^{17}$ donors/cm^3 as a function of junction depth was simulated and the results are shown in Figure 8a and Figure 8b. Figure 8a shows the simulated open circuit voltage as a function of junction depth for a short circuit photocurrent (J_{SC}) of 1 A/cm^2. Experimental values obtained from fabricated GaSb TPV cells of different junction depths show a very good agreement which validates the simulation results. Figure 8b shows the maximum power output as a function of junction depth. The maximum power is achieved at a junction depth in the range of 0.3 - 0.4 μm. This result is in agreement with the experimental results. The high doping concentration at and near the surface, shown by the SIMS profile, causes a high dark saturation current due to Auger recombination and this high doping region of the Zn diffusion has to be etched away in order to achieve a low dark saturation current. The simulations show that a way to increase the power output is by reducing the front surface recombination velocity. The results indicate that the maximum power output can be increased by approximately 8% by reducing the front surface recombination velocity from $3x10^6$ cm/s to 10^3 cm/s. This reduction could be achieved by passivating the GaSb front surface.

Subsequently, computer simulations were performed to obtain an estimate of the optimum junction depth for GaInSb. Using the SIMS profile shown in Figure 3a, the performance of GaInSb TPV cells with band gap of 0.56 eV was simulated. The maximum power output as a function of junction depth is shown in Figure 9. The optimum junction depth for the TPV cell fabricated with 0.56 eV GaInSb with the diffused Zn profile as shown in Figure 3a is around 0.2 microns. The simulations show that the short circuit photocurrent decreases as the junction depth increase since more carriers are photo-generated in the emitter region where they are lost because of the short lifetime and high surface recombination velocity. As in the case of the GaSb TPV cell, the open circuit voltage increases with increase in junction depth up to 0.25 microns beyond which the open circuit voltage decreases slightly. Because of this behavior of the short circuit photocurrent and open circuit voltage, there is a maximum in the output power as function of junction depth.

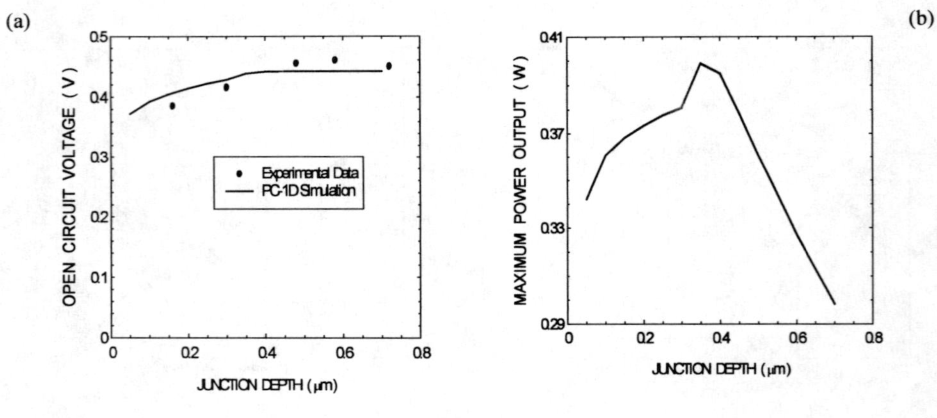

Figure 8. (a) Open Circuit Voltage and (b) Maximum power output as a function of junction depth for Zn-diffused GaSb TPV Cells

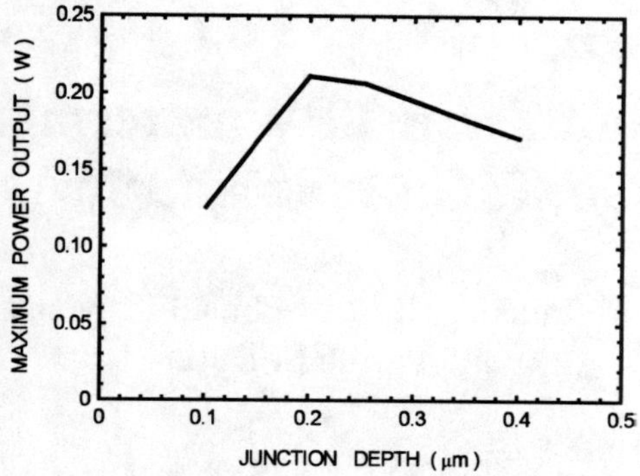

Figure 9. Maximum power output as a function of junction depth for Zn-Diffused GaInSb TPV Cells

CONCLUSION

In summary, we have been able to demonstrate high performance GaSb and GaInSb TPV cells fabricated from bulk substrates by zinc diffusion technology. Peak external quantum efficiencies exceed 50% for GaSb TPV cells and greater than 60% for InGaSb devices. The effect of junction depth on the power output has been studied experimentally as well as by theoretical simulations. Emitter depths in diffused junction cells necessary for maximum power output have been estimated for GaSb and GaInSb.

REFERENCES

1. P.S. Dutta and T.R. Miller, J. of Electronic Materials **29** (2000) 956
2. P.S. Dutta, R.J. Gutmann, D. Keller and L. Sweet, Materials Research Society Spring 2000 Proceeding, San Francisco (2001)
3. L.M. Fraas, G.R. Girard, J.E. Avery, B.A. Arau, V.S. Sundaram, A.G. Thomson and J.M. Gee, J. Appl. Phys. **66** (1989) 3866
4. V.S. Sundaram and P.E. Gruenbaum, J. Appl. Phys. **73** (1993) 3787
5. A.W. Bett, S. Keser and O.V. Sulima, J. Crystal Growth **181** (1997) 9
6. O.V. Sulima, R. Beckert, A.W. Bett, J.A. Cox and M.G. Mauk, 3rd Intern. Conf. Of Mid-Infrared Optoelectronics: Materials and Devices, Aachen, Germany (1999)
7. N.S. Reddy, M.S. Dissertation, Rensselaer Polytechnic Institute (2000)
8. C.W. Hitchcock, R.J. Gutmann, J.M. Borrego, I.B. Bhat and G.W. Charache, IEEE Trans. On Electron Devices **46,** (1999) 2154
9. PC-1D Version 5.5, Distributed by the Photovoltaics Special Research Centre at the University of South Wales, Sydney, Australia
10. M. Zierak, J.M. Borrego, I. Bhat, R.J. Gutmann, and G. Charache in Proc. of the Third NREL Conference on Thermophotovoltaic Generation of Electricity, Colorado Springs, CO (AIP Conference Proceedings **401**, New York), Eds. T.J. Coutts, C.S. Allman and J.P. Benner, (1997), p. 55
11. Private Communication, Prof. M.C. Flatte, Dept. of Physics and Astronomy, University of Iowa, Iowa City, IA

Diffusion of Zn in TPV materials: GaSb, InGaSb, InGaAsSb and InAsSbP

Oleg V. Sulima[1*], Andreas W. Bett[2], Michael G. Mauk[1], Boris Ya. Ber[3], and Partha S. Dutta[4]

[1]*AstroPower Inc., Solar Park, Newark, Delaware 19716-2000, USA*
[2]*Fraunhofer Institute for Solar Energy Systems, Heidenhofstraße 2, D-79110 Freiburg, Germany*
[3]*Ioffe Institute of Physics and Technology,26 Polytechnicheskaya, St. Petersburg 194021, Russia*
[4]*Rensselaer Polytechnic Institute, Troy, New York 12180-3590, USA*
[*]*corresponding author: sulima@astropower.com, phone: 302-366-0400 ext. 3031, fax: 302-283-0162*

Abstract. This paper reviews recent results of the study of Zn diffusion from the vapor phase in important thermophotovoltaic (TPV) materials such as GaSb-, InGaAsSb-, InGaSb- and InAsSbP. Peculiarities of Zn diffusion in each of these materials and different ways of tailoring the Zn diffusion profile for fabrication of optimized emitters in TPV cells are discussed.

INTRODUCTION

Diffusion of Zn from the vapor phase is a simple and productive method of pn-junction formation in III-V materials. Especially advantageous is the pseudo-closed box diffusion method that avoids the inconvenience of sealed ampoules [1,2]. The first papers about Zn diffusion from the vapor phase into GaSb were published at the late 1960s and early 1970s [3]. Different approaches for pseudo-closed box diffusion of Zn into GaSb are described in [4,5]. L. Fraas *et al* was the first to apply Zn diffusion into GaSb for solar cells [6]. High-efficiency GaSb solar cells with Zn-diffused emitters developed at Boeing showed a potential of this material for infrared applications and aroused a new wave of interest for thermophotovoltaics. Also, they gave a strong impetus to the study of different antimonides for PV and TPV at many institutions worldwide.

At the present stage of development of TPV systems, TPV cells do not seem to be a bottleneck. Though their efficiency and operation parameters can certainly be improved, a much larger potential for improvement lies in other parts of a TPV system. Thus, more sophisticated TPV cells are not imperative. Actually, given a high-quality substrate material and an optimized diffused emitter, TPV cells, contrary to lasers or multi-junction space solar cells, do not require complicated epitaxially grown structures. Moreover, if epitaxial growth on high-quality substrates is not perfect, as is often the case with III-V antimonides, epitaxy can degrade TPV cell parameters. Obviously, one cannot avoid growing an epitaxial layer if no substrate is currently

CP653, *Thermophotovoltaic Generation of Electricity: 5th Conference*
edited by T. J. Coutts, G. Guazzoni, and J. Luther

available for a certain wavelength range / radiator temperature. However, even in this case, a simple one-layer epitaxial structure can be sufficient for a high-efficiency TPV cell with a diffused emitter.

In this work we studied Zn diffusion from the vapor phase in important TPV materials such as GaSb, InGaAsSb, InGaSb and InAsSbP. As was shown in several papers, optimization of diffused Zn profiles in the emitters can lead to a marked improvement of device parameters [7-10]. At the same time, there are speculations in the literature that the surface recombination in GaSb is relatively low and that this factor allows the realization of good device parameters. However, comparison of experimental and modeled data of GaSb TPV cells with diffused emitters shows that the surface recombination in GaSb is high: ~10^6 cm/sec [11]. Recent data on MOCVD-grown heteroface AlGaAsSb/GaSb TPV cells without the diffused emitter also showed the same value [12]. Surface recombination velocities between 10^5-10^7 cm/sec were determined on InGaSb surfaces in MOCVD-grown InGaSb structures [13]. However, despite an observed high surface recombination in GaSb, optimizing the diffusion profiles can effectively decrease its influence. We studied the following methods of the diffused Zn profile tailoring:

- Precise thinning of the as-diffused emitter that removes a dead layer with high Zn concentration (GaSb, InGaSb and InGaAsSb)
- Additional shallow Zn diffusion after the emitter thinning (GaSb, InGaAsSb)
- Shallow diffusion followed by a high-temperature annealing under SiN_x (GaSb).

Additionally, we studied the effect of anodic oxidation of a GaSb surface on Zn diffusion.

EXPERIMENTAL

Diffusion of Zn is performed into n-type materials. In the case of GaSb and InGaSb, single crystal or polycrystalline n-type substrates are used. Single crystal n-InGaAsSb and n-InAsSbP layers are grown by liquid phase epitaxy on n-GaSb and n-InAs substrates, respectively.

A standard horizontal slide boat technique is used for the liquid-phase epitaxial growth of the InGaAsSb and InAsSbP. The graphite slide boat is situated in a sealed quartz tube placed in a microprocessor-controlled, programmable, three-zone tube furnace. The growth ambient is palladium-diffused hydrogen at atmospheric pressure. The details of InGaAsSb and InAsSbP growth are published in [14]. Bulk crystal growth of InGaSb is described in [15].

Pseudo-closed box diffusion of Zn from the vapor phase is performed into n-type substrates or layers in a H_2 atmosphere purified by a Pd cell. A specially designed multi-wafer graphite boat described in [5] is applied. Separated pure Zn and Sb vapour sources were used in more than sufficient quantities to provide the saturation of vapour pressures. In experiments with InGaAsSb and InAsSbP, an InAs- or InAs- and InP- powder is used to decrease the loss of As and P from the samples during diffusion.

It was determined for GaSb, InGaSb and InGaAsSb that the maximum (surface) Zn concentration in these materials does not depend on the diffusion temperature or Zn vapor pressure over a broad range of values. This means that the maximum Zn solubility in these materials is achieved at low temperatures/pressures. Therefore, even though it is almost certain that a Zn vapor gradient exists in a pseudo-closed diffusion boat, the same emitter profiles are obtained in different parts of the boat.

RESULTS AND DISCUSSION

GaSb

Zn diffusion from the vapor phase is performed into polished 2.5x2.5 cm^2 n-GaSb:Te ($n=2\text{-}3\times10^{17}$ cm^{-3}) (100) substrates. Prior to diffusion the substrates are degreased in acetone. Afterwards, they are treated in diluted HF or HCl to remove native oxides on the surface and dried with nitrogen.

Figure 1 shows the dependence of the Zn surface concentration on the diffusion temperature for different diffusion times.

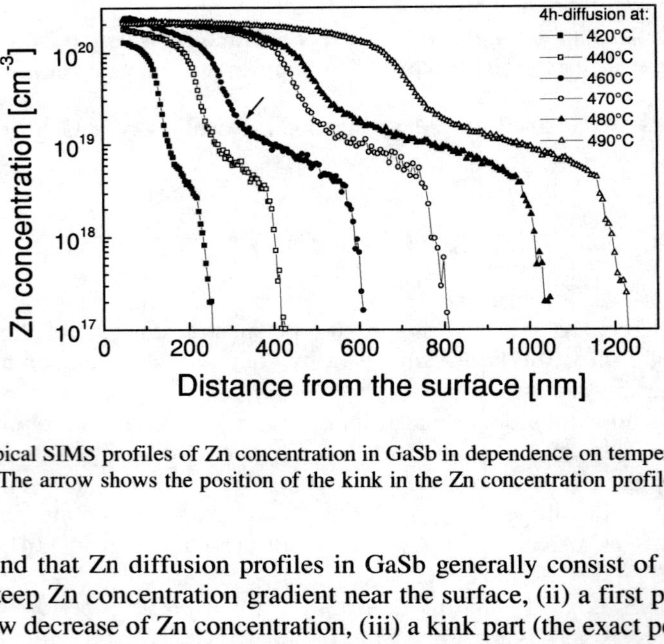

FIGURE 1. Typical SIMS profiles of Zn concentration in GaSb in dependence on temperature after a 4-hour diffusion. The arrow shows the position of the kink in the Zn concentration profile for a 460°C – diffusion.

It was found that Zn diffusion profiles in GaSb generally consist of 5 parts: (i) a part with a steep Zn concentration gradient near the surface, (ii) a first plateau with a relatively slow decrease of Zn concentration, (iii) a kink part (the exact position of the kink is defined as the point where the second derivative of the Zn concentration profile is at maximum), (iv) a second plateau with a relatively slow change of Zn concentration and (v) a steep diffusion front. The first part was usually thinner than 50 nm and it is not shown in Figure 1. We consider that this part of the SIMS measurements is not reliable enough; therefore, we did not take this part into further consideration. Hereafter we define as the surface concentration the maximum concentration at the first plateau region. Our experiments have shown that Zn concentration at part (ii) - close to $1x10^{20}$ cm^{-3} - is too high for TPV cells. Obviously, one can expect small diffusion lengths and a strong bandgap narrowing in this part of the diffused layer and consider it to be a dead layer. Therefore, the relation of the dead layer depth to a total pn-junction depth is of interest. Figures 2 and 3 show the dependencies of the both parameters on diffusion temperature (T) and time (t), respectively. One can see that the kink depth is close to the half of the pn-junction depth.

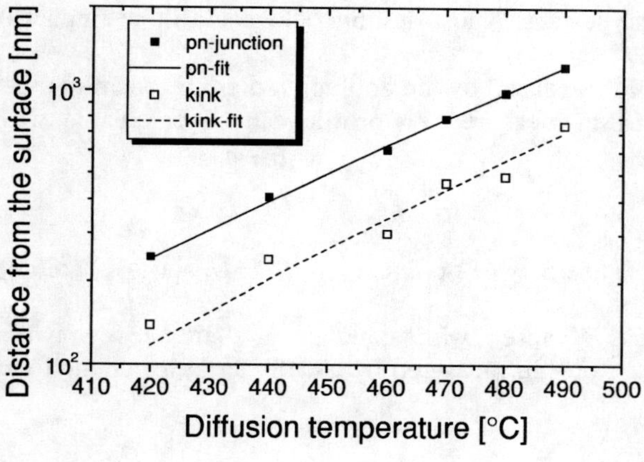

FIGURE 2. pn-junction depth (n = $3x10^{17}$ cm^{-3}) and kink depth for 4-h diffusions of Zn into GaSb in dependence on diffusion temperature.

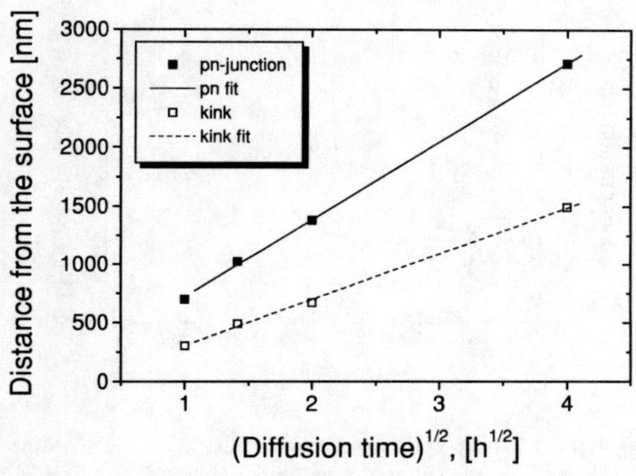

FIGURE 3. pn-junction depth (n = $3x10^{17}$ cm^{-3}) and kink depth for 480°C -diffusions of Zn into GaSb in dependence on diffusion time.

As a post-diffusion tailoring of diffused Zn emitter layers in GaSb, precise thinning through anodic oxidation followed by selective etching of anodic oxide is usually performed. An anodic oxidation regime with a GaSb consumption rate about 2.3 nm/V is used. The detailed description of the anodic oxidation process of GaSb is published elsewhere [16]. Through thinning of the diffused emitter in a TPV cell, a highly doped ("dead") part of the diffused layer can be removed with a high accuracy. Moreover, a second (shallow) diffusion of Zn can be performed resulting in an optimised Zn profile shown in Figure 4.

Figure 4 also shows the calculated profile of the net built-in electric field comprised of the (i) electric field E_1, caused by the doping gradient, and (ii) the electric field E_2 due to the band gap gradient caused by the effect of band gap narrowing.

The field E_1 moves photogenerated electrons to the p-n junction, while the field E_2 moves them to the surface. The resulting field has a large impact on the TPV cell parameters.

The electric field E_1 caused by the doping gradient between two neighbour points i and $i+1$ of the SIMS-measured Zn profile can be calculated with the following expression:

$$E_1 = \frac{kT}{qd_i} \ln \frac{N_i}{N_{i+1}}, \qquad (1)$$

where d_i is the distance between points i and $i+1$; N_i and N_{i+1} are doping densities at the corresponding points.

The electric field E_2 caused by the band gap gradient between two neighbour points i and $i+1$ of the Zn profile measured by SIMS can be calculated with the following expression:

$$E_2 = \frac{A(N_i^{1/3} - N_{i+1}^{1/3})}{qd_i}, \qquad (2)$$

where the coefficient $A = 10^{-8}$ eV·cm is taken from [17].

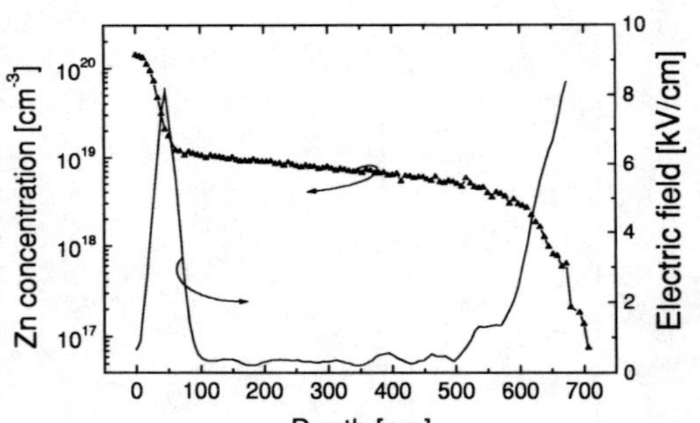

FIGURE 4. SIMS-measured Zn profile of the optimised p-GaSb emitter obtained as a result of three successive operations: (a) deep diffusion, (b) controlled thinning, and (c) shallow diffusion. Also shown is a calculated profile of the net built-in electric field.

One can see from Fig. 4, that if minority electrons are considered, the sweeping field is especially strong near the surface (up to 8 kV/cm), thus decreasing the influence of surface recombination on separation of photogenerated carriers.

Also, we performed a post-diffusion annealing to determine its influence on Zn profiles. After diffusion at low temperature (for example, 420ºC), the surface of the diffused layer is covered with 20-50 nm thick SiN_x. Afterwards, annealing in a no-Zn atmosphere at higher temperature (480-520ºC) is performed. Figure 5 illustrates changes of Zn profile in an as-diffused sample after annealing. With an increase of annealing temperature and time, the maximum concentration of Zn near the surface drops, while the depth of Zn penetration increases. Such a post-diffusion treatment can be very useful for materials containing volatile components such as P and As (InAsSbP or InGaAsSb). To avoid non-congruent evaporation of P or As from unprotected semiconductor surface, one can use low-temperature initial diffusion followed by applying of a protective SiN_x layer and high-temperature annealing.

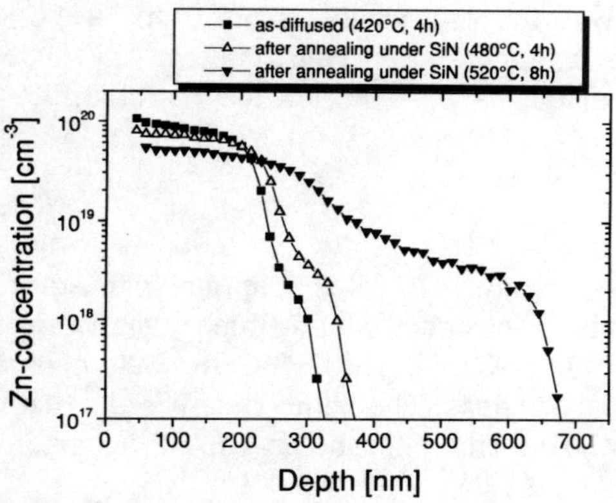

FIGURE 5. SIMS profiles of Zn concentration in a GaSb sample diffused at 420ºC (4 h) measured just after the diffusion and after a post-diffusion annealing at 480ºC (4 h) and 520ºC (8 h). SiN_x layer was deposited on the sample before annealing.

As we have found in this work, another method to accelerate Zn diffusion and thus to avoid high diffusion temperatures and long times, is to perform Zn diffusion through the preliminary formed surface oxide. To form the oxide, we used anodic oxidation providing an oxide layer with precisely controlled parameters [16].

Figure 6 shows SIMS profiles of Zn concentration in a GaSb sample diffused through anodic oxide of different thickness. All the samples are diffused in one process, thus all other diffusion parameters are the same. One can see that the diffusion depth rapidly increases with the thickness of anodic oxide. An assumption can be made that anodic oxide increases surface absorption of Zn from the vapor phase. Similar enhanced diffusion has been observed for diffusion of Zn into GaAs through an oxidized AlAs layer [18].

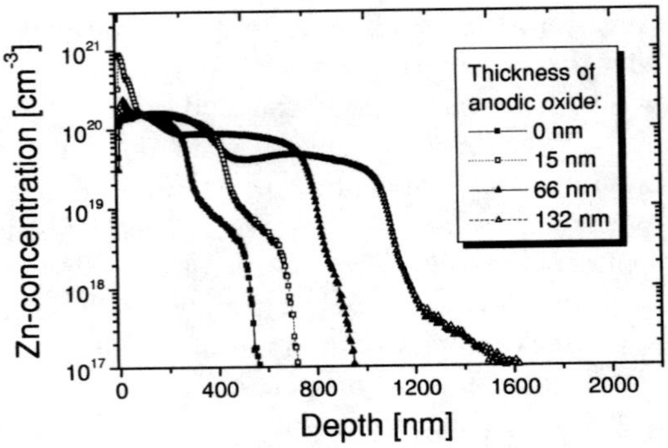

FIGURE 6. SIMS profiles of Zn concentration in a GaSb sample diffused through anodic oxide of different thickness.

In a qualitative sense, the dependence of diffusion on the anodic oxide thickness might be also true for a native oxide. Thus, one can speculate that for reproducible results the thickness of the native oxide on the GaSb surface should be minimized and/or kept constant.

InGaSb

InGaSb with a band gap of 0.5-0.7 eV is of special interest for TPV applications. The availability of ternary bulk substrates would simplify the fabrication of TPV cells with band gaps lower than that of GaSb. The TPV cells could be fabricated using diffusion without expensive epitaxial steps. The antimonide-based system (InGaSb) is preferred over the arsenic (InGaAs) and phosphorus (InAsP) counterparts due to technical simplicity during growth [15].

Zn diffusion from the vapor phase is performed into polished 2 cm^2 n-InGaSb:Te ($n \approx$ 2-3x10^{17} cm^{-3}) substrates. These InGaSb substrates are cut from a polycrystalline boule material grown at the Rensselaer Polytechnic Institute. Prior to diffusion the substrates are degreased in acetone. Afterwards, they were treated in diluted HF or HCl to remove native oxides on the surface and dried with nitrogen.

The form of diffusion profiles of Zn into $In_xGa_{1-x}Sb$ (Fig. 7) is similar to that in GaSb. All the major features described above for the Zn profile in GaSb can be also found in the ternary material. However, in addition to diffusion time and temperature, a new parameter – InSb content in $In_xGa_{1-x}Sb$ – affects the Zn diffusion. As it is shown in Figure 7, an increase of InSb content in $In_xGa_{1-x}Sb$ strongly enlarges the diffusion depth of Zn in this material. Thus, for fabrication of $In_xGa_{1-x}Sb$ TPV cells diffusion temperatures should be essentially lower than in GaSb. Polycrystalline $In_xGa_{1-x}Sb$ samples with $x = 0.12$ - 0.19 ($E_g = 0.61$-0.55 eV) and $x = 0.26$ ($E_g = 0.5$ eV) grown as a bulk material are studied in this work. Temperatures in the range of 430-450ºC for $In_xGa_{1-x}Sb$ samples with $x = 0.12$ - 0.19 are used for fabrication of InGaSb TPV cells. No TPV cells have been fabricated of $In_xGa_{1-x}Sb$ with $x = 0.26$ ($E_g = 0.5$ eV) yet.

However, it is obvious that Zn diffusion temperatures lower than 350ºC should be applied for the diffused emitter in this material.

For the tailoring of the diffused profiles anodic oxidation of InGaSb is studied. The anodic oxidation of InGaSb is performed in an electrolyte with the similar basic components as in the process developed for GaSb [16]:

(i) A component providing a suitable ionic conduction in the electrolyte. In this work we used a 5 % tartaric acid aqueous solution (AAS) with a pH value ≈ 6.0

(ii) A component decreasing the dissolution of the anodic oxide. In this work ethylene glycol (EG) with a volume ratio to the component (i) of 20 is used.

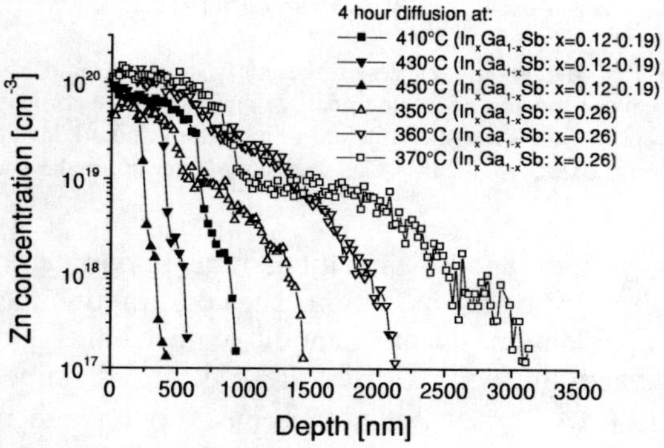

FIGURE 7. SIMS profiles of Zn concentration in $In_xGa_{1-x}Sb$ with different x in dependence on temperature after a 4-hour diffusion.

A high ratio of EG/AAS = 20 is used for InGaSb samples. In comparison, GaSb sample require EG/AAS = 5 [16]. This difference is caused by a relatively high value of the minimum current that was measured for InGaSb in electrolytes with EG/AAS = 5. Thus, it is not possible to have current densities lower than 0.3 mA/cm^2 for $In_{0.12}Ga_{0.88}Sb$ and lower than 0.7 mA/cm^2 for $In_{0.26}Ga_{0.74}Sb$ at an applied voltage of 90V. For comparison, minimum current densities as low as 0.01 mA/cm^2 are measured for GaSb at the same oxidation conditions. It is not quite clear what causes the relatively high minimum currents by the oxidation of InGaSb: higher dissolution rates of these materials, the presence of surface defects or both of them. High-quality single crystal InGaSb wafers are necessary to determine that. Since low final current densities are important for the homogeneity of anodic oxides [16], electrolytes with the lower dissolution of anodic oxides (EG/AAS = 20) are chosen.

Ellipsometry measurements of the anodic oxide thickness and refractive index are performed within an area of 5 x 5 mm^2 in 100 points. Figure 8 shows thickness of the anodic oxide of the $In_{0.12}Ga_{0.88}Sb$ sample versus applied voltage for an electrolyte EG:AAS=20:1 and final current density J_f = 0.1 mA/cm^2. The mean values are shown together with the standard deviation. One can see that the dependence of the oxide thickness is linear. The linear fit of the measured data gives the value of the oxidation ratio equal to 2.53 nm/V. It is noteworthy that practically the same dependence of the oxide thickness on the applied voltage was observed for the $In_{0.26}Ga_{0.74}Sb$ sample.

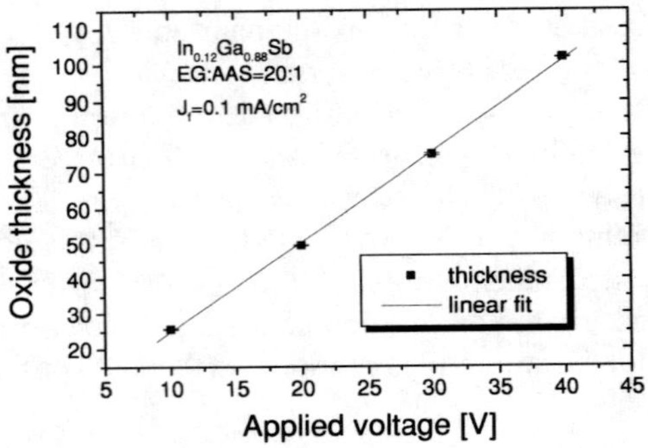

FIGURE 8. Thickness of the anodic oxide of the $In_{0.12}Ga_{0.88}Sb$ sample versus applied voltage. Ellipsometry measurements of the anodic oxide thickness and *n* were performed within an area of 5 x 5 mm^2 in 100 points. The mean values are shown together with the standard deviation. The dependence of the thickness of the anodic oxide of the $In_{0.26}Ga_{0.74}Sb$ sample on applied voltage was practically the same.

For the practical use of anodic oxidation, it is important to know how much semiconductor is consumed by the oxidation. The consumption rate can be determined by performing a local oxidation, selective anodic oxide etching and measuring the step height between oxidized and non-oxidized areas. It is determined for an electrolyte with pH = 6 and EG:AAS = 20:1 that the consumption rate is 60 ± 3 % of the oxidation rate for the $In_{0.12}Ga_{0.88}Sb$ sample and 58 ± 3 % for the $In_{0.26}Ga_{0.74}Sb$ sample.

InGaAsSb

InGaAsSb as a quaternary alloy provides a broad variation of band gaps. For the TPV application only InGaAsSb compositions lattice matched to a GaSb substrate, for example $In_{0.15}Ga_{0.85}As_{0.17}Sb_{0.83}$ that corresponds to a bandgap $E_g \approx 0.55$ eV, are considered. The first InGaAsSb cell was fabricated at AstroPower [19]. Fully epitaxial approach was used for the fabrication of that cell. Later, in a joint work of Fraunhofer ISE and AstroPower the epitaxial emitter was substituted by the diffused one [7, 8] and higher open-circuit voltage was achieved as a result.

Zn diffusion from the vapor phase is performed into epitaxial InGaAsSb:Te (n ≈ 2-3x10^{17} cm^{-3}) layers grown on polished 1.5 x 1.5 cm^2 n-GaSb:Te (100) substrates. Prior to diffusion the wafers are treated in diluted HF or HCl to remove native oxides on the surface and dried with nitrogen.

As Zn profiles in this material have a form with a kink described for GaSb, Figure 9 shows pn-junction depth, kink depth and their difference for 4-h diffusions of Zn into GaSb in dependence on diffusion temperature. Similar to InGaSb, lower than for GaSb diffusion temperatures are necessary for InGaAsSb TPV cell fabrication.

For the tailoring of the diffused layers, precise thinning through anodic oxidation and selective oxide etching, as well as additional shallow Zn diffusion after the emitter thinning is used. As it was determined in [8], the oxidation and consumption rates of $In_{0.15}Ga_{0.85}As_{0.17}Sb_{0.83}$ are 2.26 nm/V and 2.51 nm/V, respectively.

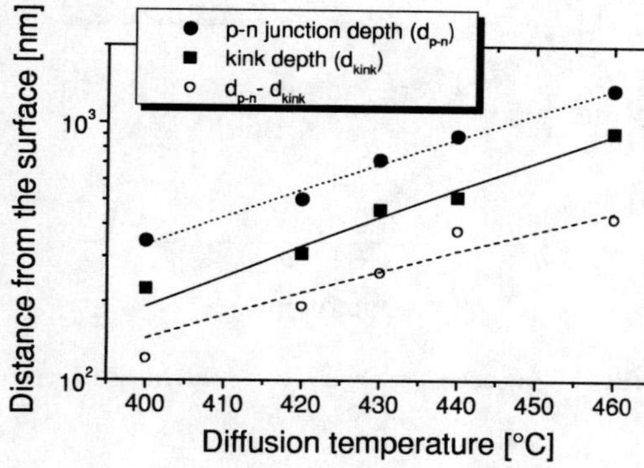

FIGURE 9. pn-junction depth (n = $3x10^{17}$ cm^{-3}), kink depth and their difference for 4-h diffusions of Zn into InGaAsSb in dependence on diffusion temperature.

InAsSbP

InAsSbP materials lattice matched to InAs or GaSb are of special interest for the TPV application. Materials with a bandgap down 0.35 eV are available in this system. There is only one publication of AstroPower reporting of an LPE-grown p-InAsSbP/n-InAsSbP/n-InAs TPV cell [20]. In [21] we fabricated a similar TPV cell with a diffused p-InAsSbP instead of an epitaxial one.

Zn diffusion from the vapor phase is performed into epitaxial InAsSbP doped with Sn or Te (n ≈ $2\text{-}3x10^{17}$ cm^{-3}) layers grown on polished 1.5 x 1.5 cm^2 n-InAs (100) substrates.

Fig. 10 shows an essential difference between Zn profiles in InAsSbP and those in GaSb, InGaAsSb or InGaSb described above. First, Zn profiles in InAsSbP exhibit no kink and relatively low Zn concentration of $2x10^{18}$ cm^{-3}. The high concentration measured at the surface may be an artifact of SIMS. Nevertheless, and similar to GaSb and its alloys, it can be removed by anodic oxidation. In other words, there is no dead layer in Zn diffused InAsSbP (or it is very thin) and thus the design of an optimal Zn-diffused emitter might be easier.

Also, Fig. 10 shows a relatively strong dependence of Zn diffusion profiles on variation of As, Sb and P. More experiments are necessary to determine which element plays the main role in the observed dependence.

For the tailoring of the diffused profiles, anodic oxidation of InAsSbP is studied. The anodic oxidation of InAsSbP is performed in the same electrolyte that was developed for the oxidation of InGaSb (see above). A consumption rate of 1.5 nm/V was determined for InAsSbP.

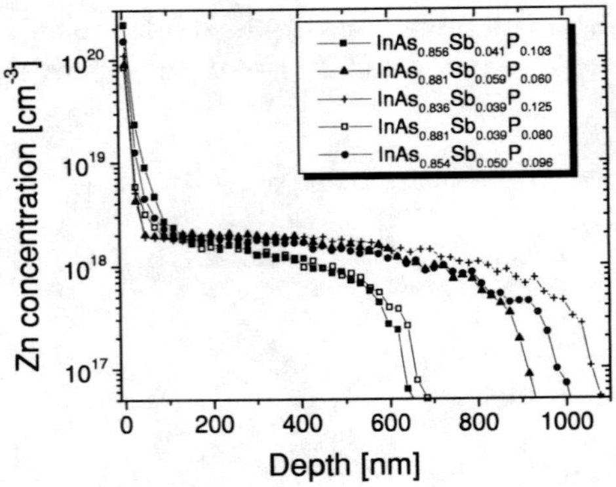

FIGURE 10. SIMS profiles of Zn concentration in $In_xGa_{1-x}Sb$ with different x after 4-hour 340°C diffusion.

CONCLUSIONS

Diffusion of Zn from the vapor phase into as GaSb-, InGaAsSb-, InGaSb- and InAsSbP using pseudo-closed system is studied.

Diffusion profiles in GaSb, InGaSb and InGaAsSb generally consist of 5 parts: (i) a part with a steep Zn concentration gradient near the surface, (ii) a first plateau with a high Zn concentration and its relatively slow decrease, (iii) a kink part, (iv) a second plateau with a relatively slow change of Zn concentration and (v) a steep diffusion front.

A tailoring of the diffused profiles in GaSb, InGaSb and InGaAsSb is necessary to use the diffused emitter in high-efficiency TPV cell. Precise thinning of the as-diffused emitter that removes a dead layer with high Zn concentration (GaSb, InGaSb and InGaAsSb) and additional shallow Zn diffusion after the emitter thinning (GaSb, InGaAsSb) are studied and found to be advantageous. Additionally, a shallow diffusion followed by a high-temperature annealing under SiN_x and the influence of anodically oxidized surface on Zn profiles are studied in GaSb.

Contrary to GaSb, InGaSb and InGaAsSb, practically no dead layer is found in Zn diffused InAsSbP. Thus the fabrication of an optimal Zn-diffused emitter InAsSbP might be simpler.

The description and comparison of GaSb-, InGaSb-, InGaAsSb- and InAsSbP-based TPV cells with Zn diffused emitters are published elsewhere [21].

ACKNOWLEDGEMENTS

The authors wish to thank A. Kovarsky from Ioffe Institute of Physics and Technology for his assistance and advice in performing SIMS measurements, C.

Schetter from Fraunhofer Institute for Solar Energy Systems for deposition of SiN_x layers and P. Sims from AstroPower for useful discussions.

REFERENCES

1. Andreev, V. M., and Sulima, O.V., Sov. Tech. Phys. Lett., 8 , 187- 188 (1982).
2. Spring Thorpe, A. J., and Svilans, M. L., "Low-temperature zinc-diffusion in GaAs, InP and InGaAs using the box-diffusion technique" in Inst. Phys. Conf. Ser., No. 65, 1982, pp.589-596.
3. Blashku, A. I., Boltaks B. I., Burdiyan, I. I., Dzhafarov,T. D, and Rzaev, M. A., Sov. Phys.-Semiconductors, 6, 402-406 (1972).
4. Sundaram, V. S., and Gruenbaum P. E., J. Appl. Phys., 73, 3787-3789 (1993).
5. Bett, A.W., Keser, S., and Sulima, O.V., J. of Cryst. Growth, 181, 9-16 (1997).
6. Fraas, L. M., Avery, J. E., Martin, J., Sundaram, V. S., Girard, G., Dinh, V.T., Davenport, T. M., Yerkes, J. W., and O'Neil, M. J., IEEE Transactions on Electron Devices 37, 443-449, (1990).
7. Sulima, O.V., Beckert, R, Bett, A.W., Cox J.A, and Mauk, M.G., IEE Proc.-Optoelectron., 147, 199-204 (2000).
8. Bett, .W., Ber, B.Ya., Mauk, M.G., South, J.T., and Sulima, O.V., " Pseudo-closed box diffusion of Zn into InGaAsSb and AlGaSb for TPV devices", ", in Thermophotovoltaic Generation of Electricity –1998, edited by T.J. Coutts et al., AIP Conference Proceedings 460, American Institute of Physics, New York, pp. 237-246.
9. Bett, A.W., Keser, S., Stollwerck, G., and Sulima, O.V., "Large-Area GaSb Photovoltaic Cells", in Thermophotovoltaic Generation of Electricity –1997, edited by T.J. Coutts et al., AIP Conference Proceedings 401, American Institute of Physics, New York, pp. 41-53.
10. Sulima, O.V., Faleev, N.N., Kazantsev, A.B., Mintairov, A.M., and Namazov, A., "Low-temperature Zn diffusion for GaSb solar cell structures fabrication" in Proceedings of the 4th European Space Power Conference, Poitiers, France, 1995, v.2, pp. 641-644.
11. Sulima, O.V., and Bett, A.W, Solar Energy Materials & Solar Cells, 66, 533-540, 2001.
12. V. Hinkov, personal communication.
13. Hitchcock, C.W., Gutmann, R.J., Borrego, J.M., Bhat, I.B., and Charache, G.W., IEEE Transactions on Electron Devices, 46, 2154-2161 (1999).
14. Mauk, M. G., Shellenbarger, Z. A., Cox, J. A., Sulima, O.V., Bett, A.W., Mueller, R. L., Sims, P. E., McNeely, J. B., DiNetta, L. C., J. of Cryst. Growth, 211, 189-193 (2000).
15. Dutta, P.S., Ostrogorsky, A.G., and Gutmann, R.J., "Bulk crystal growth of antimonide based III-V compounds for TPV applications", in Thermophotovoltaic Generation of Electricity –1998, edited by T.J. Coutts et al., AIP Conference Proceedings 460, American Institute of Physics, New York, pp. 227-236.
16. Sulima, O.V., Bett, A.W., and Wagner, J., Journal of The Electrochemical Society, 147, 1910-1914 (2000).
17. Stollwerck, G., Sulima, O.V., and Bett, A.W., IEEE Transactions on Electron Devices, 47, 448-457 (2000).
18. Andreev, V.M., Pitroff, V., and Sulima, O.V., Sov. Phys. Tech. Phys., 30, 1080-1082 (1986).
19. Shellenbarger, Z.A., Mauk, M.G., and DiNetta, L.C., " Recent Progress in InGaAsSb/GaSb TPV Devices" in Proceedings of the 25th IEEE Photovoltaic Specialists Conf., 1996, pp. 81-84.
20. Mauk, M.G., Shellenbarger, Z.A., Cox, J.A., Tata, A.N., Warden, T.G., and DiNetta, L.C., "Advances in low-bandgap InAsSbP/InAs and GaInAsSb/GaSb thermophotovoltaics", " in Proceedings of the 28th IEEE Photovoltaic Specialists Conf., 2000, pp. 1028-1031.
21. Sulima, O.V., Bett, A.W., Mauk, M.G., Dimroth, F., Dutta, P.S., and Mueller, R.L., "GaSb-, InGaAsSb-, InGaSb-, InAsSbP- and Ge-TPV cells for low-temperature TPV applications", this conference.

20% Efficient InGaAs/InPAs Thermophotovoltaic Cells

R. R. Siergiej[a], B. Wernsman[a], S. A. Derry[a], R. G. Mahorter[a], R. J. Wehrer[a], S. D. Link[a], M. N. Palmisiano[a], R. L. Messham[a], S. Murray[b], C. S. Murray[b], F. Newman[b], J. Hills[b], and D. Taylor[c]

[a]*Solid State Electronic Materials, Bechtel Bettis, Inc., West Mifflin, PA 15122, USA*
[b]*EMCORE Photovoltaics, Albuquerque, NM 87123, USA*
[c]*Bandwidth Semiconductor, LLC, Hudson, NH 03051, USA*

Abstract. Recent improvements in large area, high efficiency, monolithic interconnected modules (MIMs) represent a significant step in the development of thermophotovoltaic (TPV) technology for various power producing applications. The MIM architecture with transmissive integrated spectral control offers a desirable high-voltage, low-current output, front-side contacts for simplified packaging, high spectral utilization due to a metallic, highly reflective and specular back surface reflector, and a practical method for scale-up to full wafer devices. The n/p/n MIM TPV devices described in this work utilize a tunnel junction and a double heterostructure for improved performance. Lattice-mismatched 0.6 eV, epitaxially grown InGaAs diodes form the power-producing element. A power conversion efficiency of 20.6% and a power density of 0.90 W/cm^2 with a silicon carbide radiator operating at 1058°C is achieved for a 4 cm^2 (die area) TPV cell operating at 26.7°C.

INTRODUCTION

Monolithic interconnected module (MIM) devices for thermophotovoltaic (TPV) applications have been investigated previously [1-6]. In a MIM, several power-producing devices are connected in series using an insulating technique (e.g. semi-insulating substrate or reverse biased diodes) to isolate the devices from each other. This MIM configuration of the power-producing elements offers several key device and system advantages. The design provides a high-voltage, low-current output per MIM, thereby minimizing resistive losses due to conduction of large currents in larger arrays. A MIM design also allows the device to continue functioning if a single diode fails in a shorted mode. Having both cell electrodes available on the topside of the wafer and soldering the substrate directly to the cold plate for heat-sinking simplifies array construction. The material choice which comprise the MIM can enable an integrated spectral control strategy whereby a highly specular and reflective metallic back side reflector is employed to recuperate below bandgap light to the radiator. It

CP653, *Thermophotovoltaic Generation of Electricity: 5th Conference*
edited by T. J. Coutts, G. Guazzoni, and J. Luther

also allows a second pass of above bandgap photons through the active layers of the device, which results in an increase in the spectral utilization [7].

The choice of semiconducting material for the power-producing diode depends primarily on the peak of the spectral power density curve of the radiating material. In this work, we focus on the lattice-mismatched 0.6 eV InGaAs system which is preferred for the 1058°C graybody radiator spectral response. Lattice-mismatched diode growth requires a buffering scheme in order to reduce epitaxial layer defects that degrade device performance.

Lattice-mismatched 0.6 eV InGaAs MIM TPV devices grown on semi-insulating InP substrates have been fabricated and tested. Employing techniques to reduce shunts induced by growth and processing, in addition to improving the device design, a power conversion efficiency of 20.6% and a power density of 0.90 W/cm^2 at a cell temperature of 26.7°C using a 1058°C silicon carbide radiator have been measured. This result represents a major step in demonstrating the viability of this technology.

MIM EPITAXIAL DESIGN

The 0.6 eV InGaAs structure, seen in Figure 1, is grown on semi-insulating InP substrates by metalorganic chemical vapor deposition (MOCVD). The devices used in this study were grown using an Emcore D180 multi-wafer reactor with a maximum

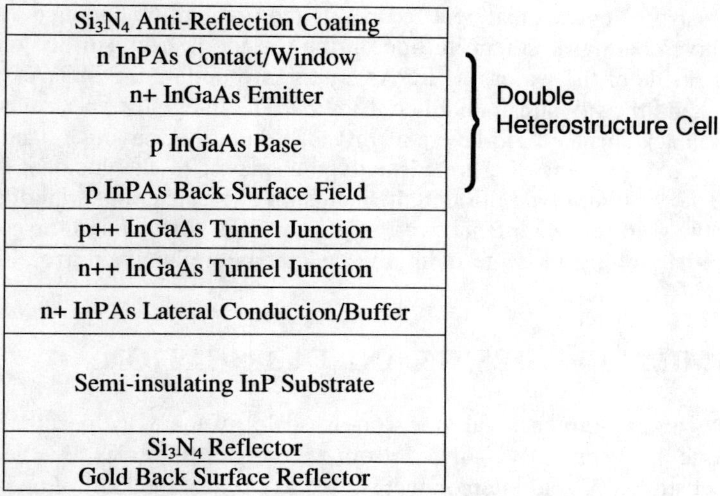

Figure 1. Cross-section of the double heterostructure (InGaAs/InPAs) 0.6 eV MIM device grown on a semi-insulating InP substrate.

capacity of three, 3-inch substrates. The substrates used were iron-doped, semi-insulating InP, with a diameter of 76 mm and a thickness of 600 μm. The growth rates of the InPAs and InGaAs layers were between 5 – 10 Å/s. Care has been exercised to minimize growth induced shunt mechanisms such as particulates, which adversely affect the electrical properties of the MIM device.

The structure considered in this work is an improved epitaxial design compared to previous work [1-6], with reduced parasitic absorption due to the minimization of non-power producing InGaAs layers. Lattice mismatched growth of the InGaAs active layers is achieved by initially growing a three step, compositionally graded InPAs buffer. The buffer is doped n-type to serve as a lateral conduction layer (LCL) to provide the back contact for the cell. In order to connect the p-type base region with the n+ lateral conduction layer, a p++/n++ tunnel diode is implemented. The active region of the cell consists of a p/n junction (base/emitter) of InGaAs surrounded by layers of InPAs (top contact/window and back surface field (BSF)) that form a double heterostructure [8]. The top InPAs layer also serves as the contact to the topside of the cell.

Because of the transmissive integrated spectral control strategy used, the design is a continual trade between the optical and electrical performance in this MIM device. The objective of this strategy is to minimize light absorption outside the p/n junction, obtain multiple passes through the junction, and return unwanted long wavelength light to the radiator. To minimize free carrier absorption effects in the semiconductor layers, it is desirable to use low doping densities as much as possible, but this leads to more electrical resistive losses, creating a competing requirement on doping density.

An anti-reflective coating of silicon nitride on the topside, a back surface reflector (BSR) on the backside of the wafer, the InPAs layers surrounding the InGaAs power-producing diode, and the InP substrate all contribute to the integrated spectral control. The window and back surface field layers of InPAs and the InP, having a bandgap of 1.0 eV and 1.35 eV, respectively, are virtually transparent to the incident photon energies. The BSR was improved through the inclusion of a barrier material of silicon nitride between the gold and the InP substrate to prevent inter-diffusion of the gold and indium species, maintaining a long-term highly specular and reflective mirror.

MIM CELL DESIGN AND FABRICATION

Figure 2 shows a three-dimensional view (not to scale) of the MIM design. In this figure each diode is electrically isolated from the adjacent diodes by the semi-insulating InP substrate. A gold interconnect is used to connect the LCL of one device to the top contact layer of the next device. The dielectric for the electrical isolation between the interconnect and the side wall of the diode also acts as the AR coating. The MIM can consist of an arbitrary number of diodes connected in series using gold interconnects. At the extreme ends of this interconnected structure, wide cathode and anode busbars are used to connect this MIM to adjacent MIMs in a large area array.

The layout of the 2 cm by 2 cm MIM design used in this study is shown in Figure 3a. The device is comprised of 30 cells sized 582 μm by 19920 μm, and four devices of this style are included in the 3 inch wafer layout seen in Figure 3b. Included with the 2 cm by 2 cm design are eight 1 cm by 1 cm MIMs, eight single junction devices, and four dual junction MIM devices. A test structure area contains devices to determine the quality of the ohmic contacts, the resistivity of the LCL, the quantum efficiency, and the electrical integrity of the AR/dielectric coating.

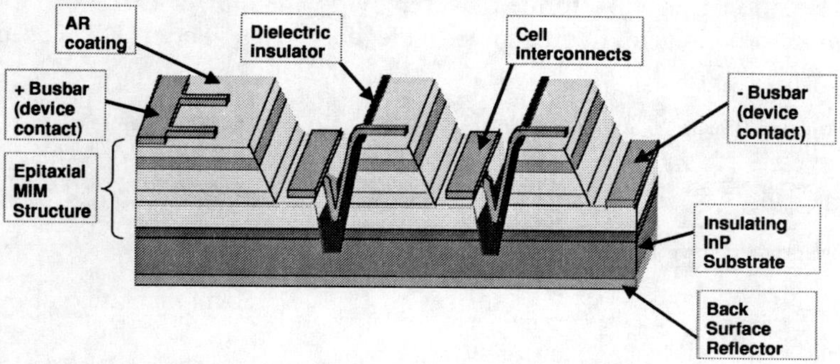

Figure 2. A conceptual drawing of the high efficiency MIM device.

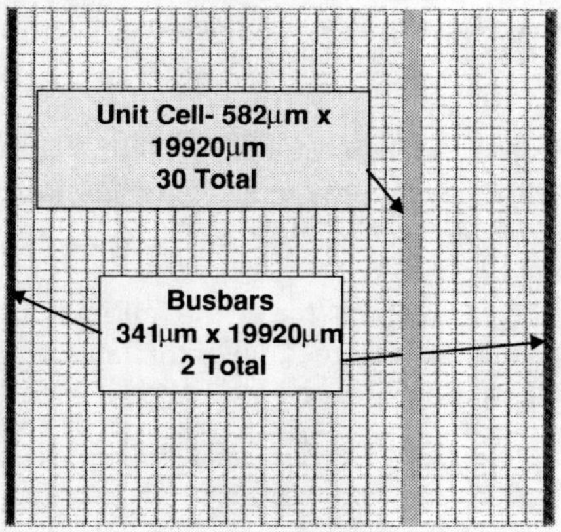

a.

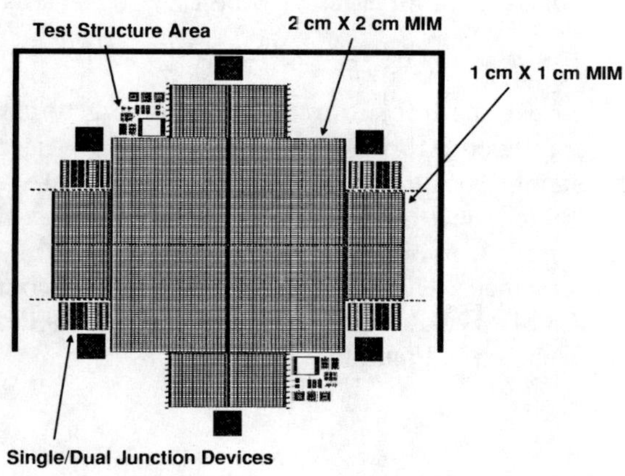

b.

Figure 3. a) The layout of the 2 cm by 2 cm, 30 junction MIM that utilizes a busbar interconnect scheme with 400 μm grid spacing. b) A three inch wafer layout includes four 2 cm by 2 cm MIMs, eight 1 cm by 1 cm MIMs and various test structures.

The MIM fabrication sequence is based on wet chemical etching techniques described previously [9] and is modified to account for our epitaxial layer design. As shown in Figure 4, the back contact etch exposes the top of the LCL. This is accomplished with an etch to remove the InGaAs and InPAs layers above the LCL. The InPAs contact/window, InGaAs power-producing p/n junction, and the InGaAs tunnel p/n junction are all etched with a solution of phosphoric acid, hydrogen peroxide, and water.

Next, the isolation trench is formed by removing the InPAs LCL/buffer with a solution of hydrobromic acid, hydrogen peroxide, and water. Then, HCl is used to V-

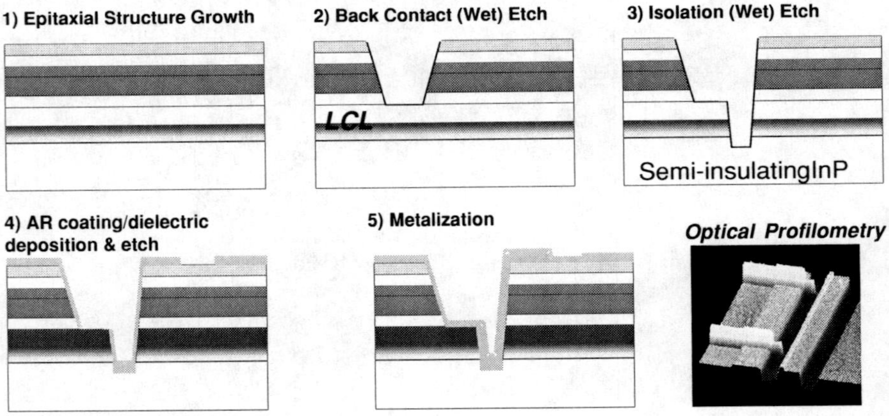

Figure 4. The MIM fabrication uses standard III-V processing techniques and features a PECVD silicon nitride AR/dielectric, wet chemical etches and e-beam evaporated Ti/Au interconnects.

groove the InP substrate a few microns to complete the isolation trench feature. The combined AR/dielectric coating of Si_3N_4 is deposited by plasma enhanced chemical vapor deposition (PECVD) and patterned for the LCL and top contact areas. A Ti/Au interconnect is achieved using a lift-off technique to complete the device. As seen in Figure 4, an optical profilometer is used to check the etch depths during the fabrication sequence. The measured depths are then compared to the designed layer thicknesses. An SEM photograph of the V-groove isolation/LCL contact area, on a finished device, is shown in Figure 5.

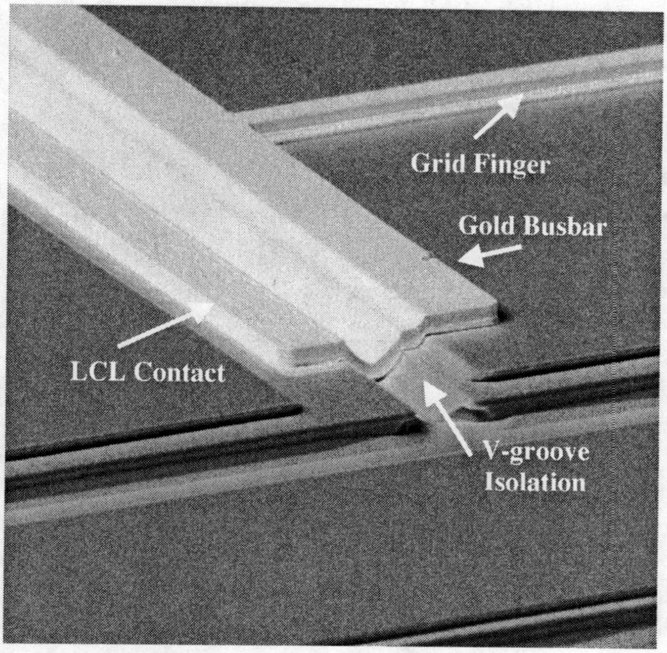

Figure 5. An SEM photograph of a finished MIM device showing the isolation V-groove, LCL contact area, busbar and a grid finger.

ELECTRICAL/OPTICAL MEASUREMENTS

Completed cells are electrically screened using a quartz-halogen lamp assembly for open circuit voltage (V_{oc}), short circuit current (I_{sc}), fill factor (FF) and maximum power ($P_{max}=V_{oc}*I_{sc}*FF$). The J-V characteristics for various illumination levels are shown in Figure 6. At a short circuit current of 136 mA and a cell temperature of 25°C, this high efficiency, 30-diode cell produces a V_{oc} of 11.6 V, FF of 67% and a P_{max} of 1.06 W. From the plot of I_{sc} versus V_{oc} values, a linear extrapolation in the n=1 region of the curve[10] gives a reverse saturation current density of 0.4 $\mu A/cm^2$, as seen in Figure 7.

In order to investigate cell-to-cell consistency in generating power, a device is forward biased under dark conditions and an IR camera is used to capture the electroluminescence. Such an IR camera image is shown in Figure 8 and clearly indicates the presence of under-performing regions. While it is difficult to quantitatively resolve diode-to-diode uniformity data from such a scan, this tool rapidly identifies diodes that have some material or process aberration.

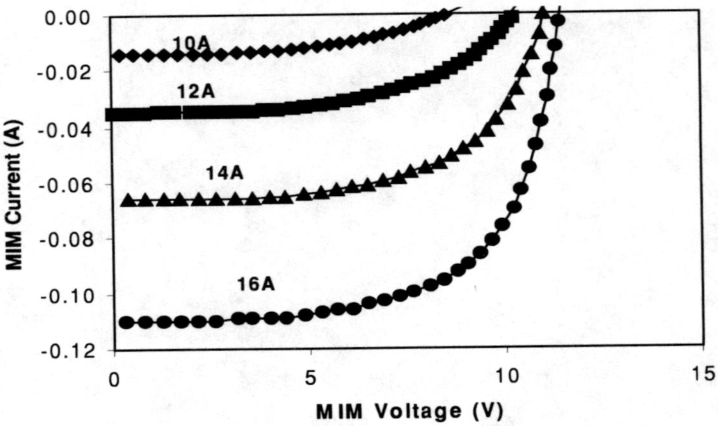

Figure 6. J-V characteristics for the 2 cm by 2 cm MIM device under different illumination levels (10, 12, 14, and 16 A) from a quartz-halogen lamp assembly.

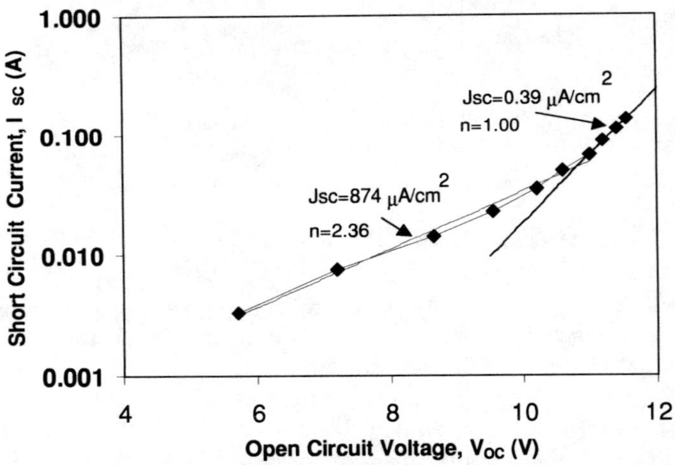

Figure 7. Short circuit current density versus open circuit voltage to enable extraction of reverse saturation and ideality values.

A measure of the external quantum efficiency is made using a single junction, 3000 μm by 3000 μm diode with no internal grid pattern. Figure 9 shows a peak external quantum efficiency of 92% at approximately 1.5 μm.

The measured reflectance of a completed MIM is shown in Figure 10. Values of reflectance > 90% are seen for a large portion of wavelengths greater than the bandgap of the diode. The dip in the reflectance around 4.5 μm is caused by absorption in the silicon nitride AR coating.

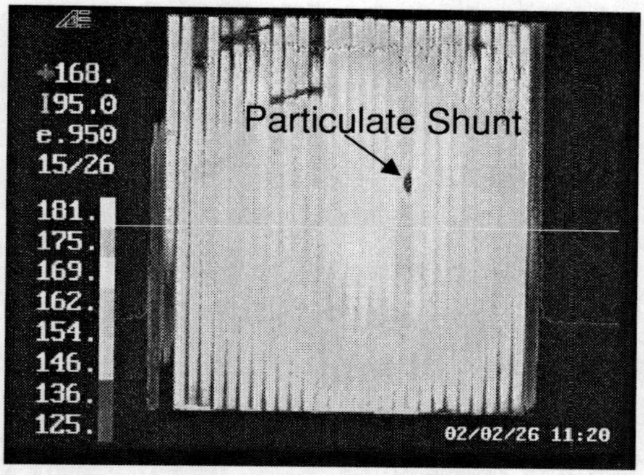

Figure 8. Electroluminescence scan of the 2 cm by 2 cm MIM devices clearly shows under-performing regions.

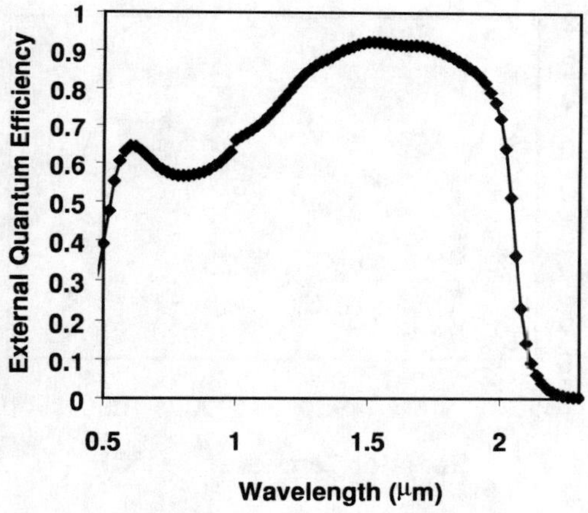

Figure 9. Measured external quantum efficiency showing approximately 92% near 1.5 μm.

Individual MIMs are mounted in ceramic saddles for in-vacuum efficiency measurements using a silicon carbide graybody radiator heated to temperatures as high as 1058°C. The ceramic saddle containing the cell is affixed atop a thermoelectric cooled copper pedestal permitting the absorbed power (P_{abs}) to be measured by monitoring a drop in temperature across the pedestal[11]. Thus, the efficiency, defined as the percentage of P_{max}/P_{abs}, may be calculated for various radiator and cell temperatures.

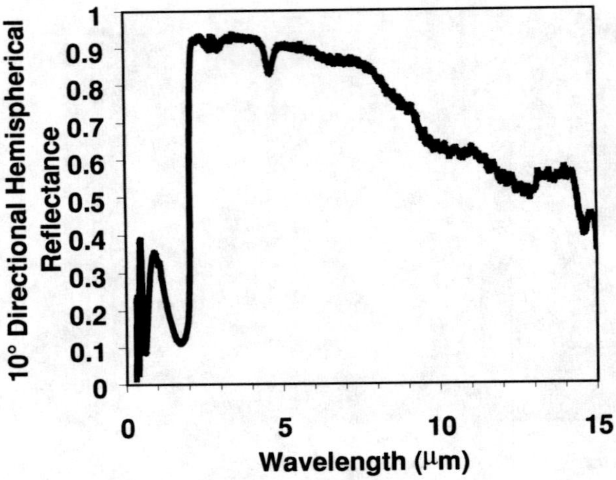

Figure 10. Measured reflectance of the MIM showing values greater than 90% for a portion of wavelengths greater than the bandgap of the diode.

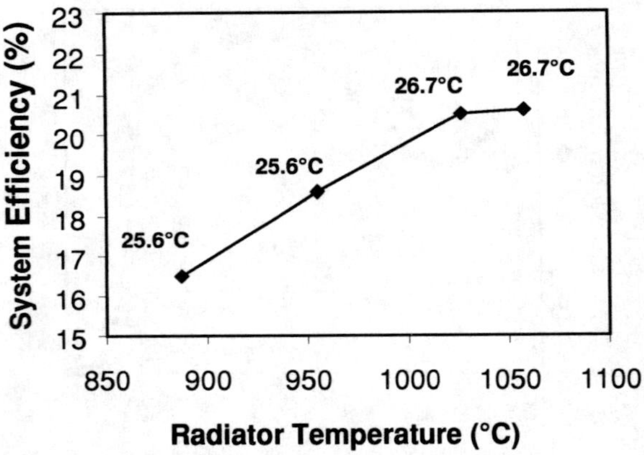

Figure 11. Measured in-vacuum MIM efficiency using a silicon carbide radiator. The cell temperatures are indicated on the graph.

Figure 12 shows the efficiency measurements using a silicon carbide graybody radiator. At a radiator temperature of 887°C and a cell temperature of 25.6°C, the TPV cells were 16.5% efficient and produced a power density of 0.38 W/cm^2. Increasing the radiator temperature to 1058°C, a power conversion efficiency of 20.6% and a power density of 0.9 W/cm^2 were measured for a cell operating at 26.7°C.

SUMMARY

In this work we describe high performance 0.6 eV lattice-mismatched InGaAs MIM devices developed for TPV applications. These devices were made utilizing standard III-V MOCVD growth and processing techniques. Large strides in improving the power density and efficiency of these devices have been made through a series of material growth and device design improvements. During the growth of the epitaxial structure, techniques have been used to minimize particulate formation, reducing shunt paths in the MIM device. Non-power producing InGaAs regions have been minimized in this design as compared to those reported previously. A barrier material of silicon nitride has been used to prevent the interdiffusion of indium and gold components, resulting in a long-term specular back surface reflector. With these improvements, a maximum efficiency of 20.6% and a power density of 0.9 W/cm^2 are measured for a device operating at 26.7°C in a system with a 1058°C silicon carbide radiator. It is believed that further improvements in material quality and device structure optimization will continue to increase device performance.

REFERENCES

1. N. S. Fatemi, D. M. Wilt, R. W. Hoffman, Jr., M. A. Stan, V. G. Weizer, P. P. Jenkins, O. S. Khan, C. S. Murray, D. Scheiman, and D. Brinker, "High-Performance, Lattice-Mismatched InGaAs/InP Monolithic Interconnected Modules (MIMs)," Proc., 4th NREL TPV Conf., AIP CP460, pp. 121-131, 1999.
2. M. W. Wanlass, J. J. Carapella, A. Duda, K. Emery, L. Gedvilas, T. Moriarty, S. Ward, J. D. Webb, X. Wu, and C. S. Murray, "High Performance, 0.6-eV, $Ga_{0.32}In_{0.68}As/InAs_{0.32}P_{0.68}$ Thermophotovoltaic Converters and Monolithically Interconnected Modules," Proc., 4th NREL TPV Conf., AIP CP460, pp. 132-141, 1999.
3. D. M. Wilt, C. S. Murray, N. S. Fatemi, and V. Weizer, "n/p/n Tunnel Junction InGaAs Monolithic Interconnected Module (MIM)," Proc., 4th NREL TPV Conf., AIP CP460, pp. 152-160, 1999.
4. C. S. Murray, N. Fatemi, M. Stan, B. Wernsman, and R. J. Wehrer, "Growth, Processing and Characterization of 0.55-eV n/p/n Monolithic Interconnected Modules," Proc. IEEE 28th Photovoltaic Specialists Conf., pp. 1238-1241, 2000.
5. D. M. Wilt, E. B. Clark, M. B. Clevenger, C. S. Murray, and R. J. Wehrer, "0.55eV n/p/n MIM TPV Cell Development," Proc. IEEE 28th Photovoltaic Specialists Conf., pp. 1024-1027, 2000.
6. C. Murray, R. Newman, S. Murray, J. Hills, D. Aiken, R. Siergiej, B. Wernsman, and D. Taylor, "Multi-Wafer Growth and Processing of 0.6-eV InGaAs Monolithic Interconnected Modules," Proc. IEEE 29th Photovoltaic Specialists Conf., in publication, 2002.
7. G. W. Charache, D. M. DePoy, P. F. Baldasaro, and B. C. Campbell, "Thermophotovoltaic Devices Utilizing a Back Surface Reflector for Spectral Control, " Proc. 2nd NREL TPV Conf., AIP CP358, pp. 339-350, 1995.
8. M. W. Wanlass, "Low Bandgap Double Heterostructure InAsP/GaInAs Photovoltaic Convert," U.S. Patent 6,300,557.
9. N. S. Fatemi, D. M. Wilt, P. P. Jenkins, R. W. Hoffman, Jr., V. G. Weizer, C. S. Murray, and D. Riley, "Materials and Process Development for the Monolithic Interconnected Module (MIM) InGaAs/InP TPV Devices," Proc. 3rd NREL TPV Conf., AIP CP401, pp. 249-262, 1997.
10. C. K. Gethers, C. T. Ballinger, and D. M. Depoy, "Lessons Learned on Closed Cavity TPV System Efficiency Measurements," Proc., 4th NREL TPV Conf., AIP CP460, pp. 335-348, 1999.
11. S. M. Sze, *Physics of Semiconductor Devices, 2nd Edition*, John Wiley and Sons, 1981, p. 264.

Effect of Metal Coverage on the Performance of 0.6-eV InGaAs Monolithic Interconnected Modules

Susan L. Murray[a], Christopher S. Murray[a], Mark A. Stan[a], Frederick D. Newman[a], Jenifer Hills[a], Richard Siergiej[b] and Bernard Wernsman[b]

[a]Emcore Photovoltaics, 10420 Research Road, Albuquerque, NM 87123
[b]Bechtel Bettis Inc., 814 Pittsburgh-McKeesport Blvd., West Mifflin, PA 15122

Abstract. With the device performance of 0.6eV InGaAs monolithic interconnected modules (MIMs) reaching open circuit voltages of 400 mV/junction and achieving excellent quantum efficiency, the next step to improve performance focuses on controlling the parasitic optical absorption in these MIMs. With an integrated spectral control approach, the design of grid finger and interconnect metallization affects both the output power and the optical absorption of the MIM. The effect of metal coverage on the optical and electrical performance of MIMs processed in a multi-wafer environment is presented.

INTRODUCTION

Increasing the efficiency of TPV devices involves increasing the output electrical power while minimizing parasitically absorbed optical power. Parasitically absorbed photons are defined as absorbed photons that do not create an electron-hole pair in a photovoltaically active device region, but act to degrade the TPV conversion process. To put this into perspective, approximately 83% of the incident photons from a 1000°C blackbody source are sub-bandgap in energy when using a 0.6-eV TPV converter. Minimizing absorption of these "parasitic" photons is therefore critical to achieving high efficiency device performance.

Figure 1 shows the normalized spectrum of a 1000°C blackbody radiator overlaid with a 0.6-eV device. The area under the curve to the right of ~2 μm is non-convertible light. The reflectance of an InP substrate with a Back Surface Reflector is shown to be 97% in this region. In an epitaxial MIM structure, reflectance decreases to 95% due to absorption in the epitaxial layers (Epi Material Reflectance with ARC), and is further reduced to 90% after processing of a MIM, due to light trapping issues associated with the metal coverage and interconnect metallization. Parasitic losses resulting from MIM processing are also seen in the convertible region, as the MIM reflectance is much higher than the epitaxial material alone.

CP653, *Thermophotovoltaic Generation of Electricity: 5th Conference*
edited by T. J. Coutts, G. Guazzoni, and J. Luther

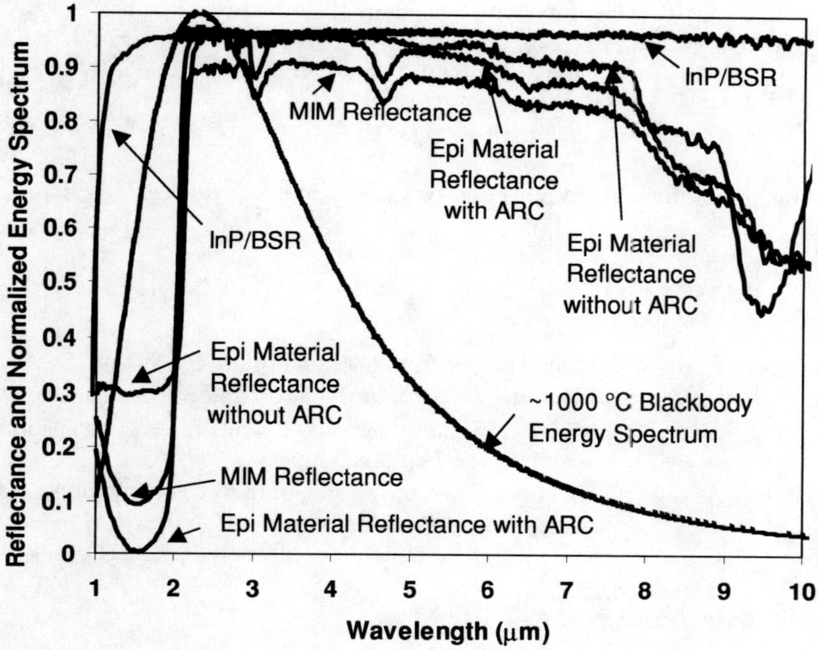

FIGURE 1. Optical reflectance curves for an InP substrate with a back surface reflector (InP/BSR), an epitaxial MIM structure without an ARC, an epitaxial MIM structure with an ARC, and a processed MIM overlaid on the 1000°C black body energy spectrum. The curves show how the MIM epitaxial layers and device processing degrade optical reflectance.

Reduction of parasitics has been approached in several ways. The approach used here is to build a monolithic interconnected module or MIM, which has been discussed by several authors [1-4].

To gain an understanding of the electrical and optical losses associated with a MIM and how the interaction between electrical and optical performance affects the overall device efficiency, parametrics are used to study the changes in device performance. For example, an estimate of the maximum power, P_{max}, can be calculated based on measurable parameters as shown below:

$$P_{max} = \varepsilon * P_{rad} * A_{MIM} * \tau * F_V * Q_E * F_O * FF \quad (1)$$

Where:

ε	=	Emissivity of the radiator
P_{rad}	=	Radiated power from the radiator ($E>E_g$), W/cm^2
A_{MIM}	=	Active area of the MIM, cm^2
τ	=	Energy weighted transmission for $E>E_g$
F_V	=	Voltage factor, V_{OC}/E_g
Q_E	=	Energy weighted quantum efficiency

F_O = Over-excitation or thermalization factor
FF = Fill factor
E = Photon energy, eV
E_g = Bandgap energy, eV
V_{OC} = Open circuit voltage, V

In addition, an estimate of the power absorbed, P_{abs} in the device can be calculated by:

$$P_{abs} = \varepsilon * P_{radT} * A_{MIM} * \tau_{abs} * (1-R_{BSR}) * \Delta R_{FCA} * \Delta R_{MIM} \qquad (2)$$

Where:
P_{radT} = Radiated power from the radiator (all E), W/cm^2
R_{BSR} = Reflection of the gold/InP back surface reflector
ΔR_{FCA} = Reduction in the reflectance due to free carrier absorption in the n- and p-doped epitaxial layers
ΔR_{MIM} = Reduction in the reflectance due to light trapping associated with the grid fingers and interconnects in the MIM.
τ_{abs} = Energy weighted transmission for all of the light.

Finally, the device efficiency, η, is defined as:

$$\eta = P_{max}/P_{abs} \qquad (3)$$

Some of the parameters defined above are affected by the array of grid fingers and the spacing between the individual diodes connected in series to form the MIM. The following sections discuss the photomask design, and how the parameters sensitive to grid finger and diode spacing affect device efficiency.

PHOTOMASK DESIGN

In order to understand the role MIM design has on reducing the device efficiency by contributing to optical and electrical losses, a photomask was designed with varying grid finger spacings and cell widths. The device being used to study the effect of grid fingers on light trapping is a 2x2 cm^2 cell, as shown in the photomask layout in Figure 2.

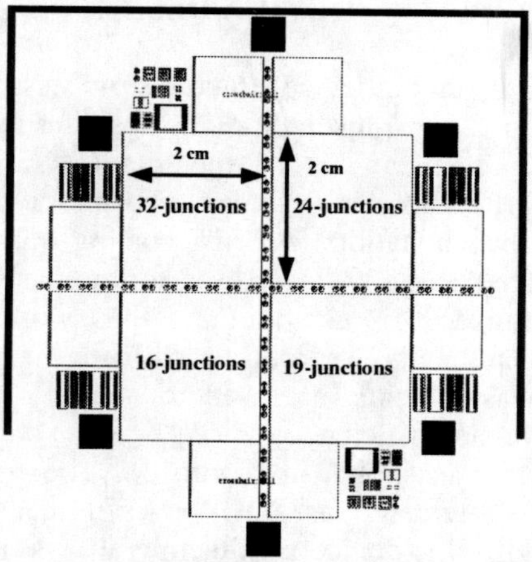

FIGURE 2. Photomask with four designs of 2x2cm MIMs. The devices have 16 junctions, 19 junctions, 24 junctions or 32 junctions.

Table 1 summarizes the active area of the individual diodes and MIMs, the grid finger spacing, spacing between individual diodes, and the device packing factor for each of the four MIM designs. Note that the active cell area doubles between the 16 junction and 32 junction MIM design. Also notice that the packing factor increases from 86% to 91% between the 32- and 16-junction devices respectively. In other words, there is a 5% increase in the device area available for conversion due to a 5% decrease in surface features on the MIM. From these observations, the 16-junction device would be expected to have better performance than the 32 junction design. In practice device performance is also governed by the electrical impact of the cell layout (eg series resistance), and neither of these junction designs is optimal for this device structure.

TABLE 1. Grid spacing and areas for mask design with 16, 19,24 and 32 junction MIMs.

Number of Junctions	Cell Width (μm)	Grid Finger Spacing (μm)	Cell Active Area (cm2)	MIM Area (cm2)	Packing Factor (%)
32	600	600	0.107	3.44	86.0
24	800	800	0.147	3.54	88.5
19	1000	1000	0.170	3.57	89.3
16	1200	1200	0.227	3.64	91.0

DEVICE DESIGN

The 0.60-eV InGaAs n/p/n device structures processed with varying metallization designs were grown on 3" semi-insulating InP substrates using an Emcore D180 multi-wafer Metal-Organic Chemical Vapor Deposition (MOCVD) reactor. The processing of the devices was performed using standard phosphoric based wet chemical etchants in an automated wet bench station. A Si_xN_y coating was used for both the dielectric and anti-reflection layers and Ti/Au was used for the metallization. The details of the growth and processing are discussed in previous work [5]. Illuminated and dark current-voltage measurements, as well as diffuse hemispherical reflectance and spectral response measurements, are used to quantify the effects of metallization pattern on MIM conversion efficiency.

The MIM structure is shown in Figure 3. The n/p/n device incorporates an InGaAs/InPAs double heterostructure (DH) on a p/n InGaAs tunnel junction to reverse the polarity of the cell. This device architecture allows the use of an n-doped lateral conduction layer, to allow lateral current flow to the back of the device to take place in more conductive n-type layers. The n-doped InPAs buffer layer, which also serves as the lateral conduction layer, is a three step compositionally graded layer used to accommodate the lattice mismatch between the active InGaAs device layers and the semi-insulating InP substrate. On the back of the InP substrate, a gold back surface reflector is used to reflect convertible light not absorbed on the first pass for a second pass through the active device layers, and to recuperate non-convertible light at the radiator. The front surface is coated with a 2000Å of Si_3N_4, used both as a dielectric and as a anti-reflection coating.

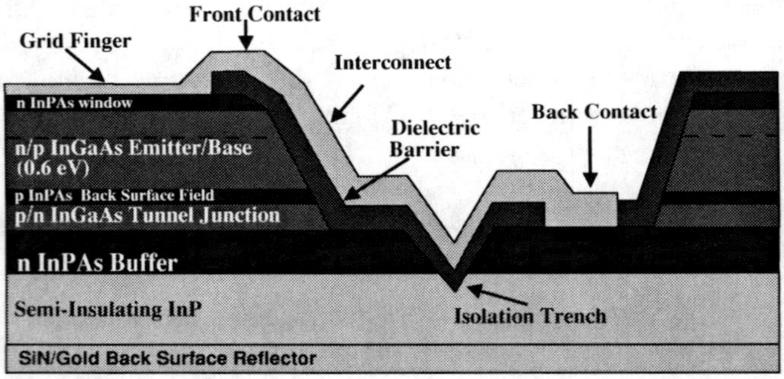

FIGURE 3. Cross section of MIM structure showing the epitaxial layers, isolation & contact designs.

RESULTS

The optical and electrical results discuss the output power, power absorbed and device efficiency as they relate to the cell design parameters shown in Equations (1-3). Figure 4 shows the measured reflectance for a) an InP wafer with a Au back surface reflector, b) grown epitaxial material with a BSR and anti-reflection coating (no processing), and c) the four MIM designs with different grid finger spacings and cell widths. The transition from low reflectance to high reflectance, for the epitaxial material curves, corresponds to the bandgap of the device. The reflectance for wavelengths less than the bandgap decreases from epitaxial material without an anti-reflection coating (approximately 30%, see Figure 1, Epi Material Reflectance without ARC) to a wafer measured with a Si_3N_4 anti-reflection coating (< 10%). The epitaxial material curve, which has a minimum reflectance of approximately 1% is indicative of the anti-reflective coating effectiveness. The increase in reflectance in the MIM curves corresponds to the increased metal coverage from the 16 to the 32 junction design.

The reflectance for wavelengths greater than the bandgap includes several loss mechanisms:

a) reflectance of the back surface reflector,
b) free carrier absorption in the n- and p-doped epitaxial layers,
c) light trapping and diffraction losses associated with the surface features of the MIM (i.e. grid fingers, etched trenches, interconnect metal, etc.)

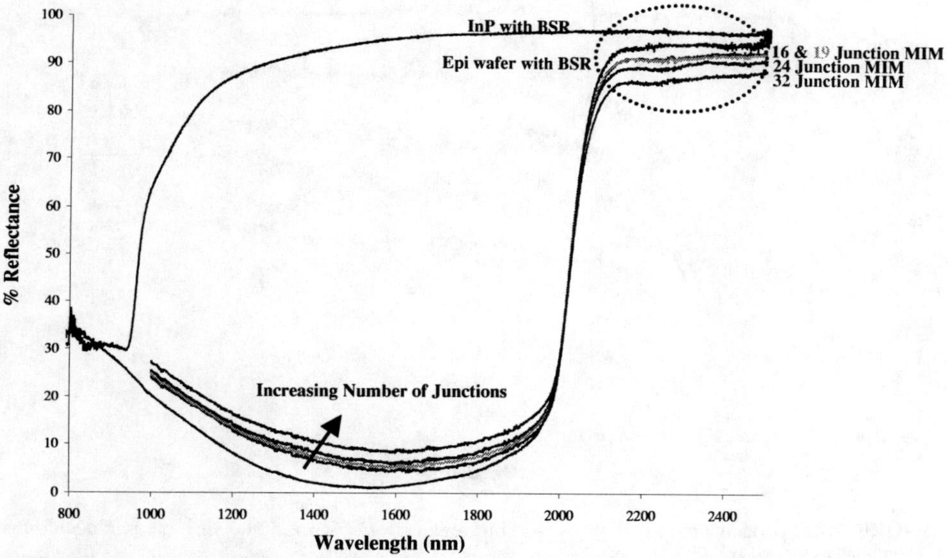

FIGURE 4. Reflection Measurements of InP substrate with back surface reflector (BSR) which is designated as R_{BSR}, epitaxial wafer with BSR, and MIMs with BSR. The reflection in the non-convertible region is shown in a blow-up in Figure 5.

The back surface reflector which is composed mainly of a thin gold layer, has a measured reflectance of 97.4%, indicating that 2.6% of the incident light is trapped in the semiconductor material following reflectance at this interface. This loss could be due to diffuse scattering from surface features on the back of the substrate, index mismatch between the InP substrate and the gold and/or absorption losses at the gold/InP interface. As seen in Figure 5, losses due to free carrier absorption are approximately 3.4% for all of the epi-layers in the structure. The loss due to light trapping is difficult to quantify although the measured results show that fewer surface features give higher total reflectance. For example the 32 junction MIM (packing factor of 86%), has a reflectance of 88% at 2.4 μm; whereas, the 16 junction MIM (packing factor of 91%) has a reflectance of 92% at the same wavelength.

Figure 5 is an expanded graph for the reflectance at wavelengths greater than the bandgap. The 32 junction MIM has the highest loss due to the largest number of surface features. The 16 and 19 junction MIMs have almost the same total reflectance which may be due to the measurement conditions (spot size of the diffuse hemispherical reflectance light source), requiring further investigation.

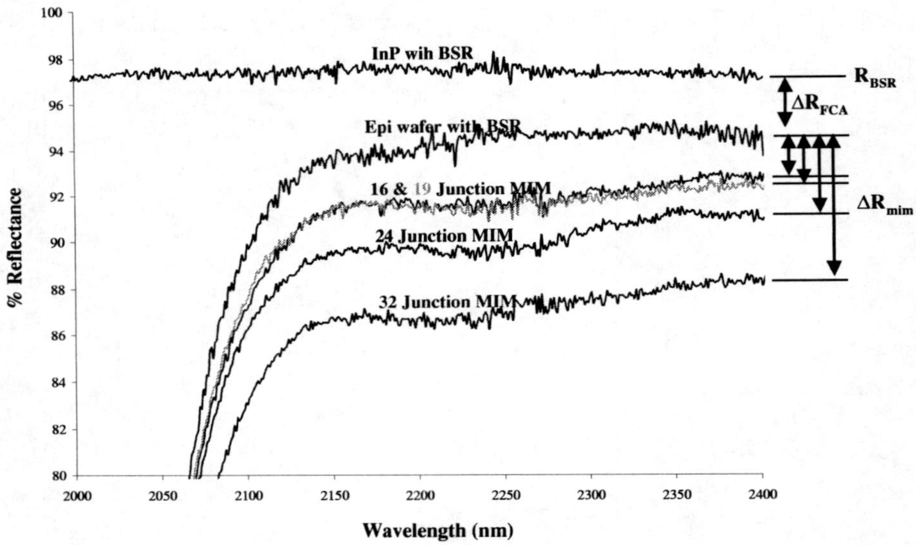

FIGURE 5. Expanded graph of non-convertible region from Figure 4 showing the parameters from Equation (2).

Unlike the optical results, which improved as the packing factor increases, the electrical performance of the MIM degrades, as shown in Figure 6. This is because as the number of junctions decreases, the conduction path length for lateral current flow increases, increasing the series resistance which decreases the fill factor (shown on the inset graph of Figure 6). The increase in the short circuit current is due to the increased cell active area as the cell width increases (see Table 1).

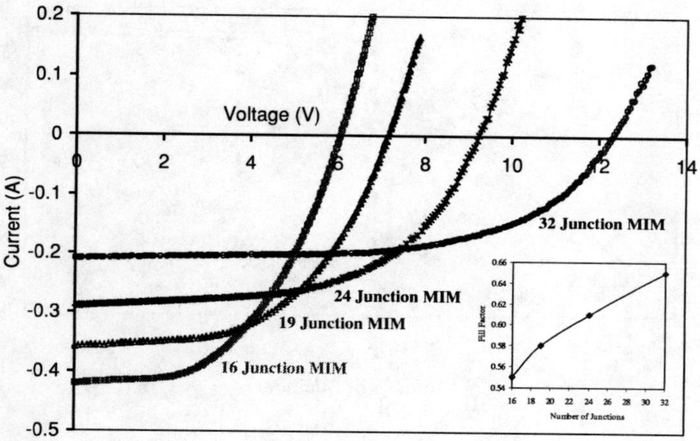

Figure 6. I-V curves for the 16, 19, 24, and 32 junction MIM designs. The inset graph shows the change in fill factor for the various designs due to increase in the effect of series resistance at larger grid spacings.

For the case of optical losses, as the grid finger spacing and cell width are increased (fewer junctions) ΔR_{MIM} decreases; however, this does not necessarily result in lower power absorbed. As the cell area increases (fewer junctions), the total power incident on the MIM area increases. It is observed that for fixed losses associated with the BSR (ΔR_{BSR} = 2.6%) and free carrier absorption (ΔR_{FCA} = 3.4%), the power absorbed in the device decreases from 32 to 19 junctions and then increases beyond a critical MIM area as shown in Figure 7. Measured values of fill factor decrease from 65% to 55% as MIM area increases due to the longer lateral conduction length (600 to 1200 μm) resulting in a corresponding decrease in the maximum power (Figure 7). The coupling of the optical and electrical losses results in an almost constant efficiency for the number of junctions considered in this work as shown in Figure 8.

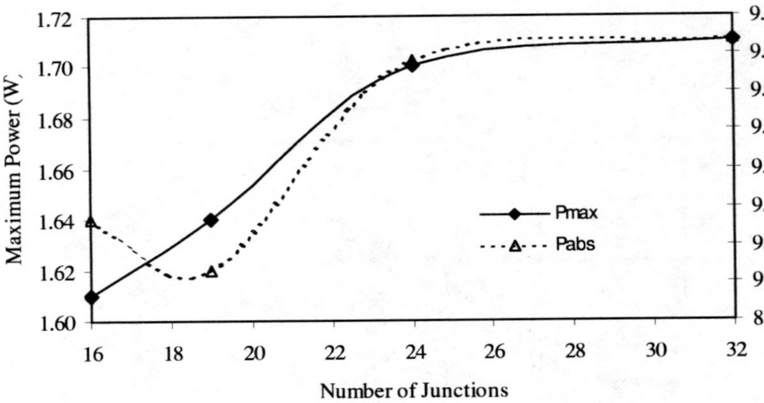

Figure 7. Projected maximum power and absorbed power as a function of number of junctions.

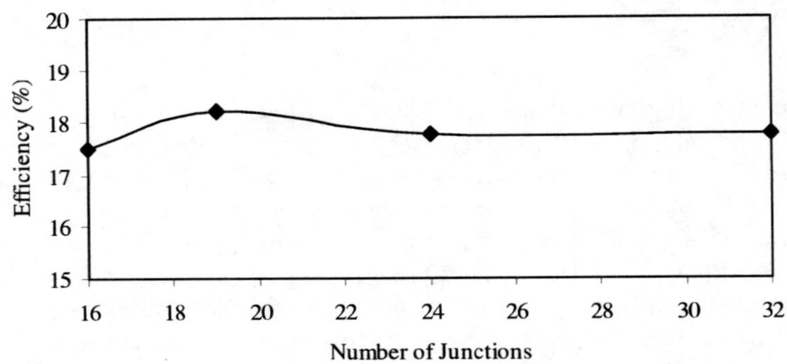

Figure 8. Projected efficiency versus junction number showing optical and electrical trade-off.

CONCLUSIONS

Optimizing the efficiency of a monolithic interconnected module using integrated spectral control is difficult because of the coupling between the electrical and optical performance of the device. As the percentage of surface features on the MIM, due to processing (i.e. etch trenches, metallization, etc.), is decreased thus improving the optical performance, the electrical performance is decreased due to longer conduction lengths for lateral current flow. In the MIM architecture considered in this work, the highest P_{max} is obtained for designs between 24 and 32 junctions with a corresponding efficiency value of ~18%.

REFERENCES

1. D. Wilt et al., "Monolithically Interconnected InGaAs TPV Module Development," *Twenty Fifth IEEE Photovoltaic Specialists Conference*, 1996, pp. 43-48.
2. N. Fatemi et al., "InGaAs Monolithic Interconnected Modules (MIMs)," Twenty Sixth *IEEE Photovoltaic Specialists Conference*, 1997, pp. 799-804.
3. M. Wanlass et al. "High Performance 0.6 eV, $Ga_{0.32}In_{0.68}As/InAs_{0.32}P$ Thermophotovoltaic Converters and Monolithic Interconnected Minimodules," *Proc. 4th NREL Thermophotovoltaic Generation of Electricity Conference*, 1998, pp. 132-141.
4. N. Fatemi et al., "High-Performance, Lattice-Mismatched InGaAs/InP Monolithic Interconnected Modules (MIMs)," *4th NREL Thermophotovoltaic Generation of Electricity Conference*, 1998, pp. 121-131.
5. C. Murray et al., "Multi-Wafer Growth and Processing of 0.6-eV InGaAs Monolithic Interconnected Modules," to be published in *Twenty Ninth IEEE Photovoltaic Specialists Conference*, New Orleans 2002.

GaSb-, InGaAsSb-, InGaSb-, InAsSbP- and Ge-TPV cells for low-temperature TPV applications

Oleg V. Sulima[1*], Andreas W. Bett[2], Michael G. Mauk[1], Frank Dimroth[2], Partha S. Dutta[3], and Robert L. Mueller[4]

[1]*AstroPower Inc., Solar Park, Newark, Delaware 19716-2000, USA*
[2]*Fraunhofer Institute for Solar Energy Systems, Heidenhofstrasse 2, D-79110 Freiburg, Germany*
[3] *Rensselaer Polytechnic Institute, Troy, New York 12180-3590, USA*
[4] *Jet Propulsion Laboratory, 4800 Oak Grove Drive, Pasadena, California 91109-8099, USA*
**corresponding author: sulima@astropower.com, phone: 302-366-0400 ext. 3031, fax: 302-283-0162*

Abstract. GaSb thermophotovoltaic (TPV) cells are the most suitable choice for modern TPV generators, both in terms of efficiency and simplicity of the diffusion technology used. Actually, TPV generators based on GaSb solar cells are the only ones available on the market. However, TPV cells with band gaps (E_g) lower than GaSb are expected to be advantageous for low-temperature (< 1000°C) non-wavelength-selective TPV radiators because they provide more effective absorption of the blackbody infrared radiation. In this work, together with GaSb (E_g = 0.72 eV), semiconductors with a lower E_g - Ge (E_g = 0.66 eV), InGaSb (E_g = 0.60 eV), InGaAsSb (E_g = 0.55 eV) and InAsSbP (E_g = 0.39 eV) - were studied for TPV cells. InGaAsSb cells seem to be the most promising candidate to replace GaSb cells in the low-temperature TPV generators.

INTRODUCTION

A simple and economic "no-epitaxy" technology of GaSb TPV cells using Zn diffusion from the vapor phase has proven to be suitable for production. Currently, only TPV generators based on GaSb solar cells are available on the market [1].

Nevertheless, materials with band gaps lower than GaSb are desirable for low-temperature (< 1000°C) non-wavelength-selective TPV radiators. Therefore, in this work, together with GaSb (E_g = 0.72 eV), semiconductors with a lower E_g - Ge (E_g = 0.66 eV), InGaSb (E_g = 0.60 eV), InGaAsSb (E_g = 0.55 eV) and InAsSbP (E_g = 0.39 eV) - are studied for TPV cells.

To form pn-junctions in III-V materials, we used the most simple and productive method – low-temperature (from 340°C for InAsSbP to 480°C for GaSb) pseudo-closed-box Zn diffusion from the vapor phase. One-layer n-type epitaxial structures were grown by liquid phase epitaxy prior to the Zn diffusion only in the case of InGaAsSb and InAsSbP.

Further ways to simplify the technology of III-V TPV cells and reduce their price include using polycrystalline substrates and/or ternary substrates such as InGaSb. In

CP653, *Thermophotovoltaic Generation of Electricity: 5th Conference*
edited by T. J. Coutts, G. Guazzoni, and J. Luther

this work, a first attempt to fabricate an InGaSb TPV cell from a polycrystalline boule material was made.

It was determined that a careful optimization of the Zn diffusion profiles in TPV cells leads to a significant improvement of device performance [2,3]. The formation of strong built-in electrical fields near the emitter surface provides a dramatic decrease in the surface recombination losses. Thus high-efficiency TPV cells can be fabricated without any passivating window.

The main obstacle to the large-scale application of III-V TPV cells is the high cost of III-V substrates (including polycrystalline). Ge TPV cells made on substrates that are 6-7 times less expensive than those of GaSb, can be a cost-effective alternative to III-V cells. In this work, Ge cells were fabricated using diffusion of P into p-type substrates.

EXPERIMENTAL

TPV cells based on single crystal GaSb, InGaAsSb, InAsSbP and Ge, as well as on polycrystalline GaSb and InGaSb structures fabricated at Fraunhofer ISE, AstroPower and RPI were tested at Fraunhofer ISE and JPL.

A schematic of a TPV cell with a diffused emitter investigated in this work is presented in Figure 1. Description of the cell parts is given in Table 1.

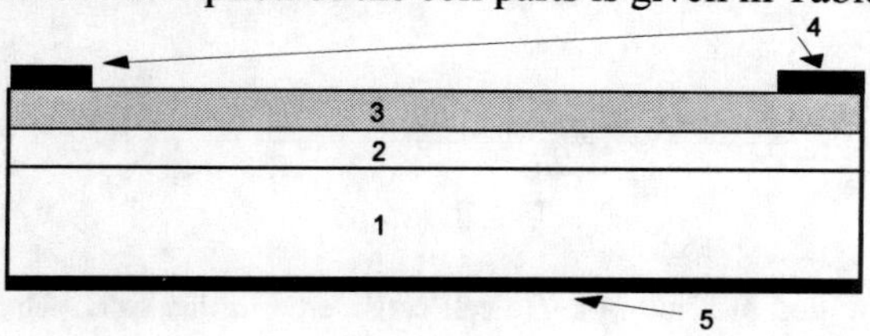

FIGURE 1. Scheme of TPV cells with a diffused emitter investigated in this work. Description of the cell parts is given in the Table 1. Numbers refer to column 1 of Table 1. Different contact grid designs used in this work are described in [4].

TABLE 1. Materials used for different parts of the studied TPV cells.

Part of the cell	GaSb-cell	InGaSb-cell	InGaAsSb-cell	InAsSbP-cell	Ge-cell
1. Substrate	n-GaSb single and polycrystalline	n-InGaSb polycrystal-line	n-GaSb single crystal	n-InAs single crystal	p-Ge single crystal
2. Epitaxial base	none	none	n-InGaAsSb	n-InAsSbP	none
3. Diffused emitter	p-GaSb	p-InGaSb	p-InGaAsSb	p-InAsSbP	n-Ge
4. Front side contact	Ti/Ni/Au				Ni/Au-Ge
5. Backside contact	Au-Ge				Ti/Pd/Ag

Details of the Zn diffusion process applied for III-V cells are published in Ref. [5]. Liquid phase epitaxy is used to grow base n-InGaAsSb or n-InAsSbP layers. Details of the epitaxial growth are described elsewhere [6].

Diffused n-type emitters in p-Ge (p = 3-5x10^{17} cm^{-3}) substrates are formed through diffusion of phosphorus. Two different methods of phosphorus diffusion are used: (i) diffusion during the MOCVD growth of InGaP on Ge, and (ii) diffusion from phosphorus spin-on dopants. In the case of method (i) the epitaxial InGaP layer is selectively removed from the Ge substrate after the diffusion. Method (i) in comparison with method (ii) seems to be too complicated. However, InGaP/Ge structures may be available as a sub-product of multi-junction solar cell fabrication.

Carrier concentration in Ge was measured by electrochemical CV profiling. Figure 2 shows the carrier concentration profile of a Ge cell that exhibited the best device parameters in this work (see below).

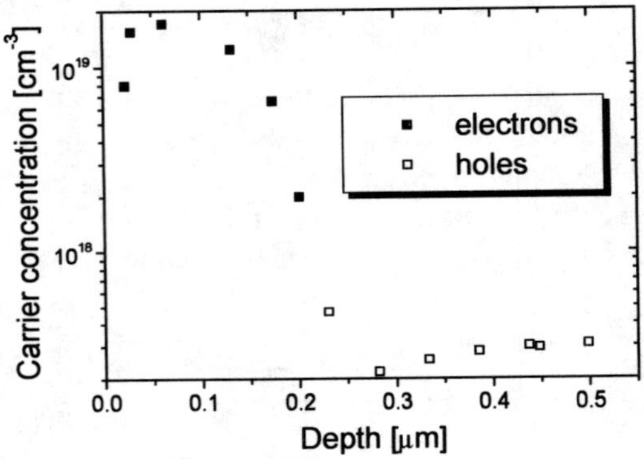

FIGURE 2. Carrier concentration profile in the Ge cell considered in this work. The profile is measured by electrochemical CV-profiling.

RESULTS AND DISCUSSION

Figure 3 shows the external quantum efficiency (EQE) spectra of a typical GaSb TPV cell together with other single crystal and polycrystalline TPV cells studied in this work.

The larger the long wavelength sensitivity of the TPV cell, the lower can be the radiator temperature. Hence if one considers only the photosensitivity range, InAsSbP cells with their sensitivity up to 3500 nm look to be the most promising for the low-temperature TPV generators. However, decrease of the bandgap inevitably causes reduction of the generated voltage. Thus, a trade-off between the increasing current and falling voltage must be found. Regarding the InAsSbP cells, problems with achieving sufficient voltage and FF impede their application. The open-circuit voltage (V_{oc}) achieved in these cells has been limited to about 0.1 V both in the cells with epitaxial [7] and diffusion emitters (this work).

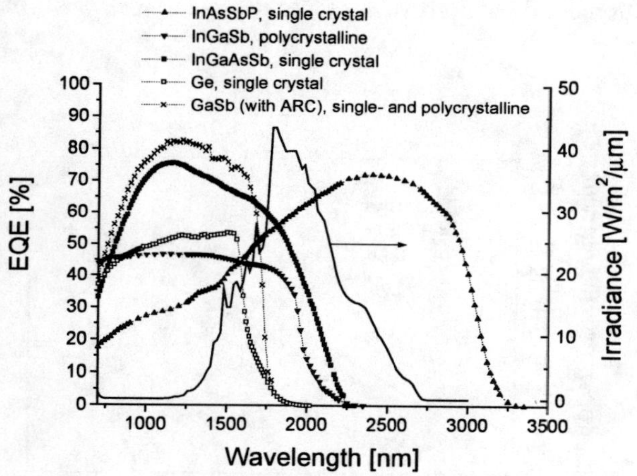

FIGURE 3. External quantum efficiency spectrum of the TPV cells (long wavelength part) measured in this work. Antireflection coating (anodic oxide) was applied only to the GaSb cell. Shadowing losses caused by the contact grid were within 15-20%. The spectrum of the JPL IR bandpass filtered large area pulsed solar simulator (LAPSS) irradiance (solid line, right axis) is shown for comparison.

Thus, the ratio between V_{oc} and the potential difference at the pn-junction in the used InAsSbP (E_g = 0.39 eV) is only 0.25. It is noteworthy that nearly the same low voltages (0.12 V) were observed in polycrystalline InGaSb cells, where the above mentioned ratio was even lower: 0.20. For comparison, this ratio equals 0.5 - 0.6 in GaSb and InGaAsSb. One can assume that it is the surface leakage that prevents obtaining a V_{oc} of around 0.2 V in InAsSbP. In the case of the polycrystalline InGaSb, both the surface and the bulk can cause the reduced voltage.

Let us now consider the InGaAsSb cell that has the next longest wavelength response up to 2250 nm (Fig. 3). In contrast to the InAsSbP cell, relatively high voltages were achieved in such a cell using an optimized diffused emitter [2]. Thus as shown below, InGaAsSb is a very promising candidate to replace GaSb in the low-temperature TPV applications.

Device parameters of GaSb, InGaAsSb and Ge TPV cells were measured under different illumination conditions including the JPL IR bandpass filtered large area pulsed solar simulator (LAPSS). The spectrum of the filtered LAPSS irradiance is shown in Figures 3 and 6. Output electric power densities of the measured devices is shown in Figure 4. The InGaAsSb cell demonstrated a superior performance under the LAPSS illumination. Ge cells showed the worst EQE (Fig. 3) and thus the lowest current. The voltage of Ge cells was relatively poor as well (Fig. 5) - lower than of the InGaAsSb ones despite the higher bandgap of Ge.

Concerning the LAPSS measurements one should note the following. It is clear from Figure 6, that even the blackbody spectrum (T = 1610 K, 3206-fold attenuated) that is the closest to the LAPSS spectrum does not match it very closely. That is the reason why additional calculations of the output power were performed for different blackbody spectra.

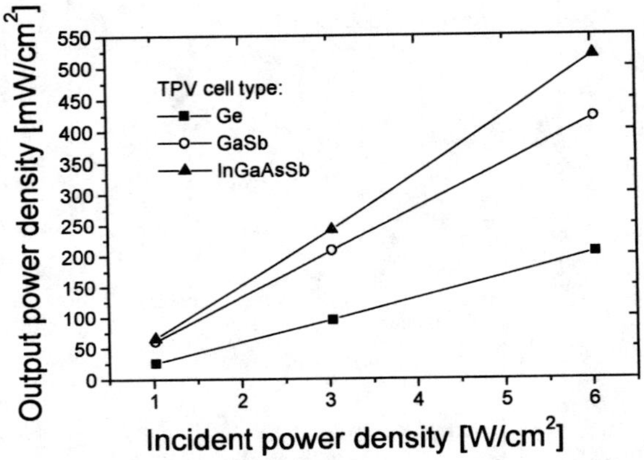

FIGURE 4. Output power density of Ge, GaSb and InGaAsSb TPV cells measured with the JPL LAPSS at different incident power densities at a cell temperature of 30°C.

Experimentally determined IV-characteristics and EQE-spectra allow an accurate calculation of TPV cell performance for any blackbody illumination at any temperature. Figure 7 shows the calculated output power of GaSb and InGaAsSb TPV cells as a function of the blackbody temperature. Taking into account optical losses in a TPV system, we assumed that the blackbody illumination of the cell is "scaled" (its intensity at all wavelengths is proportionally reduced). The scaling factor was assumed to be 0.6. Figure 7 shows that the low bandgap InGaAsSb TPV cells are advantageous for low radiator temperatures if no filters or selective emitters are used. For example, at T = 973K the output power of the InGaAsSb cell is 1.6 times higher than that of the GaSb one.

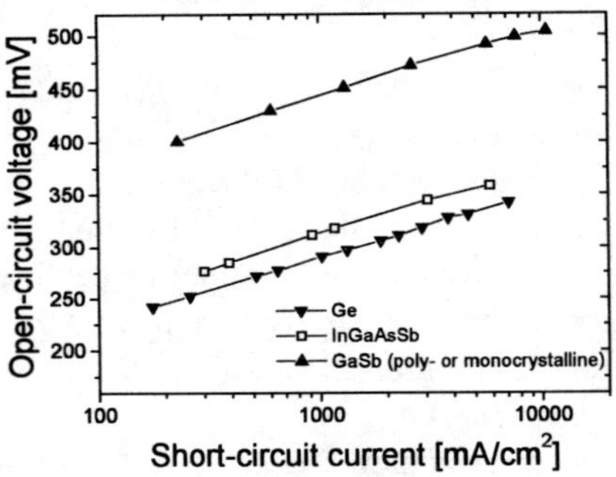

FIGURE 5. Open-circuit voltage of Ge, InGaAsSb and GaSb TPV cells vs. short-circuit current at 25°C.

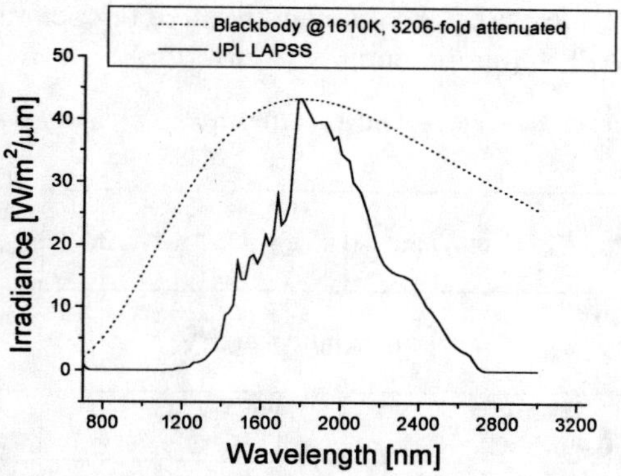

FIGURE 6. Comparison of the JPL LAPSS spectrum with the closest fit attenuated blackbody 1610 K spectrum.

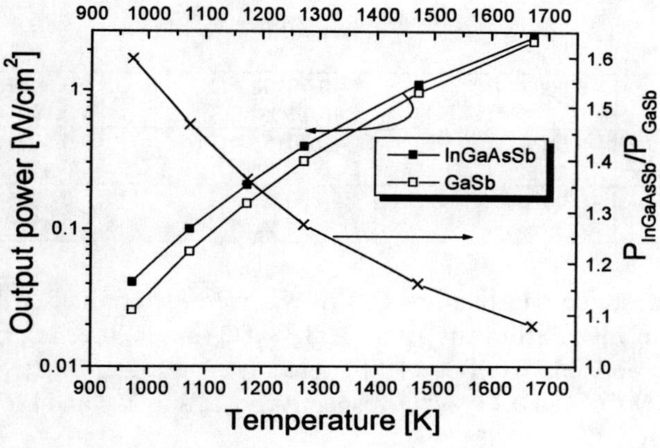

FIGURE 7. Calculated output power density of GaSb and InGaAsSb TPV cells as a function of blackbody temperature for the blackbody spectra with a scaling factor of 0.6. The calculations are performed for a cell temperature of 25°C and are based on measured device parameters (EQE, V_{oc}, FF). The ratio between the output power generated in the InGaAsSb and GaSb cells is shown on the right axis.

It is commonly believed that TPV cells operate at elevated temperatures. In this work, we determined temperature coefficients (TC) of short-circuit current (I_{sc}), V_{oc}, FF of GaSb, InGaAsSb and Ge cells using the JPL LAPSS. As the bandgap of semiconductors decreases with temperature, the spectral range of the photosensitivity widens into the long wavelength part. It is the prime reason why the TC of I_{sc} is usually positive. However, the exact value of the TC of I_{sc} is dependent on illumination spectrum and thus can be used only for the relative and not absolute cell comparison. TCs of V_{oc} and FF are not spectrum dependent and can be compared with the results of other authors. Table 2 shows the absolute and relative temperature coefficients of V_{oc} and FF at different LAPSS illumination.

TABLE 2. Absolute and relative temperature coefficients of V_{oc} and FF at different LAPSS illumination.

Incident power density / Cell material	P_{inc} = 1021 mW/cm^2	P_{inc} = 3037 mW/cm^2	P_{inc} = 6029 mW/cm^2
	Open-circuit voltage (V_{oc})		
InGaAsSb	- 1.525 mV/°C - 0.49 %/°C	- 1.496 mV/°C - 0.43 %/°C	- 1.465 mV/°C - 0.40 %/°C
GaSb	- 1.687 mV/°C - 0.39 %/°C	- 1.672 mV/°C - 0.36 %/°C	- 1.620 mV/°C - 0.33 %/°C
Ge	- 1.877 mV/°C - 0.67 %/°C	- 1.916 mV/°C - 0.62 %/°C	- 1.913 mV/°C - 0.59 %/°C
	Fill factor (FF)		
InGaAsSb	- 0.00079 /°C - 0.13 % /°C	- 0.00108 /°C - 0.17 %/°C	- 0.00128 /°C - 0.19 %/°C
GaSb	- 0.00102 /°C - 0.15 % /°C	- 0.00155 /°C - 0.22 % /°C	- 0.00216 /°C - 0.31 % /°C
Ge	- 0.00268 /°C - 0.39 % /°C	- 0.00280 /°C - 0.41 % /°C	- 0.00254 /°C - 0.41 % /°C

The lowest absolute and relative TCs of V_{OC} were determined at P_I = 6 W/cm^2: -1.465 mV/°C (in InGaAsSb-cells) and -0.33 % /°C (in GaSb-cells), respectively. The lowest absolute and relative TCs of FF were measured in InGaAsSb-cells at P_I = 1 W/cm^2: -0.00079 /°C and 0.13 %/°C, respectively. TCs of V_{OC} and FF in Ge-cells were essentially worse.

CONCLUSIONS

Device parameters of GaSb-, InGaAsSb-, InGaSb-, InAsSbP- and Ge- TPV cells were compared. GaSb- and InGaAsSb- based TPV cells showed the best performance.

Compared to GaSb, InGaAsSb cells demonstrated better parameters when measured under the JPL LAPSS or calculated on the basis of experimental results. The advantage of InGaAsSb cells is especially large if low-temperature blackbody radiators without filters are used.

Although Ge TPV cells exhibit lower efficiencies and worse temperature characteristics than GaSb and InGaAsSb ones, the relatively low cost of Ge may still make TPV Ge cells attractive.

Polycrystalline InGaSb- and single crystal InAsSbP-based cells demonstrated high photocurrent. However, their voltage and FF need to be improved.

ACKNOWLEDGEMENTS

The authors wish to thank J. Cox from AstroPower for epitaxial growth of quaternary alloys, E. Schaeffer, R. Beckert, G. Siefer, Klaus Breithaupt and H. Lautenschlager from Fraunhofer Institute for Solar Energy Systems for assistance by fabrication and testing of TPV cells, and P. Sims from AstroPower for useful discussions.

REFERENCES

1. www.realgoods.com/renew/shop/product.cfm?dp=1000&sd=1006&ts=3006374
2. Sulima, O.V., Beckert, R, Bett, A.W., Cox J.A, and Mauk, M.G., IEE Proc.-Optoelectron., 147, 199-204 (2000).
3. Sulima, O.V., and Bett, A.W, Solar Energy Materials & Solar Cells, 66, 533-540, 2001.
4. Bett, A.W., Keser, S., Stollwerck, G., and Sulima, O.V., "Large-Area GaSb Photovoltaic Cells", in Thermophotovoltaic Generation of Electricity –1997, edited by T.J. Coutts et al., AIP Conference Proceedings 401, American Institute of Physics, New York, pp. 41-53.
5. Sulima, O.V., Bett, A.W., Mauk, M.G., Ber, B.Ya., and Dutta, P.S., "Diffusion of Zn in TPV materials: GaSb, InGaSb, InGaAsSb and InAsSbP", this conference.
6. Mauk, M. G., Shellenbarger, Z. A., Cox, J. A., Sulima, O.V., Bett, A.W., Mueller, R. L., Sims, P. E., McNeely, J. B., DiNetta, L. C., J. of Cryst. Growth, 211, 189-193 (2000).
7. Mauk, M.G., Shellenbarger, Z.A., Cox, J.A., Tata, A.N., Warden, T.G., and DiNetta, L.C., "Advances in low-bandgap InAsSbP/InAs and GaInAsSb/GaSb thermophotovoltaics", in Proceedings of the 28th IEEE Photovoltaic Specialists Conf., 2000, pp. 1028-1031.

Key Issues for an Accurate Modelling of GaSb TPV Converters

Diego Martín and Carlos Algora

Instituto de Energía Solar. E.T.S.I. Telecomunicación. Universidad Politécnica de Madrid. Ciudad Universitaria s/n. 28040. Madrid. Spain.
Fax: (+34) 91.544.63.41, Phone: (+34) 91.549.57.00 ext. 281, E-mail: diego@ies-def.upm.es

Abstract. GaSb TPV devices are commonly manufactured by Zn diffusion from the vapour phase on a n-type substrate, leading to very high doping concentrations in a narrow emitter. This fact emphasizes the need of a careful modelling that must include high doping effects to simulate the optoelectronic behaviour of devices. In this work the key parameters that have strong influence on the performance of GaSb TPV devices are underlined, more reliable values are suggested and our first results on the study of the absorption coefficient dependence with p-type high doping concentration are presented.

INTRODUCTION

In contrast to the extended experimental work carried out in the manufacturing of GaSb TPV converters within the last decade, there are few publications tackling the task of developing a reliable and accurate model of GaSb devices that eventually results in an improved device design. This is primarily due to the absence of an accurate set of material and device parameters in the literature for feeding the available semiconductor simulation tools or the specific software for modelling thermophotovoltaic devices. Some worthy advances have been made in this field, as those published in [1], but there is still a long way to go.

It is well known that vapour phase Zn diffusions at the saturation level lead to highly doped emitters with a Zn concentration greater than 10^{20} cm^{-3} near the surface of the device. Since the consequences of the lack of accurate parameters are amplified by such high doping levels, it is needed a realistic modelling of minority carrier devices to take carefully into account enhanced Auger recombination, electrical band gap narrowing, variation of the absorption coefficient with the doping level, etc. These effects are not yet fully known for GaSb, compared to other photovoltaic materials like Si or GaAs. Our approach in this work consist of performing a wealth of calculations for quantum efficiency and dark I-V curves of GaSb TPV cells using different sets of published GaSb parameters. The method used to analyze these simulations by comparison with experimental data evidences the inconsistency of those reported sets of data. Hence we propose the guidelines to overcome these facts, giving a set of accurate material parameters, suggesting the careful study of the not enough accurate ones and presenting our first results in the study of the absorption changes in the semiconductor due to the high doping concentrations in the emitter.

CP653, *Thermophotovoltaic Generation of Electricity: 5th Conference*
edited by T. J. Coutts, G. Guazzoni, and J. Luther

DEVICE MODELLING INCLUDING HIGH DOPING EFFECTS

In order to include high doping effects and doping gradients in the emitter, we have built a one-dimensional non-linear set of differential equations based in the minority-carrier continuity and transport ones, solved with the help of the non-stiff Adams method aided by the shooting technique. The numerical method first uses the doping gradient and the bandgap narrowing to calculate the built-in electrostatic field in the emitter. Then this field is used to compute minority concentrations under both illumination and dark conditions, from which internal QE and J_{01} are obtained. A complete description of the device model and the set of equations can be found in [2].

The Zn doping profile depicted in figure 1 is used to perform the calculations. It has been measured by SIMS and corresponds to a 450ºC, 45 minutes process with Zn pre-loading. Complete GaSb TPV devices have been manufactured and tested with this profile, after removing the first 100 nm of the emitter surface by anodic oxidation. Extended characterization including dark I-V curves, spectral response and reflectivity measurements to obtain the internal QE has been carried out. They will be used to compare with the calculations.

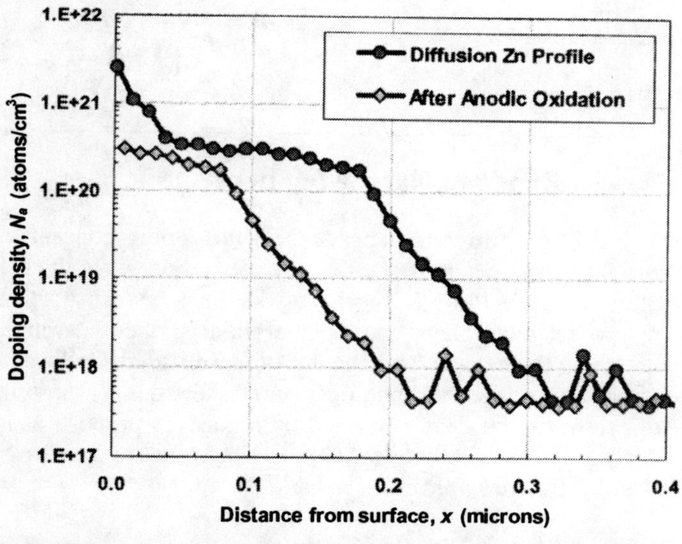

FIGURE 1. Zn doping profile as diffused and after anodic oxidation used to perform the calculations. This latter is used to manufacture complete devices.

CRITICAL REVIEW OF EXISTING DATA ON MATERIAL PARAMETERS

A complete review of the published GaSb material parameters and their effect in the device performance has been published in [1]. We have used mainly that set of parameters but taking special attention to the ones that have stronger influence at high doping concentrations. These are p-GaSb bandgap narrowing (p-BGN), Auger recombination and particularly the absorption coefficient dependence with doping concentrations. In the following sections their influence in device behaviour at high p-type doping levels will be discussed in detail. Table 1 presents a summary of the fixed GaSb parameters used for calculations.

TABLE 1. Fixed Material Parameters used for Calculations

GaSb Parameter	Numerical Value	Reference
Intrinsic carrier concentration n_i at 300 K	$1.4 \cdot 10^{12}$ cm^{-3}	[3]
Radiative coefficient, B_{opt}	$8.5 \cdot 10^{-11}$ cm^3/s	[1,3]
Photon recycling factor in n-GaSb, ϕ_n	10	[1,3]
SRH lifetime of electrons, $\tau_{SRH,e}$	10 ns	[1,3]
SRH lifetime of holes, $\tau_{SRH,h}$	10 ns	[This work]
BGN in n-GaSb at 300 K (N_d is the base doping)	$\Delta E_{gap,n} = 1.36 \cdot 10^{-8} N_d^{1/3}$ eV	[3]
Electron mobility in p-GaSb, $\mu_e = \mu_{\min,e} + \dfrac{\mu_{\max,e} - \mu_{\min,e}}{1 + \left(\dfrac{N_d + N_a}{N_{ref,e}}\right)^{a_e}}$	$\mu_{max,e} = 6600$ cm^2/V·s $\mu_{min,e} = 850$ cm^2/V·s $N_{ref,e} = 8 \cdot 10^{17}$ cm^{-3} $a_e = 1.8$	[1,3]
Hole mobility in n-GaSb, $\mu_h = \mu_{\min,h} + \dfrac{\mu_{\max,h} - \mu_{\min,h}}{1 + \left(\dfrac{N_d + N_a}{N_{ref,h}}\right)^{a_h}}$	$\mu_{max,h} = 1250$ cm^2/V·s $\mu_{min,h} = 85$ cm^2/V·s $N_{ref,h} = 4 \cdot 10^{17}$ cm^{-3} $a_h = 0.44$	[1,3]

Bandgap Narrowing in p-GaSb

The electrical bandgap narrowing dependence with doping concentration in both p-type and n-type GaSb has been published by several authors since the early stages of GaSb fundamental research [4, 5]. Recent works have shown that the bandgap narrowing in p-type material plays an essential role in the determination of the quasi-electrostatic field in the first nanometers of Zn-diffused GaSb devices [1, 3]. The need of a field that pushes the photogenerated electrons towards the depletion layer, instead of towards the device surface, is argued to propose a new bandgap narrowing value, using the well-known $N_a^{1/3}$ dependence versus p-type doping concentration. All of them are summarized in table 2.

TABLE 2. Models for p-type GaSb Bandgap Narrowing

Dependence of $\Delta E_{gap,p}$ with N_a (in eV). N_a must be expressed in cm^{-3}.	Reference
$1.8 \cdot 10^{-8} N_a^{1/3}$	[4]
$8.07 \cdot 10^{-9} N_a^{1/3} + 2.8 \cdot 10^{-7} N_a^{1/4} + 4.12 \cdot 10^{-12} N_a^{1/2}$	[5]
$1 \cdot 10^{-8} N_a^{1/3}$	[1]

Our novel approach in this work is based on using the device model to calculate the emitter contributions to the injection saturation current J_{01}, taking into account the different models for p-BGN in table 2. Then they are added to the J_{01} of the base. This allows us to determine the most reliable dependence (from those included in table 2) just by comparing the calculations obtained with actual experimental values of J_{01}. This comparison is shown in figure 2, which evidences the big influence of p-BGN in the dark characteristic of devices. Here the dependence of J_{01} with base doping N_d is essentially formal. It is justified due to the availability of experimental data of J_{01} for different base doping values. It can be seen that the model proposed in [1] shows a reasonable agreement with the measured values for complete devices. The experimental J_{01} values are relatively high due to the low SRH holes lifetime (10 ns) of the n-GaSb wafers used for manufacturing the devices.

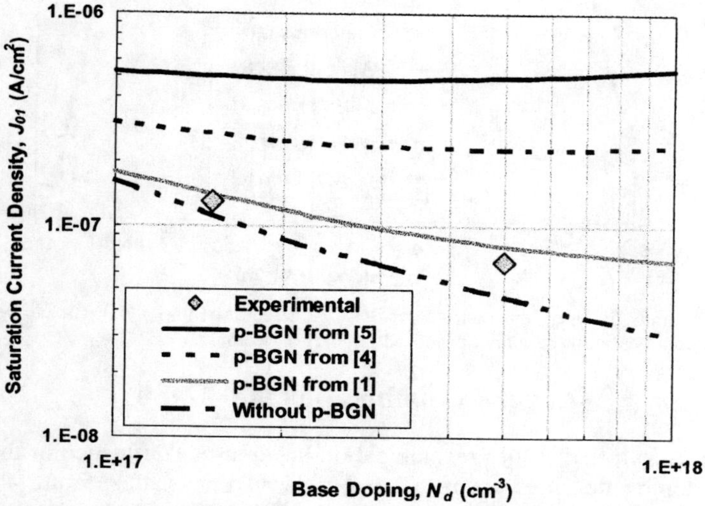

FIGURE 2. J_{01} dependence with base doping for the p-BGN models proposed in table 2 (and assuming no p-BGN). C_{AUGER} and S_n are set to their final values of $5 \cdot 10^{-30}$ cm^6/s and 10^6 cm/s, respectively.

The process that is carried out to fix the parameters and study its influence consists of the following steps: first, reasonable values of C_{AUGER} and S_n are chosen and the different p-BGN models are used to fit the J_{01} in dark conditions. Then the internal QE can be fitted too by adjusting the other material parameters that have a big influence on the device generation at high doping levels, such as Auger coefficient C_{AUGER} and surface recombination rate S_n. By doing so it is selected each GaSb parameter from the device electro-optical characteristic in which its influence is stronger. Given C_{AUGER} and S_n from internal QE fitting, J_{01} is recalculated to take into account the influence from the variations of the aforementioned parameters, that are much smaller than that due to the p-BGN. After that we are able to set the proper device parameters and dependences that fit both illumination and dark characteristics of our devices.

Regardless of that, the p-BGN has also a certain influence on the internal QE. In figure 3, the internal QE variation with p-BGN is presented when C_{AUGER} and S_n are fixed. The most important feature is that for all the p-BGN models described in table 2 a proper fitting of the QE could be reached by modifying C_{AUGER} and S_n, but the set of parameters obtained would be useless to describe the dark characteristic of the devices. The procedure discussed here circumvents this inconvenience.

To perform the internal QE calculations it has been used a novel model for the absorption coefficient that will be discussed in the following sections.

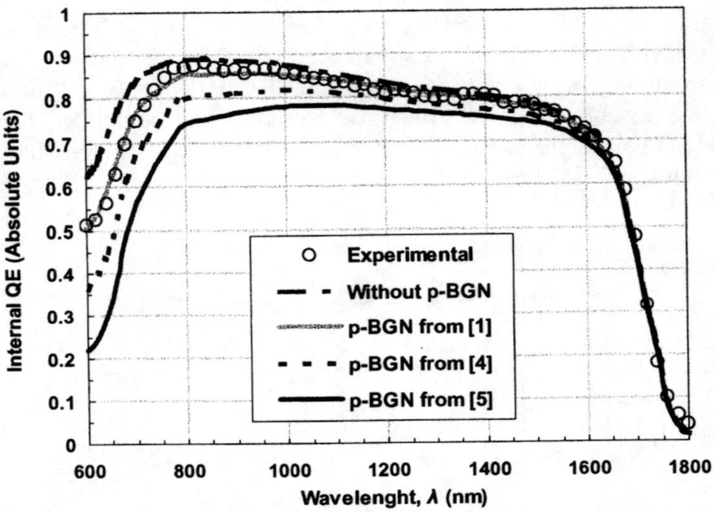

FIGURE 3. Internal QE dependence with the p-BGN model and with no p-BGN. It can be seen that an excellent fit is obtained with the model proposed in [1] (grey colour).

Auger Recombination in p-GaSb

It is well known that Auger recombination has a strong influence in the p-emitter when high doping densities are reached. To clarify this matter, figure 4 shows the Auger, radiative and SRH contributions to the total lifetime of minority carriers in the emitter as a function of the distance from the surface, as used to solve the one-dimensional equations in our model. Calculations are carried out setting C_{AUGER} to the value that will be used in the device final model. As can be seen, the Auger recombination dominates the lifetime behaviour in the first 125 nm of the 200 nm thick emitter. This corresponds to a Zn doping concentration of $2\cdot10^{19}$ cm^{-3}, from which SRH lifetime becomes the main contribution.

In order to set the C_{AUGER} that helps to fit both the internal QE and J_{01} experimental measurements, we have used the fitting method discussed above. The results for internal QE are shown in figure 5, in which the C_{AUGER} variation is ranged from $1\cdot10^{-30}$ to $1\cdot10^{-29}$ cm^6/s, which corresponds to a reasonable range that includes the experimental data for C_{AUGER} in p-GaSb [6]. The best matching is obtained with a value of $5\cdot10^{-30}$ cm^6/s, in agreement with the value published in [6] for $N_a=3\cdot10^{19}$ cm^{-3} at 300K of $5\text{-}6\cdot10^{-30}$ cm^6/s. The fitting between the calculations and the experimental QE data is very good, as can be seen in the figure 5 (in grey colour). It confirms that published values of C_{AUGER} can be used to obtain reliable GaSb device models and there is no need to change them as happened with the p-BGN. This conclusion belies the assumptions made in reference [1].

The surface recombination rate obtained from the iterative fitting is very high, about $1\text{-}2\cdot10^6$ cm/s. This result is in agreement with reference [3], and clearly in contrast with the extended S_n values used for simulations or obtained from fitting in previous works, like 103 cm/s [7], 10^3 cm/s [1] or 10^4 cm/s [8]. It can be explained by the presence of the aforementioned quasi-electrostatic field near the surface. This field acts as a passivation layer, preventing minority photogenerated carriers from recombine at the surface of the device. This is the reason to justify the use of an "effective" surface recombination rate 2 or 3 orders of magnitude lower than its real value, when no electrostatic field is considered in less accurate former device models.

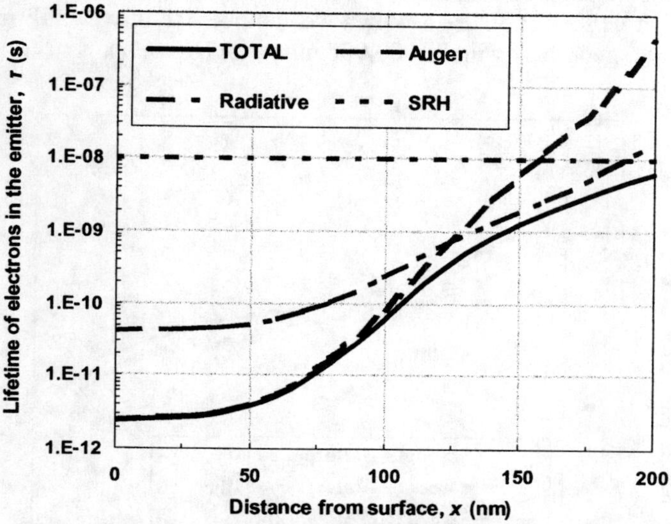

FIGURE 4. Total non-constant lifetime of minority carriers in the emitter as a function of the distance from the surface, showing its complete breakdown into Auger, radiative and SRH components.

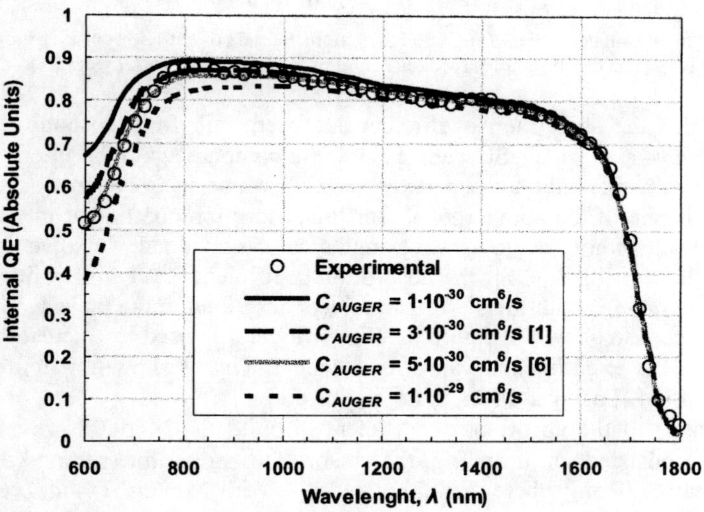

FIGURE 5. Internal QE data and calculations for different C_{AUGER} values. The best matching is obtained with $C_{AUGER} = 5 \cdot 10^{-30}$ cm^6/s [6] (grey colour).

Absorption Coefficient

One of the parameters that has a definite weight in the device modelling is the absorption coefficient. There are different sets of published data in which the absorption coefficient for GaSb is reported. First we built a model for the absorption coefficient based on the data in [9] from 600 to 800 nm, the data in [10] from 1500 to 1800 nm and the model proposed in [1] from 800 to 1500 nm. This set of data is called Model 1 and can be seen in figure 6. When we compute the Internal QE using that set of data there is no good fitting in the 700-900 nm range (figure 7).

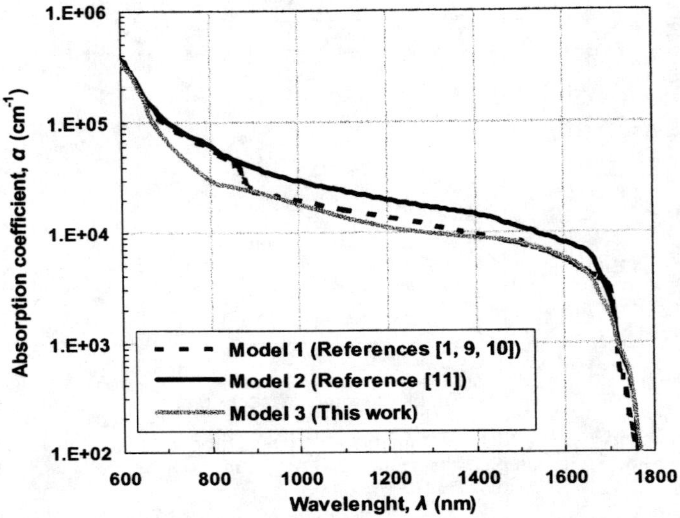

FIGURE 6. Absorption coefficient for GaSb from published sets of data (Model 1), from ellipsometric measurements of n-type GaSb samples (Model 2) and proposed in this work (Model 3).

Model 2 for the absorption coefficient has been built from spectral ellipsometric measurements of a n-type GaSb wafer [11]. If we compute again the internal QE using the same set of semiconductor parameters, even a worse fit is obtained (figure 7). We believe that if one of the above models for the absorption coefficient in GaSb is used no good fit in the whole range of wavelengths can be obtained. To solve this situation we have built a third model for the absorption coefficient that helps to fit properly the measured internal QE with the set of parameters described in the table 1. This is called Model 3 and is also shown in figure 6. When Model 3 is used to calculate the internal QE, the fit to the experimental values is excellent. This is shown in figure 7 and has been reported in [2] using a different doping profile.

The changes in the absorption coefficient to build the Model 3 are not arbitrary. They have consisted in dropping the absorption value for certain wavelengths, basically near 800 nm where the simulated QE with Models 1 and 2 have higher differences with the experimental data. It could be possible that the doping levels of the emitter (higher that 10^{20} cm^{-3}) were responsible for the absorption coefficient decrease, as has been reported for GaAs [12] at even lower doping concentrations.

Presumably the question that arises from the last considerations seems to be: is there any measurable change in the absorption coefficient caused by high doping concentrations in the emitter?.

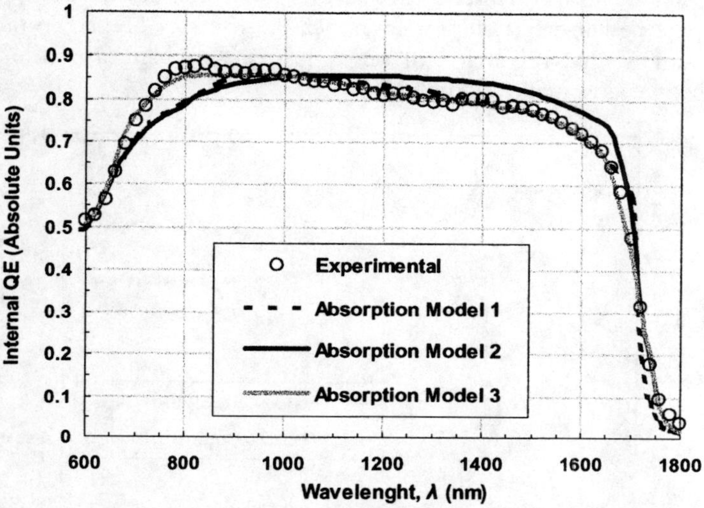

FIGURE 7. Internal QE derived from measurements and calculated using the different Models for the absorption coefficient. All other parameters are fixed to their value obtained from prior fitting.

ELLIPSOMETRY OF HIGHLY DOPED GaSb

To the authors knowledge, there are no studies dealing with the absorption properties of highly doped p-type GaSb out of the region of the fundamental gap. With the aim of addressing the problem raised, we have started some experiments with the first purpose of measuring hypothetical changes in the optical properties of GaSb after a Zn diffusion, under the same temperature and time conditions used for the manufacture of real devices. Room temperature measurements of the complex dielectric function have been performed in the range of 300-1700 nm using "Variable-Angle Spectral Ellipsometry" near the pseudo-Brewster angle. First of all we have checked the uniformity of this optical constants in a 2'' n-type Te doped GaSb wafer, to rule out initial variations that could hide some change after the diffusion step. It has been confirmed that there is good uniformity across the whole wafer, with GaSb native oxide thickness variations in our wafers between 2.24-2.48 nm.

After that, a Zn diffusion from the vapour phase at 475ºC during 45 minutes was done, and new ellipsometric measurements were carried out. It has been found a clear change in the optical constants n and k (figure 8) that can be neither explained nor fitted by the presence of a single layer of native GaSb oxide. The change produced in the effective extinction coefficient k points out a relative decrease in the absorption properties after the doping step. This can be easily seen in figure 9, where k data are translated into an effective absorption coefficient. After the doping step there is a drop in the value of *k*, in the ranges 800-1200 nm and 1400-1700 nm, but up to now this changes do not fully reproduce the absorption Model 3 suggested by us. This first measurement can be taken as a rough empirical check on the qualitative behaviour of the absorption changes with high doping densities in GaSb. In the future this variation must be understood using theoretical models such as effective medium approaches, with the help of new spectral ellipsometric measurements of deeper Zn profiles. The challenging aim of that research will be to justify the changes of the absorption coefficient that we have made building Model 3.

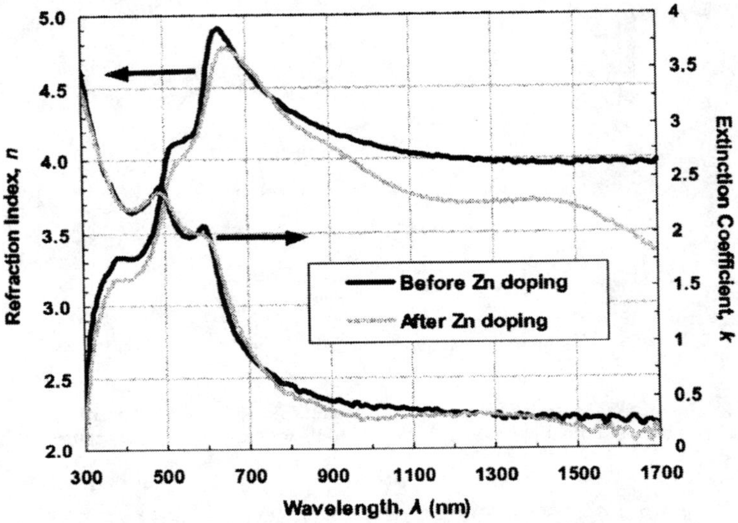

FIGURE 8. Complex Index of Refraction data before and after the Zn diffusion step

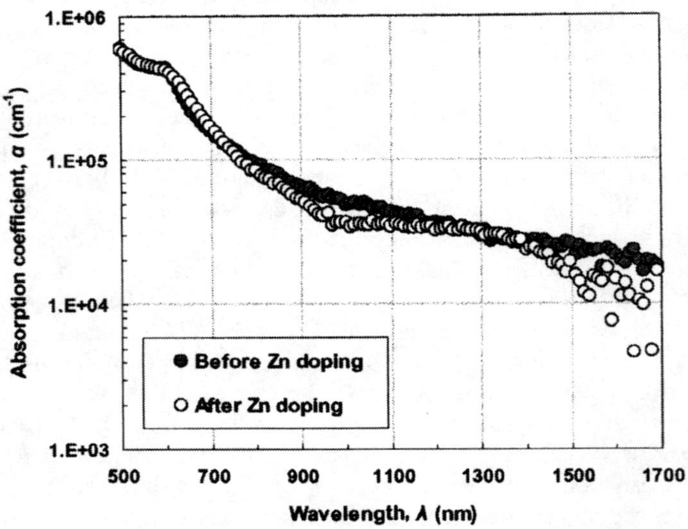

FIGURE 9. Absorption coefficient derived directly from ellipsometric measurements.

SUMMARY AND CONCLUSIONS

Zn diffusions from the vapour phase lead to more than 10^{20} cm^{-3} doping densities in the emitter of p-on-n GaSb TPV converters. In this work we have presented a new attempt to determine an accurate value for the GaSb parameters that have strong influence in the device modelling under high doping concentrations in the emitter. It is basically done by comparison of experimental data for internal QE and J_{01} with theoretical simulations where p-BGN, C_{AUGER} and S_n are modified remaining the other GaSb parameters fixed. The best fit is found when $\Delta E_{gap,p} = 1\cdot 10^{-8} N_a^{1/3}$ eV, $C_{AUGER} = 5\cdot 10^{-30}$ cm^6/s and $S_n = 1\text{-}2\cdot 10^6$ cm/s.

A new model for the absorption coefficient has been built to obtain proper matching of the internal QE data. It has been considered that high doping densities could be responsible for the proposed decreases in the absorption at certain wavelengths. Spectral ellipsometric measurements of Zn doped GaSb showing a relative drop in absorption have been presented for the first time.

ACKNOWLEDGMENTS

The authors wish to thank Dr. V. P. Khvostikov and Dr. M. Shvartzs from Ioffe Institute (St. Petersburg, Russia) for their kind help during characterization of the GaSb converters. Authors are also indebted to Dr. R. Serna from Instituto de Óptica (CSIC, Madrid) for the Spectral Ellipsometry measurements. This work has been supported by the Spanish CICYT, with the PETRI project under contract R-PETRI 95-0357-OP. Diego Martín holds a Ph.D. F.P.U. grant with reference number AP99-52539230 from the Spanish Ministry for Education & Culture.

REFERENCES

1. Stollwerck, G., Sulima, O. V., and Bett, A. W., *IEEE Transactions on Electron Devices* **47 (2)**, 448-457 (2000).
2. Algora, C., and Martin, D., "Modelling and Manufacturing of GaSb TPV Converters", *in these Proceedings*.
3. Sulima, O.V., and Bett, A.W., *Solar Energy Materials And Solar Cells* **66**, 533-540 (2001).
4. Titkov, A.N., Chaikina, E.I., Komova, E.M. and Ermakova, N.G., *Sov. Phys. Semicond.* **15**, 198-202 (1981).
5. Jain, S.C., McGregor, J.M., and Roulston, D.J., *J. Appl .Phys.* **68**, 3747-3749 (1990).
6. Yevick, D., and Bardyszewski, W., *Appl. Phys. Lett.* **51 (2)**, 124-126 (1987).
7. Fraas et al., "Fundamental Characterization Studies of GaSb Solar Cells", *22th IEEE PVSC*, Las Vegas (USA), 1991, pp.80-84.
8. Bett, A.W. et al., "Optimization of Low-Temperature Zn Diffusion for GaSb Solar Cell Structures Fabrication", in *13th European PVSEC*, Nice (France), 1995, pp. 88-91.
9. Apnes, D.E. and Studna, A.A., *Physical Review B* **27**, 985-1009 (1983).
10. Seraphin, B.O., and Bennett, H.E., *Optical Properties of III-V Compounds,* New York, R.K. Willardson and A.C. Beer, Eds., Academic, 1975.
11. Charache, G.W., "Spectral Ellipsometry of GaSb: Experiment and Modelling", *DOE Technical Report KAPL-P-000301* (1999).
12. Casey, H.C., Sell, D.D. and Wecht, K.W., *J. Appl. Phys.* **46 (1)**, 250-257 (1975).

Modelling And Manufacturing GaSb TPV Converters

Carlos Algora and Diego Martín

Instituto de Energía Solar – E.T.S.I. Telecomunicación
Universidad Politécnica de Madrid
Ciudad Universitaria s/n – 28040 Madrid (SPAIN)
Phone: 34.91.3367232; E-mail: algora@ies-def.upm.es

Abstract. A complete model for GaSb TPV converters considering multiple real conditions (spectra, temperature, ARC, size, number and characteristics of grid fingers, size of busbar, etc.) is presented. The model has been applied to GaSb TPV converters fabricated in our laboratory by single zinc diffusion. The agreement between theory and experiments is very good. Future guides of performance improvement like the passivation of the surface together with the correspondent structure optimisation and the achievement of a better GaSb quality available more widespread are suggested.

INTRODUCTION

GaSb converters are the best-known TPV devices. However, neither the GaSb material parameters are well established nor the modelling of the converter has reached a high level of accuracy. Just recently, efforts like that of Ref. [1] have tried to enhance the understanding of device performance in dependence of physical parameters. One step beyond in order to accurately know the suitable parameters of GaSb for modelling TPV converters is given in these proceedings [2].

But in addition to these valuable efforts, a detailed numerical modelling of GaSb TPV converters must overcome some challenges very similar to those of high concentrator III-V compounds solar cells. Both TPV and concentrator PV converters operate under high flux densities. Consequently, optical, thermal and electrical problems (and their related solutions) are shared by both devices. A first attempt of a real model was carried out in [3] although the emphasis was put in a concrete set of external conditions and no description of the used equations was done. Now, the purpose of this work is to present with more detail a model of GaSb TPV converters considering real conditions and to verify it with experimental measurements over manufactured devices. Additionally, the model will help in showing the improvement paths of our present device technology.

THEORETICAL MODEL

The calculation of the GaSb TPV converter performance is essentially a four step process: 1) the transmittance through the ARC as a function of photon energy is calculated, 2) the photogenerated current is determined using the minority-carrier transport equations; 3) the dark currents including the effects of recombination at base,

CP653, *Thermophotovoltaic Generation of Electricity: 5th Conference*
edited by T. J. Coutts, G. Guazzoni, and J. Luther

emitter, depletion region and perimeter are computed; 4) the I-V characteristic is achieved by using a lumped resistance model.

Although the goal of this paper is to present a device model as general as possible, several equations will be adapted to the device structure of figure 1 which is the one we manufacture.

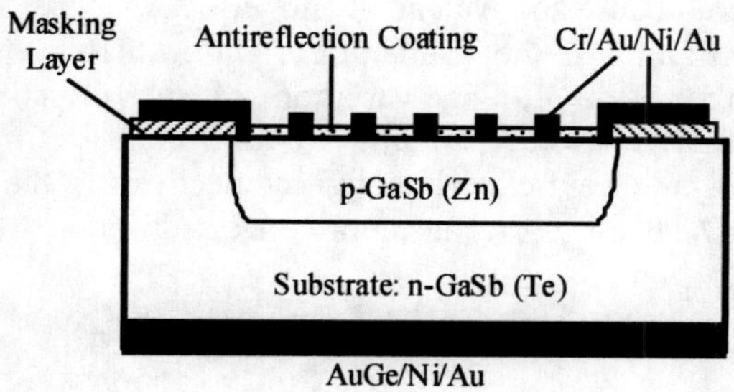

FIGURE 1. Typical structure of the GaSb TPV converter that is analysed in this work.

Light Impinging the Emitter of the Converter

The light spectrum impinging the GaSb converter, $N_{ph}(\lambda)$, (expressed in number of photons/time·area) is transmitted to the active photovoltaic surface once the ARCs have been crossed. The transmittance, $T(\lambda)$, of the stack composed of incident medium, ARCs and semiconductor layers can be calculated by using the transfer matrix method [4]. Therefore, the spectrum impinging the emitter of the converter is:

$$F(\lambda) = N_{ph}(\lambda)T(\lambda) \tag{1}$$

Photocurrent and Spectral Response

The rate at which minority carriers are generated, $G(\lambda,x)$ is

$$G(\lambda,x) = F(\lambda)\alpha(\lambda,x)\exp\left[-\int_0^x \alpha(\lambda,x)dx\right] \tag{2}$$

This generation term is introduced in the continuity equations in order to calculate the electron and hole currents. More specifically the minority-carrier equations are solved within the quasiequilibrium approximation. This does not introduce significant errors because the minority concentrations remain small when compared to the majority ones. The current equations to introduce in the continuity equations are:

$$J_e = \mu_e n\left(qE_e - \frac{kT}{N_c}\frac{dN_c}{dx}\right) + kT\mu_e\frac{dn}{dx} \tag{3a}$$

$$J_h = \mu_h p\left(qE_h - \frac{dE_g}{dx} + \frac{kT}{N_v}\frac{dN_v}{dx}\right) - kT\mu_h \frac{dp}{dx} \quad (3b)$$

Where J_e and J_h are the electron and hole currents, respectively, q is the electron charge, n and p are the electron and hole concentrations, E_g is the band-gap energy, N_c and N_v are the conduction and valence-band effective densities of states, k is the Boltzmann constant and T is the temperature. Unfortunately, there are not available data at present about the doping variation of the effective density of states. Consequently, the derivatives dN_V/dx and dN_C/dx remain zero in equations (3a, 3b). The built-in electrostatic field experimented by electrons in the p-type material, E_e is obtained by setting J_h=0 and $p=N_A$ (acceptor concentration) so,

$$E_e = \frac{1}{q}\left(\frac{kT}{N_A}\frac{dN_A}{dx} + \frac{dE_g}{dx}\right) \quad (4)$$

By using Eqs. (3b), (4) and the corresponding continuity equation, the electron density distribution in the p-material, $n(x)$, can be calculated. Similarly, for holes in the n-material.

Because in the usual configuration of standard GaSb TPV converters, the emitter is achieved by diffusion, the doping level is not a constant. Consequently, mobilities, absorption coefficient (although in this work it will be considered as constant with doping because there is not any available model of its doping dependence in spite of there are some attempts to reach it [2]), electric fields, diffusion lengths, etc. are doping dependant (that is position dependant) and there is no analytic solution for the equations. Therefore, a numerical solution implemented in a computer calculation must be carried out. In this calculation the photogenerated current is obtained by solving the continuity equation with the boundary conditions

$$n = n_0 \quad \text{at the depletion layer} \quad (5a)$$

$$J_e = qS(n - n_0) \quad \text{at the surface} \quad (5b)$$

where S is the surface recombination velocity and n_0 is the equilibrium value of n.

Analogous but simpler conditions are used in the n-type material for calculating J_h. This is because the base of the TPV device of figure 1 (substrate) is considered as constant in doping concentration and therefore, there is neither electric field nor parameter variations with x. Regarding the depletion region, all absorbed photons have been assumed to contribute to photocurrent, J_{dr}. The quantum efficiency, $QE(\lambda)$, is derived as the sum of these currents at the junction edges

$$QE(\lambda) = \frac{J_e(\lambda) + J_{dr}(\lambda) + J_h(\lambda)}{qF(\lambda)} \quad (6)$$

Finally, the total photocurrent for a given spectrum between λ_1 and λ_2 is

$$J_L = q\int_{\lambda_1}^{\lambda_2}(1-F_S)QE(\lambda)F(\lambda)d\lambda \tag{7}$$

being F_S the shadowing factor of the front metal grid.

Recombination Losses

The injected dark current is calculated by setting $G(\lambda,x)=0$ and changing the boundary condition in the depletion region border by

$$n = n_0 \exp\frac{qV_j}{kT} \tag{8}$$

being V_j the voltage across the junction. Similarly for the n-type material. This way, recombination density currents in the emitter, J_{01E}, and base, J_{01B}, can be calculated in order to give

$$J_{01} = J_{01E} + J_{01B} \tag{9}$$

To calculate the 2-kT recombination current density, the contribution of depletion region recombination plus that of the perimeter one have been assumed so,

$$J_{02}\left[\exp\left(\frac{qV}{2kT}\right)-1\right] = \left(J_{02,per} + J_{02,dr}\right)\left[\exp\left(\frac{qV}{2kT}\right)-1\right] \tag{10}$$

$$J_{02,per} = qn_iS_0L_s\frac{P}{A} \tag{11}$$

$$J_{02,dr} = \frac{\pi}{2}qn_iW_{dr}C\frac{kT/q}{V_{bi}-V_j} \tag{12}$$

Equation (11) is the proposed one by Henry, Logan and Merrit [5]. The meaning of each symbol is n_i intrinsic concentration, S_0 recombination velocity at the perimeter, L_s diffusion length at the perimeter, P length of the perimeter and A junction area. Note that in the structure of figure 1 the perimeter recombination is suppressed thanks to a diffusion mask. Consequently, in the following analysis the perimeter recombination is not taken into account.

Equation (12) is the proposed one by Hovel [6] which is based in the classical theory of Sah-Noyce-Sockley finally remodelled by Choo.

J-V Characteristic

To calculate the J-V characteristic, a model for the series resistance is necessary. Series resistance has been considered as a lumped parameter. Different components of the series resistance considering the current flowing in: 1) the emitter, r_E, 2) the front contact, r_{FC}, 3) the front metal grid, r_M, 4) the base, r_B, 5) the substrate, r_{Sub}, 6) the back contact, r_{BC}, 7) the busbar and external connections, r_{BW}, are included in the model. Their particular expressions can be found in [7] Therefore, the series resistance, r_S is

$$r_S = r_E + r_{FC} + r_M + r_B + r_{Sub} + r_{BC} + r_{BW} \tag{13}$$

The expression for the J-V characteristic including series and shunt resistance, r_p, is

$$J = J_L - J_{01}\left[\exp\left(\frac{V + Jr_S}{kT/q}\right) - 1\right] - J_{02}\left[\exp\left(\frac{V + Jr_S}{2kT/q}\right) - 1\right] - \frac{V + Jr_S}{r_p} \tag{14}$$

MANUFACTURING

Te-doped, n-type GaSb substrates, $1.5 \cdot 10^{17}$ cm^{-3} in doping density and with a mobility of 3100 $V/cm^2 \cdot s$ were covered with a pirolitic SiO_2 oxide. Windows in the oxide were opened by using photolithography. A vapour phase zinc diffusion was carried out at 450°C in order to achieve the p-region within the n-substrate.

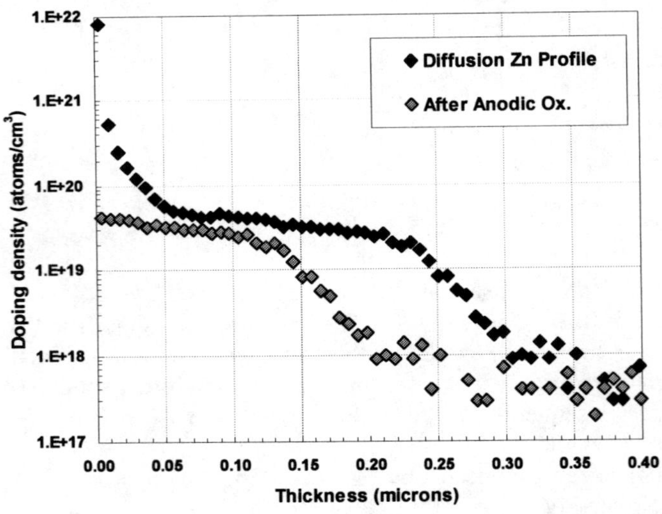

FIGURE 2. Zinc diffusion profile measured by SIMS. The profile of the same diffusion after anodic oxidation is also shown.

The zinc profile is shown in figure 2. After the diffusion, an anodic oxidation was used to remove about 100 nm of the damaged surface.

For back contact AuGe/Ni/Au was evaporated onto the n-type back surface and alloyed at 250°C for 1 minute. For front contact, Cr/Au/Ni/Au was directly evaporated and not alloyed onto the p-type GaSb while the grid was delineated by a lift-off technique. The shadowing factor was 6% and the active area was 49 mm^2.

After the mesa etching a double ZnS/MgF_2 ARC with thicknesses optimised for a 1000°C blackbody was deposited by thermal evaporation. The resulting GaSb converters were 7.4x7.4 mm in size and were finally attached by their back side to a copper disk by means of a solder paste. Wire bonding was used for their front side.

CHARACTERIZATION AND DISCUSSION

In this section we present the characterization and simulation carried out always over the same device. The dark J-V curve was measured as figure 3 shows. The fit of such curve produced the parameters set shown in Table 1. As can be seen the agreement is excellent.

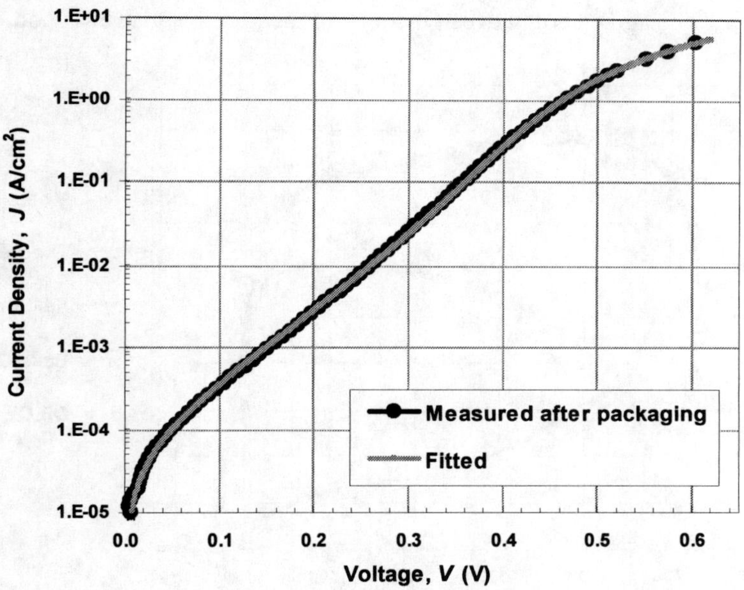

FIGURE 3. Measured dark I-V and best fit.

The behaviour of the device is very close to the ideal model of two diodes with kT and $2kT$ slopes presenting m_1=1.11 and m_2=1.98. In order to verify the physical meaning of this fit, the J_{01} was calculated by the sum of the contributions of base and emitter. The result gives $J_{01E}=3.85\cdot10^{-8}$ A/cm^2 and $J_{01B}=8.55\cdot10^{-8}$ A/cm^2. So, $J_{01}=1.24\cdot10^{-7}$ A/cm^2 which is in excellent agreement with the value of Table 1.

Table 1. Fitted parameters from experimental curve of figure 3

Parameters	Value
J_{01}	$1.3\cdot10^{-7}$ A/cm^2
m_1	1.11
J_{02}	$5.2\cdot10^{-5}$ A/cm^2
m_2	1.98
r_S	22 mΩ·cm^2
r_P	5 kΩ·cm^2

External quantum efficiency was measured at the Ioffe Institute of St. Petersburg. From this measurement the internal quantum efficiency was derived as figure 4 shows. As can be seen, the efficiency is fairly good except for high wavelengths. This is probably due to low quality of our n-substrate that is not suitable to act as a good base. Additionally, some efforts could be carried out in order to take the efficiency close to unity in some range by decreasing the emitter doping level while surface recombination is kept below 10^5 cm/s as it was stated in Ref [3].

In the same figure 4 the theoretical quantum efficiency is also presented including its different components. For this calculation the device model presented in this paper together with the set of physical parameter model described in [2] have been used with the only modification of an emitter surface recombination velocity of $2\cdot10^6$ cm/s and a $\tau_{SRH,p}$=15 ns. The agreement between theory and measurement is very good.

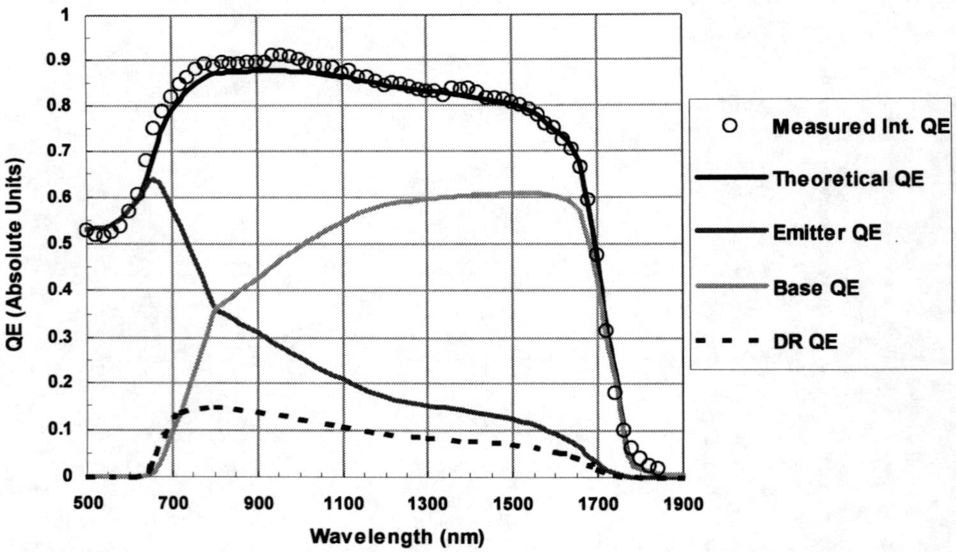

FIGURE 4. Measured and calculated internal quantum efficiency. The calculated components of the internal quantum efficiency are also shown.

By using a flash lamp at Ioffe Institute, the GaSb converter was illuminated under different power densities while its fill factor and open circuit voltage were measured, as figure 5 shows. By using the parameters of Table 1, the V_{oc} and FF as a function of

J_{sc} were calculated. Again the agreement between theory and measurement is very good.

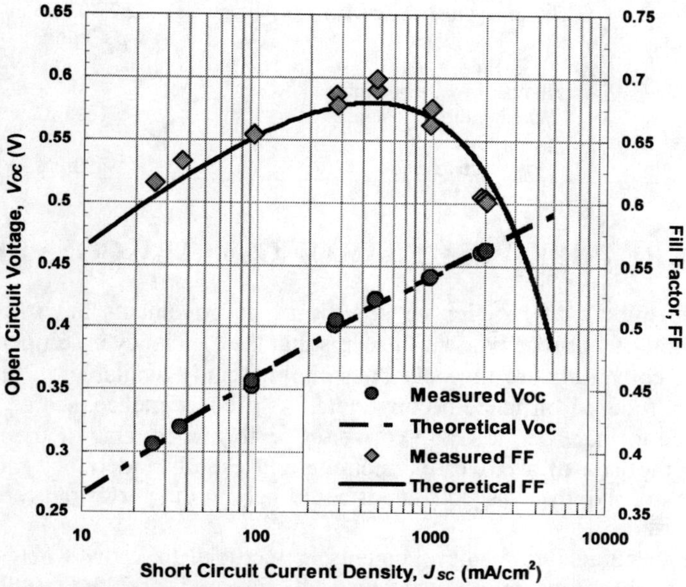

FIGURE 5. Experimental and calculated V_{oc}-J_{sc} and FF-J_{sc}

Finally, the analysis of the series resistance components was carried out. The different contributions of series resistance as a function of geometrical and technological parameters were calculated following the expressions of Ref. [7] and are presented in Table 2. In Table 3 the real parameters used in the calculations of Table 2 are depicted.

TABLE 2. Calculated components of the series resistance.

Series resistance components	Value
Emitter sheet resistance, R_E	15.5 Ω/
Emitter resistance, r_E	0.032 mΩ·cm^2
Base plus substrate resistance, r_B+r_{Sub}	3.44 mΩ·cm^2
Grid resistance, r_M	17.61 mΩ·cm^2
Front contact resistance, r_{FC}	1.67 mΩ·cm^2
Back contact resistance, r_{BC}	0.1 mΩ·cm^2
Bus and external connection resistance, r_{BW}	12.27 mΩ·cm^2
TOTAL series resistance, r_S	**34.88 mΩ·cm^2**

As can be seen, the total value of the modelled series resistance is 34.88 mΩ·cm^2 while the experimental value fitted from the dark I-V curve is 22 mΩ·cm^2 (as Table 1 shows). This is because Table 2 calculates the illumination series resistance while Table 1 shows the dark series resistance. Of course, the illumination series resistance is higher than that the illumination one.

TABLE 3. Real geometric and technological parameters used in the calculations of Table 2.

Parameter	Value
Grid metal thickness	425 nm
Grid metal resistivity (Au)	$2.2 \cdot 10^{-6}$ Ω·cm
Finger width	3 μm
Finger pitch	47 μm
Number of wire bondings	8
Wire bonding resistance	100 mΩ
Bus width	200 μm
Active area	49 mm^2

FUTURE IMPROVEMENTS OF GaSb TPV CONVERTERS

From the middle of nineties no significant improvements have appeared in the manufacturing of GaSb converters. It seems that the efficiency is stopped and although this kind of converters are the only ones commercially available all the efforts have been mainly focused on tandem converters. As a consequence, scarce attempts have been devoted to increase the GaSb converter performance. One of the reasons of this situation is the lack of a powerful, accurate and useful model. The model presented here is able to solve these aspects and it could help in the performance increase of the GaSb converters.

In order to guide future improvements it is crucial to know where the losses are produced. With the quantum efficiency of figure 4 and the 1000ºC blackbody spectrum, the efficiency of our converter has been calculated. To account the incident power, photons with an energy lower than the bandgap are not considered and a geometric factor of 0.6 is assumed. The efficiency so obtained we have called "present efficiency" in Table 4.

TABLE 4. Power losses analysis of a GaSb converter with a 1000ºC blackbody spectrum.

Item	Efficiency (%)	Increment (%)
Present efficiency	20.31	
Surface passivation	20.71	0.40
Semiconductor structure optimisation	23.98	3.27
Shadowing	25.55	1.57
Series resistance	27.71	1.82
Shunt resistance	27.71	0.00
Reflectivity	28.31	0.60
Depletion layer recombination	29.96	1.65
Emitter and base non radiative recombination	34.78	4.82
Auger recombination	35.39	0.61
Others	48.67	11.07
Radiative limit	51.58	2.31

The present p-on-n structure achieved by diffusion does not incorporate any passivation. If a perfect passivation layer were placed on the top of the converter the efficiency would increase 0.4 points. This efficiency can be increased in 3.27 points if the emitter and base were optimised in thickness and doping. Most of the following terms in the Table 4 are clear and assume a zero shadowing factor, a zero series resistance, an infinite shunt resistance and a zero reflectivity. Assuming a J_{02}=0 the improvement is of 1.65% meaning that the device is operating in the region dominated by the $2kT$ slope. The SRH recombination (that is, imperfections of base and emitter material) produces 4.82 points of losses while the Auger only 0.61 points. Finally, the item "others" includes aspects far away of the present technology that would contribute to a quantum efficiency equal to unity. The radiative limit assume incident and emission angles of 90°

Consequently, it can be said that the main realistic aspects for increasing the performance of the present GaSb converters come from an optimised structure achieved over a better GaSb material. Therefore, technological aspects like passivation of the surface together with an optimised structure and the achievement of GaSb of a more widespread improved quality should be deeply investigated.

SUMMARY AND CONCLUSIONS

A very complete model for GaSb converters considering multiple real conditions has been presented. Several GaSb converters have been manufactured and characterized. The experimental performance of the converter fits very well with the prediction of the model.

Between the capabilities of the model is the possibility of detecting the loss origins. As a consequence, the main aspects detected for improving the performance of the GaSb converters are the passivation of the surface together with an optimised structure and the achievement of GaSb of a more widespread improved quality.

ACKNOWLEDGMENTS

The authors would like to thank to Dr. V. Khvostikov and Mr. M. Shvarts of the Ioffe Institute of St. Petersbourg for several measurements presented here. This work has been supported by the Spanish Ministery of Education and Culture under contract R-PETRI-95-0357-OP. Diego Martín holds a grant of the Spanish Ministery of Education and Culture.

REFERENCES

[1] G. Stollwerck, O. V. Sulima and A. W. Bett, "Characterization and Simulation of GaSb Device-Related Properties", IEEE Transactions on Electron Devices 47 (2), 448-457 (2000).

[2] D. Martín and C. Algora, "Key issues for an accurate modelling of GaSb TPV converters" in these Proceedings.

[3] C. Algora and D. Martín, "Modelling GaSb thermophotovoltaic converters under real conditions", Proc. of the 17th European Photovoltaic Solar Energy Conference, Munich (Germany), pp.29-32, 2001

[4] M. Born and E. Wolf, "Principles of Optics", pp.51-70, Pergamon Press, UK 1980.

[5] C. H. Henry, R. a. Logan and F. R. Merrit, "The effect of surface recombination on current in $Al_xGa_{1-x}As$ heterojunctions", J. Appl. Phys. 49 (6) 3530-3542 (1978).

[6] H. J. Hovel, "The effect of depletion region recombination currents on the efficiencies of Si and GaAs solar cells", Proc. 10th IEEE PVSC, pp. 34-39 (1973).

[7] C. Algora and V. Díaz, "Influence of series resistance on guidelines for manufacture of concentrator p-on-n GaAs solar cells", Prof. Photolt. Res. Appl. 8, 211-225. 2000.

SESSION 7

Cell Technology

Chair: M. N. Palmisano

New Flexible Photocell Module for Thermophotovoltaic Applications

Bernd Bitnar[1], Wilhelm Durisch[1], Andreas Meyer[2], Günther Palfinger[1]

[1] *Paul Scherrer Institut, CH-5232 Villigen PSI, Switzerland*
phone: +41 56 310 4085, fax: +41 56 310 2646
[2] *Solaronix SA, CH-1170 Aubonne*
phone : +41 21 821 2280, fax : +41 21 821 2289
Author for correspondence: Bernd Bitnar, email: bernd.bitnar@psi.ch

Abstract. This paper describes the practical implementation of TPV systems. As a key part of these systems, we developed a new flexible photocell module for TPV applications. The module consists of commercially available monocrystalline silicon solar cells, which were cut into small stripes that are series connected to achieve a high voltage and a reduced current for operation at an illumination density of 5 – 10 suns. For an easy mounting into a heating boiler, the module is highly flexible and can be bent into a cylindrical shape. The rear side of the module is electrically isolated, but achieves a good heat coupling through a covering aluminium foil to the boiler wall. Thus, it can directly be mounted into a water-cooled boiler housing without the need of an additional water-cooling circuit. A first test module with an active cell area of 481 cm^2 achieved an electricity output of 50 W by irradiation with an Yb_2O_3 mantle emitter heated with a 20 kW methane burner.

INTRODUCTION

This paper continues our preceding paper in this volume (in the following text referred as Ref. B):

Bitnar B., Mayor J.-C., Durisch W., Meyer A., Palfinger G., von Roth F., Sigg H., „Record electricity-to-gas power efficiency of a silicon solar cell based TPV system".

One key component of a TPV system is the photocell generator. Silicon photocells have the advantages of being not expensive, consisting of a non-toxic material and being already available on the market in form of monocrystalline solar cells (Ref. B). For the application in a TPV self-powered boiler [1], the costs of the TPV system are a crucial parameter. In our existing technology, the photocell module, including the cells and the cell-cooling system, is the most expensive part of the TPV generator [1]. Actually, the cost for the cell-cooling material exceeds the cell cost itself.

TPV systems built until now, use photocell generators, which consist of single cells fixed onto cooling elements. The elements were mechanically mounted to a cylindrical unit [1, 2, 3]. Simpler systems use planar geometries, which are easier to fabricate, but suffer from a poor view factor of the cells [4].

In this work we developed a TPV silicon photocell module, which is highly flexible to be bent into a cylindrical shape. For its fabrication, modified conventional techniques used for commercially available solar modules are applied. The module

CP653, *Thermophotovoltaic Generation of Electricity: 5th Conference*
edited by T. J. Coutts, G. Guazzoni, and J. Luther

can easily be built into a heating boiler without the need of an additional cooling circuit. It is expected that the production cost of TPV systems can drastically be reduced by using the new flexible module compared to the existing mechanically assembled modules [5].

TPV PROTOTYPE SYSTEM

In this and the next chapter our TPV systems are described. Their individual components, like Yb_2O_3 emitter, filter and silicon photocell generator are presented in Ref. B. The smaller system is called prototype system and experimental results achieved with it are discussed in Ref. B and [6].

Beside a filter we used a quartz tube, which simply protects the cells from a direct contact with the hot exhaust gas. In axial direction, the system is closed by gold plated glass reflectors with a broadband reflectivity of 95 %. The upper reflector has a central hole with a diameter of 40 mm, through which the exhaust gas can escape. The mantle emitter is heated by a self-sucking tubeless butane burner (Venturi-principle), which is mounted onto a butane gas cylinder. The thermal power of the burner was determined by measuring the weight loss of the gas cylinder during the experiment. A detailed description of the system is given in [6, 7, 8, 9].

The butane burner system achieved a very high system efficiency of 2.4 % (see Ref. B) and the emitter temperature agrees well with simulations, which means that no temperature difference between the flame close to the emitter and the emitter itself occurs. But it has some major disadvantages:

- The air number λ, which is defined as the mass of combustion air to the mass of air necessary for a stochiometric combustion, can not directly be determined. In our experiments, we adjusted the air input to a maximum current output of the cells, assuming that under this condition $\lambda = 1$.
- For experiments with preheated combustion air, the airflow cannot be controlled quantitatively.
- The top reflector influences the amount of combustion air due to its flow resistance.

To circumvent these problems, a second generation TPV prototype system was built using a 0.7 – 2.5 kW methane burner. The thermal power and the air number can be set by gas flow controllers. This burner uses small burner tubes, on which the Yb_2O_3 mantle emitter is fixed. The electrical power is measured with a power tracker. Using a small furnace in the combustion air line, experiments with preheated air can be carried out with this system. The experimental results with this system and the comparison to simulations are given in Ref. B.

TPV DEMONSTRATION SYSTEM

An up-scale of the technique developed for the prototype system is a TPV demonstration system, which is shown in Fig. 1. This demonstrator was developed to achieve a sufficient high electrical output power for a self-powered residential heating system and can be mounted into a prototype heating boiler.

FIGURE 1. The TPV demonstration system. The silicon photocells, their supporting aluminium cooling-plates and the cooling water lines are clearly visible.

The demonstration system uses a 20 kW methane tube burner from Hoval AG (Vaduz, Liechtenstein), an Yb_2O_3 mantle emitter with an area of about 500 cm^2 and monocrystalline silicon photocells from ASE (Heilbronn, Germany). The 50x100 mm^2 cells are glued onto water cooled aluminium plates. For simplicity, a quartz tube without filter protects the cells from a contact with the hot exhaust gas and no axial reflectors are used. A detailed description of the system is given in [1]. We achieved an electrical output power of 121 W at 12 kW thermal power and 164 W at 20 kW. This electrical power is sufficient to self-power a residential heating system.

We projected to improve the photocell module corresponding the following requirements:

- Simplifying the cell cooling to reduce its costs (see [1, 5])
- Choose of a module design for an easy mounting into the boiler housing.
- Use of smaller photocells to reduce the loss due to the series resistance of the cell interconnection.
- Protection of the front surface of the cells against mechanical damage.

THE NEW FLEXIBLE PHOTOCELL MODULE

We developed a new flexible TPV photocell module, which fulfils all of these requirements:

- The module is highly flexible to be bent into a cylindrical shape. It can directly be glued onto the inner surface of the furnace housing, which is water-cooled.
- An aluminium foil forms the rear side of the module and couples the cells thermally to the boiler wall. Therefore, no additional cooling circuit is necessary.
- The cells are electrically isolated from the aluminium foil with a thin isolating layer.
- 12 – 16 mm wide cell stripes were cut and series connected with soldered metal tabs for a high mechanical flexibility and a low resistive interconnection.
- The front side is protected by a special ethylvinylacetate (EVA) foil with a thickness of about 100 μm.

Fig. 2 shows schematically a drawing of the module structure.

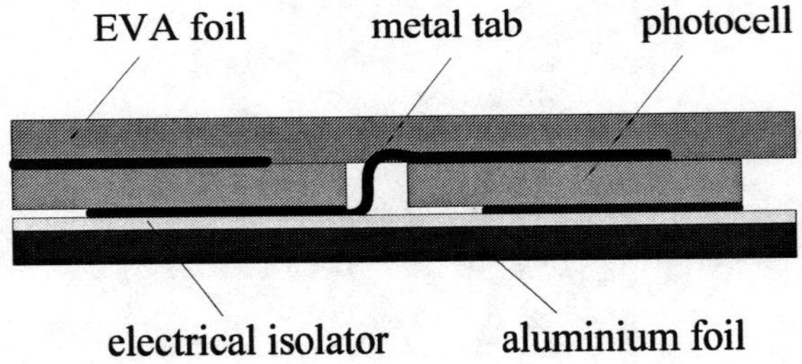

FIGURE 2. Schematically drawing of the structure of the new flexible photocell module.

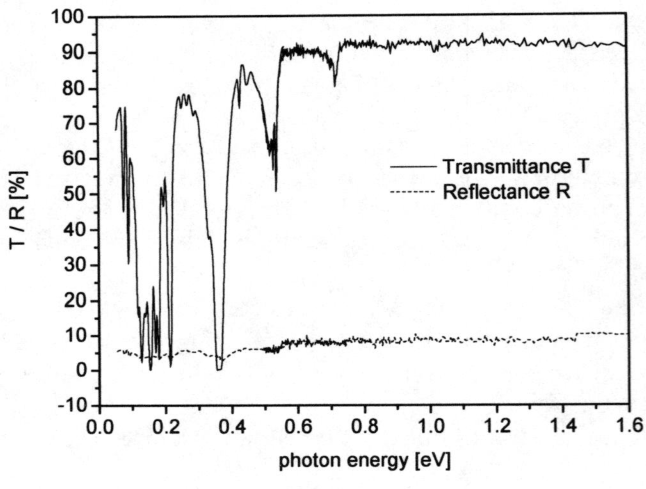

FIGURE 3. Transmittance T and reflectance R spectra of the special EVA foil.

We measured the transmittance and reflectance of the EVA laminate foil to study, if the foil is useful for TPV applications. Fig. 3 shows the result of the measurement. A high transmittance of more than 90 % and the absence of absorption for convertible radiation can be seen. From the reflectance we determined the refractive index of the foil to n = 1.56. Strong absorption bands are visible at 0.54 eV, 0.36 eV and below 0.21 eV. When compared to the emitter spectrum (see Fig. 3 in Ref. B), fortunately, no strong emission occurs in these bands and they are located far below the silicon bandgap. Only the continuous emission at E < 0.2 eV can be absorbed in the foil. Therefore, the foil will not be heated too strong from the Yb_2O_3 irradiation, even without SnO_2 filter, and seems to be suitable for the encapsulation of silicon TPV cells.

Two test modules with this structure were fabricated. Both modules consist of SH2 photocells from ASE (Heilbronn, Germany) (see Ref. B). The smaller module is shown in Fig. 4 (top). Due to its high flexibility it can easily be bent into a cylindrical shape. We plan to use similar modules for small TPV systems for camping gas cookers with a thermal power of about 1 kW. The module can be mounted with an uncovered rear side and is simply cooled by natural air convection.

Fig. 4 (bottom) shows a larger module, which is designed to be built into a TPV boiler prototype. The cell stripes have a width of 16 mm and the total cell area of the module is 481 cm^2.

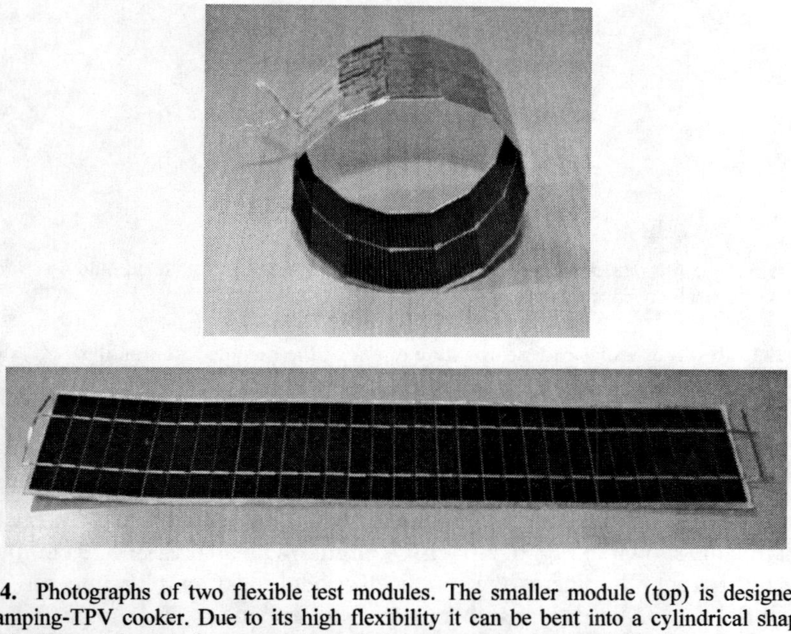

FIGURE 4. Photographs of two flexible test modules. The smaller module (top) is designed for a portable camping-TPV cooker. Due to its high flexibility it can be bent into a cylindrical shape. The larger module (bottom) is part of an array to be built into a prototype heating boiler.

MODULE CHARACTERISATION

To measure IV-characteristics of the flexible module shown in Fig. 4 (bottom), it was mounted onto a sun tracker. Fig. 5 shows a photograph of the sun tracker with six modules of the photocell module array, which is planned to be built into a TPV boiler. One module is electrically connected for the IV-measurement. At the top of the picture, one of seven pyranometers can be seen, which are used to measure the sun light intensity. The whole IV-characterisation setup is described in [10].

The result of the IV-measurement is given in Table 1. I_{SC} is the short circuit current and V_{OC} the open circuit voltage of the module. The electrical power P_{max} of 7.5 W at 91.5 mW/cm^2 global irradiation power gives a module efficiency (referred to the active cell area) of 17.1 %. The high efficiency and the fillfactor FF of 77 % validate the low resistivity cell interconnection in the module.

FIGURE 5. The 6 modules of the flexible TPV photocell array are mounted onto a sun tracker. One module is electrically connected to the IV-measurement system.

TABLE 1. Result of an IV-characterisation of one of the flexible test modules

radiation	I_{SC} [A]	V_{OC} [V]	FF [%]	P_{max} [W]	T [°C]
sun AM2.0	0.53	18.6	77	7.5	-
Yb_2O_3 P_{th}=12 kW	3.6	19.9	59	42	34
Yb_2O_3 P_{th}=20 kW	5.0	19.5	51	50	46

The test of the module under TPV conditions was performed with our TPV test setup, which is shown in Fig. 6. An Yb_2O_3 emitter with an area of 500 cm^2 heated with a methane burner illuminates the photocell module. To simplify the cell cooling, a water filter [2, 5] is used to absorb sub-bandgap radiation. The photocell module was bent around the filter and fixed with tape. No additional cooling was applied. The electrical power P_{max} was measured using a power tracker.

FIGURE 6. Test setup for the new flexible module. In the centre the shining Yb_2O_3 selective emitter is visible. The photocell module is heat protected by a water filter.

Table 1 gives the result of the IV-characterisations for 12 kW and 20 kW thermal power of the burner. The module achieved a maximum electrical power of 50 W, which is by a factor of 7 higher than with solar irradiation. The fillfactor is low, especially at 20 kW burner power, because the cells are designed for unconcentrated solar irradiance. V_{OC} drops for increasing burner power, because the cell temperature T increases.

CONCLUSION AND OUTLOOK

The flexible photocell module fulfils all our requirements to a TPV module in a residential heating system. It delivers sufficient electrical power under Yb_2O_3 irradiation, although the water filter in the test setup (see Fig. 6) absorbed about 10 % of convertible radiation power. The low fillfactor is caused by the series resistance of the cells itself. Cutting smaller cell stripes, which reduces the module current, does not affect the FF. That means that the FF is not limited by the cell interconnection.

The next step will be to mount the modules into a furnace housing by clogging them with heat conducting adhesive directly to the water-cooled boiler wall. A test of the whole module array under real operation conditions of the boiler should follow. Then it can be judged, whether the new flexible module array is able to achieve a comparable - or higher - electrical power than the conventional photocell module with cooling water circuit.

Certainly, there are also various solar applications for the new module. It can easily be mounted on curved surfaces like on caravan roofs or boats.

We expect that the development of the flexible module array is - together with the techniques of the Yb_2O_3 emitter and the SnO_2 filter - an important step towards a TPV heating system prototype. Because commercially available silicon solar cells are used, no major technical difficulties for an industrial production should occur.

ACKNOWLEDGMENTS

We thank Stefan Stutz for the FTIR measurement of the front side encapsulation foil. The project was supported by the Swiss "Kommission für Technologie und Innovation" KTI, contract No. 5692.2 EBS.

REFERENCES

1. Palfinger G., Bitnar B., Durisch W., Mayor J.C., Grützmacher D., Gobrecht J., "Cost estimates of electricity from a TPV residential heating system", Proc. 17th Europ. Photov. Solar Energy Conf. & Exh. 2001, pp. 117 - 120.
2. Bitnar B., Durisch W., Grützmacher D., Mayor J.-C., Müller C., von Roth F., Anna Selvan J.A., Sigg H., Tschudi H.R., Gobrecht J., „A TPV system with silicon photocells and a selective emitter", Proc. 28th IEEE Photov. Spec. Conf. 2000, pp. 1218 – 1221.
3. Fraas L.M., Samaras J.E., Huang H.X., Minkin L.M., Avery J.E., Daniels W.E., Hui S., "TPV generators using the radiant tube burner configuration", Proc. 17th Europ. Photov. Solar Energy Conf. & Exh. 2001, pp. 2308-2311.
4. Fraas L.M., Ballantyne R., Hui S., Ye S.-Z., Gregory S., Keyes J., Avery J., Lamson D., Daniels B., "Commercial GsSb cell and circuit development for the midnight sun TPV stove", in *Thermophotovoltaic Generation of Electricity: Fourth NREL Conference*, edited by Coutts/ Benner/Allman, AIP Conference Proceedings 460, New York: American Institute of Physics, 1999, pp. 480-487.
5. Palfinger G., Bitnar B., Durisch W., Mayor J.C., Grützmacher D., Gobrecht J., "Cost estimates of electricity from a TPV residential heating system", this volume.
6. Durisch W., Bitnar B., Palfinger G., von Roth F., "Small TPV prototype systems", this volume.
7. Durisch W., Bitnar B., Mayor J.-C., von Roth F., Sig H., Tschudi H.R., Palfinger G., „Progress in the development of a small thermophotovoltaic prototype system", Proc. 17th Europ. Photov. Solar Energy Conf. & Exh. 2001, pp. 2296 – 2299.
8. Bitnar B., Durisch W., Mayor J.-C., Sigg H., Tschudi H.R., Palfinger G., Gobrecht J., „Development of a small TPV prototype system with an efficiency > 2%", Proc. 17th Europ. Photov. Solar Energy Conf. & Exh. 2001, pp. 33 – 36.
9. Bitnar B., Durisch W., Mayor J.-C., Palfinger G., Sigg H., Grützmacher D., Gobrecht J., „Record electricity-to-gas-power efficiency of a silicon solar cell based TPV system", Proc. 29th IEEE Photov. Spec. Conf., in print.
10. Durisch W., Struss O., Robert K., "Efficiency of selected photovoltaic modules under varying climatic conditions", *Renewable Energy*, Special Edition WREC VI 2000, pp. 779 – 788 (2000).

Wafer-Bonded Internal Back-Surface Reflectors for Enhanced TPV Performance

C.A. Wang*, P.G. Murphy*, P.W. O'Brien*, D.A. Shiau*, A. C. Anderson*, Z.L Liau*, D.M. DePoy#, G. Nichols#

**Lincoln Laboratory, Massachusetts Institute of Technology, Lexington, MA 02420-9108*

#Lockheed Martin Corporation, Schenectady, NY 12301

Abstract. This paper discusses recent efforts to realize GaInAsSb/GaSb TPV cells with an internal back-surface reflector (BSR). The cells are fabricated by wafer-bonding GaInAsSb/GaSb device layers to GaAs substrates with a dielectric/Au reflector, removing the GaSb substrate, and subsequently processing the layers using standard photolithographic techniques. The internal BSR enhances optical absorption within the device, while the dielectric layer provides electrical isolation. This approach is compatible with monolithic integration of series-connected TPV cells and can mitigate the requirements of filters used for front-surface spectral control.

INTRODUCTION

High-efficiency thermophotovoltaic (TPV) systems require highly efficient TPV cells as well as excellent spectral control to recuperate below bandgap photons back to the radiator [1]. A back-surface reflector (BSR), which is a highly reflecting mirror deposited on the backside of the substrate, was used to provide both spectral control and device performance enhancement for InGaAs/InP TPV systems [2-5]. Spectral control is achieved by reflection of below band-gap photons that pass through the device and substrate back to the radiator, while device improvement results from the increased probability of above bandgap photons that are not absorbed in the first pass to be reflected back into the active layers for a second-pass absorption. This approach, however, has not been pursued for GaSb-based TPV systems since the effectiveness of the BSR is reduced by free-carrier absorption in the GaSb substrate [6]. Consequently, front-surface spectral control techniques with a filter placed between the radiator and the front surface of the TPV cell are being developed for GaSb-based TPV systems [7-9].

CP653, *Thermophotovoltaic Generation of Electricity: 5th Conference*
edited by T. J. Coutts, G. Guazzoni, and J. Luther

This paper describes a new approach for GaInAsSb/GaSb TPV cells with an internal BSR, which is formed by wafer-bonding the epitaxial layers directly to a GaAs handle wafer. After bonding the epitaxial layers, the GaSb substrate is completely removed. In this manner, the reflector is located directly below the epitaxial device layers. This concept of wafer-bonding using metal as an adhesive layer is similar to that developed for integration of dissimilar materials [10] and for fabrication of internal mirrors for photovoltaic cells [11], light emitting diodes [12], and vertical-cavity surface-emitting lasers [13]. The resulting TPV device structure consists of a broad-band high-reflectivity dielectric/Au mirror sandwiched between the GaInAsSb/GaSb device layers and the GaAs handle substrate.

Wafer-bonded GaInAsSb/GaSb TPV cells with an internal BSR have the potential of improving TPV cell performance; providing spectral control; and allowing series connections of TPV devices. Specifically, the internal BSR enhances photon recycling effects, and as mentioned above, increases the effective optical thickness so that the active layers can be thinner, which is advantageous for reducing dark current and increasing open circuit voltage. In addition, reflection of below bandgap photons back to the radiator can aid in spectral control. Furthermore, since the epitaxial layers are electrically isolated from the handle substrate by the dielectric layer, TPV cells can be connected in series to fabricate monolithically interconnected modules (MIMs) [2-5].

EXPERIMENTAL PROCEDURES

GaInAsSb/GaSb TPV device structures were grown lattice matched to 5-cm-diam GaSb substrates by organometallic vapor phase epitaxy [14]. Epitaxial layers were grown in a reverse sequence compared to conventional TPV structures in order to obtain a p-on-n structure after wafer bonding. In addition, an InAsSb layer was grown as the etch-stop layer. The as-grown inverted TPV structure shown in Figure 1a consists of the following layers grown on a (001) n-GaSb substrate miscut 6° toward (1-11)B: u-GaSb buffer layer, u-InAsSb etch-stop layer, p-GaSb contact layer, p-AlGaAsSb window layer, p-GaInAsSb emitter layer, n-GaInAsSb base layer, and n-GaSb lateral conduction layer (LCL). (In some structures, the p-AlGaAsSb window layer was omitted.) Although the InAsSb etch-stop layer is eventually removed during fabrication of wafer-bonded epitaxial TPV structures, it is critical that this layer is also lattice matched to GaSb since the GaInAsSb active layers are grown on top of the InAsSb etch-stop layer.

The inverted TPV structure on the GaSb substrate and the handle semi-insulating GaAs wafer were prepared for bonding with a dielectric/reflector. First, both GaSb and GaAs surfaces were solvent cleaned and etched in HCl and NH_4OH to remove native oxides. The GaSb epitaxial TPV structure was sputter-coated with a three-layer reflector consisting of SiO_x/Ti/Au, while the GaAs was sputter-coated with Ti/Au. SiO_x not only provides electrical isolation of the epitaxial layers but also increases above band-gap reflectivity. Although Ti decreases the reflectivity, it improves the adhesion of Au to both the SiO_x and GaAs. To bond the wafers, the two Au surfaces were placed in contact, and the wafers were heated under vacuum to a temperature of 250 °C and a mechanical pressure of 250 psi was applied. The low bonding temperature is compatible with subsequent wafer processing. In a separate set of experiments to study reflector properties, GaAs wafers were sputter-coated with SiO_x of various thicknesses ranging from 500 to 3600 nm and a 0.5-μm-thick Au layer.

The 5-cm-diam bonded wafer was cut into four quarters for ease of handling and processing, and the GaSb substrate was spin-etched with H_2O_2:H_2O:NaK tartrate tetrahydrate to remove the bulk of the substrate, then chemomechanically polished. The remaining GaSb substrate and GaSb buffer layer were selectively etched using CrO_3:HF:H_2O until the InAsSb layer was exposed. Finally, the InAsSb etch-stop layer was removed with H_2O_2 saturated with citric acid. The resulting wafer-bonded epitaxy can then be processed into TPV devices using standard photolithographic processes and chemical etching to define the TPV cells. Figure 1b schematically shows the wafer-bonded GaInAsSb/GaSb TPV device with the internal BSR.

The surface flatness of bonded wafers was measured with an interferometer. The bonded epitaxial structures, as well as unbonded control structures, were characterized by high-resolution x-ray diffraction (HRXRD) and photoluminescence (PL) measurements at room temperature. Reflectivity measurements were performed at near normal incidence over the spectral range from 1 to 20 μm.

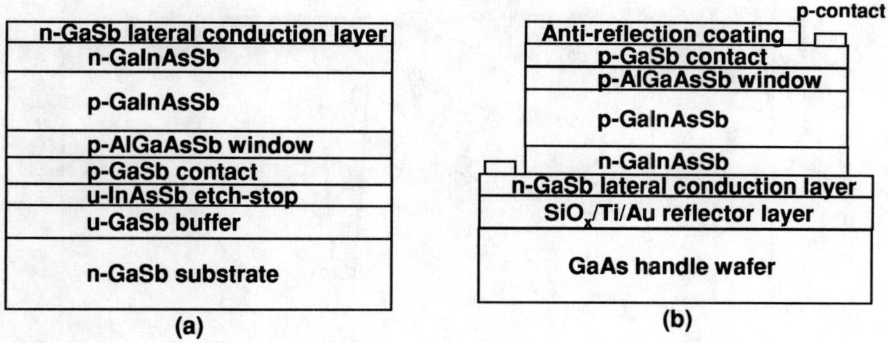

FIGURE 1. Schematic structure of (a) inverted GaInAsSb/GaSb TPV device structure grown on n-GaSb substrate for wafer bonding and (b) wafer-bonded GaInAsSb/GaSb TPV structure on GaAs handle wafer after removal of GaSb substrate and InAsSb etch-stop layer.

RESULTS AND DISCUSSION

Figure 2 shows the measured reflectivity of the SiO_x/Au reflector, with various SiO_x thicknesses of 50, 200, and 360 nm, as measured through the GaAs wafer. For wavelengths below ~5 μm, the reflectivity increases with thickness. Above 5 μm, absorption in SiO_x decreases the reflectivity, especially in the wavelength range between 7.5 and 10 μm. Below 1.4 μm, the reflectivity decreases due to absorption in the GaAs substrate. High reflectivity below 2.5 μm is important for enhancing GaInAsSb TPV device performance, while above 2.5 μm it is important for spectral control. The integrated reflectivity from 1.5 to 2.5 μm (since GaAs is absorbing below 1.4 μm) is 0.962, 0.97, and 0.981 for SiO_x thickness of 50, 200, and 360 nm, respectively. For 2.5 to 15 μm, the integrated reflectivity is 0.952, 0.943, and 0.936 for SiO_x thickness of 50, 200, and 360 nm, respectively. The high reflectivity over the wide wavelength range is extremely attractive for GaSb-based TPV systems, although, there is a trade-off between above- and below-bandgap reflector performances when using SiO_x. Further improvements in reflective properties of the internal BSR are expected by replacement of SiO_x with an alternative low-index dielectric with low absorption above 2.5 μm, such as MgF_2, AlF_2, or other metal-fluorides [15]. As a starting point in this study, the SiO_x thickness of the BSR layer used for wafer-bonded samples was 200 nm. In these structures, Ti was deposited on the SiO_x to improve Au adhesion, and R decreased only by a few percent.

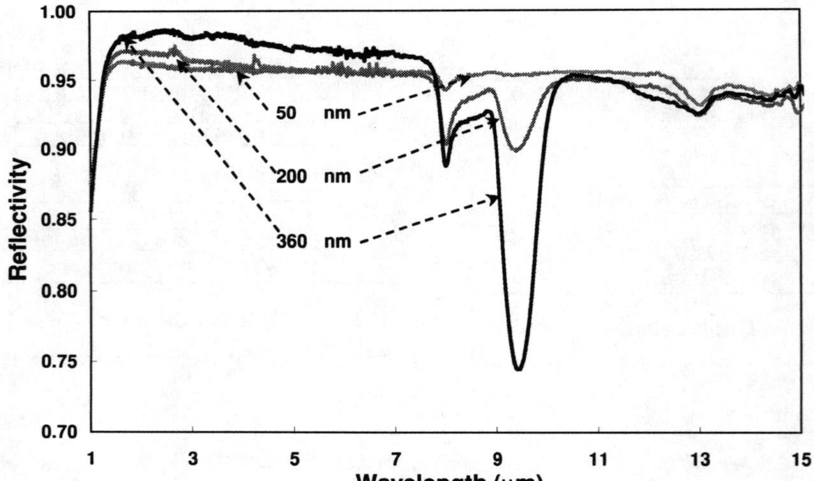

***FIGURE 2**. Measured reflectivity of SiO_x/Au reflector with various SiO_x thicknesses on GaAs.*

Interferograms of the GaInAsSb/GaSb and GaAs bonded wafers are shown in Figure 3. The GaSb wafer is concave and exhibits an average wafer bow of 22 μm, while the GaAs is convex with bow of 24 μm. The circular fringes and nearly symmetrical wafer bow are indicative that bonding of the TPV device wafer and GaAs wafer was achieved over the entire 5-cm-diameter area. The bow is due to residual stress in the bonded wafers, and results from different thermal expansion coefficients of the SiO_x, Au, GaSb, and GaAs alloys. If this residual stress exceeds the mechanical bond strength of the sputtered and bonded layers, debonding can occur. Furthermore, adhesion of the sputtered layers and the bonded interface must be able to withstand substrate removal and multiple processes during TPV fabrication. In numerous tests, the integrity of the Au-bonded epitaxial layers was maintained throughout all processing steps.

FIGURE 3. Interferograms of (a) back side of GaSb substrate bonded to GaAs handle substrate; (b) back side of GaAs handle substrate. One fringe corresponds to 4.8 μm.

The HRXRD rocking curves for the lattice-matched control GaInAsSb TPV device structure on a GaSb substrate and a wafer-bonded GaInAsSb TPV structure after removal of the GaSb substrate and InAsSb etch-stop layer are shown in Figures 4a and 4b, respectively. The diffraction intensity is lower for the wafer-bonded TPV device structure since the GaSb substrate has been removed and only the epitaxial layers that were grown lattice-matched to GaSb are remaining. The slightly broadened peak of the wafer-bonded structure is due to residual stress.

Figure 5 shows 300 K PL spectra for the control and wafer-bonded GaInAsSb TPV structures. Multiple peaks in the PL spectrum of the wafer-bonded sample are due to thin-layer interference effects discussed below. The PL peak intensity of the wafer-bonded epitaxy is over two times greater than that of the control TPV structure. These results suggest that the optical efficiency is enhanced as a result of the internal BSR, and there is negligible material degradation after wafer bonding and substrate removal.

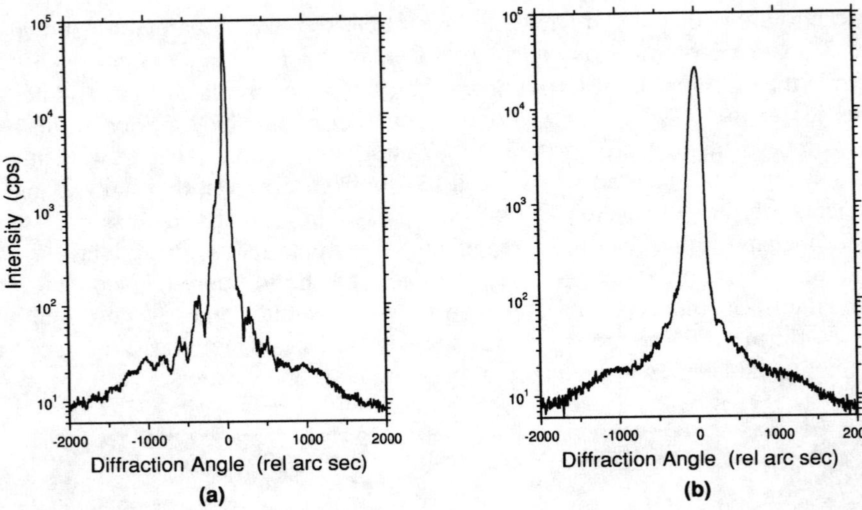

FIGURE 4. X-ray diffraction of (a) control GaInAsSb/GaSb TPV wafer and (b) wafer-bonded GaInAsSb/GaSb TPV structure.

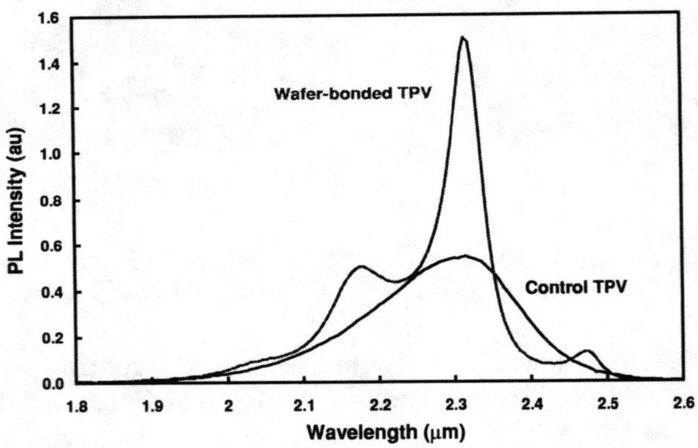

FIGURE 5. 300 K photoluminescence of control and wafer-bonded GaInAsSb/GaSb TPV structures.

Figure 6 shows the reflectivity of the wafer-bonded GaInAsSb TPV device structure without an anti-reflection (AR) coating. Without the AR coating, resonant absorption is observed across the whole wavelength range due to destructive interference between reflections from the air/GaSb (R~34%) and GaSb/SiO_x/Ti/Au interfaces. Such effects introduce absorption losses and will reduce the efficiency of the internal BSR [16]. There are, however, approaches that could minimize this affect. One is to utilize an AR coating on the front surface of the TPV cell to minimize resonant absorption of above-bandgap photons, or to use a thinner TPV device layer structure, which increases the resonant fringe spacing. Alternatively, epitaxial transfer of the GaInAsSb/GaSb TPV device layers can be achieved by direct fusion to the GaAs substrate, as shown in Figure 7, with the BSR deposited on the back side of the semi-insulating GaAs wafer.

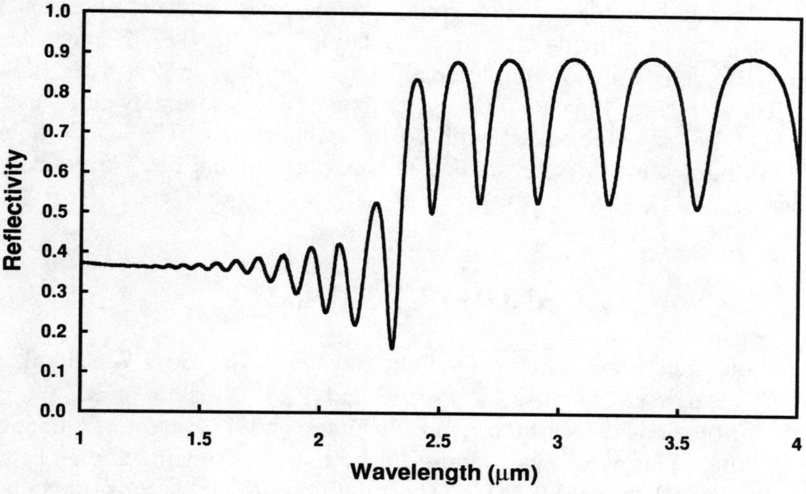

FIGURE 6*. Reflectivity spectrum of uncoated wafer-bonded GaInAsSb/GaSb TPV structure. The TPV device structure is 4 μm thick. Thinner device layers will increase the resonant fringe spacing.*

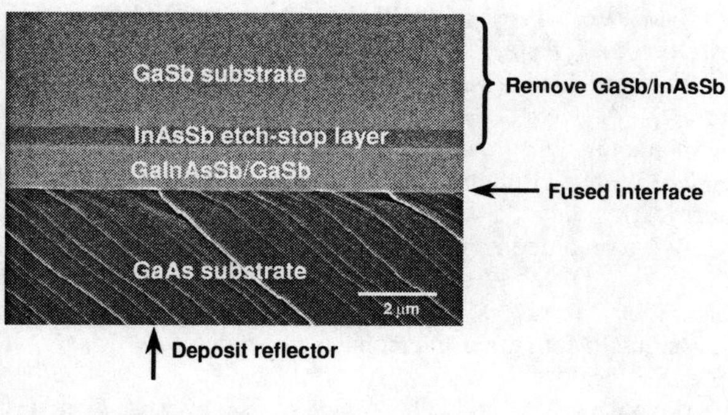

FIGURE 7*. Cross-section scanning electron micrograph of GaInAsSb/GaSb fused to directly to GaAs substrate. The GaSb substrate and InAsSb etch-stop layer can be removed and a BSR deposited on the back-side of the GaAs substrate.*

CONCLUSIONS

In conclusion, wafer bonding of epitaxial GaInAsSb/GaSb TPV device layers to GaAs substrates with a high-reflectivity broad-band mirror was demonstrated. The structural and optical properties of the wafer-bonded TPV device layers were maintained after fabrication, and improvement in TPV cell performance due to photon recycling and double-pass absorption is anticipated. Furthermore, since the BSR is designed to provide electrical isolation of the TPV device layers, GaInAsSb TPV cells can be monolithically interconnected in series for building voltage. These results indicate that wafer bonding is extremely attractive for GaSb-based TPV cells fabricated with MIM architecture. The wafer bonding process is not limited to GaAs substrates, and alternative handle substrates can also be considered.

ACKNOWLEDGMENTS

The authors gratefully acknowledge D.R. Calawa, J. Caissie, J.W. Chludzinski, P. Foti, C.T. Harris, S. Hoyt, J.M. Porter, and P.M. Nitishin of MIT Lincoln Laboratory for technical assistance and P. Dutta and I. Bhat for guidance in selective etching. This work was sponsored by the Department of Energy under AF Contract No. F19628-00-C-0002. The opinions, interpretations, conclusions, and recommendations are those of the author and are not necessarily endorsed by the United States Government.

REFERENCES

1. P.F. Baldasaro, J.E. Raynolds, G.W. Charache, D.M. DePoy, C.T. Ballinger, T. Donovan, J.M. Borrego, J. Appl. Phys. 89, 3319-3327 (2001).
2. J.S. Ward, A. Duda, M.W. Wanlass, J.J. Carapella, X. Wu, R.J. Matson, T.J. Coutts, T. Moriaty, 'A Novel Design for Monolithically Interconnected Modules (MIMs) for Thermophotovoltaic Power Conversion,' AIP Conf. Proc. 401, 227-236 (1997).
3. D. M. Wilt, N.S. Fatemi, P.P. Jenkins, V.G. Weizer, R.W. Hoffman, Jr., Raj K. Jain, C.S. Murray, D. Riley, 'Electrical and Optical Performance Characteristics of 0.74 eV p/n InGaAs Monolithic Interconnected Modules' AIP Conf. Proc. 401, 237-247 (1997).
4. N.S. Fatemi, D. M. Wilt, R.W. Hoffman, Jr., M.A. Stan, V.G. Weizer, P.P. Jenkins, O.S. Khan, C.S. Murray, D. Scheiman, D. Brinker, 'High-Performance, Lattice-Mismatched InGaAs/InP Monolithic Interconnected Modules (MIMs),' AIP Conf. Proc. 460, 121-131 (1999).
5. M.W. Wanlass, J.J. Carapella, A.Duda, K. Emery, L. Gedvilas, T. Moriaty, S. Ward, J.D. Webb, X. Wu, 'High-Performance 0.6 eV, $Ga_{0.32}In_{0.68}As/InAs_{0.32}P_{0.68}$ Thermophotovoltaic Converters and Monolithically Interconnected Modules, AIP Conf. Proc 460, 132-141 (1999).

6. G.W. Charache, D.M. DePoy, P.F. Baldasaro, B.C. Campbell, 'Thermophotovoltaic Devices Utilizing a Back Surface Reflector for Spectral Control,' AIP Conf. Proc. 321, 339-350 (1996).
7. G.W. Charache, J.L. Egley, D.M. DePoy, L.R. Danielson, M.J. Freeman, R.J. Dziendziel, J.F. Moynihan, P.F. Baldasaro, B.C. Campbell, C.A. Wang, H.K. Choi, G.W. Turner, S.J. Wojtczuk, P. Colter, P. Sharps, M. Timmons, R.E. Fahey, K. Zhang, J. Electron. Mater. 27, 1038-1042 (1998).
8. G.W. Charache, D.M. DePoy, J.E. Raynolds, P.F. Baldasaro, K.E. Miyano, T. Holden, F.H. Pollak, P.R. Sharps, M.L. Timmons, C.B. Geller, W. Mannstadt, R. Asahi, A.J. Freeman, W. Wolf, J. Appl. Phys. 86, 452-458 (1999).
9. S.J. Spector, D.K. Astolfi, S.P. Doran, T.M. Lyszczarz, J.E. Raynolds, J. Vac. Sci. Technol. B, 2757-2760 (2001).
10. I.-H. Tan, C. Reaves, A. Holmes, Jr., E.L. Hu, J.E. Bowers, S. DenBaars, Electron. Lett. 31, 588-589 (1995).
11. F. Omnes, J.-C. Guillaume, G. Nataf, G. Jager-Waldau, P. Vennegues, P. Gibart, IEEE Trans. Electron. Dev. 43, 1806-1811 (1996).
12. R.H. Horng, D.S. Wuu, S.C. Wei, C.Y. Tseng, M.F. Huang, K.H. Chang, P.H. Liu, K.C. Lin, Appl. Phys. Lett. 75, 3054-3056 (1999).
13. H.C. Lin, W.H. Wang, K.C. Hsieh, K.Y. Cheng, Electron. Lett. 38, 516-517 (2002).
14. C.A. Wang, H.K. Choi, G.W. Charache, IEE Proc.-Optoelectron. 147, 193-198 (2000).
15. X. Wu, A. Duda, J.J. Carapella, J.S. Ward, J.D. Webb, M.W. Wanlass, 'A Study of Contacts and Back-Surface Reflectors for 0.6-eV $Ga_{0.32}In_{0.68}As/InAs_{0.32}P_{0.68}$ Thermophotovoltaic Monolithically Interconnected Modules,' AIP Conf. Proc. 460, 517-524 (1999).
16. M.B. Clevenger, C.S. Murray, S.A. Ringel, R.N. Sacks, L. Qin, G.W. Charache, D.M. DePoy, 'Optical Properties of Thin Semiconductor Device Structures with Reflective Back-Surface Layers,' AIP Conf. Proc. 460, 327-334 (1999).

A Broadband antireflection for GaSb by means of subwavelength grating (SWG) structures

Hiroo Yugami, Ken-ichi Kobayashi and Hitoshi Sai

Department of Machine Intelligence and Systems Engineering, Tohoku University, Aoba01, Aramaki, Aoba-ku, Sendai, 980-8579, Japan

Yoshiaki Kanamori and Kazuhiro Hane

Department of Mechatronics and Precision Engineering, Tohoku University, Aoba01, Aramaki, Aoba-ku, Sendai, 980-8579, Japan

Abstract. Broadband antirefraction property is demonstrated by using subwavelength grating structures on GaSb wafers, which are the base material of GaSb PV cells. Surface nano-structure with 350nm periods is fabricated by means of electron beam lithography and fast atom beam (FAB) etching. Since FAB is electrically neutral atomic or molecular beam, it is possible to obtain fine patterns with nanometer order. The reflectivity of this sample is strongly suppressed from the visible to near IR region. The experimental data is compared with numerical simulations by using the rigorous coupled wave analysis. The thermal stability of SWG structures are also studied by measuring reflection spectra of heated samples.

INTRODUCTION

Antireflection coating on photovoltaic (PV) cells is very important for high efficiency thermophotovoltaic (TPV) systems. Multilayered coating of dielectric materials with different refractive indices is the most popular technique for PV cells. This technique has been well established and can be applied to wide surface area of PV cells. However, in some cases, the thermal stability of such multiplayer coating limits the performance of TPV, because the surface temperature of TPV cells is higher than that of the conventional PV cells.

A subwavelength grating (SWG), which is the surface-relief grating with the period smaller than the light wavelength, behaves as an antireflection surface [1,2]. The SWG with a deep tapered shape grating especially suppresses the reflection over a wide spectral bandwidth and a large field of view. The SWG is excellent in heat resistance because consisted of no materials except a substrate material. The schematic view of antireflection SWG is shown in Fig. 1. In order to achieve high performance SWG, the structural periodicity (Λ) must be smaller than the incident radiation wavelength (λ). A deep tapered

CP653, *Thermophotovoltaic Generation of Electricity: 5th Conference*
edited by T. J. Coutts, G. Guazzoni, and J. Luther

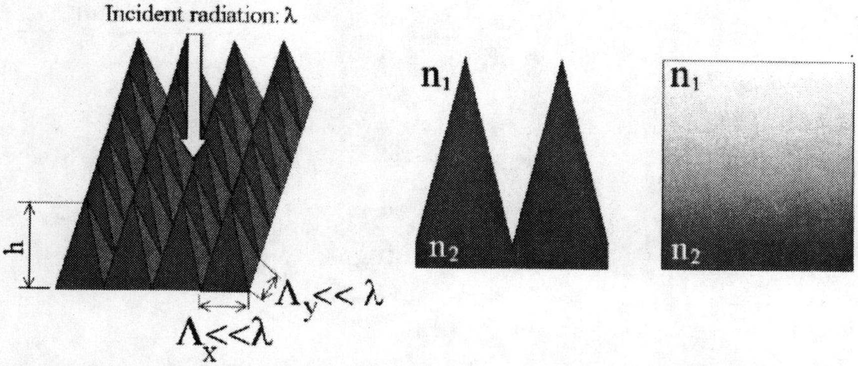

FIGURE 1. Schematic drawing of antireflection SWG (a), and the equivalent refractive index distribution along the light propagation direction (b).

shape is also important to obtain a broadband antireflection. The spectral properties of antireflection SWG can be explained by the graded refractive index model.

In this study, we have fabricated a two-dimensional antireflection SWG on a GaSb substrate. The SWG has consisted of 350 nm-period tapered grating. In the fabrication, electron beam (EB) lithography and fast atom beam (FAB) etching have been used. Since FAB is electrically neutral atomic or molecular beam, it is possible to obtain fine patterns with nanometer order. The reflectivity of the sample is strongly suppressed from the visible to near infrared region. The experimental data is compared with numerical simulations using the rigorous coupled analysis (RCWA). We have also demonstrated that the SWG works well for reducing the reflection even at the temperature of 240 °C.

FABRICATION PROCESS

We are studying the fabrication process of high performance selective emitter or absorber with surface microstructures[3-5]. In the case of selective emitters with surface microstructures, the structural periodicity is nearly equal the wavelength of thermal radiation, and the aspect ratio of unity is enough to obtain the sufficient performance in most cases. On the other hand, the most important properties of antireflection SWG are high aspect ratios and short periodicity of the grating. It is also necessary the two dimensional grating to obtain the characteristics independent on the polarization of radiation, The aspect ratio of 1 obtained by Han et al. using replication process had been the highest value in the two dimensional SWG [6]. Recently, two dimensional SWG with 150nm period and 350nm depth is fabricated on by Kanamori et al. on Si substrates [7] . Up to now, SWG has been fabricated on SiO_2 and Si in most cases.

The periodicity of SWG must be less than 450nm in order to match the spectral response of GaSb PV cells. In this study, the periodicity of SWG is fixed to 350nm.

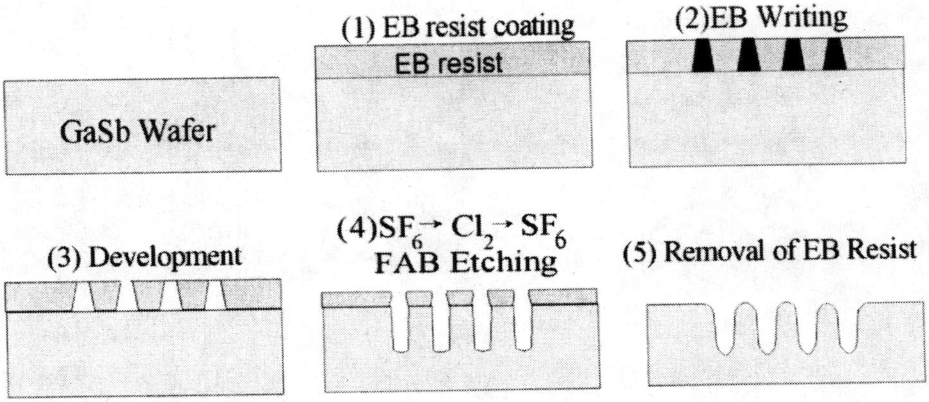

FIGURE 2. Fabrication process steps used in the fabrication of SWG.

This value was determined from both of fabrication reliability and performance of SWG.

The process steps used in the fabrication of the SWG are shown in Fig.2. A 500mm thick GaSb (100) wafers (p-type, carrier density;1~2 x 10^{17} cm^{-3}) is used as substrates. EB positive resist (Nippon Zeon Co., Ltd.; ZEP-520) is coated on the GaSb substrate by the spin coating technique. The thickness of EB resist is about 350nm. The SWG pattern is drawn on the resist by EB machine (ELIONIX Co., Ltd.; ELS3700). The EB resist is exposed by drawing the grating lines vertically and horizontally.

After the development, the resist grating is used as a mask for the FAB etching. In order to obtain the low spectral reflectivity, the gratings must have tapered shape as shown in Fig. 1. The tapered structure is fabricated by alternative FAB etching with two different kinds of gases. First, the GaSb substrate is etched by the FAB (EBARA Co., Ltd.; FAB-60ML) of SF_6 gas. In the second FAB etching, the Cl_2 gas is used. In the third FAB etching, the SF_6 gas is used again. In the SF_6 etching, the tapered grating shape is generated mainly by the physical spattering. On the other hand, for Cl_2 gas, the GaSb is etched vertically due to chemical reaction. By using the two etching gases alternately, the tapered grating with high aspect ratio can be fabricated. The EB resist is finally removed with tetrahydrofuran.

RESULTS AND DISCUSSION

Figure 3 shows the SEM images of fabricated SWG on a GaSb substrate. The grating with 350nm periodicity is fabricated in 1 x1 mm^2. From the analysis of SEM images, the depth of grating is estimated to be 1 μm. This result shows that the aspect ratio of fabricated SWG is about 2.86. This is the highest value of aspect ratio of two-dimensional SWG fabricated on III-V compound semiconductors. Furthermore, the fabricated SWG consists of a smooth tapered profile. Figure 4 shows the reflectivities of

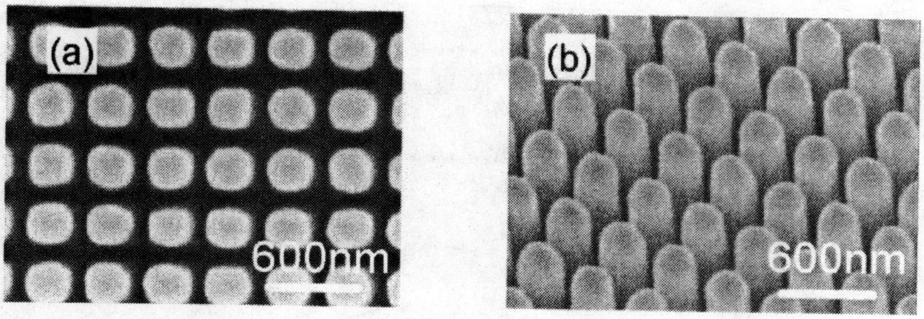

FIGURE 3. Top (a) and oblique view (b) of SEM images of fabricated SWG on GaSb

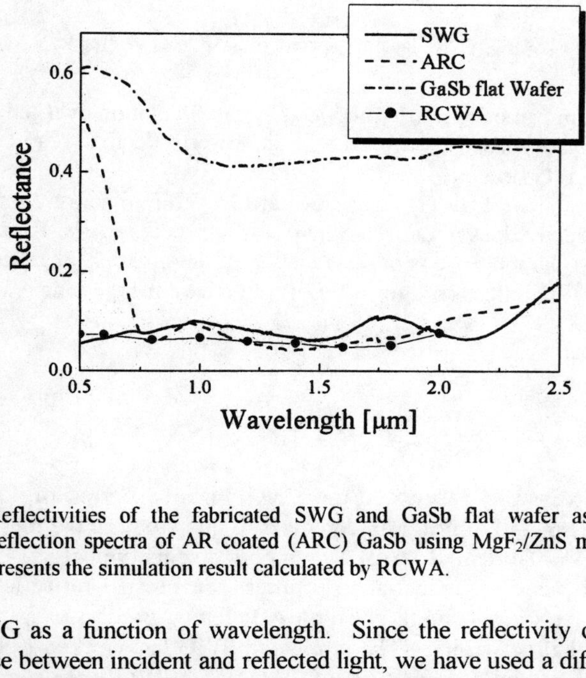

FIGURE 4. Reflectivities of the fabricated SWG and GaSb flat wafer as a function of wavelength. Reflection spectra of AR coated (ARC) GaSb using MgF_2/ZnS multilayers. The marked line represents the simulation result calculated by RCWA.

the fabricated SWG as a function of wavelength. Since the reflectivity of SWG shows angular dependence between incident and reflected light, we have used a diffused reflection geometry to measure the spectra. The reflected light within the angle of ±40° from the normal direction is collected in the measurement. The reflectivity of an antireflection coating (ARC) layer coated on the GaSb substrate is also measured. As shown in Fig.4, the SWG decreases the reflection drastically compared with that of the polished GaSb substrate. The reflectivity of the SWG is decreased as much as that of the ARC layer. At the wavelengths shorter than 0.75 μm, the reflectivity of the SWG is drastically decreased, while that of the ARC layer is increased with decreasing the wavelengths. The reflectivity of SWG is simulated by RCWA using literature data of optical constants of GaSb [8]. The

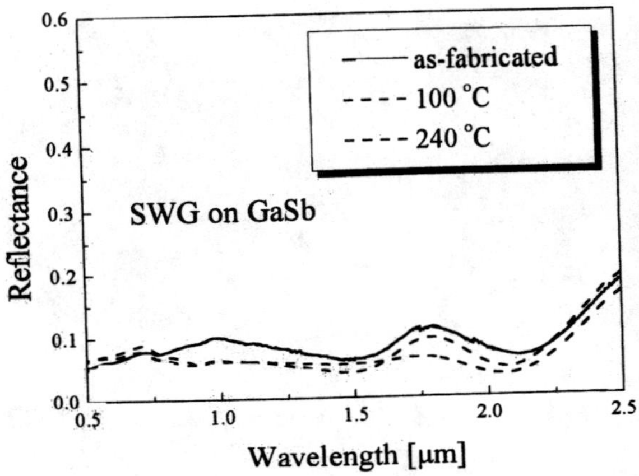

FIGURE 5. Reflectance spectra of SWS after heating tests.

reflection spectrum can be almost reproduced by the simulation over the measured spectral range. The simulation analysis reveals that the lower reflectivity can be achieved by the optimization of SWG structure.

The thermal stability of SWG was studied at 100°C for 36 hours. As shown in Fig.5, the antireflection property does not affected by this heat process. After the subsequent heating up to 240°C for 6 hours, there is no degradation of the property of reflectivity. This result indicates that the antireflection using SWG is effective for high temperature application of PV cells such as TPV.

CONCLUSIONS

We have successfully fabricated the SWG with the period of 350 nm on a GaSb substrate. The aspect ratio of 2.86 achieved in this study is the highest value of two-dimensional SWG fabricated on III-V compound semiconductors. For fabricating the tapered grating, which is essential to suppress the Fresnel reflection at the GaSb-air interface, we have introduced three-step FAB etching process using SF_6 and Cl_2 gases combined with EB patterning. The fabricated SWG has worked well for reducing the reflection even at the temperature of 240 °C. SWG shows good antireflection properties even for radiation with large incident angle. This property is very useful for TPV generators, because PV cells for TPV system are irradiated by thermal radiation with large incident angle range. It is confirmed that the GaSb PV cell with SWG will be powerful for improving the efficiency of TPV generation system.

ACKNOWLEDGMENTS

The authors appreciate the monetary supports by the Ministry of Education, Science, Sports and Culture, Japan, Japan Space Forum, and the Asahi Glass Foundation. The authors also thank the workers of the Venture Business Laboratory in Tohoku University for their thchnical supports.

REFERENCES

1. Clapham, P.B., and Hutley, M.C., Nature **244**, 281-282 (1973).
2. Kanamori, Y., Sasaki, M., and Hane, K, Opt. Lett., **24**, 1422-1424 (1999).
3. Sai, H., Yugami, H., Akiyama, Y., Kanamori, Y., and Hane, K., *J. Opt. Soc. Am. A*, **18**, 1471-1476 (2001).
4. Maruyama, S., Kashiwa, T., Yugami, H., and Esashi, E., *Appl. Phys. Lett.*, **79**, 1393-1395 (2001).
5. Sai, H., Yugami, Y., Kanamori, Y., and Hane, *Solar Energy Materials and Solar Cells,* in press (2002).
6. Han, S.T., Tsao, Y-L., Walser, R.M., and Becker, M.F., *Appl. Opt*. **31**, 2343-2352 (1992).
7. Kanamori, Y., Hane, K Sai, H. and, Yugami, H., Appl. Phys. Lett., **78**, 142-143 (2001).
8. Edwards, David F., and White, Richard H., in *Handbook of Optical Constants of Solids II,* edited by E.D. Palik, New York: Academic Press, 1991, pp.597-606.

Effect of Graded Base p-Doping on MIM Performance

B. Wernsman[a], R. J. Wehrer[a], D. M. Wilt[b], R. W. Smith[a], R. R. Siergiej[a], G. P. Schmuck[a], I. Ju[a], and C. B. Geller[a]

[a]Bechtel Bettis, Inc., West Mifflin, PA 15122
[b]NASA Glenn Research Center, Cleveland, OH 44135

Abstract. 0.6 eV InGaAs/InPAs double heterostructure n/p/n tunnel junction monolithic interconnected modules (TJ-MIMs) having p-type bases doped uniformly at $2x10^{17}$ cm^{-3} and inhomogeneously graded from $4x10^{16}$ cm^{-3} to $4x10^{17}$ cm^{-3} (10:1 doping ratio) were grown, processed, and characterized. The electrical performance of the devices with inhomogeneously graded bases was higher than that of the uniformly doped case due to the built-in electric field in the base region, making QE_{int} and J_{SC} larger. The free carrier absorption increased due to the relatively higher doping levels for the linearly doped case but remained unchanged for the exponential profile. Therefore, results showed that exponentially grading the base p-type doping increased the MIM performance by ~ 8% (relative).

INTRODUCTION

InGaAs/InPAs double heterostructure n/p/n tunnel junction (TJ) monolithic interconnected modules (MIMs) have demonstrated higher performance compared to InGaAs/InPAs double heterostructure p/n MIMs [1]. This increase occurs because the emitter and lateral conduction layer (LCL) are comprised of n-type materials with a relatively low sheet resistance, yielding MIMs with reduced series resistance. Also, due to the relatively large lifetimes and mobilities of electrons in p-type layers, lowly doped, thick p-type bases can be used to increase device performance without significantly increasing parasitic heating. Using this device structure, 0.6 eV InGaAs/InPAs double heterostructure TJ-MIMs have demonstrated system efficiencies [2] greater than 20% at a cell and etched silicon carbide radiator temperature of ~ 25°C and ~ 1000°C, respectively [3]. These results are approaching the theoretical performance limit of MIMs having the structure given in Reference [3]. However, it is still possible to increase the performance of these MIMs.

CP653, *Thermophotovoltaic Generation of Electricity: 5th Conference*
edited by T. J. Coutts, G. Guazzoni, and J. Luther

One possible method of increasing 0.6 eV InGaAs/InPAs double heterostructure n/p/n TJ-MIM performance is to use inhomogeneous doping and/or composition in the device structure. These types of inhomogeneous structures have been theoretically and experimentally shown to improve the efficiency of solar cells by creating an electric field in the inhomogeneous layer [4-7]. These built-in electric fields sweep the excess minority charge carriers toward the depletion region of the p/n junction, increasing the short circuit current density (J_{SC}). They have also been shown to decrease the reverse saturation current density (J_o), increasing the device open circuit voltage (V_{OC}) and fill factor (FF). Thus, the use of inhomogeneous layers tends to increase the maximum output power (Q_e^{MAX}) and system efficiency (η). To understand how using inhomogeneous layers affects the performance of 0.6 eV InGaAs n/p/n TJ-MIM devices, a theoretical and experimental evaluation was made.

THEORETICAL PREDICTIONS

Theoretical performance modeling of the 0.6 eV InGaAs/InPAs double heterostructure n/p/n TJ-MIM is used to guide the device growth. The model is based on the work of Hovel [8] where analytic solutions of the transport equations are used. These solutions are only achieved for either a uniformly doped or an exponentially doped InGaAs layer where the geometric mean doping is used to calculate the mobility and effective mass of the charge carriers. It is important to note that the exponentially graded layers build a constant electric field (E). A correction was made to the generated photo-current ($J_{h\nu}$) equation of Hovel [8] for the constant electric field case so that it would reduce to the uniform doping equation when E was set to zero. This corrected $J_{h\nu}$ term is given by:

$$J_{h\nu} = \frac{qD\alpha\tau F_o}{\alpha L^2(\alpha - 2E') - 1}\left[\frac{H}{L'G} - E' + \alpha\right] = J_{SC}, \tag{1}$$

where

$$H = L'\left(\frac{S}{D} + 2E'\right)e^{(E'-\alpha)t} - L'\left(\frac{S}{D} + E'\right)\cosh\frac{t}{L'} - \sinh\frac{t}{L'} - \alpha L' e^{(E'-\alpha)t}, \tag{2}$$

$$G = \cosh\frac{t}{L'} + L'\left(\frac{S}{D} + E'\right)\sinh\frac{t}{L'}, \tag{3}$$

$$\frac{1}{L'} = \sqrt{E'^2 + \frac{1}{L^2}}, \tag{4}$$

$$E' = \frac{E}{kT}, \tag{5}$$

q is the electronic charge, D is the electron diffusion coefficient, α is the photon absorption coefficient, τ is the minority carrier lifetime, F_o is the photon flux, t is the layer thickness, S is the surface recombination velocity, L is the diffusion length, k is the Boltzmann constant, and T is the temperature.

Figure 1 shows the diode structure (electrically active layers) used in the performance predictions of the 0.6 eV n/p/n TJ-MIM. The p-type base of the converter is ~ 8 times thicker than the emitter to ensure that the majority of the photon absorption and electron-hole pair production occur in this region. By allowing the electron-hole pair production to occur here, the output current is governed mainly by the more mobile minority electrons, making the output power larger. Thus, the inhomogeneous layer (built-in electric field) was incorporated into the base of the converter to enhance the output power. Rather than grading the composition (variation in composition will vary the lattice parameter and lead to strain), inhomogeneous InGaAs p-type bases were modeled by grading the p-type doping from a minimum near the n/p junction to a maximum at the base/BSF interface.

Table 1 shows the theoretical performance of 0.6 eV n/p/n TJ-MIMs having uniform, 10:1 exponentially, and 100:1 exponentially graded p-doped bases. The average p-doping for these cases was $8x10^{16}$ cm^{-3}. As shown in Table 1, the system efficiency of the MIMs having graded p-type bases was larger than that of the MIM with uniform base doping, where the 100:1 exponentially graded case had the highest efficiency. Since no increase in J_{SC} was predicted, the predicted efficiency enhancement was due to V_{OC} and FF increasing. V_{OC} and FF increased because J_o decreased from 4.37 $\mu A/cm^2$ to 1.28 $\mu A/cm^2$ as the p-type base changed from uniform doping to 100:1 exponential doping. To verify that inhomogeneous bases improve performance, structures incorporating this layer were grown and processed.

Window; n-InPAs
Emitter; n-InGAs
Base; p-InGaAs,
Back Surface Field (BSF); p-InPAs
Tunnel Junction; InGaAs
Lateral Conduction Layer; n-InGaAs

FIGURE 1. Simplified Block Diagram of MIM Structure with Buffers and Substrate not Shown.

TABLE 1. MIM Performance Predictions with a Cell Temperature of 25°C and Radiator Temperature and Emissivity of 1000°C and 0.90, respectively.

Base p-Doping Profile	Doping Range (cm^{-3})	J_{SC} (A/cm^2)	V_{OC} (V)	FF (%)	Q_e^{MAX} (W/cm^2)	Q_{abs} (W/cm^2)	η (%)
Uniform	$8x10^{16}$	2.91	0.344	67.6	0.605	3.56	17.1
10:1 Exponential	$2.1x10^{16}$–$2.1x10^{17}$	2.91	0.365	68.9	0.655	3.56	18.5
100:1 Exponential	$3.7x10^{15}$–$3.7x10^{17}$	2.91	0.376	69.5	0.680	3.56	19.3

Q_{abs} is the total heat absorbed by the device.

MIM GROWTH AND PROCESSING

The n/p/n TJ-MIM structures were produced using a low-pressure, horizontal organo-metallic vapor phase epitaxial reactor described previously [9]. A double dilution system was installed onto this reactor to allow for low p-type Zn doping. Calibrations of the system showed that Zn doping levels as low as $4x10^{16}$ cm^{-3} were easily achievable. However, this value of Zn doping is higher than the minimum values given in Table 1. To observe the effect of an inhomogeneous base on the performance of the n/p/n TJ-MIM and to minimize free carrier absorption, 10:1 p-type doping profiles were grown with a minimum doping of $4x10^{16}$ cm^{-3}. Table 2 shows the predicted electrical performance of the uniformly and 10:1 exponentially graded MIMs having an average doping of $1.6x10^{17}$ cm^{-3}, keeping Q_{abs} constant. Again, as with the case given in Table 1, the electrical performance of the device having a 10:1 doped base was better than the uniform case.

TABLE 2. MIM Electrical Performance Predictions with a Cell Temperature of 25°C and Radiator Temperature and Emissivity of 1000°C and 0.90, respectively.

Base p-Doping Profile	Doping Range (cm^{-3})	J_{SC} (A/cm^2)	V_{OC} (V)	J_o ($\mu A/cm^2$)	FF (%)	Q_e^{MAX} (W/cm^2)	Q_{abs} (W/cm^2)	η (%)
Uniform	$1.6x10^{17}$	2.90	0.345	4.14	67.7	0.608	3.61	17.0
10:1 Exp.	$4x10^{16}$–$4x10^{17}$	2.91	0.366	1.90	68.9	0.655	3.61	18.4

Along with the uniformly and exponentially graded bases, a linearly graded base was also grown to help understand the effect of different grading profiles. Figure 2 shows the SIMS analysis from the three different growths (uniform, exponential, and linear). The grown wafers were processed into MIMs using a previously described wet chemical etching technique [10].

RESULTS AND DISCUSSION

The electrical and optical performances of the MIMs were evaluated separately. By evaluating these independent performance criteria separately, the effect of the inhomogeneous layer on device performance could be better understood, allowing for future MIM design optimization.

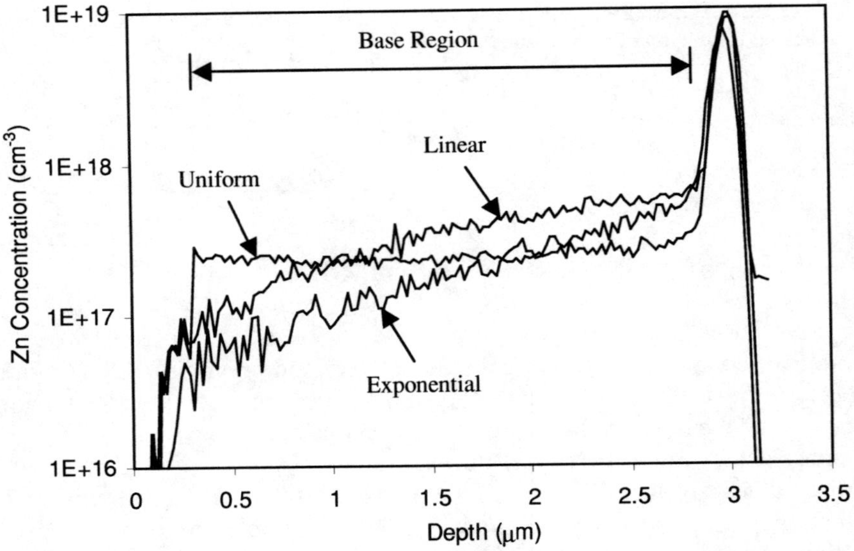

FIGURE 2. SIMS Analysis of Grown Structures Showing Zn (p-Type Doping) Profiles.

Electrical Performance

The J_{SC}, V_{OC} and FF are used to determine the electrical performance (Q_e^{MAX}) of the MIM. Figures 3-5 show the J_{SC}-V_{OC}, FF-J_{SC} and Q_e^{MAX}-J_{SC} curves, respectively, for the three different doping profiles. As shown in Fig. 3 at 1 A/cm^2, V_{OC} was the highest for the exponential profile, lower for the linear

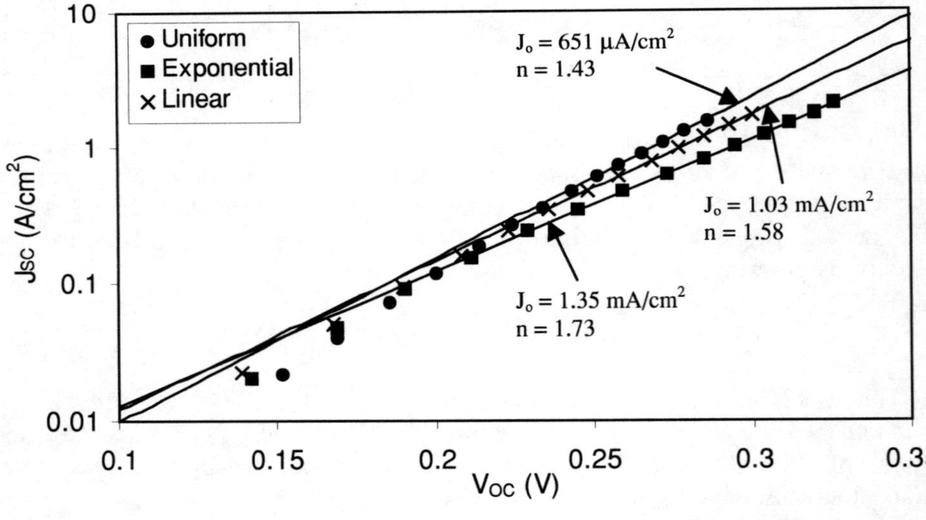

FIGURE 3. J_{SC}-V_{OC} Curves for MIMs Having Different Base p-Type Doping Profiles.

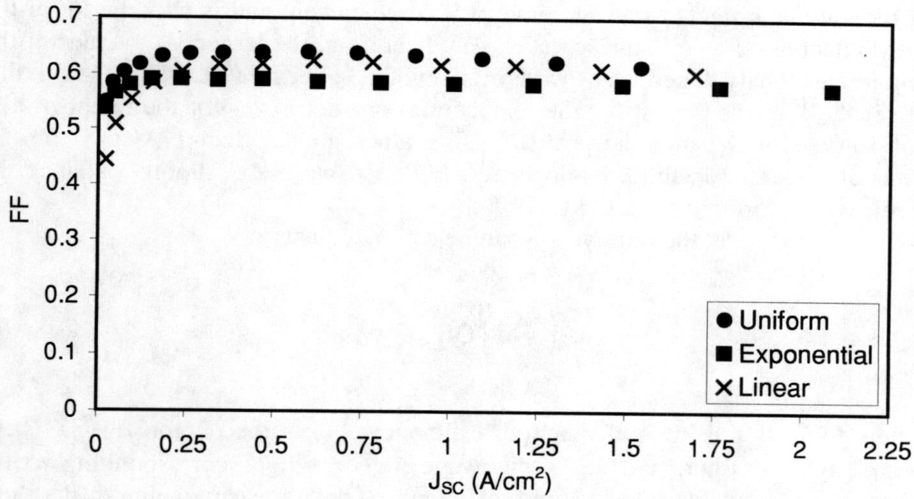

FIGURE 4. FF-J_{SC} Curves for the MIMs Having Different Base p-Type Doping Profiles.

profile, and the lowest for the uniform profile. However, the higher V_{OC} values did not correspond to smaller J_o values (Fig. 3). This apparent discrepancy between the J_o and V_{OC} values was due to the different ideality factors (n) [11] for the linear, exponential, and BSF profiles where the larger n values correspond to the larger V_{OC} values (Fig. 3). It is believed that the larger n values were due to the relatively low base doping level near the n/p junction that makes the depletion zone large, allowing for more recombination to occur in the zone. This increase

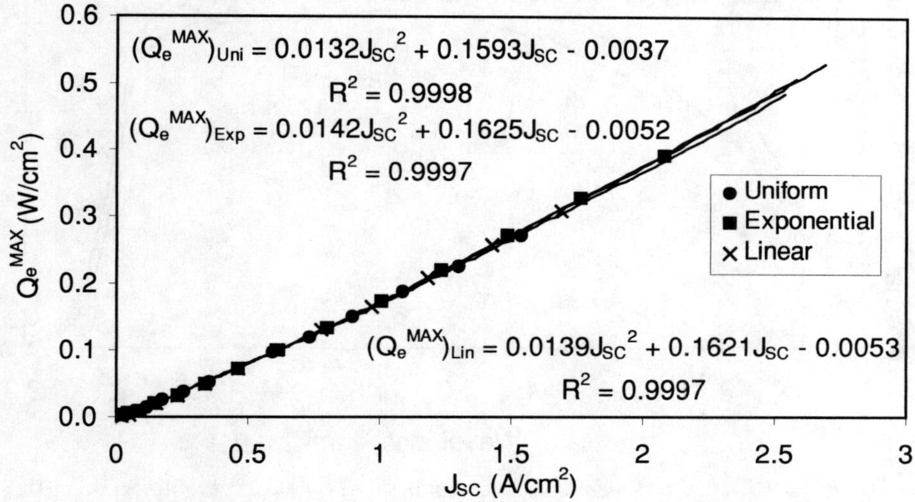

FIGURE 5. Q_e^{MAX}-J_{SC} Curves for the MIMs Having Different Base p-Type Doping Profiles.

in recombination acts as an increase in leakage current and is why the FF of the MIMs decreased as n increased (Fig. 4). Therefore, the larger V_{OC} values of the inhomogeneously doped bases were offset by the lower FF values. Even with this offset, Q_e^{MAX} was largest for the exponential profile, lower for the linear profile and lowest for the uniform profile at J_{SC} values greater than 1 A/cm^2 (Fig. 5). Thus at a given J_{SC}, the inhomogeneously doped bases did slightly enhance the electrical performance of the MIMs.

For each device, the relative J_{SC} can be estimated using:

$$J_{SC} = q\int_0^{\lambda_g} QE_{int}\Gamma_{hv}d\lambda\,, \tag{6}$$

where QE_{int} is the internal quantum efficiency, Γ_{hv} is the photon flux, λ is the wavelength of light, and λ_g is the wavelength of light corresponding to the bandgap energy. Internal quantum efficiency (QE_{int}) measurements of the three different devices were made to study the change in J_{SC} and are given in Fig. 6. Using these QE_{int} curves and Eq. (6), estimates of the J_{SC} for each device illuminated with a 1000°C, 0.90 emissivity blackbody were made and are also given in Fig. 6.

As shown in Fig. 6, QE_{int} was the highest for the linear profile, lower for the exponential profile and lowest for the uniform case. Thus, the calculated J_{SC}

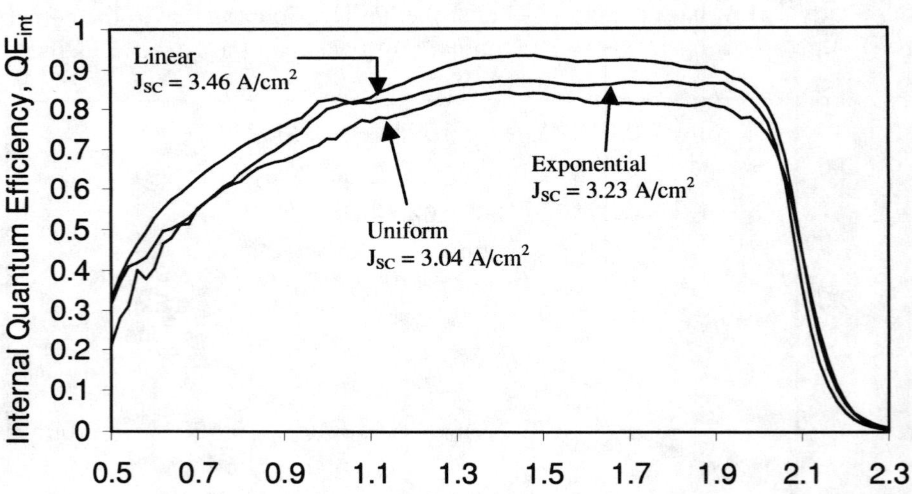

FIGURE 6. QE_{int} of the MIMs Having Different Base p-Type Doping Profiles Showing Their Relative J_{SC} Values when Illuminated by a 1000°C, 0.90 Emissivity Radiator.

(using Eq. (6)) for the linear, exponential and uniform cases were 3.46 A/cm^2, 3.23 A/cm^2 and 3.04 A/cm^2, respectively. As discussed earlier, the exponential profile builds a constant electric field in the base region making QE_{int} larger than that of the uniform case. The electric field for the linear profile is proportional to 1/x where x is the distance into the base region from the n/p junction. Therefore, near the n/p junction where x is small, the electric field for the linear case is very large, making QE_{int} increase above the uniform and exponential cases. Thus, the increase in QE_{int} and thus J_{SC} above that given by the uniform case is due to the built in electric field in the base region.

Optical Performance

The heat absorbed for a MIM (Q_{abs}) is given by:

$$Q_{abs} = \int_0^\infty (1-R)\rho_{hv} d\lambda , \qquad (7)$$

where R is its reflectance and ρ_{hv} is the spectral heat flux. Thus, the reflectance of the MIM is used to determine its optical performance. Figure 7 shows the measured 10° directional hemispherical reflectance of the grown MIM structures with the uniform, exponential and linear doped bases. As shown, the reflectance of the light for wavelengths $> \lambda_g$ ($\lambda_g \sim 2.1$ μm) for the uniform and exponential

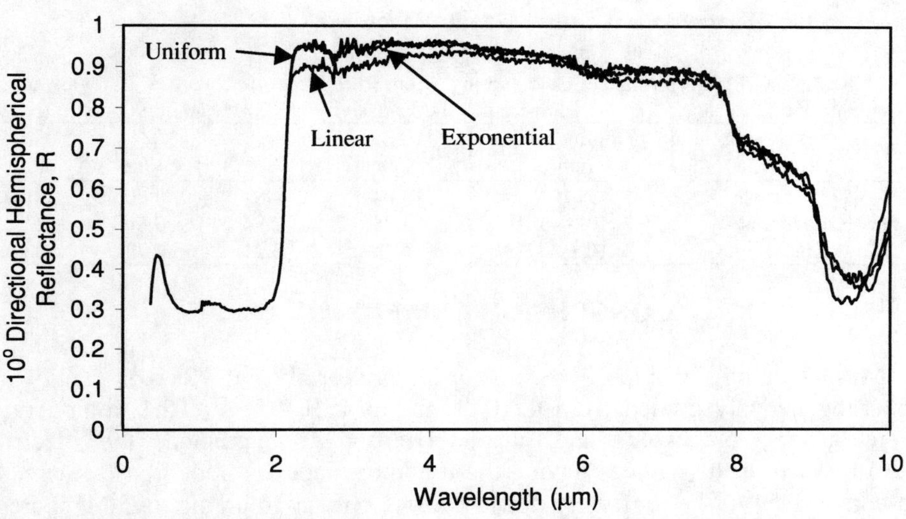

FIGURE 7. 10° Directional Hemispherical Reflectance of Grown MIM Structures Having Different p-Type Doping Profiles.

cases were relatively equal while that of the linear case was lower due to an increase in free carrier absorption. Since free carrier absorption increases as the p doping increases, free carrier absorption for the linear profile was larger because a greater portion of the base region was doped higher than $2x10^{17}$ cm^{-3} (Fig. 2). Thus, the parasitic heating of the MIM is larger for the linear case.

MIM Performance

Table 3 shows the effect of the measured electrical and optical performance on calculated system efficiency (η) using a 1000°C, 0.90 emissivity radiator. As shown, η for both inhomogeneously doped base cases is larger than that for the uniform case where the highest η occurred for the exponential case. It is important to note that although the linear Q_e^{MAX} is the largest, this large Q_e^{MAX} value is offset by the larger Q_{abs} value due to its increased free carrier absorption. The 8.1% (relative) increase in η due to exponentially grading the p-type dopant in the base (21.0% to 22.7%) corresponds well with that predicted in Table 2 (8.2% (relative); 17.0% to 18.4%). However, contrary to that shown in Table 2, the experimental results suggest that η should increase due to increasing J_{SC} and not due to increasing V_{OC} and FF (decreasing J_o). Also, η calculated from the experimental results is larger than that given in Table 2. This contradiction between experiment and prediction may be due to the material properties (i.e. effective mass, mobility etc.) used in the predictive tool, requiring further evaluation of these properties. Also, the predictive model based on Hovel [8] may need to be modified to account for variable material properties across the base and the observed variability in ideality factor (n).

TABLE 3. MIM Performance Calculations From the Experimental Results with a Cell Temperature of 25°C and Radiator Temperature and Emissivity of 1000°C and 0.90, respectively.

Base p-Doping Profile	Doping Range (cm^{-3})	J_{SC} (A/cm^2)	Q_{abs} (W/cm^2)	Q_e^{MAX} (W/cm^2)	η (%)
Uniform	$2x10^{17}$	3.04	2.87	0.603	21.0
Exponential	$4x10^{16}$–$4x10^{17}$	3.23	2.94	0.668	22.7
Linear	$4x10^{16}$–$4x10^{17}$	3.46	3.21	0.722	22.5

CONCLUSION AND FUTURE WORK

MIMs having p-type bases doped uniformly at $2x10^{17}$ cm^{-3} and inhomogeneously graded from $4x10^{16}$ cm^{-3} to $4x10^{17}$ cm^{-3} (10:1 doping ratio) were grown, processed, and characterized. As predicted, the electrical performance of the devices having the inhomogeneously graded bases increased in comparison to the uniformly doped case. Contrary to prediction, this increase in performance for the devices with inhomogeneous bases was due to the built-in electric field that increased QE_{int}, making J_{SC} larger. The optical performance of the exponential case was relatively the same as that for the uniform case. However, free carrier absorption for the linear profile was larger due to the relatively high doping levels. By using the exponentially p-doped base, an ~ 8% (relative) increase in device performance using a 1000°C, 0.90 emissivity radiator can be obtained. Thus, inhomogeneously p-doped bases can increase MIM performance.

Efforts are underway to determine improved material properties. Also, to better understand the effect of inhomogeneous layers on device performance, more sophisticated models are being developed. These models will be able to predict the performance of MIMs using inhomogeneous material properties and doping profiles.

REFERENCES

1. Wilt, D. M., Murray, C. S., Fatemi, N. S., and Weizer, V., "n/p/n Tunnel Junction InGaAs Monolithic Interconnected Module (MIM)," in *Thermophotovoltaic Generation of Electricity: Fourth NREL Conf*erence, edited by T. J. Coutts et al., AIP Conference Proceedings 460, New York, 1999, pp. 152-160.
2. Gethers, C. K., Ballinger, C. T., and DePoy, D. M., "Lessons Learned on Closed Cavity TPV System Efficiency Measurements," in *Thermophotovoltaic Generation of Electricity: Fourth NREL Conf*erence, edited by T. J. Coutts et al., AIP Conference Proceedings 460, New York, 1999, pp. 335-348.
3. Siergiej, R. R., Wernsman, B., Derry, S. A., Mahorter, R. G., Wehrer, R. J., Link, S. D., Palmisiano, M. N., Messham, R. L., Murray, S., Murray, C. S., Newman, F., Hills, J., and Taylor, D., "20% Efficient InGaAs/InPAs Thermophotovoltaic Cells," in *Thermophotovoltaic Generation of Electricity: Fifith NREL Conf*erence, to be published.
4. Ellis, B., and Moss, T. S., *Solid-State Electronics* **13**, 1-24 (1970).
5. Hamaker, H. C., *J. Appl. Phys.* **58**, 2344-2351 (1985).
6. Virshup, G. F., Ford, C. W., and Werthen, J. G., *Appl. Phys. Lett.* **47**, 1319-1321 (1985).
7. Sassi, G., *J. Appl. Phys.* **65**, 322-328 (1989).
8. Hovel, H. J., *Semiconductors and Semimetals: Volume 11, Solar Cells*, Academic Press, Orlando, FL, 1975.
9. Wilt, D. M. et al., "Lattice Mismatched InGaAs PV Devices for TPV Power Systems," in *Proceedings of First World Conference on PV Energy Conversion*, 1994, p. 1738.
10. Fatemi, N. S., Wilt. D. M., Jenkins, P. P., Hoffman, R. W., Jr., Weizer, V. G., Murray, C. S., and Riley, D., "Materials and Process Development for the Monolithic Interconnected Module (MIM) InGaAs/InP TPV Devices," in *Thermophotovoltaic Generation of Electricity: Third NREL Conf*erence, edited by T. J. Coutts et al., AIP Conference Proceedings 401, New York, 1997, pp. 249-262.
11. Sze, S. M., *Semiconductor Devices; Physics and Technology*, John Wiley & Sons, New York, p. 95, 1985.

Performance Limits of Low Bandgap Thermophotovoltaic Antimonide-Based Cells for Low Temperature Radiators

J.M. Borrego [1,2], C.A.Wang[3], P.S. Dutta [1,2], G.Rajagopalan[1], R.J. Gutmann [1,2], I.B. Bhat [1,2], H. Ehsani [4], J.F.Beausang[4], G. Nichols [4], and P.F. Baldasaro [4]

[1] *Center for Integrated Electronics and Electronics Manufacturing*
[2] *Department of Electrical, Computer and Systems Engineering*
Rensselaer Polytechnic Institute, Troy, NY – 12180
[3] *Lincoln Laboratory, Massachusetts Institute of Technology, Lexington, MA -02420*
[4] *Lockheed Martin Inc., Schenectady, NY – 12301*

Abstract. This paper assesses the performance of antimonide-based thermophotovoltaic cells fabricated by different technologies. In particular, the paper compares the performance of lattice matched quaternary (GaInAsSb) cells epitaxially grown on GaSb substrates to the performance of ternary (GaInSb) and binary (GaSb) cells fabricated by Zn diffusion on bulk substrates. The focus of the paper is to delineate the key performance advantages of the highest performance-to-date of the quaternary cells to the performance of the alternative ternary and binary antimonide-based diffusion technology. The performance characteristics of the cells considered are obtained from PC-1D simulations using appropriate material parameters.

INTRODUCTION

In order for Thermophotovoltaic (TPV) cells to be commercially viable it is necessary that they can be fabricated by processes that give cells with relatively good performance and with reproducible characteristics in semiconductor materials with energy band gaps in the range between 0.7 eV and 0.5 eV. At the present time the following fabrication process are being used when the material used is Sb based: molecular beam epitaxy (MBE), organometallic vapor phase epitaxy OMVPE and bulk diffusion. Recent work in our laboratory has shown that it is possible to fabricate GaSb and ternary GaInSb TPV cells in which the material has been grown by a Bridgman type technique and then the cell emitter is obtained by low temperature Zn diffusion. This fabrication process is inexpensive and has shown to produce cells with very reproducible characteristics. MBE and OMVPE are complex fabrication processes that have the advantage of giving cells

CP653, *Thermophotovoltaic Generation of Electricity: 5th Conference*
edited by T. J. Coutts, G. Guazzoni, and J. Luther

in which the physical structure and the doping concentration of the material in several regions can be tailored to achieve high performance cells. It is the purpose of this paper to present the results of computer simulations using PC-1D of the performance of Zn bulk diffused GaSb and GaInSb TPV cells and compare them to the performance of quaternary OMVPE grown cells as function of radiator temperature.

MATERIAL PARAMETERS

Computer simulations using PC-1D require a large number of material parameters that can be grouped as energy band parameters, carrier mobility parameters, optical parameters and recombination parameters. The energy band parameters of the III-V compounds have been compiled by Madelung (1). The mobility dependence upon carrier concentration for III-V binaries GaAs, GaSb, InAs and InSb has been reported in several papers (2,3,4). Optical constants of all the binary III-V compounds have been recently compiled (5). Experimental values of similar material parameters for the ternaries GaInSb and quaternaries GaInAsSb have not been reported. They can be estimated with confidence, from the corresponding material parameters of the binaries GaAs, InAs, GaSb and InSb, using equations developed by Adachi (6) and we have used that procedure for estimating those parameters for GaInAs and GaInAsSb alloys with band gaps in the range of 0.60 eV to 0.50 eV.

Figure 1 shows the measured mobility of electrons and holes in OMVPE grown quaternary material GaInAsSb with a bandgap of 0.55 eV and compares it with the mobility calculated using Matthiessen rule starting from the experimental mobility of the binaries GaAs, InAs, GaSb and InSb.

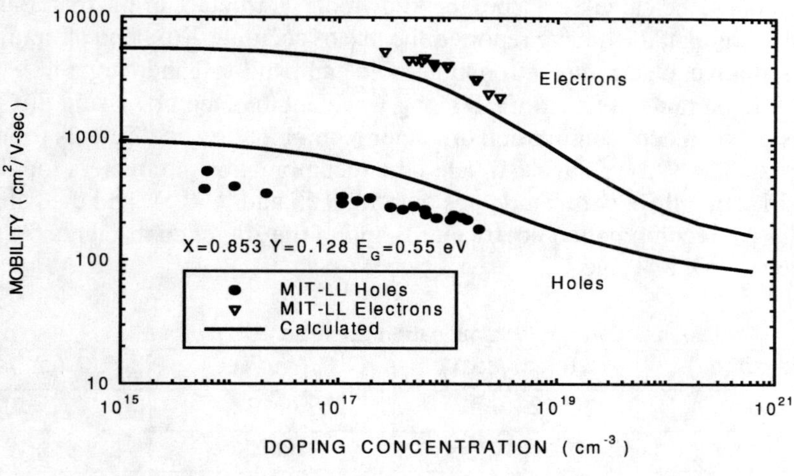

FIGURE 1. Electron and Hole Mobility of GaInAsSb Alloy.

The agreement between the experimental values and the calculated values is rather good considering the many assumptions used in obtaining the mobility of the alloy using Matthiessen rule. Similarly good agreement has been found between the measured optical absorption constant of the quaternary alloy GaInAsSb and the one calculated from the absorption constant of the binaries (7). Figure 2 shows a comparison between the calculated and the measured absorption constant for a GaInAsSb quaternary alloy of 0.53 eV bandgap (7).

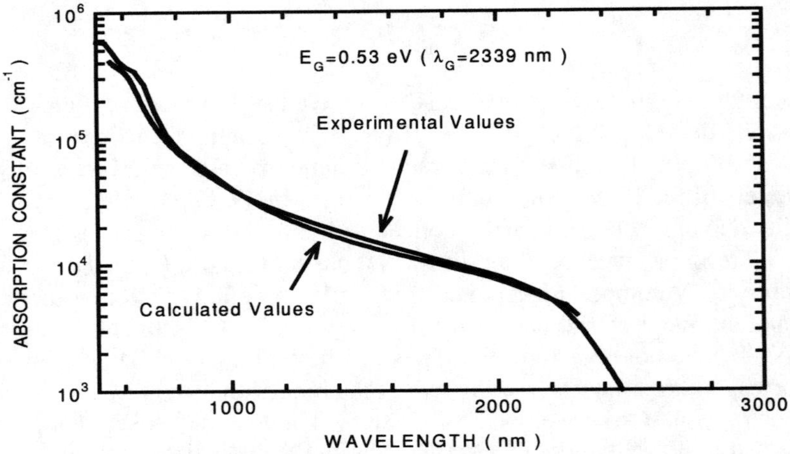

FIGURE 2. Optical Absorption Constant of GaInAsSb Alloy.

The recombination parameters are SRH, radiative or band-to-band and Auger recombination. Of these recombination processes, SRH cannot be estimated a priori since it is dependent upon the recombination centers introduced during the fabrication process. On the other hand, radiative and Auger recombination parameters can be calculated from the known band structure of the material. Recently Grein et al (8) have reported the use of accurate **K·p** band structures to obtain radiative, electron-electron, hole-hole and band-to-band Auger recombination rates in InAs/InGaSb long wavelength detectors. Using this paper as basis plus a recent compilation of Auger recombination coefficients for many semiconductors (9), we have estimated the recombination parameters for GaInSb and GaInAsSb alloys with band gaps of 0.50, 0.55 and 0.60 eV and the values of the radiative recombination coefficient B and of the Auger coefficients C_N and C_P are given below in Table 1.

TABLE 1. Radiative and Auger Recombination Coefficients

Bandgap (eV)	B (cm^3/sec)	C_N (cm^6/sec)	C_P (cm^6/sec)
0.60	8.2×10^{-11}	7.4×10^{-28}	7.5×10^{-28}
0.55	7.6×10^{-11}	1×10^{-27}	1.3×10^{-27}
0.50	6.9×10^{-11}	1.5×10^{-27}	2.6×10^{-27}

There is a question about the accuracy of these values as well as if they agree with measured values in these alloys, but at the present there are not reliable measurements and work is in progress to measure these coefficients. We believe the values of the Auger coefficients in Table 1 are accurate within a factor of 3 or 4 and have used them in simulations using PC-1D.

STRUCTURE OF SIMULATED Zn DIFFUSED CELLS

The physical structure of the simulated cells is very simple; it consists of a bulk wafer of binary GaSb or ternary GaInSb n-type material in which a p-type emitter has been formed by Zn diffusion at low temperature. A typical SIMS Zn diffusion profile in bulk GaInSb is shown in Fig. 3.

The profile shows that there is a high Zn concentration at the surface, around 10^{19} cm^{-3}, which is followed by a relatively slow exponential decay in the doping concentration with distance to $5x10^{18}$ cm^{-3} at 0.25 microns and then it decreases very sharply to a concentration to around 10^{16} cm^{-3} at 0.3 microns. Figure 3 shows that the doping concentration in the emitter is larger than 10^{18} cm^{-3} so the recombination in the emitter is dominated by Auger recombination and this cannot be changed as long as the emitter is formed by Zn diffusion.

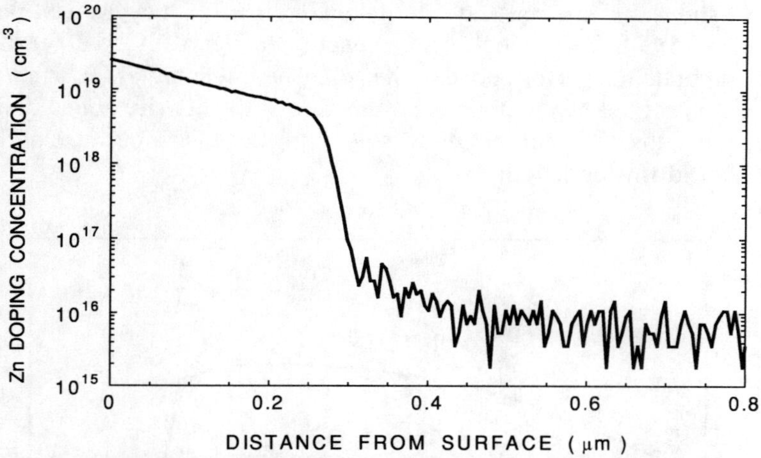

FIGURE 3. SIMS Emitter Doping Profile of Zn Bulk Diffused Cells

In order to achieve a high open circuit voltage, the doping concentration in the base should be chosen such that the recombination is mainly radiative. This implies low doping concentration but high enough so the PN junction has an adequate built-in voltage that allows carrier collection near open circuit conditions. This last condition can be achieved by doping the N-base at a concentration comparable to the conduction band density of states. In these low band gap materials the conduction band density of states is of the order of 10^{17}

cm^{-3} so the doping in the N-base should be within the range of 10^{16} cm^{-3} to 10^{17} cm^{-3}. As a starting design point we have chosen a doping concentration of $1x10^{17}$ cm^{-3} for the N-base of all the simulated cells since it is very difficult to obtain GaSb and GaInSb bulk material with lower doping concentration due to the native defects present.

For simulation purposes we have chosen the TPV cells to have a total area of 1 cm^2 and a thickness of 100 microns. The surface recombination in the front free surface is assumed to be 10^6 cm/sec. We have found that this value matches reasonably well the measured external quantum efficiency at short wavelengths. In the simulation we have assumed the metallic grid to form a perfect ohmic contact with the P-type emitter with a 10% obscuration ratio. Since the cell thickness is 100 microns and the diffusion length in the base is a few microns, the surface recombination velocity in the back surface is not relevant and for simulation purposes we have assumed to have a value of 10^7 cm/sec. Besides the above assumptions, the simulations do not include any effects due to photon recycling nor effects due to series resistance of the metallic grid in the TPV cell.

SIMULATION RESULTS FOR GaSb AND GaInSb TPV CELLS

Figure 4 shows the internal and external quantum efficiency, as well as the reflectance, of a GaInSb TPV cell with an energy bandgap of 0.55 eV using PC-1D to simulate the performance of the cell. The low internal QE at short wavelengths is caused by the high recombination in the heavily doped emitter . At longer wavelengths, the internal QE does not rise above 70% because of the long base and short diffusion length.

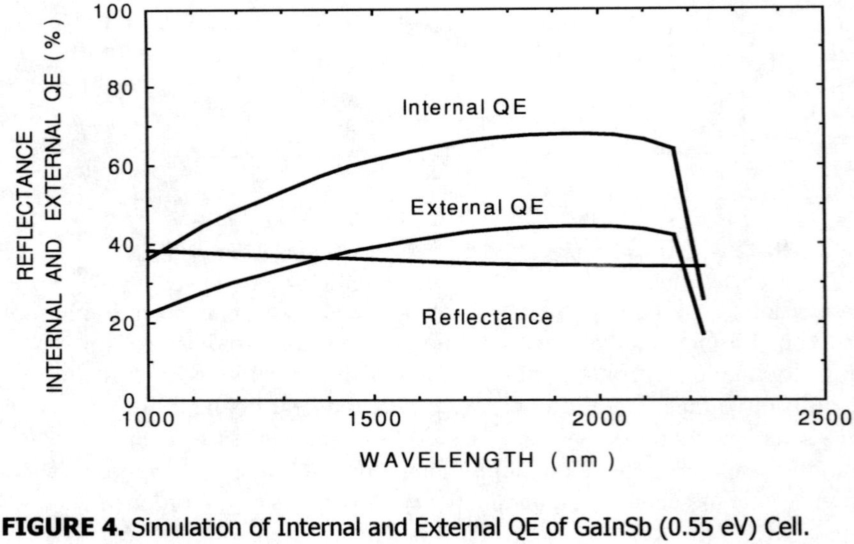

FIGURE 4. Simulation of Internal and External QE of GaInSb (0.55 eV) Cell.

Minority carrier's photogenerated in the base may diffuse toward the junction and collected or they diffuse away from the junction and are lost by bulk recombination that is predominantly Auger recombination. Besides the above simulations, we have determined the maximum power output performance of the Zn diffused cells as function of radiator temperature. The maximum power as function of Black Body radiator temperature for the several Zn diffused GaSb and GaInSb TPV cells is shown in Fig.5.

Figure 5 shows that in spite of the low radiator temperatures, the GaSb TPV cell always produces the largest output power. The reason for this surprising result is that the Auger recombination coefficients increase exponentially as the band gap decreases which causes an exponential increase in the dark current and a large decrease in the open circuit voltage. The increase in short circuit current with decreasing band gap does not compensate for the large decrease in open circuit voltage. In order to increase the performance of the low band gap GaInSb TPV cells it is necessary to lower the doping concentration in the N-base to a value in the range between 10^{16} cm^{-3} to 10^{17} cm^{-3} which may or may not be possible. This makes the recombination process in the base to be mainly radiative and also increases the value of the lifetime and diffusion length in the base bringing an increase in the QE and also lowering the cell dark current.

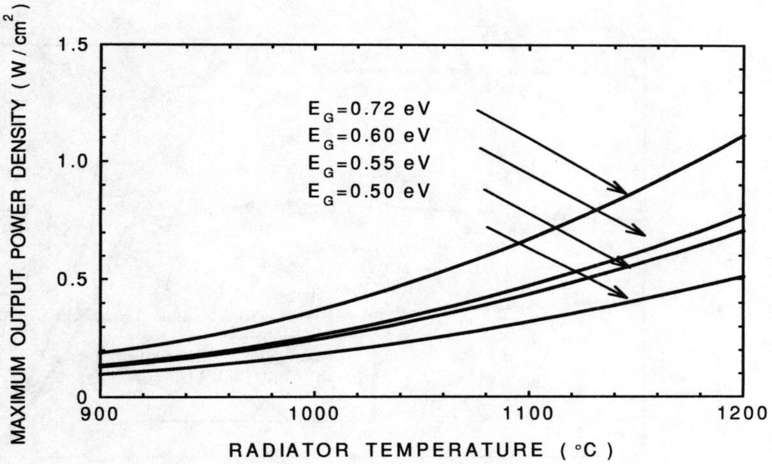

FIGURE 5. Maximum Output Power Density vs. Radiator Temperature

STRUCTURE OF QUATERNARY GaInAsSb TPV CELLS

The quaternary material was OMVPE grown lattice matched to GaSb.The typical structure of the TPV cell is a p-on-n configuration with a 1μm thick n-type GaInAsSb base layer doped to ~5x10^{17} cm^{-3}, a 4 μm-thick GaInAsSb p-emitter layer doped to ~$2x10^{17}$ cm^{-3} and a 0.05 μm GaSb window layer doped p-type to $2x10^{18}$ cm^{-3}. One of the advantages of the above structure is that there are window GaSb layers both at the top of the emitter and at the bottom of the base. Since the GaSb layers have a higher bandgap and doping concentration than the emitter and base layers, the main effect is to reduce the effective recombination velocity at the window layer-active layer interface.

SIMULATION OF GaInAsSb TPV CELLS

We have used PC-1D to simulate the performance of cells with the physical structure described above. In addition we have assumed that the GaSb layer below the cell base layer is 10 μm thick and we have assumed that the surface recombination velocity at the top and bottom GaSb window layers is 10^6 cm/sec. Also we have assumed that the lifetime of excess carriers due to SRH recombination is 1 millisec. These are the only two assumptions of the simulation, all the parameters of the GaInAsSb quaternary alloy have been obtained from the material parameters of the binaries GaAs, InAs, GaSb and InSb as described at the beginning of the paper. Figure 6 shows the results of the simulation and compares the experimentally measured external QE to the simulated one.

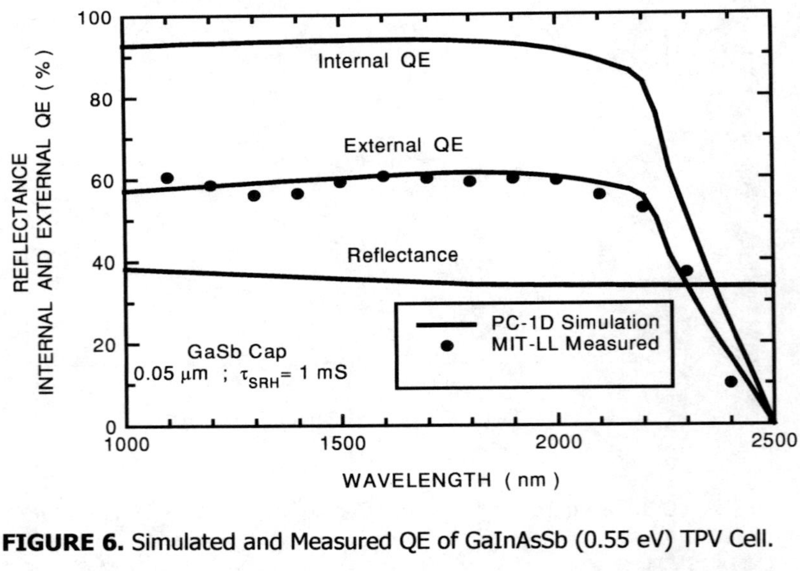

FIGURE 6. Simulated and Measured QE of GaInAsSb (0.55 eV) TPV Cell.

The agreement between the simulated and measured external QE is very good considering that the only parameters used in the simulation were the physical dimensions and doping concentration of the base and emitter layers. The excellent agreement at long wavelengths, above 2200 nanometers, is because we have augmented the absorption constant calculated from the binaries by the absorption due to the Ürbach edge experimentally measured in this alloy (7).

Comparing this result with the external QE shown in Fig. 4 for a GaInSb TPV cell, we see that the advantage of the OMVPE grown TPV cell is to reduce the effect of the surface recombination velocity. This is caused by the presence of the window layers that because of their higher doping and larger energy bandgap effectively reduce surface recombination effects. Since the doping concentrations in the emitter and base are of the order of 10^{17} cm^{-3}, the lifetime is long enough that any carrier generated in the emitter or base is collected by the junction and not lost by bulk recombination. This makes the internal QE reach values close to 95 % and it could reach 100 % if the doping is lowered to ~ 10^{16} cm^{-3}.

In addition we have simulated the illuminated I-V characteristics of the above cell and are shown in Fig. 7 for a maximum short circuit current density of 3.5 A/cm^2 in order to compare them to the measured characteristics. The simulation gives an open circuit voltage of 0.279 V and the measured open circuit voltage for a current density of 3.5 A/cm^2 was 0.310 V. The difference of 31 mV could be to either inaccurate values used of the intrinsic concentration for the 0.55 eV quaternary alloy, or in the Auger and radiative recombination coefficients or to wrong values assumed for the doping concentrations. Considering that all the quaternary alloy parameters were obtained from the parameters for the binaries, the agreement is rather remarkable.

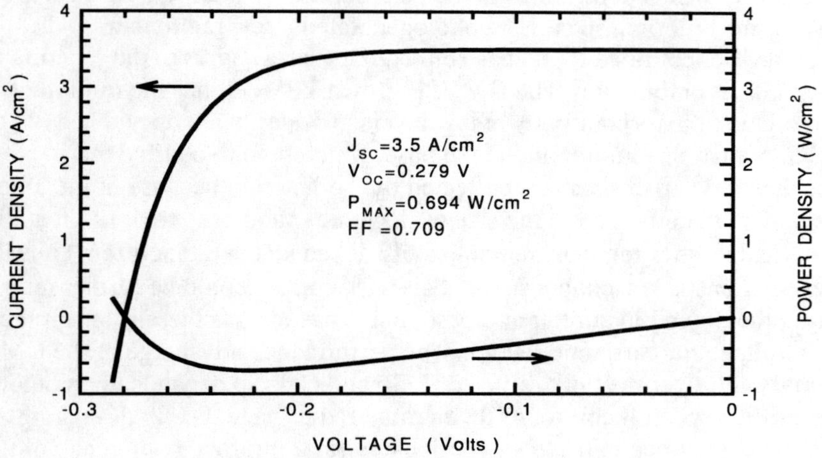

FIGURE 7. Simulated Light I-V Characteristics of GaInAsSb TPV Cell.

In addition we have simulated the performance of the above OMVPE quaternary cell as function of radiator temperature, as shown in Fig.5 for bulk diffused GaSb and GaInSb cells, and the results are given in Fig. 8. The results in the above figure show that the OMVPE quaternary cell outperforms all the other cells at all radiator temperatures simulated.

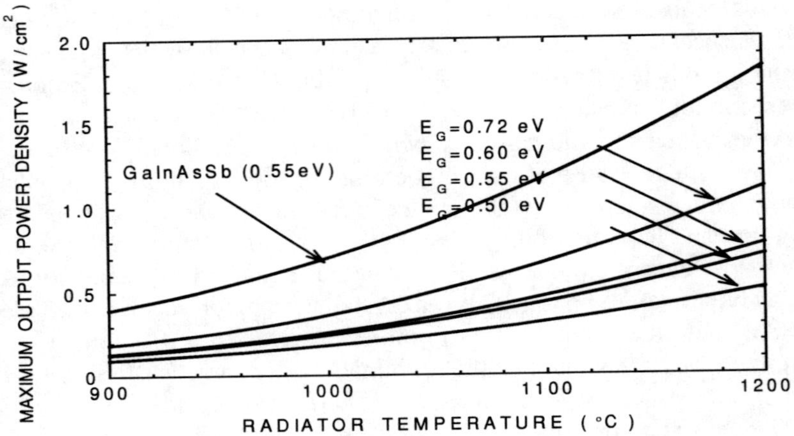

FIGURE 8. Maximum Output Power Density vs. Radiator Temperature

We believe the main reason for the performance advantage of the OMVPE GaInAsSb grown cell compared to the bulk diffused GaInSb is not so much to the material parameters since they have similar energy bandgap but to the structure of the cells. The bulk-diffused cells have a heavily doped emitter that does not collect photogenerated carriers very efficiently and the emitter behaves rather like a "dead layer" which contributes to the dark current. In addition the base is very long with short lifetime so carriers photogenerated a few microns from the junction are not confined to the base but can diffuse away from the junction and lost by bulk recombination. The OMVPE grown TPV cell has the important characteristic of incorporating window layers of wider bandgap and heavily doped material, both at the emitter and at the base, which help to confine the photogenerated carriers and are collected by the junction because of the long diffusion length compared to the size of the emitter and base regions. In addition the base and emitter regions are not heavily doped so the dark current contributed by these regions is reasonable small. This dark current could be further reduce by lowering the doping in emitter and base regions as long as there is an appreciable built in voltage for carrier collection. The performance advantage of the OMVPE quaternary cell over the bulk diffused GaSb and GaInSb depends upon other variables, like spectral control of the thermal radiation and cell AR coating, which has not been included in these simulations. Further improvement in any of the considered cells will change the performance advantage of the OMVPE quaternary cell over the bulk Zn diffused GaSb and GaInSb cells.

CONCLUSIONS

In this paper we have compared the performance of GaSb, GaInSb TPV cells fabricated by bulk diffusion to the performance of quaternary cells GaInAsSb grown by OMVPE with similar bandgap as function of radiator temperature. The performance was obtained from simulation using PC-1D. The material parameters for the ternary and the quaternary alloys were derived from the material parameters of the binary materials GaAs, InAs, GaSb and InSb using well-known rules. The calculated mobility and optical absorption constants were compared with experimental measurements and reasonable agreement was obtained which gives confidence to the method employed for calculating the parameters of the alloys. The results of the simulation of the cells as function of radiator temperature show the superior performance of the quaternary OMVPE GaInAsSb TPV cell for radiator temperatures between 900°C and 1200°C .We believe this is due to the window layers in the emitter and base of the OMVPE quaternary cell that help to confine the photogenerated carriers.

REFERENCES

1. Madelung, O., Editor *Semiconductors Group IV and III-V Compounds*, Springer-Verlag 1991.

2. Wiley, J.D., in *Semiconductors and Semimetals*, Willarsdon, R.K and Beer, A.C. eds, vol. 10, Academic Press, New York 1975.

3. Hilsum, C. and Rose-Innes, A.C., *Semiconducting III-V Compounds*, Pergamon Press, New York 1961.

4. Milnes, A.G. and Polakov, A.Y., *Solid State Electronics*, **36**, 803-818, 1993.

5. -Adachi, S., *Optical Constants of Crystalline and Amorphous Semiconductors*, Kluwer Academic Publishers, Boston 1999.

6. Adachi, S., *Physical Properties of III-V Semiconductor Compounds*, John Wiley and Sons, New York 1992.

7. Munoz, M., Wei, K., Pollak, Fred H., Freeouf, J.L., Wang, C.A., Charache,G.W., *J. of App. Physics,* **87**, 1780-1787, (2000).

8.Grein, C.H., Young, P.M., Flattė, M.E. and Ehrenreich, H, *J. of App. Physics*, **78**,7143-7152, (1995).

9.Charache,G.W.,Baldasaro,P.F.,Danielson,L.R.,DePoy,D.M.,Freeman,M.J., Wang,A.C.,Choi,H.K.,Garbuzov,D.V.,Martinelli,R.U.,Khalfin,V.,Saroop,S., Borrego,J.M. and Gutmann,R.J.,*J.of App.Physics*,**85**,2247-2252,(1999).

Author Index

A

B

C

D

E

F

G

H

J

K

L

M

N

O

P

Q